STATISTICAL SHAPE AND DEFORMATION ANALYSIS

Computer Vision and Pattern Recognition Series

Series Editors

Horst Bischof Institute for Computer Graphics and Vision, Graz University of Technology, Austria

Kyoung Mu Department of Electrical and Computer Engineering, Seoul National University, Republic of Korea

Sudeep Sarkar Department of Computer Science and Engineering, University of South Florida, Tampa, United States

Also in the Series:

Lin and Zhang, Low-Rank Models in Visual Analysis: Theories, Algorithms and Applications, 2017, 9780128127315

Murino et al., Group and Crowd Behavior for Computer Vision, 2017, 9780128092767

De Marsico et al., Human Recognition in Unconstrained Environments: Using Computer Vision, Pattern Recognition and Machine Learning Methods for Biometrics, 2017, 9780081007051

Saha et al., Skeletonization: Theory, Methods and Applications, 2017, 9780081012918

STATISTICAL SHAPE AND DEFORMATION ANALYSIS

Methods, Implementation and Applications

Edited by

GUOYAN ZHENG

SHUO LI

GABOR SZÉKELY

ACADEMIC PRESS

An imprint of Elsevier

Academic Press is an imprint of Elsevier
125 London Wall, London EC2Y 5AS, United Kingdom
525 B Street, Suite 1800, San Diego, CA 92101-4495, United States
50 Hampshire Street, 5th Floor, Cambridge, MA 02139, United States
The Boulevard, Langford Lane, Kidlington, Oxford OX5 1GB, United Kingdom

Notices

Knowledge and best practice in this field are constantly changing. As new research and experience broaden our
understanding, changes in research methods, professional practices, or medical treatment may become necessary.

Practitioners and researchers must always rely on their own experience and knowledge in evaluating and using any
information, methods, compounds, or experiments described herein. In using such information or methods they
should be mindful of their own safety and the safety of others, including parties for whom they have a professional
responsibility.

To the fullest extent of the law, neither the Publisher nor the authors, contributors, or editors, assume any liability for
any injury and/or damage to persons or property as a matter of products liability, negligence or otherwise, or from
any use or operation of any methods, products, instructions, or ideas contained in the material herein.

Library of Congress Cataloging-in-Publication Data
A catalog record for this book is available from the Library of Congress

British Library Cataloguing-in-Publication Data
A catalogue record for this book is available from the British Library

ISBN: 978-0-12-810493-4

For information on all Academic Press publications
visit our website at https://www.elsevier.com

Working together
to grow libraries in
developing countries

www.elsevier.com • www.bookaid.org

Publisher: Joe Hayton
Acquisition Editor: Tim Pitts
Editorial Project Manager: Charlotte Kent
Production Project Manager: Kiruthika Govindaraju
Designer: Vicky Pearson Esser

Typeset by VTeX

CONTENTS

List of Contributors   xi
About the Editors   xv
Preface   xvii
Acknowledgment   xix

I. Basic Concepts, Methods and Algorithms

1. Automated Image Interpretation Using Statistical Shape Models   3
Claudia Lindner

   1.1. Introduction   3
   1.2. Statistical Shape Analysis   4
   1.3. Feature Point Detection Using Shape Model Matching   10
   1.4. Fully Automated Image Analysis via Shape Model Matching   23
   1.5. Automated Image Interpretation and Its Applications   26
   1.6. Limitations of Statistical Shape Models for Image Interpretation   27
   1.7. Conclusion   28
   Acknowledgements   28
   References   29

2. Statistical Deformation Model: Theory and Methods   33
Xiahai Zhuang, Yipeng Hu

   2.1. Introduction   34
   2.2. Deformation Representation   35
   2.3. Statistical Approaches   40
   2.4. General-Purpose Deformation Models   48
   2.5. Biophysics-Based Deformation   58
   References   64

3. Correspondence Establishment in Statistical Shape Modeling: Optimization and Evaluation   67
Song Wang, Brent Munsell, Theodor Richardson

   3.1. Introduction   67
   3.2. PDM and Shape Correspondence   69
   3.3. Landmark Sliding for Shape Correspondence   69
   3.4. Groupwise Shape Correspondence   75
   3.5. Performance Evaluation of Shape Correspondence   79
   3.6. Experiments   82
   3.7. Conclusions and 3D Shape Correspondence   85
   References   86

4. Landmark-Based Statistical Shape Representations   89
Bulat Ibragimov, Tomaž Vrtovec

4.1. Introduction   89
4.2. Landmark-Based Shape Representation   91
4.3. Shape-Based Landmark Detection   101
4.4. Conclusion   111
References   112

5. Probabilistic Morphable Models   115
Bernhard Egger, Sandro Schönborn, Clemens Blumer, Thomas Vetter

5.1. Introduction   115
5.2. Methods   120
5.3. Applications and Results   129
5.4. Conclusion   131
References   133

6. Object Statistics on Curved Manifolds   137
Stephen M. Pizer, J.S. Marron

6.1. Objectives of Object Statistics   138
6.2. Objects Live on Curved Manifolds   139
6.3. Statistical Analysis Background   144
6.4. Advanced Statistical Methods for Manifold Data   148
6.5. Correspondence   153
6.6. How to Compare Representations and Statistical Methods   155
6.7. Results of Classification, Hypothesis Testing, and Probability Distribution Estimation 157
6.8. Conclusions   161
Acknowledgements   162
References   162

7. Shape Modeling Using Gaussian Process Morphable Models   165
Marcel Lüthi, Andreas Forster, Thomas Gerig, Thomas Vetter

7.1. Introduction   165
7.2. Shape Modeling Using Gaussian Processes   166
7.3. Non-Rigid Registration Using Gaussian Process Priors   172
7.4. Case Study: Building a Statistical Shape Model of the Skull   174
7.5. Modeling and Analyzing Pathologies   184
7.6. Conclusion   186
Appendix 7.A   188
References   190

8. Bayesian Statistics in Computational Anatomy   **193**
Christof Seiler

    8.1. Introduction   193
    8.2. Parametric Bayesian Statistics   194
    8.3. Nonparametric Bayesian Statistics   204
    8.4. Conclusions and Open Problems   212
    References   212

II. Open Source Implementation Examples

9. Morpho and Rvcg – Shape Analysis in R   **217**
Stefan Schlager

    9.1. Introduction   218
    9.2. Preliminaries and Installation   218
    9.3. Landmark Based Shape Analysis with Morpho   219
    9.4. Manipulations on Triangular Meshes Using Rvcg (and Morpho)   244
    9.5. Beyond CRAN   250
    9.6. Final Remarks   253
    References   255

10. ShapeWorks   **257**
Joshua Cates, Shireen Elhabian, Ross Whitaker

    10.1. Introducing ShapeWorks   258
    10.2. Particle-Based Modeling   262
    10.3. PBM Extensions   273
    10.4. ShapeWorks Software Implementation and Workflow   284
    10.5. ShapeWorks in Biomedical Applications   290
    10.6. Conclusions and Future Work   292
    References   292

III. Applications

11. Applications of Statistical Deformation Model   **301**
Yipeng Hu, Xiahai Zhuang

    11.1. Image-Guided Prostate Intervention   301
    11.2. Whole Heart Segmentation   319
    References   327

12. Statistical Shape and Deformation Models Based 2D–3D Reconstruction   **329**
Guoyan Zheng, Weimin Yu

    12.1. Introduction   329

12.2. Statistical Shape Model Based 2D–3D Reconstruction and Its Application in THA   331
12.3. Statistical Deformation Model Based 2D–3D Reconstruction   338
12.4. Final Remarks   346
References   347

13. Statistical Shape Analysis for Brain Structures   351
Li Shen, Shan Cong, Mark Inlow

13.1. Introduction   351
13.2. Surface Modeling and Registration   353
13.3. Statistical Inference on the Surface   362
13.4. An Example Application   369
13.5. Conclusions   372
Acknowledgments   373
References   373

14. Statistical Respiratory Models for Motion Estimation   379
Christoph Jud, Philippe C. Cattin, Frank Preiswerk

14.1. Background   379
14.2. 4-Dimensional MR Imaging   381
14.3. Motion Model Building   384
14.4. Establishment of Correspondence   388
14.5. Statistical Motion Modeling   391
14.6. Bayesian Reconstruction from Sparse Data   392
14.7. Applications of Population-Based Statistical Motion Models to Motion Reconstruction   394
14.8. Reconstruction by Regression   401
14.9. Conclusion   404
Acknowledgments   404
References   404

15. Statistical Shape and Appearance Models for Bone Quality Assessment   409
Patrik Raudaschl, Karl Fritscher

15.1. Introduction   410
15.2. Fundamentals of Statistical Shape and Appearance Models   414
15.3. Approaches for Bone Quality Assessment   419
15.4. Discussion and Conclusion   432
References   436

16. Statistical Shape Models of the Heart: Applications to Cardiac Imaging   445
Concetta Piazzese, M. Chiara Carminati, Mauro Pepi, Enrico G. Caiani

16.1. Introduction   445
16.2. The heart   446

16.3. Cardiac Imaging Techniques   449
16.4. Statistical Shape Models   453
16.5. Discussion   462
References   473

Index   *481*

LIST OF CONTRIBUTORS

Clemens Blumer

Department of Mathematics and Computer Science, University of Basel, Switzerland

Enrico G. Caiani

Dipartimento di Elettronica, Informazione e Bioingegneria, Politecnico di Milano, Italy

M. Chiara Carminati

Paul Scherrer Institut, Villigen, Switzerland

Joshua Cates

Scientific Computing and Imaging Institute, University of Utah, Salt Lake City, United States

Biomedical Image and Data Analysis Core, University of Utah, Salt Lake City, United States

Comprehensive Arrhythmia Research and Management Center, University of Utah, Salt Lake City, United States

Philippe C. Cattin

Department of Biomedical Engineering, University of Basel, Allschwil, Switzerland

Shan Cong

Department of Radiology and Imaging Sciences, Center for Neuroimaging, Indiana University School of Medicine, Indianapolis, IN, United States

Department of Electrical and Computer Engineering, Indiana University–Purdue University Indianapolis, Indianapolis, IN, United States

Bernhard Egger

Department of Mathematics and Computer Science, University of Basel, Switzerland

Shireen Elhabian

Scientific Computing and Imaging Institute, University of Utah, Salt Lake City, United States

Biomedical Image and Data Analysis Core, University of Utah, Salt Lake City, United States

Faculty of Computers and Information, Cairo University, Cairo, Egypt

Andreas Forster

Department of Mathematics and Computer Science, University of Basel, Switzerland

Karl Fritscher

Institute for Biomedical Image Analysis (IBIA), University for Health Sciences, Medical Informatics and Technology (UMIT), Hall in Tirol, Austria

Thomas Gerig
Department of Mathematics and Computer Science, University of Basel, Switzerland

Yipeng Hu
Centre for Medical Image Computing, Department of Medical Physics and Biomedical Engineering, University College London, London, United Kingdom

Bulat Ibragimov
Stanford University School of Medicine, Department of Radiation Oncology, United States

Mark Inlow
Department of Radiology and Imaging Sciences, Center for Neuroimaging, Indiana University School of Medicine, Indianapolis, IN, United States

Christoph Jud
Department of Biomedical Engineering, University of Basel, Allschwil, Switzerland

Claudia Lindner
The University of Manchester, Centre for Imaging Sciences, Manchester, United Kingdom

Marcel Lüthi
Department of Mathematics and Computer Science, University of Basel, Switzerland

J.S. Marron
University of North Carolina at Chapel Hill, Department of Statistics and Operations Research, United States

Brent Munsell
College of Charleston, Department of Computer Science, Charleston, SC 29401, United States

Mauro Pepi
Centro Cardiologico Monzino IRCCS, Milan, Italy

Concetta Piazzese
Centro Cardiologico Monzino IRCCS, Milan, Italy

Stephen M. Pizer
University of North Carolina at Chapel Hill, Department of Computer Science, United States

Frank Preiswerk
Department of Biomedical Engineering, University of Basel, Allschwil, Switzerland
Department of Radiology, Brigham and Women's Hospital, Harvard Medical School, Boston, MA, United States

Patrik Raudaschl

Institute for Biomedical Image Analysis (IBIA), University for Health Sciences, Medical Informatics and Technology (UMIT), Hall in Tirol, Austria

Theodor Richardson

South University, College of Business, Savannah, GA 31405, United States

Stefan Schlager

Biological Anthropology, Albert-Ludwigs University Freiburg, Freiburg, Germany

Sandro Schönborn

Department of Mathematics and Computer Science, University of Basel, Switzerland

Christof Seiler

Department of Statistics, Stanford University, United States

Li Shen

Department of Radiology and Imaging Sciences, Center for Neuroimaging, Indiana University School of Medicine, Indianapolis, IN, United States

Center for Computational Biology and Bioinformatics, Indiana University School of Medicine, Indianapolis, IN, United States

Thomas Vetter

Department of Mathematics and Computer Science, University of Basel, Switzerland

Tomaž Vrtovec

University of Ljubljana, Faculty of Electrical Engineering, Slovenia

Song Wang

University of South Carolina, Department of Computer Science and Engineering, Columbia, SC 29208, United States

Ross Whitaker

Scientific Computing and Imaging Institute, University of Utah, Salt Lake City, United States

School of Computing, University of Utah, Salt Lake City, United States

Weimin Yu

Institute for Surgical Technology and Biomechanics, University of Bern, Bern, Switzerland

Guoyan Zheng

Institute for Surgical Technology and Biomechanics, University of Bern, Bern, Switzerland

Xiahai Zhuang

School of Data Science, Fudan University, Shanghai, China

ABOUT THE EDITORS

Prof. Dr. Guoyan Zheng, University of Bern, Bern, Switzerland. Prof. Zheng is the Head of the Information Processing in Medical Interventions Group, the Institute for Surgical Technology and Biomechanics, University of Bern. In 2010, he did his habilitation and was awarded the title 'Privatdozent' from the same university. His research interests include medical image computing, machine learning, computer assisted interventions, medical robotics, and multi-modality image analysis. He has published over 160 peer-reviewed journal and conference papers and was granted 6 US and European patents. He has won over ten national and international awards/prizes including the best basic science paper published in the Journal of Laryngology and Otology in year 2011, the 2009 Ypsomed Innovation Prize, and the best technical paper award in the 2006 annual conference of the International Society of Computer Assisted Orthopaedic Surgery. He is on the program committee of the 16th, the 18th and the 19th International Conference on Medical Image Computing and Computer Assisted Interventions (MICCAI 2013, 2015 and 2016). He is the general chair of the 7th International Workshop on Medical Imaging and Augmented Reality (MIAR 2016).

Prof. Dr. Shuo Li, University of Western Ontario, London, ON, Canada. Dr. Li is the director of Digital Imaging Group (DIG) of London, an associate professor in the department of medical imaging and medical biophysics at the University of Western Ontario and a scientist in Lawson Health Research Institute. Before this position, he was a research scientist and project manager in General Electric (GE) healthcare for 9 years. He funds and directs the DIG (http://digitalimaginggroup.ca/) since 2006, which is a highly dynamic and multiple disciplinary group. He received his Ph.D. degree in computer science from Concordia University 2006, where his Ph.D. thesis won the doctoral prize given to the most deserving graduating student in the faculty of engineering and computer science. He has published over 100 publications. He is the recipient of several GE internal awards. He serves as a guest editor and associate editor in several prestigious journals in the field. He serves as a program committee member in highly influential conferences. He is the editor of six books. His current interest is development of intelligent analytic tools to help physicians and hospital administrative to handle the big medical data, centered on medical images.

Prof. Dr. Gabor Székely, Computer Vision Lab, ETHZ, Switzerland. Prof. Székely was born in 1951 in Budapest, Hungary and graduated from the Technical University of Budapest in chemical engineering in 1974 and from the Eötvös Loránd University of Budapest in Applied Mathematics in 1981. He obtained his Ph.D. in analytical chemistry in 1985 from the Technical University of Budapest. Between 1974 and 1986 he has been working at the Computer Department of the Institute of Isotopes of the Hungarian Academy of Sciences, since 1985 as the Head of the Department, focusing on automatic structure elucidation of organic compounds. Between 1986 and 1990 he developed computer support systems for the chemical and biomedical applications of magnetic resonance at Bruker Spectrospin. In 1991 he joined the Computer Vision Laboratory of the ETH Zurich as a senior researcher. In 2002 he has been elected as Associate Professor at the Department of Information Technology and Electrical Engineering and founded the Medical Image Analysis and Visualization research group. He has been promoted to Full Professor in 2008. Between 2001 and 2013 he has been the Director of the Swiss National Center for Competence in Research on Computer Aided and Image Guided Medical Interventions. In 2007 he co-founded the spinoff company Virtamed, producing virtual reality based system for surgical training. His major research interest is developing algorithms and clinical systems for the optimal computer support of medical diagnosis, therapy, training and education.

PREFACE

The importance of statistical shape and deformation analysis as a research tool in medical image analysis, computational radiology, computational anatomy, computer graphics, biometry, and physical anthropology is recognized today by most researchers in those related fields. Appearance and structure, which are also known as form, have long been used by natural scientists to perceive and classify organisms. Statistical shape and deformation analysis is generally associated with quantitative study of form and form change. Although the problem of mathematically characterizing biological shapes was studied by D'Arcy Thompson one hundred years ago, it was not before the early nineties of the last century when this issue has become a major research focus within the emerging new fields of medical image computing and geometric morphometrics. Since then, a broad spectrum of methods has been developed by the different research communities, increasing the utility of shape modeling and introducing it in various applications.

Several good text-books exist on statistical shape analysis. However, in spite of the highly visual nature of statistical shape and deformation analysis, and the ready availability of open source software tools, most of these books approach the subject from a pure mathematical perspective. We do believe that a book accompanied with readily available, open source tools will be a major driver in the uptake of statistical shape and deformation analysis methods by researchers in different fields. To this end we organized the book in three main parts. Part I deals with various concepts, methods, algorithms and techniques that have been developed in the past three decades. Part II focuses on actual implementation examples, building on open source packages that have been used in different research communities. Finally, Part III discusses the applications of statistical shape and deformation analysis in biometry, anthropology, medical image analysis, and clinical practice.

Statistical shape and deformation analysis is a field potentially as vast and diverse as shape and deformation analysis itself. In an attempt to capture this diversity, each chapter of this book is contributed by different authors. Each subject is presented and discussed by the experts who first defined the problem, presented solutions to the problems, and formulated the principles underlying the solutions. Principles and methods of statistical shape and deformation analysis are explained through direct examples of recent or ongoing research. Unlike other text-books on the same topic, with this book understanding the methods and algorithms in Part I can be facilitated by the open source implementation examples in Part II and the real life applications in Part III. More details on downloading and installing open source software as well as invaluable supplementary resources can be found in the book's companion website at: http://ssda-book.weebly.com/.

This book may be used as a text-book by readers interested in learning the basic concepts, methods, algorithms, techniques and implementations of statistical shape and deformation analysis or geometric morphometrics, or as a unique reference to consult for key topic (both conceptual and technical) in these new areas of investigation. Prime intended audience will be active researchers and graduate students working in the field of medical image analysis. They should be able to understand the chapters, just like when they read papers published in high-quality scientific journals. Advanced under-graduate students with a background in biomedical engineering or computer science will also find this book highly accessible. Due to the broad applications of statistical shape and deformation analysis in different fields from medical image analysis, it is ex-pected that the book will also generate interests in researchers and graduate students in many different communities.

ACKNOWLEDGMENT

We wish to extend our gratitude to all the people who made this book possible. We sincerely thank our colleagues for their hard work in making their contributions. We also deeply appreciated the efforts of the publisher, Elsevier, to make this book possible. The help provided by Mr. Tim Pitts and Ms. Charlie Kent from Elsevier in all the phases of organization and editing was outstanding.

PART I

Basic Concepts, Methods and Algorithms

CHAPTER 1

Automated Image Interpretation Using Statistical Shape Models

Claudia Lindner
The University of Manchester, Centre for Imaging Sciences, Manchester, United Kingdom

Contents

1.1 Introduction	3
1.2 Statistical Shape Analysis	4
1.2.1 Defining Shape	4
1.2.2 Statistical Shape Models	5
1.2.3 Automated Shape Annotation	9
1.3 Feature Point Detection Using Shape Model Matching	10
1.3.1 Active Shape Models	11
1.3.2 Active Appearance Models	14
1.3.3 Constrained Local Models	18
1.3.4 Random Forest Regression-Voting Constrained Local Models	20
1.4 Fully Automated Image Analysis via Shape Model Matching	23
1.4.1 Object Detection Methods	24
1.4.2 Combining Object Detection with Shape Model Matching	25
1.5 Automated Image Interpretation and Its Applications	26
1.6 Limitations of Statistical Shape Models for Image Interpretation	27
1.7 Conclusion	28
Acknowledgements	28
References	29

1.1 INTRODUCTION

Image interpretation covers a wide range of techniques to extract meaningful information from raw image data. Statistical Shape Models (SSMs) provide a means to describe the variation in shape of an object class across a set of images, allowing the qualitative and quantitative analysis of image data. For the detailed analysis of the shape of an object using SSMs, the object of interest needs to be annotated using many points in every image. Depending on the object type (e.g. faces or skeletal structures), this annotation may not only be tedious and time-consuming but may also require significant expertise to correctly place the points. In this chapter, we are going to introduce the basic methodology behind SSMs, and describe a number of automated annotation methods to enable fast and objective shape analyses with a wide variety of applications.

Statistical Shape and Deformation Analysis
DOI: 10.1016/B978-0-12-810493-4.00002-X

1.2 STATISTICAL SHAPE ANALYSIS

An object in an image can be characterized by both *shape* and *texture*. While shape provides geometric information about the object, texture gives its intensity appearance. Hence, shape-based image interpretation utilizes the contour of an object while texture-based image interpretation utilizes its gray-level (or color-level) variation. Within an object class (e.g. hands, faces) the shape and texture may vary widely across images. For example with reference to medical images, these differences may be related to one or more clinical variables such as age and gender, anatomical variation between individuals, or disease progression. Furthermore, differences in the shape and texture of an object may also result from the usage of different imaging techniques or image acquisition protocols.

A detailed introduction to *statistical shape analysis* can be found in [30]. Below, we provide a summary of the main aspects of statistical shape analysis as deemed relevant to the remainder of this chapter.

1.2.1 Defining Shape

Traditionally, morphometric analyses were based on a set of predefined measurements such as lengths and angles. Though predefined measurements are often easily obtainable, they do not capture the overall shape of the object and predominantly measure size rather than shape.

> *"If we are not interested in the location, orientation or scale ..., then we find ourselves working with ... change of shape."* [44, p. 428]

According to Kendall [44], *shape* describes all geometric information of an object disregarding location, orientation and scale of the data. Modern morphometrics, thus, suggests to analyze the contour of an object so as to capture its overall shape. This is commonly done by placing a number of landmark points along the object's contour. A landmark[1] is considered to be a point of correspondence that marks a specific part of the contour, or structure, of an object across images of the same object class. The shape of the object is then expressed by the combination of all landmark points. When identifying landmarks, the following criteria should be taken into account [74]. Landmarks should:

- provide adequate coverage of the morphology of the object;
- be chosen so as to quantify any significant shape change;
- be placed such that they can be found repeatedly and reliably;
- mark consistent positions relative to other landmarks.

---

[1] Depending on application area and information content of the point, several synonyms for landmark are used in the literature, including anchor point, fiducial point, model point, key point, and feature point.

In studies involving images of organisms, landmarks are often manually chosen to mark points of anatomical significance (e.g. finger tips in images of the hands) in combination with evenly spaced points along the object's contour. When landmarks are evenly spaced it is not always obvious how to select the (number of) points required such that the variation in shape is well represented and that all points are in correspondence across the object class. The latter applies to both object classes with and without anatomical landmarks or key points of interest. Furthermore, this holds true for shape surfaces in 3D where manual identification of suitable landmarks is often difficult and impractical. Given a representative set of objects for an object class of interest, a number of automated methods are available to establish correspondence in these cases; see [26] for an overview. A technique referred to as Minimum Description Length (MDL) has emerged as the state-of-the-art in this field [25]. MDL methods establish point correspondences automatically by minimizing the description length of a shape model created from the point placements, aiming for models with high generalization ability, specificity and compactness. However, automated methods to establish point correspondences often require the shape to be readily defined as a curve i.e. the shape of the object of interest to be outlined in every image – which is not always available.

A set of landmark points placed at key features of an object provides a description of the geometry of the object and allows the detailed statistical analysis of the *global* shape of the object. Though various statistical analyses could be used for this purpose, Principal Component Analysis (PCA) is commonly applied in this context.

1.2.2 Statistical Shape Models

Statistical Shape Models (SSMs)[2] are commonly used to study the morphometrics of a deformable object. They describe the shape of the object by applying PCA to a set of landmark points, and are based on the assumption that each shape is a deformed version of a reference shape. SSMs aim to establish the shape variation of an object class in order to create a statistical model that gives a parameterized representation of this variation in shape.

Building an SSM is based on a set of landmark points capturing the shape of the object in every image. The types of shape deformation present across the data are identified and redundancies in the shape distribution of the object class are removed. SSMs yield a compact representation of the shape of the object class, and can be used in a variety of application areas. They are not only useful for the analysis of shape differences in the data, on which the model was built, but also to analyze the shape of new data and to synthesize new shapes that are similar to the original data.

A prerequisite for building an SSM is a set of annotated images. The dataset that the model is built on is commonly referred to as the *training* dataset. Given a set of

[2] In the literature, SSMs are also known as Point Distribution Models.

n training images, for each image the shape of the object of interest is assumed to be described by ℓ landmark points placed at key features and/or along the contour of the object. Note that the landmark points may be in any dimension but in this chapter we focus on the 2D case.[3] The landmark points of any image I are denoted by shape vector $\mathbf{x}^I = [x_1, x_2, \ldots, x_\ell, y_1, y_2, \ldots, y_\ell]^T \in 2\ell \times 1$. For boundary landmarks (i.e. landmark points along the contour of the object), all landmarks are given in the order that they are connected with each other (if applicable).

A detailed description of SSMs can be found in Cootes et al. [16] and Davies et al. [26]. The following provides an overview of the ideas and algorithms behind building SSMs.

Shape Alignment via Generalized Procrustes Analysis

The training images can vary widely in the way they present the object of interest, i.e. in terms of rotation, scale and position of the object. Therefore, the annotated landmark points of all training images need to be aligned into a common coordinate frame. This can be achieved by applying a modification of Procrustes Analysis to remove all information that is unrelated to shape (i.e. translational, rotational and scaling information). The aim is to find transformation T_θ with parameters θ that aligns all shape vectors into a common reference frame such that $\mathbf{x}^I = T_\theta(\mathbf{x})$ with $\mathbf{x}$ being the aligned shape vector. To do so, the landmark sets of all images need to be centered to remove translational differences. This is done by determining and aligning their centers of gravity. Generalized Procrustes Analysis then describes the process of accounting for rotation and scale by finding transformation T_i such that each shape $\mathbf{x}_i$ best aligns with a target mean shape $\bar{\mathbf{x}}$. *Best* here is defined in terms of the minimized sum of square distances over equivalent landmark points in the image coordinate frame: $\min \sum_i |\mathbf{x}_i^I - T_i(\bar{\mathbf{x}})|^2$. The alignment process is conducted in an iterative manner, where the shape of each training image is repeatedly aligned with the respective mean over all aligned shapes until the mean shape converges. To initialize the alignment, the mean shape is set to the landmark points of the first shape in the training set.

After all shapes have been aligned, PCA is applied to build the SSM.

Modeling Shape Variation via Principal Component Analysis

PCA is a dimensionality reduction method that can be used to represent the shape variation of n aligned shape vectors $\mathbf{x}_i$, $1 \leq i \leq n$, in a simplified way. To start with, the mean shape vector $\bar{\mathbf{x}} = \frac{1}{n} \sum_{i=1}^{n} \mathbf{x}_i$ is calculated over all training shapes. This is then used to analyze how the elements of the different shape vectors vary together by calculating

---

[3] For a review of medical imaging studies that applied SSMs in 3D, the interested reader is referred to [42].

the covariance matrix over all training data

$$\mathbf{C} = \frac{1}{n-1} \sum_{i=1}^{n} (\mathbf{x}_i - \bar{\mathbf{x}})(\mathbf{x}_i - \bar{\mathbf{x}})^T. \qquad (1.1)$$

Covariance matrix $\mathbf{C}$ provides the basis to compute the principal components, its 2ℓ eigenvectors $\mathbf{p}_j$, and the corresponding eigenvalues λ_j with $\mathbf{C}\mathbf{p}_j = \lambda_j \mathbf{p}_j$. Without loss of generality, it is assumed that all eigenvalues and eigenvectors are sorted according to $\lambda_j \geq \lambda_{j+1}$. Applying PCA to the set of training shape vectors, thus, provides the matrix $\mathbf{P} = (\mathbf{p}_1 \mid \mathbf{p}_2 \mid \ldots \mid \mathbf{p}_{2\ell})$ containing the 2ℓ orthogonal shape eigenvectors of the covariance matrix $\mathbf{C}$ in its columns as well as a 2ℓ-dimensional vector $\mathbf{b} = \mathbf{P}^T(\mathbf{x} - \bar{\mathbf{x}})$. The landmarks of any aligned shape can then be defined by the mean shape vector plus a linear combination of eigenvectors of $\mathbf{C}$

$$\mathbf{x} = \bar{\mathbf{x}} + \mathbf{P}\mathbf{b}. \qquad (1.2)$$

Matrix $\mathbf{P}$, shape mode vector $\mathbf{b}$ and mean shape $\bar{\mathbf{x}}$ define an SSM where $\mathbf{b}$ gives the deviation from the mean shape and allows for the representation of different shapes. The variance of every eigenvector $\mathbf{p}_j$ is given by the corresponding eigenvalue λ_j, and the sum of all eigenvalues gives the overall variance of the training data. Eigenvectors with larger eigenvalues describe more significant shape variations whereas those with lower eigenvalues express smaller, often more local, variations. In general, most of the overall variation can be explained by a subset of the 2ℓ eigenvectors and eigenvalues. To reduce dimensionality while still preserving a sufficient large proportion of the overall shape variance of the training set, it is usually satisfactory to only consider the t largest eigenvalues and their corresponding eigenvectors such that the sum of these eigenvalues $\sum_{j=1}^{t} \lambda_j$ captures a certain proportion (e. g. 95%) of the total shape variance $\sum_{j=1}^{2\ell} \lambda_j$. An alternative approach to defining t, the number of shape modes to be included, is to choose enough modes so that the model can approximate any training shape to within a given accuracy (e.g. within one pixel). Based on the above, every aligned shape vector $\mathbf{x}$ can be approximated by

$$\mathbf{x} \approx \bar{\mathbf{x}} + \mathbf{P}'\mathbf{b}' \qquad (1.3)$$

with $\mathbf{P}'$ containing the t eigenvectors corresponding to the t largest eigenvalues that describe the most significant shape variations, and $\mathbf{b}'$ being a t-dimensional vector representing the t shape mode parameters. To accurately describe shape vector $\mathbf{x}$ after dimensionality reduction, a residual term needs to be introduced such that

$$\mathbf{x} = \bar{\mathbf{x}} + \mathbf{P}'\mathbf{b}' + \mathbf{r} \qquad (1.4)$$

where residuals $\mathbf{r}$ can be used to estimate how well the SSM fits to a dataset. The larger the residuals the more the model deviates from the data.

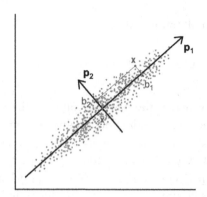

Figure 1.1 Applying PCA to 2D vectors. Every vector can be described by $\mathbf{x} = \bar{\mathbf{x}} + b_1\mathbf{p}_1 + b_2\mathbf{p}_2$ and can be approximated by $\mathbf{x} \approx \bar{\mathbf{x}} + b_1\mathbf{p}_1$, where $\mathbf{p}_1$ and $\mathbf{p}_2$ are the two principle axes.

The shape represented by Eq. (1.3) can be altered by varying the elements of $\mathbf{b}'$. Each shape mode represents a pattern of shape variation learned from the training set. The variance of an element b'_j of $\mathbf{b}'$ is described by λ_j, $1 \leq j \leq t$. To guarantee that the SSM only captures shapes that are sufficiently similar to the shapes given by the training set, limits can be applied to the values of elements b'_j. Originally it was suggested to limit elements b'_j with respect to the standard deviation (SD) from the mean value, e.g. $-3\sqrt{\lambda_j} \leq b'_j \leq 3\sqrt{\lambda_j}$. However, assuming the distribution of elements b'_j to be Gaussian in general, it was later recommended to use an upper bound on the Mahalanobis distance instead [16]. For a *new* aligned shape $\mathbf{x}^I$, the squared Mahalanobis distance to the mean $\bar{\mathbf{x}}$ of the given distribution of shape variation is defined by $d_M^2 = (\mathbf{x}^I - \bar{\mathbf{x}})^T \mathbf{C}^{-1}(\mathbf{x}^I - \bar{\mathbf{x}})$ with $\mathbf{C}$ being the covariance matrix of the distribution. When relating the squared Mahalanobis distance to the PCA results, it is given by $d_M^2 = \sum_{j=1}^{2\ell} \frac{b_j^2}{\lambda_j}$ and corresponds to the probability that $\mathbf{x}^I$ belongs to the distribution of shape variation given by the training set.

Fig. 1.1 gives an example of applying PCA to a set of 2D vectors. In this example, the data are correlated and hence contain redundant information. PCA removes this redundancy by identifying orthogonal directions corresponding to the main variance of the data. The two principle components, $\mathbf{p}_1$ and $\mathbf{p}_2$, point in these directions and provide the axes of the new coordinate system where every vector $\mathbf{x}$ can be described by $\mathbf{x} = \bar{\mathbf{x}} + b_1\mathbf{p}_1 + b_2\mathbf{p}_2$. To reduce the dimensionality of the data, every vector can be approximated by the nearest neighbor on the first principle axis $\mathbf{x} \approx \bar{\mathbf{x}} + b_1\mathbf{p}_1$. Hence, every 2D vector in the dataset can be approximated by an SSM that uses only a single parameter, b_1. Similarly, a few parameters (or shape modes) may be sufficient to represent high dimensional data as, for example, given by a dataset that uses many landmarks to capture the contour of an object.

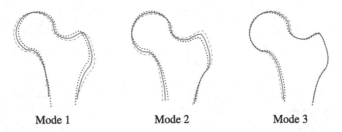

Mode 1   Mode 2   Mode 3

Figure 1.2 First three shape modes of variation in proximal femur morphology and the effect of varying the PCA parameters between ±2.5 SD for an SSM trained on 2516 female AP pelvic radiographs from the Osteoarthritis Initiative dataset (https://oai.epi-ucsf.org/). Each figure shows the average (solid line) and ±2.5 SD.

Fig. 1.2 shows the first three shape modes of a proximal femur shape model that was trained on 2516 female AP pelvic radiographs from the Osteoarthritis Initiative dataset (https://oai.epi-ucsf.org/). These three modes together explain 70% of the variance in proximal femur morphology across all radiographs analyzed. This figure demonstrates that an SSM mode does not simply reflect one morphometric measurement such as the femoral neck width but rather represents a combination of several related differences between proximal femurs of the study population.

Applying PCA to the aligned shape vectors allows the transformation of the image landmark coordinates into a new coordinate system. To project the data from the reference coordinate frame back to the image coordinate frame, transformation T_θ needs to be applied

$$\mathbf{x}^I = T_\theta(\bar{\mathbf{x}} + \mathbf{P}'\mathbf{b}' + \mathbf{r}). \tag{1.5}$$

To interpret an image using an SSM, model parameters $\mathbf{b}'$ that best match the model to the image (i.e. such that residuals $\mathbf{r}$ are small) need to be identified.

1.2.3 Automated Shape Annotation

SSMs provide a method to quantify the shape of an object using only a limited number of parameters. Applying SSMs in the setting of morphometric analyses (e.g. of skeletal structures in medical images) has the advantage of capturing the *global* shape of the object of interest rather than reducing it to a set of fixed geometric measurements such as lengths and angles.

Annotating images manually is prone to inconsistencies, time-consuming and subjective. This poses a problem in particular for application areas with large datasets and high accuracy requirements. Over the years, several methods have been developed to automate the annotation process. These methods broadly fall into two categories – feature point detection and groupwise image registration. Feature point detection aims to directly predict the position of every landmark in every image, and is often combined

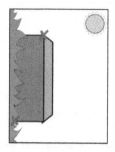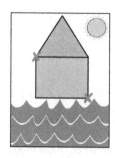

Figure 1.3 Example dataset where the object of interest, the house, is not sufficiently aligned across images for groupwise image registration to be applied without some form of initialization.

with some kind of shape model to regularize the output of the detection. Groupwise image registration, in contrast, aims to bring a set of *related* images into spatial alignment which can then be used to find sets of corresponding points across images using a single annotated reference image. Furthermore, several hybrid methods combining feature point detection and image registration have been presented (see e.g. Zhang and Brady [76] or Guo et al. [41]). An alternative approach to generating an SSM, without the need for annotating new images, would be to directly predict the parameters of the shape model. For more details, see for example the literature on Marginal Space Learning [77] or Shape Particle Filtering [27].

Groupwise image registration requires all images to be similar in terms of the position, orientation and scale of the object of interest in the image (e.g. as in MRI or CT datasets). If this is not the case, a semi-automatic registration approach can be followed by guiding the registration with a sparse set of manually annotated points in every image. For example, the three images in Fig. 1.3 are not well aligned if the object of interest was the house. However, if one was to manually annotate two corners of the house (crosses) then these could be used to initialize and guide the registration. Depending on the application area and the object of interest, the number of points required to guide the registration may significantly increase. More details on groupwise image registration can be found elsewhere (see e.g. Cootes et al. [17] and Sotiras et al. [66]). In this chapter, the focus is on feature point detection methods to generate the annotations as these provide the opportunity to fully automate the annotation process – even if the images are not aligned and vary significantly in the position, orientation and scale of the object of interest.

1.3 FEATURE POINT DETECTION USING SHAPE MODEL MATCHING

For an automated annotation system to be able to reliably deal with a range of intra-class variation and to provide robust results, it needs to be both general and specific. To accomplish this, it is inevitable for the system to incorporate prior knowledge about the expected shape of the object and/or its texture. Prior knowledge is particularly useful in cases of outliers (e.g. due to occlusions) and noise as well as when there is significant variation in shape and/or texture across images. Most recent feature point detection methods include prior knowledge via machine learning approaches. While this allows the annotation system to learn about the expected shape and/or texture variation, it also means that these methods will only perform well if the shape/texture variation in the training data is representative of the variation in the given object class.[4]

Texture-based feature point detection techniques focus on identifying the best position for every landmark by analyzing the texture in a region of interest of a given image (see e.g. Reinders et al. [61], Feris et al. [32] or Patil et al. [60]). Shape-based approaches, in contrast, aim to identify the shape that best matches the search image. Hence, in the latter a single point is *not* necessarily placed at its best position based on the texture in the region around the landmark point but at its best position given the overall shape in the image. Shape-based methods will only identify point positions that provide an overall *valid* shape example of the object class, which can be a big advantage in some application areas. A variety of shape models can be used in this context, ranging from simple geometric relationships between sets of points [5,58] over graphical representations of shape [28,78] to SSMs. We refer to methods that regularize the output of a feature point candidate detector using an SSM as *shape model matching* methods. Of note is that shape model matching methods often have a local texture-based feature point detection step, and hence tend to combine shape with texture information.

In the following, we describe a range of shape model matching methods.

1.3.1 Active Shape Models

Active Shape Models (ASMs) [16] are early forms of shape model matching methods where, based on a trained SSM and an initial estimate of the position of all landmarks $\mathbf{x}^I$ in the search image, model parameters $\mathbf{b}'$ (Eq. (1.3)) and transformation T_θ are iteratively updated for the SSM to best match the image.[5] Starting from the SSM mean shape (i.e. with $\mathbf{b}'$ initialized to zero) at the estimated position, orientation and scale, a model instance $\mathbf{x}$ is created and the following model matching steps are applied:

1. Each landmark point in $\mathbf{x}$ is optimized by moving it to the locally *best* position $\mathbf{x}_o$;
2. Transformation T_θ and shape model parameters $\mathbf{b}'$ are updated to best fit the optimized positions $\mathbf{x}_o$ of all landmarks. Limits put on the values of $\mathbf{b}'$ ensure that the

---

[4] In this context, the term training data describes the annotated images used to develop the annotation system.

[5] In the literature, the term ASM is often confused with SSM such that ASM often refers to any methodology that combines the output of a set of feature point detectors with the global shape constraints of an SSM – rather than ASM referring to the specific SSM-based shape model matching method described in Cootes et al. [16].

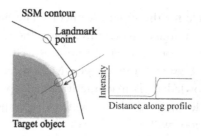

Figure 1.4 Applying edge detection to identify the best landmark point position.

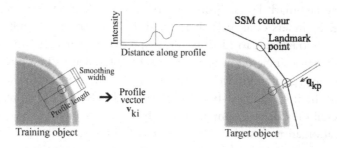

Figure 1.5 Modeling local gray-value profiles to identify the best landmark point position.

optimized model instance is consistent with the shape variation learned from the training images. Model instance **x** is re-calculated;

3. Steps 1 and 2 are repeated until the shape model instance converges to the shape in the search image.

The very first ASM algorithms used edge detection methods based on the profile normal to the shape model contour to identify the locally *best* position for each landmark (see Fig. 1.4). This required all landmarks to lie on the locally strongest edge. However, depending on the dataset, the locally strongest edge may not be the edge representing the contour of the object of interest. Therefore, later ASM versions improved upon this by using statistical models of the gray-value structure along the profile normal to the SSM model contour.

Local Gray-Value Models

ASMs that use local gray-value models learn the nearby structure of every landmark point and hence do not require landmarks to lie on the locally strongest edge but only on locally distinctive structures [15].

Model Building. For every landmark point k, $1 \le k \le \ell$ and every training image i, $1 \le i \le n$, a profile vector $\mathbf{v}_{ki}$ normal to the shape model contour is sampled as illustrated in Fig. 1.5. Either side of the landmark, s pixels are sampled into $\mathbf{v}_{ki}$ of length $2s + 1$. Depending on the dataset, applying a smoothing operation perpendicular to the profile may increase the robustness of the local gray-value model. In addition, sampling the derivatives rather than the smoothed intensity values may help to overcome global gray-value variations. All profile vectors $\mathbf{v}_{ki}$ are normalized such that $|\mathbf{v}_{ki}| = 1$. For every landmark point k, the distribution of the normalized profile vectors $\mathbf{v}_{ki}$ of the training set is assumed to be multivariate Gaussian defined by the covariance matrix and mean over vectors $\mathbf{v}_{ki}$. The probability of the corresponding landmark point in the search image to fit to this distribution is then related to the squared Mahalanobis distance between the normalized profile vector at that point and the mean of the landmark profile model. Minimizing the latter optimizes the quality-of-fit of the landmark point position to the learned profile model for that landmark.

Model Matching. To find the *best* fit to a search image based on an initial estimate, for every landmark point k, a profile of length $2m + 1$, $m \gg s$, centered at the estimated landmark point position is sampled (Fig. 1.5). For each of the possible positions of the learned profile model along this profile sample, quality-of-fit $\mathbf{q}_{kp}$, $1 \leq p \leq 2(m - s) + 1$, is calculated. Landmark point k is then moved to the center of the profile model position that optimizes $\mathbf{q}_{kp}$. This is an iterative process which is done independently for every landmark point. Once the positions of all landmark points have been optimized, transformation T_θ and shape model parameters $\mathbf{b}'$ are updated to best match these positions. As with the strongest edge approach, this procedure is repeated iteratively until the shape model instance converges to the shape in the search image.

To further increase the robustness of ASMs, a multi-resolution framework following a Gaussian pyramid approach can be used. This requires several ASMs to be trained, one for each stage of the framework. During shape model matching this approach involves first a search for the point locations in a coarse, low resolution version of the image and then a series of refinement searches at stages of increasing resolutions. Keeping the number of pixels constant across stages allows the coarse model to take a larger area of the image into account, guiding the model towards the correct position. For more details on how to efficiently match an SSM to a search image using ASMs the reader is referred to [12,14,16].

Despite these improvements to increase the robustness of ASMs, they are very sensitive to the landmark positions used for initialization. This limits their application areas. Particularly in medical imaging, anatomical landmarks may not always be placed on an edge while there may be edges of non-relevant structures close by. Furthermore, the patterns of intensities around anatomical landmarks are not always sufficiently distinctive across subjects. Therefore, ASMs are likely to drift off or get stuck if not initialized appropriately (see Fig. 1.6). Over the years, several extensions to the traditional ASM algorithm have been suggested aiming to overcome these shortcomings [23,50,52,56, 68,75].

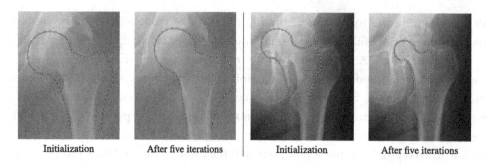

| Initialization | After five iterations | Initialization | After five iterations |

Figure 1.6 Two examples of where an insufficient initialization in position, orientation or scale causes a gray-level model based ASM to get stuck in a local optimum; showing the initialization of the model as well as the result after five search iterations (additional iterations did not improve the performance).

1.3.2 Active Appearance Models

Statistical Appearance Models (SAMs) combine SSMs as described above with a model of texture variation retrieved from a shape-free image patch. Active Appearance Models (AAMs) have been proposed as a shape model matching method that uses a joint shape and texture SAM for feature point detection [13].

Texture Models

Similar to SSMs, texture models use eigenanalysis to build a statistical model of the variation in texture of a target object across a set of annotated training images. To allow for comparison of textures across images, texture variations that can be solely attributed to variations in shape need to be removed. This is achieved by performing shape-normalization to all training images via warping the object of every image into a shape-free patch according to mean shape $\bar{\mathbf{x}}$. Warping is conducted by applying a triangulation algorithm (e.g. Delaunay triangulation), and the gray-value information of the synthesized shape-free patch is then sampled into a texture vector $\mathbf{g}'$.

In addition to texture variations caused by shape change, there may also be texture variations that are due to other reason such as global lighting variations. To account for this, the texture vector in the image coordinate frame needs to be normalized. Projections between image coordinate frame and reference coordinate frame, i.e. normalization and de-normalization, are transformation-based.

Transformation T_ψ that is used to project the object from the model coordinate frame to the image coordinate frame can be defined by

$$T_\psi(\mathbf{g}) = \alpha\mathbf{g} + \beta\mathbf{1} \quad \text{with} \quad \alpha = \mathbf{g}' \cdot \bar{\mathbf{g}} \quad \text{and} \quad \beta = (\mathbf{g}' \cdot \mathbf{1})/k. \tag{1.6}$$

Using the above, texture training vector $\mathbf{g}'$ can be normalized according to

$$\mathbf{g} = T_\psi^{-1}(\mathbf{g}') = (\mathbf{g}' - \beta \mathbf{1})/\alpha \qquad (1.7)$$

where α represents a scaling factor, β gives the mean pixel value as an offset factor, $\mathbf{1}$ is a vector of ones, and k is the number of elements of $\mathbf{g}$, $\mathbf{g}'$ and $\mathbf{1}$. The mean texture vector over all normalized texture training vectors $\mathbf{g}_i$, $1 \leq i \leq n$, is defined by $\bar{\mathbf{g}}$. The latter is recursively calculated and re-estimated using α and β as in Eq. (1.6), where one of the texture training vectors is chosen as the first estimate of the mean.

Based on the set of normalized texture vectors $\mathbf{g}_i$, PCA can then be applied to generate a shape-free texture model

$$\mathbf{g} = \bar{\mathbf{g}} + \mathbf{P}_\tau \mathbf{b}_\tau \qquad (1.8)$$

where the columns of matrix $\mathbf{P}_\tau$ consist of the set of texture eigenvectors describing the texture spectrum given by the data, and $\mathbf{b}_\tau = \mathbf{P}_\tau^T (\mathbf{g} - \bar{\mathbf{g}})$ defines the variation of the gray-value patterns. The texture model described in Eq. (1.8) is defined in the reference coordinate frame. To project the texture model from the reference coordinate frame back to the image coordinate frame, transformation T_ψ can be applied

$$\mathbf{g}' = T_\psi(\mathbf{g}) = \alpha(\bar{\mathbf{g}} + \mathbf{P}_\tau \mathbf{b}_\tau) + \beta \mathbf{1}. \qquad (1.9)$$

As for SSMs, dimensionality reduction can be achieved by only including the first t texture modes that capture a certain proportion of the overall texture variation.

Statistical Appearance Models

SAMs define combined shape and texture models that also capture correlations between shape and texture. According to the above, the shape and appearance of an object of interest can be summarized by shape model parameters $\mathbf{b}_\sigma$[6] and texture model parameters $\mathbf{b}_\tau$. To reflect correlations that are present between shape and texture, appearance vector $\mathbf{b}_\alpha$ can be defined by

$$\mathbf{b}_\alpha = \begin{pmatrix} \mathbf{W}_\sigma \mathbf{b}_\sigma \\ \mathbf{b}_\tau \end{pmatrix} \qquad (1.10)$$

where $\mathbf{W}_\sigma$ is a diagonal matrix of weights to account for the difference in units between the shape and texture models (e.g. coordinates versus gray-values). Applying PCA to the combined appearance vector $\mathbf{b}_\alpha$ defines an SAM

$$\mathbf{b}_\alpha = \mathbf{Q}_\alpha \mathbf{c}_\alpha = \begin{pmatrix} \mathbf{Q}_\sigma \\ \mathbf{Q}_\tau \end{pmatrix} \mathbf{c}_\alpha \qquad (1.11)$$

---

[6] In this section, $\mathbf{P}_\sigma$ and $\mathbf{b}_\sigma$ correspond to $\mathbf{P}$ and $\mathbf{b}$, respectively, as defined in Eq. (1.2).

where $\mathbf{Q}_\alpha$ is a matrix with the combined appearance eigenvectors in the columns, and $\mathbf{c}_\alpha$ is a vector of appearance coefficients describing both the shape and texture of the object. Compared to the shape and texture models defined above, SAMs do not include a mean appearance vector as both the shape and the texture coefficients have zero mean by construction. The shape and texture of an SAM can also be defined directly via

$$\mathbf{x} = \bar{\mathbf{x}} + \mathbf{P}_\sigma \mathbf{W}_\sigma^{-1} \mathbf{Q}_\sigma \mathbf{c}_\alpha \quad \text{and} \quad \mathbf{g} = \bar{\mathbf{g}} + \mathbf{P}_\tau \mathbf{Q}_\tau \mathbf{c}_\alpha. \tag{1.12}$$

Active Appearance Models

Active Appearance Models (AAMs) describe an optimization problem to minimize the difference between an appearance model instance and the object of interest in an image. Given an SAM as described above, the aim is to adjust the model parameters such that the appearance model instance matches the target object as closely as possible.

In contrast to ASMs, instead of optimizing the position of individual landmarks, AAMs aim to match the entire object simultaneously. An AAM is matched to an image by minimizing the residual between the object's texture in the image and that synthesized by the SAM. As defined in Eq. (1.12), both the shape and texture variability of an SAM are regulated by appearance vector $\mathbf{c}_\alpha$. Given a trained SAM, based on an initial estimate of $\mathbf{c}_\alpha$ as well as of the position of all landmarks $\mathbf{x}^I$ in the search image (i.e. an estimate of transformation parameters $\boldsymbol{\theta}$ to account for translation, rotation and scale) a shape model instance in the image coordinate frame can be approximated by $\mathbf{x}^I = T_\theta(\bar{\mathbf{x}} + \mathbf{P}_\sigma \mathbf{W}_\sigma^{-1} \mathbf{Q}_\sigma \mathbf{c}_\alpha)$. To capture the texture variation of the search image, all pixels in the region of shape model instance $\mathbf{x}^I$ are sampled into texture vector $\mathbf{g}'_I$. Transformation $T_\psi^{-1}(\mathbf{g}'_I)$ is then applied to obtain the sampled texture vector in the reference coordinate frame, $\mathbf{g}^I$, where ψ is the vector of transformation parameters α and β. Initial values for $\mathbf{c}_\alpha$ and ψ are learned from the training set. Initial values for $\boldsymbol{\theta}$ can be learned from the training set or obtained from a manual initialization of the pose of the object in the image.

Texture vector $\mathbf{g}^I$ defines the texture of a patch of the object of interest in the search image. The texture of an object patch synthesized by the texture model can be described by texture vector $\mathbf{g}^M = \bar{\mathbf{g}} + \mathbf{P}_\tau \mathbf{Q}_\tau \mathbf{c}_\alpha$. Both the image and the synthesized texture vectors are given in the reference coordinate frame and the difference can be easily determined by

$$\mathbf{r}(\mathbf{p}) = \mathbf{g}^I - \mathbf{g}^M \tag{1.13}$$

with $\mathbf{p}$ defining all required parameters, i.e. appearance model parameters $\mathbf{c}_\alpha$ as well as transformation parameters $\boldsymbol{\theta}$ (translation, rotation and scale) and ψ (scale and offset). To minimize the difference given by vector $\mathbf{r}$, parameters $\mathbf{p}$ need to be incrementally adjusted. Without loss of generality, it is assumed that there is an approximately linear relationship between changes in the parameters and the difference between the image patch and the synthesized model patch.

The difference vector $\mathbf{r}$ is iteratively optimized by minimizing its sum of squares

$$E = \mathbf{r}^T \mathbf{r} \tag{1.14}$$

via incrementally modifying parameter vector $\mathbf{p}$ by some small additive variance $\delta\mathbf{p}$. The latter can be calculated by use of a gradient matrix $\mathbf{R}$ which reflects the linear relationship between changes in $\mathbf{p}$ and the difference between $\mathbf{g}^I$ and $\mathbf{g}^M$

$$\delta\mathbf{p} = -\mathbf{R}\mathbf{r}(\mathbf{p}). \tag{1.15}$$

If $\mathbf{R}$ was needed to be computed at every step then this would be an expensive operation. However, since E is computed in the reference frame, which is consistent across training and search images, $\mathbf{R}$ can be pre-computed from the training data.[7]

To iteratively optimize E and match the SAM instance to the search image, the following steps are applied:

1. Given a normalized image texture vector $\mathbf{g}^I$ and a synthesized model texture vector $\mathbf{g}^M$, difference vector $\mathbf{r}$, error E and update vector $\delta\mathbf{p}$ are calculated.
2. Model parameters $\mathbf{p}$ are updated by adding $\delta\mathbf{p}$, i.e. according to $\mathbf{p} \to \mathbf{p} + \delta\mathbf{p}$.
3. Using the updated model parameters $\mathbf{p}$, shape model instance $\mathbf{x}^I$ and texture model instance $\mathbf{g}^M$ are re-calculated.
4. The updated shape model instance $\mathbf{x}^I$ is used to re-sample the texture of the search image into $\mathbf{g}^{I\prime}$. Transformation T_{ψ}^{-1} is then applied to obtain the updated texture vector $\mathbf{g}^I$ in the reference coordinate frame.
5. Difference vector $\mathbf{r}$ and error value E are re-calculated and re-evaluated based on updated texture vector $\mathbf{g}^I$. If error E has decreased as a result of the re-calculations then update $\delta\mathbf{p}$ of model parameters $\mathbf{p}$ (as in Step 2) is accepted. If, however, the optimization was not successful, then all calculations in Steps 3 to 5 are repeated with some smaller parameter update (e.g. by adding $\frac{1}{2}\delta\mathbf{p}$ instead of adding $\delta\mathbf{p}$).

This procedure is iteratively repeated as long as error E can be minimized, and hence it converges to a locally optimal appearance model instance.

The above only provides a summary of AAMs, more details on AAMs and their variants can be found in [13,35,55].

As with ASMs, the model matching performance of AAMs depends on the local initialization of the model. Fig. 1.7 gives two examples of AAM model matching results where the model initialization and the training data were the same as for Fig. 1.6. This shows that AAMs, compared to ASMs, show improved robustness to the local

---

[7] Though this approach works well for many application areas, several suggestions to improve efficiency and robustness have been proposed. For example, Batur and Hayes [4] suggest to gradually adapt the entries of the gradient matrix, and Matthews and Baker [55] present a modified fitting algorithm that is based on image alignment.

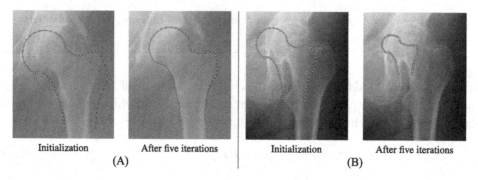

Initialization	After five iterations	Initialization	After five iterations
(A)		(B)	

Figure 1.7 Two examples of AAM model matching results (using standard AAMs as in [13]) highlighting the sensitivity to the initialization in position, orientation or scale; showing the initialization of the model as well as the result after five search iterations (additional iterations did not improve the performance). In (A) the AAM accurately outlined the contour of the proximal femur whereas in (B) it failed completely to match the proximal femur in the image.

initialization in some cases but similar performance for cases where the initialization pose varies significantly from the true pose in the search image. Furthermore, due to the minimization of the difference between the entire object in the search image and the texture model of the object, the AAM model matching process is very sensitive to texture variations between the training images and the search image. Specifically illumination variations pose a major challenge to AAMs (see e.g. [35] for a review of AAM variants addressing this).

1.3.3 Constrained Local Models

Constrained Local Models (CLMs) combine a *global* shape model with *local* texture models for every landmark point, and have been introduced as a more accurate and robust method to combine shape and texture information [21]. They can be considered as a hybrid feature point detection method that models the local texture around each landmark point to identify its best position as well as regularizes the overall output using an SSM. CLMs are very similar to AAMs in the way the combined shape-texture model is built, but CLMs use *localized* texture patches and apply a more robust non-linear search. In the following, the key facts are summarized. The reader interested in more details is referred to [21].

Model Building. As for AAMs, shape and texture models can be defined by

$$\mathbf{x} = \bar{\mathbf{x}} + \mathbf{P}_\sigma \mathbf{b}_\sigma \quad \text{and} \quad \mathbf{g} = \bar{\mathbf{g}} + \mathbf{P}_\tau \mathbf{b}_\tau. \tag{1.16}$$

In contrast to AAMs, for CLMs texture vector $\mathbf{g}'$ is obtained by sampling a rectangular patch from the region around each landmark point. The intensity values of all patches are normalized and concatenated to form a single texture vector per image. The shape, texture and appearance models are then built in the same way as for AAMs such that

$$\mathbf{x} = \bar{\mathbf{x}} + \mathbf{P}_\sigma \mathbf{b}_\sigma = \bar{\mathbf{x}} + \mathbf{P}_\sigma \mathbf{W}_\sigma^{-1} \mathbf{Q}_\sigma \mathbf{c}_\alpha \quad \text{and} \quad \mathbf{g} = \bar{\mathbf{g}} + \mathbf{P}_\tau \mathbf{b}_\tau = \bar{\mathbf{g}} + \mathbf{P}_\tau \mathbf{Q}_\tau \mathbf{c}_\alpha. \qquad (1.17)$$

Model Matching. The combined shape-texture model can be used to search for the object of interest in a new image using the following optimization procedure.

Based on an initial estimate of the position of all landmarks $\mathbf{x}^I$ in the search image, transformation T_θ and shape parameters $\mathbf{b}_\sigma$ can be approximated by fitting the shape model instance to the estimated positions of all landmark points. Texture parameters $\mathbf{b}_\tau$ can be approximated by sampling a patch at the estimated position of every landmark point in the reference frame and fitting the texture model to all sampled patches. After estimating combined parameters $\mathbf{c}_\alpha$, a texture template for each landmark point can be generated.

Using the generated templates, a set of response images (one for every landmark point) is obtained via sampling patches from the area around the current estimate of every landmark point's position and determining their similarity to the texture template for this landmark. Every response image $\mathbf{R}_k$, $1 \leq k \leq \ell$, corresponds to the area around the estimated position of landmark k, and at every position $\mathbf{R}_k$ contains the value of a similarity measure. The latter indicates the quality-of-fit of the local texture template model to the texture of the image at that position.

Based on the obtained response images, the positions of all landmarks k can be updated via searching for peaks in all response images $\mathbf{R}_k$ while also taking the learned shape variation of the object class into account. This is done via fitting the shape model instance so as to maximize the overall quality-of-fit

$$Q(\mathbf{b}_\sigma, \boldsymbol{\theta}) = \omega \sum_{k=1}^{\ell} \mathbf{R}_k(x_k, y_k) - \sum_{j=1}^{t} \frac{b_j^2}{\lambda_j} \qquad (1.18)$$

where b_j are the elements of $\mathbf{b}_\sigma$ and λ_j are the corresponding eigenvalues of the SSM. The first term gives the image matching information while the second term specifies the shape regularization using the squared Mahalanobis distance. Parameter ω is a weight used to balance the relative importance of similarity and shape constraints, and a suitable value can be calculated from the training set. In [21], it was suggested to optimize objective function $Q(\mathbf{b}_\sigma, \boldsymbol{\theta})$ using the Nelder–Mead simplex algorithm [57] but alternative optimization methods could be used.

After the optimization, the positions of all landmark points are updated accordingly and the procedure of generating the texture templates, obtaining the response images and optimizing the model matching are iteratively repeated until convergence.

Different measures of similarity could be used in the above CLM framework. In [21] normalized correlation was used to measure the similarity between the intensities of the

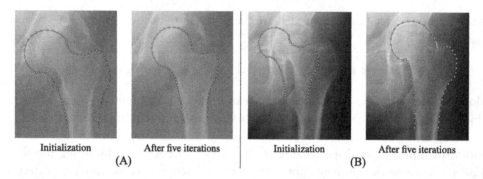

Initialization	After five iterations	Initialization	After five iterations
(A)		(B)	

Figure 1.8 Two examples of CLM model matching results (using normalized correlation as similarity measure) highlighting the sensitivity to the initialization in position, orientation or scale; showing the initialization of the model as well as the result after five search iterations (additional iterations did not improve the performance). In (A) the CLM accurately outlined the contour of the proximal femur whereas in (B) it failed to correctly capture the superior femoral head–neck junction.

local texture templates and the sampled image patches. However, the term "Constrained Local Model" is now used more broadly to describe any method in which local response images are generated and then combined using global shape constraints to optimize the overall fitting. For example, in Saragih et al. [64] this term was used to describe a generic framework for local landmark point models and any kind of shape regularization rather than specifically SSMs.

As for ASMs and AAMs, the model matching performance of CLMs depends on the local initialization of the model. Fig. 1.8 shows two examples of CLM model matching results where the model initialization and the training data were the same as for Figs. 1.6 and 1.7. This demonstrates the improved performance of CLMs over ASMs and AAMs. With similar performance to AAMs in some cases as shown in Fig. 1.8A and more robust performance for cases with a less accurate initialization as in Fig. 1.8B. However, Fig. 1.8B also demonstrates that CLMs may have difficulties in correctly outlining the full contour of an object of interest.

1.3.4 Random Forest Regression-Voting Constrained Local Models

The generation of response images for CLMs can take various forms. A recent approach that has emerged as a powerful and promising technique is to apply Random Forest regression-voting (RFRV) in the CLM framework. RFRV-CLMs use Random Forests [6] to vote for the best position of each landmark point. They combine predictions from a number of independent decision trees and accumulate predictions from multiple regions of the image. Full details are given in [48], the following provides a summary of the approach.

Model Building. As above, to describe the shape variation of the object of interest across all training images, an SSM is built such that the position of each landmark point k, $1 \leq k \leq \ell$, is given by

$$\mathbf{x}_k^I = T_\theta(\bar{\mathbf{x}}_k + \mathbf{P}_k' \mathbf{b}' + \mathbf{r}_k) \tag{1.19}$$

where $\bar{\mathbf{x}}_k$ is the mean position of landmark point k in the reference frame, $\mathbf{P}_k$ gives the modes of shape variation, $\mathbf{b}$ are the shape model parameters, $\mathbf{r}_k$ are the residuals used to allow small deviations from the model, and T_θ applies a global transformation.

Random Forests (RFs) are composed of multiple independent decision trees that are trained independently on a random subset of data. The region of interest of the image that captures all landmark points of the object is re-sampled into a standardized reference frame by applying the inverse of transformation T_θ as described above. One RF is trained for each of the landmark points in $\mathbf{x}$. For every landmark point, a set of patches is sampled at a number of random displacements $\mathbf{d}_j$ from the landmark's true position in the reference frame (with $\mathbf{d}_j$ drawn from a flat distribution $[-d_{max}, +d_{max}]$). A set of features $\mathbf{f}_j$ is then randomly sampled from every patch, and all features per landmark (i.e. over all sampled patches) are used to train a RF regressor on pairs $\{(\mathbf{f}_j, \mathbf{d}_j)\}$ to learn to predict the position of the landmark point. For each tree leaf, both the mean and the standard deviation of the displacements of all training samples arriving at that leaf are stored. In [48], Haar features are used as they have been shown to be effective and can be calculated efficiently from integral images [69]. However, other features could be used within the same RFRV-CLM framework.

Similar as for ASMs, the robustness of RFRV-CLMs can be improved further by applying them in a multi-resolution framework. Here, this is achieved by building a sequence of models with varying resolutions of the reference frame. A common approach is to apply a coarse-to-fine search strategy, i.e. to have a RFRV-CLM with a low resolution reference frame followed by one with a high resolution reference frame.

Model Matching. Given a search image and an initial estimate of the position of all landmarks $\mathbf{x}^I$ in the image, transformation T_θ and shape parameters $\mathbf{b}_\sigma'$ are estimated. To compute all response images $\mathbf{R}_k$ in the reference frame, the region of interest capturing all ℓ landmark points is re-sampled by applying the inverse of the estimated transformation T_θ.

For every landmark point k, a grid over the area around the estimated point position is searched and the relevant feature values (according to the trained RF) from patches at every position within the grid are extracted. The feature values are then used for the trees in the RF to independently vote for the best position of the landmark in an accumulated 2D histogram of votes, yielding response image $\mathbf{R}_k$ (see Fig. 1.9). This is done independently for each of the landmark points.

All response images are then combined with the global shape constraints given by the trained SSM by fitting the shape model instance to all response images (i.e. landmark

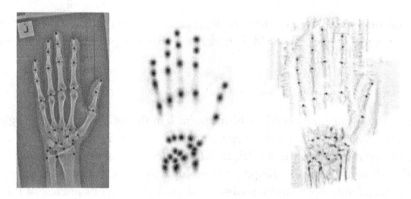

Figure 1.9 Example hand radiograph showing the manual annotation of the joints of the hand with 37 points (left) as well as the superimposed histograms of votes (i.e. response images) for applying a coarse RFRV-CLM (middle) and a fine RFRV-CLM (right).

points) such that the number of votes is maximized over all landmarks. This is done by seeking shape model and pose parameters $\{\mathbf{b}', \boldsymbol{\theta}\}$ that optimize

$$Q(\{\mathbf{b}', \boldsymbol{\theta}\}) = \sum_{k=1}^{\ell} \mathbf{R}_k(T_{\boldsymbol{\theta}}(\bar{\mathbf{x}}_k + \mathbf{P}'_k\mathbf{b}' + \mathbf{r}_k)) \qquad (1.20)$$

$$\text{s.t. } \mathbf{b}'^T\mathbf{S}_b^{-1}\mathbf{b}' \le M_t \text{ and } |\mathbf{r}_k| < r_t$$

where $\mathbf{S}_b$ is the covariance matrix of shape model parameters $\mathbf{b}'$, M_t is a threshold on the Mahalanobis distance, and r_t is a threshold on the residuals. Lindner et al. [48] describe an efficient technique on how to solve this optimization problem.

RFRV-CLM model matching is an iterative process, several iterations of the above can be applied to optimize the result.

Combining the CLM approach with RF regression–voting from the area around the estimated position of a landmark point significantly improves both the model matching performance and the robustness to the initialization in position, orientation and scale. Fig. 1.10 demonstrates the superior model matching performance of RFRV-CLMs over ASMs, AAMs and CLMs when using the same model initialization and training data. This shows that RFRV-CLMs achieve high accuracy in outlining the contour of the proximal femur in both the cases shown in Figs. 1.10A and 1.10B.

In [48], it was shown that RFRV-CLMs perform equally well on non-medical data such as images of the face. Fig. 1.11 shows some example images of applying the RFRV-CLM described in [48] to images from the AFLW dataset [45]. This example application also demonstrates that shape model matching methods do not require the landmark points to be placed along the contour of an object but can also be used to mark key features of an object.

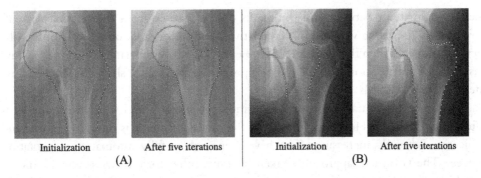

Initialization   After five iterations   Initialization   After five iterations

(A)   (B)

Figure 1.10 Two examples of RFRV-CLM model matching results highlighting their robustness to the initialization in position, orientation or scale; showing the initialization of the model as well as the result after five search iterations.

Figure 1.11 Example images from the AFLW dataset [45] showing the automatically obtained annotations using the RFRV-CLM as described in [48].

RF regression has been shown to be effective for a range of applications. Whereas in the RFRV-CLM framework it is combined with SSMs, there is also various other work where it is combined with alternative ways of describing shape. For some examples, the interested reader is referred to [10,29,43,54]. Furthermore, the application and effectiveness of RF regression is not limited to feature point detection but has also been applied to predict objects and image properties [20,33,37,59,72]. More details on *regression forests* can be found in [19].

The latest developments in the field of feature point detection are exploring how *deep learning* [46] can be applied for accurate and robust shape model matching. The reader is referred to [1,53] for examples of first steps towards this direction.

1.4 FULLY AUTOMATED IMAGE ANALYSIS VIA SHAPE MODEL MATCHING

The above introduced shape model matching techniques are all semi-automatic and require an estimated position of all landmark points to initialize the search. If the position, orientation and scale of the object in the image do not vary significantly across images then an estimated position of all landmarks can be learned from the training data. If, however, there is large variation in the relative position, orientation or scale of the object across images then the model needs to be initialized appropriately on a case-by-case basis. A *good* initialization is all the more important for less robust shape model matching methods such as ASMs.

In practice the initialization of the position, orientation and scale of the model is often done manually. However, initializing the model manually is time-consuming and subjective. Recently, increasing efforts have been made to fully automate the annotation process. The two main approaches taken here are either to initialize the shape model matching step by means of an object detection step, or to directly predict landmark point candidates from the whole image. For examples of the latter, the reader is referred to [10,38,67]. In the following, we provide details on object detection methods, and describe how to apply object detection for fully automated shape model matching.

1.4.1 Object Detection Methods

Early methods of detecting an object, or segmenting its outline, were based on thresholding or edge detection (see e.g. [7,63]). Although these methods can be successful in cases where the object of interest is well-defined in the image (e.g. detecting a black ball in front of a clear blue sky), they are very sensitive to background noise and complex structures. Reasons for failure include varying intensity values for the same object or the same intensity values for different objects as well as additional edges in the image other than the contour of the object of interest.

Most recent developments in object detection follow a sliding window approach where rectangular patches are systematically sampled at multiple positions across the search image at a range of orientations and scales. In the simplest case, the object is then detected by comparing a template of the object to every sampled patch and identifying the patch that best matches the template according to some similarity measure; yielding an estimate of the position, orientation and scale of the object within the image. However, this approach only works well if the object in the image is similar enough to the template, e.g. if there is little intra-class appearance variation.

Various improvements to the above have been introduced to achieve more robust object detection, specifically in the case of large intra-class appearance variation or in the presence of noise and occlusions. A common and effective approach is to combine the sliding window framework with machine learning methods for template matching. Well-known classification-based techniques are, for example, described in Dalal and Triggs [22], Felzenszwalb et al. [31] or Viola and Jones [69]. To also take *global* image context into account, regression-based methods can be applied to *vote* for the best pose of the object in a sliding window framework.

The foundations of regression-based object detection methods were laid by the introduction of the Generalized Hough Transform [3] which uses a gradient orientation-based voting scheme to detect arbitrary shapes. The latter was taken one step further by Leibe et al. [47] and their Implicit Shape Models that match code book entries of local appearance to cast votes for possible positions of the center of the object. More recently, Hough Forests [33] were proposed to overcome the computationally expensive generation of large codebooks and their matching. Hough Forests apply the Generalized Hough Transform in the RF framework and combine classification with regression such that patches containing parts of the object of interest cast votes for possible positions of the center of the object. In the field of medical imaging, a similar approach utilizing multi-class regression forests was used by Criminisi et al. [20] to vote for the positions of the sides of bounding boxes around organs in CT images. Lindner et al. [49] used class-independent regression forests to allow all image structures to vote for the position, orientation and scale of a reference frame that captures the object of interest.

1.4.2 Combining Object Detection with Shape Model Matching

When aiming at the fully automated annotation of an object by combining object detection with shape model matching, the overall performance depends on both the estimated position, orientation and scale of the object in the image as well as the convergence range of the model matching technique.

For the shape model matching methods described in Section 1.3, a reference frame is used to give a standardized bounding box for all landmark points. To utilize an object detection step for initialization of any of these methods, it is therefore sufficient to estimate the position, orientation and scale of the bounding box. The latter will then provide an estimate of the position of all landmark points within the reference frame based on the SSM mean shape. Any object detector could be trained and applied to predict the bounding box but ideally the detector would provide not only an estimate of the position but also of the orientation and scale of the bounding box. The more accurate the pose of the bounding box can be predicted the better the initialization of the shape model matching technique will be. In [49], regression forests were used to initialize a RFRV-CLM, and it was shown that the automatically initialized search performed equally well to a search initialized with the mean shape at the correct pose (as given by manual annotations).

Fig. 1.12 demonstrates the performance of fully automatically annotating the left proximal femur in *full* pelvic images by combining object detection with shape model matching. All searches were initialized with the result of a class-independent regression forest to predict the position, orientation and scale of the reference frame [49], followed by five iterations of shape model matching per method.

The results show that the overall performance of a fully automated annotation system is dependent on both the performance of the object detector as well as the accuracy and

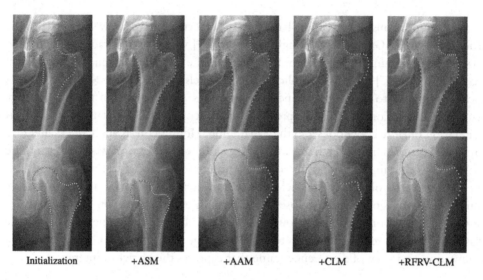

Initialization   +ASM   +AAM   +CLM   +RFRV-CLM

Figure 1.12 Two examples (top and bottom) of fully automatically annotating the proximal femur by combining object detection with shape model matching. All searches were run on full pelvic images. The initialization was given by fitting the mean SSM shape to the result of a class-independent regression forest that predicted the position, orientation and scale of the reference frame [49]. Five search iterations of each of the shape model matching methods were applied to the initialized mean shape to obtain the shown results.

robustness of the shape model matching method. The initialization of the mean shape from the object detector is more accurate in the top example, leading to overall better annotation results (i.e. across shape model matching methods). In the bottom example, the initialization from the object detector is not sufficient for some of the investigated shape model matching methods, causing them to get trapped in a local optimum away from the true contour of the object. In these examples, the fully automated annotation system that combined object detection with a RFRV-CLM performed best and achieved an average point–to–curve error of 0.5 mm when compared to the manual ground truth.

1.5 AUTOMATED IMAGE INTERPRETATION AND ITS APPLICATIONS

Once an automated landmark point annotation system is available, SSMs as well as SAMs can be easily obtained and used for further analysis of the object in the image. There are various fields and applications where this would be of benefit, and the following provides a brief overview of some of the application areas.

Medical imaging is one of the fields with a substantial body of research being undertaken to develop automated techniques for improved interpretation and understanding of medical images. To name just a few, SSM- and/or SAM-based methods are developed to improve the diagnosis and treatment of musculoskeletal disorders such as

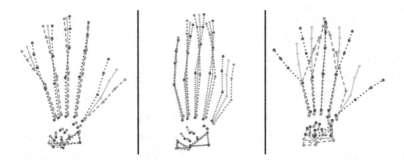

Figure 1.13 Example to demonstrate the importance of the shape variation expressed by the training data: Main mode of shape variation in hand radiographs (mean ± 2.5 SD) for three SSMs each trained on (left) images with the fingers spread; (middle) images with the fingers close together; and (right) both images with the fingers spread and images with the fingers close together. In each image the joints of the hand were annotated with 37 points.

osteoporosis [9,39] and osteoarthritis [8,40,51,70] as well as for orthopaedic implants and surgery planning [65]. They are further applied to areas such as the analysis of body decomposition [71] in tackling obesity, or facial expression interpretation in assessing depression [11].

Furthermore, SSMs and SAMs provide many opportunities in areas outside of medical imaging. For example, analyzing and interpreting images of the face can be used for avatar animation in the film and gaming industry [2,73] as well as for human face identification [24] and age estimation [36] in forensics. Face tracking is also commonly applied in advanced driver-assistance systems to ensure that the driver's attention is directed at the road [62]. In addition, SSM- and/or SAM-based methods show promise for their application in industrial quality assurance when applied as part of an automated industrial inspection procedure [34]. An everyday example of the application of statistical shape and appearance models is given by devices such as smartphones or digital cameras where the models are used to offer the user various ways of retouching their photos such as improving the lighting or slimming people's faces [18].

1.6 LIMITATIONS OF STATISTICAL SHAPE MODELS FOR IMAGE INTERPRETATION

When working with or applying SSMs, an important factor that needs to be kept in mind is that the shape variation represented by the model is based on the shape variation expressed by the data that was used to train the model. Therefore, a model resulting from a non-representative training dataset may not match all shapes of an object class. Fig. 1.13 demonstrates this by visualizing the first mode of shape variation for three different hand SSMs, one trained on images with the fingers spread, a second one trained on images with the fingers close together, and a third one trained on both images with the fingers spread and images with the fingers close together. The first two SSMs will not fit each other's training data very well, and will result in large residuals and insufficient shape model matching performance (when applied as part of a fully automated annotation system). The third model, however, has encountered a larger variation of shapes during training and provides a better representation of the object class. This model is therefore likely to fit to a range of hand images, independently of whether the fingers are spread or close together. While the shape variation exhibited by an SSM can be useful to restrict the shape model matching to valid shapes only, it also carries the danger of being too restrictive to sufficiently represent the object class of interest. This equally applies to automated methods that apply SSMs for shape model matching. An SSM should, hence, always be trained using representative data for the expected shape variation across the object class.

In addition, SSMs are based on linear PCA. Therefore, they are aimed at shape distributions that can be described as multivariate Gaussian. Their application may be inadequate for non-linear distributions where they may overestimate the number of shape parameters needed to approximate the data, or where the model may not be able to fully explain the non-linear shape variation.

1.7 CONCLUSION

In this chapter, we have described the core methodology behind Statistical Shape Models and have demonstrated how they can be applied to automatically analyze a range of image types. One of the big advantages of these models is that they provide a way of quantitatively describing the overall shape of an object rather than reducing it to a limited set of measurements, while at the same time allowing for the reduction of the complexity of the shape representation. In addition, they allow for the synthesis of new shapes. Many interesting extensions to the core methodology could not be presented here, simply for reasons of limited space, and we refer the reader to the references for further details. We hope that this chapter has conveyed the main concepts of building and matching SSMs, demonstrated their usefulness and inspired the reader to read on.

ACKNOWLEDGEMENTS

The author is grateful to Prof. Tim Cootes for fruitful discussions, to Dr. Paul Bromiley for supplying the data for Fig. 1.11, and to Jessie Thomson for helpful comments on the chapter. The author acknowledges funding from the Engineering and Physical Sciences Research Council, UK (EP/M012611/1).

REFERENCES

[1] B. Aubert, C. Vazquez, T. Cresson, S. Parent, J. De Guise, Automatic spine and pelvis detection in frontal X-rays using deep neural networks for patch displacement learning, in: 2016 IEEE 13th International Symposium on Biomedical Imaging (ISBI), 2016, pp. 1426–1429.

[2] I. Bacivarov, P. Corcoran, Facial expression modeling using component AAM models – gaming applications, in: 2009 International IEEE Consumer Electronics Society's Games Innovations Conference, 2009, pp. 1–16.

[3] D. Ballard, Generalizing the Hough transform to detect arbitrary shapes, Pattern Recognit. 13 (1981) 111–122.

[4] A. Batur, M. Hayes, Adaptive active appearance models, IEEE Trans. Image Process. 14 (2005) 1707–1721.

[5] P. Belhumeur, D. Jacobs, D. Kriegman, N. Kumar, Localizing parts of faces using a consensus of exemplars, in: Proceedings CVPR 2011, IEEE Press, 2011, pp. 545–552.

[6] L. Breiman, Random forests, Mach. Learn. 45 (2001) 5–32.

[7] J. Canny, A computational approach to edge detection, IEEE Trans. Pattern Anal. Mach. Intell. 8 (1986) 679–698.

[8] M. Castaño-Betancourt, J. Van Meurs, S. Bierma-Zeinstra, F. Rivadeneira, A. Hofman, H. Weinans, A. Uitterlinden, J. Waarsing, The contribution of hip geometry to the prediction of hip osteoarthritis, Osteoarthr. Cartil. 21 (2013) 1530–1536.

[9] I. Castro-Mateos, J. Pozo, T. Cootes, M. Wilkinson, R. Eastell, A. Frangi, Statistical shape and appearance models in osteoporosis, Curr. Osteoporos. Rep. 12 (2014) 163–173.

[10] C. Chen, W. Xie, J. Franke, P. Grutzner, L.P. Nolte, G. Zheng, Automatic X-ray landmark detection and shape segmentation via data-driven joint estimation of image displacements, Med. Image Anal. 18 (2014) 487–499.

[11] J. Cohn, T. Kruez, I. Matthews, Y. Yang, M. Nguyen, M. Padilla, F. Zhou, F. De la Torre, Detecting depression from facial actions and vocal prosody, in: 2009 3rd International Conference on Affective Computing and Intelligent Interaction and Workshops, 2009, pp. 1–7.

[12] T. Cootes, Model-based methods in analysis of biomedical images, in: R. Baldock, J. Graham (Eds.), Image Processing and Analysis: A Practical Approach, Oxford University Press, 2000, pp. 223–248.

[13] T. Cootes, G. Edwards, C. Taylor, Active appearance models, IEEE Trans. Pattern Anal. Mach. Intell. 23 (2001) 681–685.

[14] T. Cootes, A. Hill, C. Taylor, L. Haslam, The use of active shape models for locating structures in medical images, Image Vis. Comput. 12 (1994) 355–366.

[15] T. Cootes, C. Taylor, Statistical models of appearance for medical image analysis and computer vision, in: M. Sonka, K. Hanson (Eds.), Proceedings SPIE 2001, 2001, pp. 236–248.

[16] T. Cootes, C. Taylor, D. Cooper, J. Graham, Active shape models – their training and application, Comput. Vis. Image Underst. 61 (1995) 38–59.

[17] T. Cootes, C. Twining, V. Petrovic, K. Babalola, C. Taylor, Computing accurate correspondences across groups of images, IEEE Trans. Pattern Anal. Mach. Intell. 32 (2010) 1994–2005.

[18] P. Corcoran, C. Stan, C. Florea, M. Ciuc, P. Bigioi, Digital beauty: the good, the bad, and the (not-so) ugly, IEEE Consum. Electron. Mag. 3 (2014) 55–62.

[19] A. Criminisi, J. Shotton (Eds.), Decision Forests for Computer Vision and Medical Image Analysis, Springer, 2013.

[20] A. Criminisi, J. Shotton, D. Robertson, E. Konukoglu, Regression forests for efficient anatomy detection and localization in CT studies, in: B. Menze, G. Langs, Z. Tu, A. Criminisi (Eds.), Proceedings MICCAI 2010 – Workshop MCV, Springer, 2010, pp. 106–117.

[21] D. Cristinacce, T. Cootes, Automatic feature localisation with constrained local models, J. Pattern Recognit. 41 (2008) 3054–3067.

[22] N. Dalal, B. Triggs, Histograms of oriented gradients for human detection, in: Proceedings CVPR 2005, 2005, pp. 886–893.

[23] L. Dang, F. Kong, Facial feature point extraction using a new improved active shape model, in: Third International Congress on Image and Signal Processing (CISP), 2010, pp. 944–948.

[24] A. Dantcheva, P. Elia, A. Ross, What else does your biometric data reveal? A survey on soft biometrics, IEEE Trans. Inf. Forensics Secur. 11 (2016) 441–467.

[25] R. Davies, C. Twining, T. Cootes, J. Waterton, C. Taylor, A minimum description length approach to statistical shape modelling, IEEE Trans. Med. Imaging 21 (2002) 525–537.

[26] R. Davies, C. Twining, C. Taylor, Statistical Models of Shape: Optimisation and Evaluation, Springer, London, 2008.

[27] M. de Bruijne, M. Nielsen, Shape particle filtering for image segmentation, in: Proceedings MICCAI 2004, Springer, 2004, pp. 168–175.

[28] R. Donner, B. Menze, H. Bischof, G. Langs, Fast anatomical structure localization using top-down image patch regression, in: Proceedings MICCAI 2012 – Workshop MCV, Springer, 2013, pp. 133–141.

[29] R. Donner, B. Menze, H. Bischof, G. Langs, Global localization of 3D anatomical structures by pre-filtered Hough forests and discrete optimization, Med. Image Anal. 17 (2013) 1304–1314.

[30] I. Dryden, K. Mardia, Statistical Shape Analysis, Wiley, New York, 1998.

[31] P.F. Felzenszwalb, R. Girshick, D. McAllester, D. Ramanan, Object detection with discriminatively trained part based models, IEEE Trans. Pattern Anal. Mach. Intell. 32 (2010) 1627–1645.

[32] R. Feris, J. Gemmell, K. Toyama, V. Kruger, Hierarchical wavelet networks for facial feature localization, in: Proceedings of the Fifth International Conference on Automatic Face and Gesture Recognition, 2002, pp. 118–123.

[33] J. Gall, V. Lempitsky, Class-specific hough forests for object detection, in: Proceedings CVPR 2009, IEEE Press, 2009, pp. 1022–1029.

[34] H. Gao, C. Ding, C. Song, J. Mei, Automated inspection of E-shaped magnetic core elements using K-tSL-center clustering and active shape models, IEEE Trans. Ind. Inform. 9 (2013) 1782–1789.

[35] X. Gao, Y. Su, X. Li, D. Tao, A review of active appearance models, IEEE Trans. Syst. Man Cybern. 40 (2010) 145–158.

[36] X. Geng, Z. Zhou, K. Smith-Miles, Automatic age estimation based on facial aging patterns, IEEE Trans. Pattern Anal. Mach. Intell. 29 (2007) 2234–2240.

[37] R. Girshick, J. Shotton, P. Kohli, A. Criminisi, A. Fitzgibbon, Efficient regression of general-activity human poses from depth images, in: Proceedings ICCV 2011, IEEE Press, 2011, pp. 415–422.

[38] B. Glocker, D. Zikic, E. Konukoglu, D. Haynor, A. Criminisi, Vertebrae localization in pathological spine CT via dense classification from sparse annotations, in: Proceedings MICCAI 2013 – Part II, Springer, 2013, pp. 262–270.

[39] S. Goodyear, R. Barr, E. McCloskey, S. Alesci, R. Aspden, D. Reid, J. Gregory, Can we improve the prediction of hip fracture by assessing bone structure using shape and appearance modelling? Bone 53 (2013) 188–193.

[40] J. Gregory, J. Waarsing, J. Day, H. Pols, M. Reijman, H. Weinans, R. Aspden, Early identification of radiographic osteoarthritis of the hip using an active shape model to quantify changes in bone morphometric features, Arthritis Rheum. 56 (2007) 3634–3643.

[41] J. Guo, X. Mei, K. Tang, Automatic landmark annotation and dense correspondence registration for 3D human facial images, BMC Bioinform. 14 (2013) 232.

[42] T. Heimann, H. Meinzer, Statistical shape models for 3D medical image segmentation: a review, Med. Image Anal. 13 (2009) 543–563.

[43] B. Ibragimov, B. Likar, F. Pernus, T. Vrtovec, A game-theoretic framework for landmark-based image segmentation, IEEE Trans. Med. Imaging 31 (2012) 1761–1776.

[44] D. Kendall, The diffusion of shape, Adv. Appl. Probab. 9 (1977) 428–430.

[45] M. Kostinger, P. Wohlhart, P. Roth, H. Bischof, Annotated facial landmarks in the wild: a large-scale, real-world database for facial landmark localization, in: Proceedings ICCV 2011 – Computer Vision Workshop, IEEE Press, 2011, pp. 2144–2151.

[46] Y. LeCun, Y. Bengio, G. Hinton, Deep learning, Nature 521 (2015) 436–444.

[47] B. Leibe, A. Leonardis, B. Schiele, Combined object categorization and segmentation with an implicit shape model, in: Proceedings ECCV 2004 – Workshop on Statistical Learning in Computer Vision, 2004, pp. 17–32.

[48] C. Lindner, P. Bromiley, M. Ionita, T. Cootes, Robust and accurate shape model matching using random forest regression-voting, IEEE Trans. Pattern Anal. Mach. Intell. 37 (2015) 1862–1874.

[49] C. Lindner, S. Thiagarajah, M. Wilkinson, The arcOGEN Consortium, G. Wallis, T. Cootes, Fully automatic segmentation of the proximal femur using random forest regression voting, IEEE Trans. Med. Imaging 32 (2013) 1462–1472.

[50] J. Liu, J. Udupa, Oriented active shape models, IEEE Trans. Med. Imaging 28 (2009) 571–584.

[51] J. Lynch, N. Parimi, R. Chaganti, M. Nevitt, N. Lane, The association of proximal femoral shape and incident radiographic hip OA in elderly women, Osteoarthr. Cartil. 17 (2009) 1313–1318.

[52] M. Mahoor, M. Abdel-Mottaleb, Facial features extraction in color images using enhanced active shape model, in: Proceedings of the Seventh International Conference on Automatic Face and Gesture Recognition, 2006, pp. 144–148.

[53] A. Mansoor, J. Cerrolaza, R. Idrees, E. Biggs, M. Alsharid, R. Avery, M. Linguraru, Deep learning guided partitioned shape model for anterior visual pathway segmentation, IEEE Trans. Med. Imaging 35 (2016) 1856–1865.

[54] B. Martinez, M. Valstar, X. Binefa, M. Pantic, Local evidence aggregation for regression-based facial point detection, IEEE Trans. Pattern Anal. Mach. Intell. 35 (2013) 1149–1163.

[55] I. Matthews, S. Baker, Active appearance models revisited, Int. J. Comput. Vis. 60 (2004) 135–164.

[56] S. Milborrow, F. Nicolls, Locating facial features with an extended active shape model, in: Proceedings ECCV 2008, Springer, 2008, pp. 504–513.

[57] J. Nelder, R. Mead, A simplex method for function minimization, Comput. J. 7 (1965) 308–313.

[58] J. Nicolle, K. Bailly, V. Rapp, M. Chetouani, Locating facial landmarks with binary map cross-correlations, in: Proceedings ICIP 2013, IEEE Press, 2013, pp. 501–508.

[59] M. Oda, N. Shimizu, K. Karasawa, Y. Nimura, T. Kitasaka, K. Misawa, M. Fujiwara, D. Rueckert, K. Mori, Regression forest-based atlas localization and direction specific atlas generation for pancreas segmentation, in: Proceedings MICCAI 2016 – Part II, Springer, 2016, pp. 556–563.

[60] R. Patil, V. Sahula, S. Mandal, Automatic detection of facial feature points in image sequences, in: Proceedings ICIIP 2011, 2011, pp. 1–5.

[61] M. Reinders, R. Koch, J. Gerbrands, Locating facial features in image sequences using neural networks, in: Proceedings of the Second International Conference on Automatic Face and Gesture Recognition, 1996, pp. 230–235.

[62] M. Rezaei, R. Klette, Look at the driver, look at the road: no distraction! No accident!, in: Proceedings CVPR 2014, 2014, pp. 129–136.

[63] P. Sahoo, S. Soltani, A. Wong, A survey of thresholding techniques, Comput. Vis. Graph. Image Process. 41 (1988) 233–260.

[64] J. Saragih, S. Lucey, J. Cohn, Deformable model fitting by regularized landmark mean-shift, Int. J. Comput. Vis. 91 (2011) 200–215.

[65] N. Sarkalkan, H. Weinans, A. Zadpoor, Statistical shape and appearance models of bones, Bone 60 (2014) 129–140.

[66] A. Sotiras, C. Davatzikos, N. Paragios, Deformable medical image registration: a survey, IEEE Trans. Med. Imaging 32 (2013) 1153–1190.

[67] D. Stern, T. Ebner, M. Urschler, From local to global random regression forests: exploring anatomical landmark localization, in: Proceedings MICCAI 2016 – Part II, Springer, 2016, pp. 221–229.

[68] B. van Ginneken, A. Frangi, J. Staal, B. ter Haar Romeny, M. Viergever, Active shape model segmentation with optimal features, IEEE Trans. Med. Imaging 21 (2002) 924–933.

[69] P. Viola, M. Jones, Rapid object detection using a boosted cascade of simple features, in: Proceedings CVPR 2001, IEEE Press, 2001, pp. 511–518.

[70] J. Waarsing, R. Rozendaal, J. Verhaar, S. Bierma-Zeinstra, H. Weinans, A statistical model of shape and density of the proximal femur in relation to radiological and clinical OA of the hip, Osteoarthr. Cartil. 18 (2010) 787–794.

[71] B. Xie, J. Avila, B. Ng, B. Fan, V. Loo, V. Gilsanz, T. Hangartner, H. Kalkwarf, J. Lappe, S. Oberfield, K. Winer, B. Zemel, J. Shepherd, Accurate body composition measures from whole-body silhouettes, Med. Phys. 42 (2015) 4668–4677.

[72] C. Xu, A. Nanjappa, X. Zhang, L. Cheng, Estimate hand poses efficiently from single depth images, Int. J. Comput. Vis. 116 (2016) 21–45.

[73] L. Zalewski, S. Gong, 2D statistical models of facial expressions for realistic 3D avatar animation, in: Proceedings CVPR 2005, 2005, pp. 217–222.

[74] M. Zelditch, D. Swiderski, H. Sheets, W. Fink, Geometric Morphometrics for Biologists, Elsevier Academic Press, New York, 2004.

[75] F. Zhang, R. Shen, L. Xu, G. Chen, J. Wang, Real time facial feature localization based on improved active shape model, in: Fifth International Congress on Image and Signal Processing (CISP), 2012, pp. 921–925.

[76] W. Zhang, S. Brady, Feature point detection for non-rigid registration of digital breast tomosynthesis images, in: Digital Mammography, Springer, 2010, pp. 296–303.

[77] Y. Zheng, B. Georgescu, D. Comaniciu, Marginal space learning for efficient detection of 2D/3D anatomical structures in medical images, in: Proceedings IPMI 2009, Springer, 2009, pp. 411–422.

[78] F. Zhou, J. Brandt, Z. Lin, Exemplar-based graph matching for robust facial landmark localization, in: Proceedings ICCV 2013, IEEE Press, 2013, pp. 1025–1032.

CHAPTER 2

Statistical Deformation Model: Theory and Methods

Xiahai Zhuang*, Yipeng Hu†
*School of Data Science, Fudan University, Shanghai, China
†Centre for Medical Image Computing, Department of Medical Physics and Biomedical Engineering, University College London, London, United Kingdom

Contents

2.1	Introduction	34
2.2	Deformation Representation	35
	2.2.1 Ubiquitous Displacement Models	36
	2.2.2 Rigid and Nonrigid Models	37
	2.2.3 Global and Local Models	39
	2.2.4 Heuristic, Biophysical and Statistical Deformation Models	40
2.3	Statistical Approaches	40
	2.3.1 Principal Component Analysis	41
	2.3.2 Multiple Regression Analysis	42
	2.3.3 Kernel Regression of Deformations	44
	2.3.3.1 Computation on Deformation Manifold	44
	2.3.3.2 Regression of Subject-Specific Pathological Condition	46
	2.3.4 Common Space and Model Generalization	47
	2.3.5 A Note on Population	48
2.4	General-Purpose Deformation Models	48
	2.4.1 A Kernel-Based Approach	48
	2.4.1.1 Smoothing and Approximation	50
	2.4.1.2 Thine-Plate Spline and Multi-Quadrics Splines	51
	2.4.1.3 Gaussian Kernels and Other Locally Supported Models	51
	2.4.1.4 Elastic Body Spline	52
	2.4.2 Local Deformation Models	53
	2.4.2.1 Free Form Deformation (FFD)	53
	2.4.2.2 Nonrigid Registration	54
	2.4.2.3 Locally Affine Transformation (LAT)	54
	2.4.2.4 Locally Affine Registration Method Using Mutual Information	56
2.5	Biophysics-Based Deformation	58
	2.5.1 Biomechanical Modeling with Finite Element Analysis	58
	2.5.1.1 Meshing	59
	2.5.1.2 Material Properties	61
	2.5.1.3 Boundary Conditions	62
	2.5.1.4 Statistical Modeling of Finite Element Analysis	63

Statistical Shape and Deformation Analysis
DOI: 10.1016/B978-0-12-810493-4.00003-1

2.5.2   Modeling Intervention-Induced Deformation and Diffusion-Based Modeling of
           Biological Changes   63
References   64

2.1 INTRODUCTION

There is no shortage of literatures which survey the deformation models in medical image analysis and also in several other related fields such as computer vision. Most of them deal with elaborate mathematical framework which does not take into account the details of implementation and, quite rightly, they usually list the relevant applications without discussing practical experience in greater depths for the purpose of technical review. Therefore, it is very difficult to serve as a tutorial to help beginners to start. For instance, B-spline-based free-form deformation (which is covered in Section 2.4.2.1) has gained its great popularity since 1990s, later has been trialled out with virtually all imaging modalities from many potential applications. To the best of the author's knowledge, many existing image guidance systems have not formally adopted this approach. This may reflect the fact that there has been a short of papers reporting negative results. We feel that this experience, albeit somewhat intuitive and empirical, needs to be clarified for practitioners in this field. A good approach to achieve this goal is to go through implementing and experimenting.

The purpose of this chapter is to introduce some basic concepts, together with the implementation details using common programming languages. This will not only help to the reading of this chapter but also to the screening of the existing literatures. Although no official code goes along with the chapter. The outlook of this textbook style chapter is to enable readers with basic programming experience to make their own working applications to implement or adapt applied methods to the application to hand. As a slight difference to formal research papers, we only list the reference complementing the content of this chapter in terms of implementation, in order to minimize the overhead in reading this chapter.

We discuss a class of statistical models of deformation which, by its literal definition, is a subset of statistical models that are used to describe, summarize and predict deformation encountered in medical image analysis. *Deformation* of an organ, a tumor or certain piece of biological tissue of interest within the same patient, is usually caused by gross patient movement, periodical respiratory and cardiac motions, interventions (e.g. by force exerted from surgical instruments), or natural process (e.g. growth, response to treatment). *Motion* is often used to refer to this description, changes within the same subject. In a broader definition, however, deformation can also refer to the change between two subjects, i.e. difference of the same organ between image data from two or more patients. This potential ambiguity can be avoided if one always consider what the *population* is of interest when the deformation is modeled. We will deal with this statistical concept. In this chapter, we do not limit ourselves to either type but distinguishing these two does help in a number of practical aspects, such as in choosing the more appropriate model given a problem.

Moreover, in both scenarios when biological tissue deforms, different biological processes may occur, the state-of-the-art deformation models only deal primarily with morphological changes, i.e. position, size, shape etc., while some pioneering study investigated more fundamental changes in tissue deformation, therefore changes in image intensity values cannot be obtained by interpolation. The models described in this chapter may or may not be suitable to handle but some potentially relevant methodologies are discussed in Section 2.5.

A note on the terminology is to point out: although the term "statistical deformation model" has been popular since the contributions from [17], where dense displacement fields (DDF) produced by image registration were considered as representation of the deformation, researchers have been using this term in different contexts, most of which were covered by the description in the above paragraph. Statistical motion models, for instance, is an example effort to distinguish the deformation caused by patient-specific motion from other types of deformation.

2.2 DEFORMATION REPRESENTATION

From a practical perspective, we classify the deformation models based on underlying representation that links the human perception of how things of interest deform to some mathematical formulation so it can be quantified. Examples range from the most basic point cloud, which samples displacement of a set of points, to the more complex spline-based parametric models that have built-in smoothness constraints. This can be presented, following the notations used in [7], as follows:

$$\mathbf{x}_A = \mathrm{T}(\mathbf{x}_B) \qquad (2.1)$$

where $\mathbf{x}_A$ and $\mathbf{x}_B$ are *positions* (position vectors) from images A and B, respectively. The position vectors are represented by the spatial coordinates in 3D.[1] In the context of medical image, the positions are the spatial locations of voxels, of a resampled grids (e.g. free-form transformation) or other regions of interest. An inherently related problem is medical image registration, in which the goal of registering two images is to find a spatial transformation T that best aligns *corresponding locations*[2] using image intensities or features

---

[1] We only consider 3D in this chapter as it is widely accepted as standard in medical imaging community. Most algorithms and methods can be reduced to 2D, whilst the projective transformation is usually not considered as deformation in the special case of modeling 2D-to-3D, although some of the discussion in this chapter may still apply.

[2] The term *corresponding features* may have different meanings besides locations, *corresponding location* is used here to distinguish between the features used in *feature registration*.

$A(\mathbf{x}_A)$, $\mathbf{x}_A \in \Omega_A$ and $B(\mathbf{x}_B)$, $\mathbf{x}_B \in \Omega_B$, which are functions of spatial locations $\mathbf{x}_A$ and $\mathbf{x}_B$, and Ω_A and Ω_B are domains over which the intensities/features representing image A and B are sampled, respectively. In this chapter, the primary goal is to investigate various forms of the transformation T, in particular, a parametric representation $\mathrm{T}(\boldsymbol{\theta})$ that is governed by the *transformation parameters*, in vector $\boldsymbol{\theta}$. A registration or other equivalent tasks, such as segmentation, becomes an optimization for $\boldsymbol{\theta}$. We focus on the parametric models in this chapter. The primary reason is that the statistical approaches applied on these representations are much better developed with parametric models. For statistical modeling analysis, nonparametric models, in general, can be re-parameterized.

2.2.1 Ubiquitous Displacement Models

T can take in theory any form to the extent that every image voxel or feature retains completely separate displacement components to transform B to A. It first looks like a lazy option but nonetheless has been adopted in a wide range of applications. Two examples are dense displacement field (DDF) and point cloud representation for image voxels and features respectively.

The DDF has x-, y- and z-displacement components assigned to each voxel, re-sulting in three volumes of the same size of the original scalar[3] image to represent displacement in all directions. In practice, it can be written in the original image file format with three separate files, or a single file having a "4D" data.

Similarly, a point cloud is a collection of 3-tuple vectors representing displacement for all points sampled from regions of interest. The point cloud itself may represent a surface of an organ or a more "solid" form, by which the internal points are also sampled. One example that will be discussed further in the Application chapter is the vertices of a tetrahedron mesh from finite element analysis. Therefore, for the mth $(m = 1, \ldots, M)$ location:

$$\mathrm{T}(\mathbf{x}_m) = \mathbf{d}_m, \quad m \in [1, M] \tag{2.2}$$

where $\mathbf{d}_m$, $m = 1, \ldots, M$ are the displacement vectors and M is the number of points. This representation is efficient for forward- or backward transformation, although in the latter, an inverse transformation is usually computed so the displacement will be defined on the voxels or features on A:

$$\mathbf{x}_B = \mathrm{T}^{-1}(\mathbf{x}_A) \tag{2.3}$$

If the transformation is linear, one can take the inverse of the transformation matrix directly. Otherwise, the inverse transformation can be computed numerically using a number of algorithms, such as the Dynamic Resampling And distance Weight-ing (DRAW) interpolation [25,26]. It is a simple interpolation scheme that assigns a

---

[3] Scalar image where the intensity values are only available information.

weighted sum of displacements to each voxel, with the weights being inverse distances from K closest points $\mathbf{x}_k$ with forward displacement vector $\mathbf{d}_k$:

$$\mathrm{T}^{-1}(\mathbf{x}_m, K) = -\frac{1}{\sum_{k=1}^{K} \|\mathbf{x}_k - \mathbf{x}_m\|_2^{-e}} [\mathbf{d}_k^T]_K^T \cdot [\|\mathbf{x}_k - \mathbf{x}_m\|_2^{-e}]_K \qquad (2.4)$$

where $\|\mathbf{x}_k - \mathbf{x}_m\|_2$ indicates the Euclidean distance between $\mathbf{x}_m$ and the kth neighbor point and e determines the smoothness of the resulting deformation. Please notice that the superscript T indicates transpose of a matrix or vector in this chapter. The matrix multiplication calculates the inverse-distance-weighted sum for all x-, y- and z-components of the displacement matrix. The matrix form is also handy once an algebra numerical computing package is available. The distances can usually be efficiently and partially pre-computed using many well-packaged K-Nearest Neighbor (k-NN) algorithms, such as K-D Tree. It is also a useful algorithm in some regression methods and kernel-based spline calculations. Although not necessary for some transformation models, this interpolation scheme generally applies to compute displacement field from locations where no *control points*[4] are defined.

It is with some tasks when the exhaustive displacement representation encounters problems. One of such scenario is transformation estimation, i.e. registration. The little constraint imposed by the individual displacement vectors, in general, leads to an ill-posed problem, having only a third of information from the image intensity data. The other potential problem is that analysis of the transformation having high dimensionality can be challenging. The reminder of the chapter is looking at these problems with the aims of reducing the dimensionality and/or adding constraints to the displacement.

2.2.2 Rigid and Nonrigid Models

Nonrigid transformations are discussed in this chapter, hence the word *deformation* in the chapter title. However, it is important to acknowledge that the rigid-body transformation is not only a basis of several nonrigid deformation models but also is often combined with other nonrigid transformation models that excluding the rigid component. In most of the real clinical scenarios, significant rigid transformations are usually present due to patient movement and change of coordinate systems. It represents the rotation and translation as follows:

$$\mathbf{x}_A = \mathrm{T}_{r6}(\mathbf{x}_B) = \mathbf{R}\mathbf{x}_B + \mathbf{t} \qquad (2.5)$$

where $\mathbf{x}_A$ and $\mathbf{x}_B$ are two 3-dimensional position vectors, containing 3 elements to represent the spatial co-ordinates (of points or of pixels/voxels). $\mathbf{t}$ is a translation vector

---

[4] Control points are the locations where transformation is explicitly defined. These are usually anatomical landmarks, fiducials and regular grid points such as those in the free-form deformations.

$\mathbf{t} = [t_x, t_y, t_z]^T$. $\mathbf{R}$ is the 3×3 rotation matrix constructed by 3 rotation angles. For instance, $\mathbf{R}$ is defined in 3D as:

$$\mathbf{R} = \begin{bmatrix} 1 & 0 & 0 \\ 0 & \cos\theta_x & -\sin\theta_x \\ 0 & \sin\theta_x & \cos\theta_x \end{bmatrix} \cdot \begin{bmatrix} \cos\theta_y & 0 & \sin\theta_y \\ 0 & 1 & 0 \\ -\sin\theta_y & 0 & \cos\theta_y \end{bmatrix} \cdot \begin{bmatrix} \cos\theta_z & -\sin\theta_z & 0 \\ \sin\theta_z & \cos\theta_z & 0 \\ 0 & 0 & 1 \end{bmatrix} \tag{2.6}$$

where θ_x, θ_y and θ_z are Euler angles representing the rotation about x-, y- and z-axis, respectively. This is not a unique formulation as it depends on the order in which these matrices are multiplied and the direction of rotation. Therefore, the constrained matrix $\mathbf{R}$ is preferred to generally represent rotation, subject to $\mathbf{R}^T\mathbf{R} = \mathbf{I}$, $|\mathbf{R}| = 1$, where $\mathbf{I}$ is the identity matrix and $|\mathbf{R}|$ is the determinant of the rotation matrix. An isotropic scaling factor s may be included to extend the rigid transformation as follows:

$$\mathbf{x}_A = T_{r7}(\mathbf{x}_B) = s\mathbf{R}\mathbf{x}_B + \mathbf{t} \tag{2.7}$$

More generally, an affine transformation describes a group of transformations preserving straight lines, including scaling, shearing, reflection and rigid transformations. It can be defined as:

$$\mathbf{x}_A = T_{affine}(\mathbf{x}_B) = \mathbf{A}\mathbf{x}_B + \mathbf{t} \tag{2.8}$$

where $\mathbf{A}$ is an "unconstrained" 3×3 matrix. Due to similar properties in representation, with higher DOFs (up to 12 with additional 3 DOF from translation $\mathbf{t}$), the affine transformation is sometimes referred to as a generalized rigid transformation [1]. In most cases, image re-positioning, scaling and/or distortion can be approximated by an affine transformation.

The rigid transformation and its variants all enjoy decades of development since the prevalence of personal computers. The physical meaning of each parameter is clearly distinguished. Perhaps more importantly, the solutions are well established. Given correspondent location vectors $\mathbf{x}_A$ and $\mathbf{x}_B$, $\mathbf{R}$ and $\mathbf{t}$ may be solved using singular value decomposition (SVD) as follows:

$$\mathbf{R} = \mathbf{V} \begin{bmatrix} 1 & 0 & 0 \\ 0 & 1 & 0 \\ 0 & 0 & \det(\mathbf{V}\mathbf{U}^T) \end{bmatrix} \mathbf{U}^T \tag{2.9}$$

$$\mathbf{t} = \bar{\mathbf{x}}_A - \mathbf{R}\bar{\mathbf{x}}_B \tag{2.10}$$

where $\mathbf{U}$ and $\mathbf{V}$ can be computed by decomposing the matrix $\mathbf{X}_B\mathbf{X}_A^T$ containing mean-subtracted data, where, for A and B, we have:

$$\mathbf{X} = \left[(\mathbf{x}_n - \overline{\mathbf{x}})^T \right]_{m=1,\dots,M}^T \tag{2.11}$$

and

$$\overline{\mathbf{x}} = \frac{1}{M} \sum_{m=1}^{M} \mathbf{x}_m \tag{2.12}$$

as:

$$\mathbf{X}_B\mathbf{X}_A^T = \mathbf{U\Sigma V}^T \tag{2.13}$$

The determinant is useful to correct the reflection so the resulting transformation does not include such physically implausible flipping. Many numerical packages have efficient SVD algorithms implemented. SVD again can become useful in linear regression and principal component analysis, both of which are discussed in this chapter. The estimate of the *pose*, i.e. rotation and translation, is also known as absolute orientation problem or equivalent Procrustes problem. A least square solution also exists based on a different formulation for rotation using quaternions [9]. The existence of a direct, fast and convergence guaranteed solution is practically desirable for a transformation model.

The less constrained affined transformation can be solved by a linear least-squares approach:

$$\mathbf{A} = \mathbf{X}_A\mathbf{X}_B^T \left(\mathbf{X}_B\mathbf{X}_B^T \right)^{-1} \tag{2.14}$$

$$\mathbf{t} = \overline{\mathbf{x}}_A - \mathbf{A}\overline{\mathbf{x}}_B \tag{2.15}$$

Inclusion of any additional parameters, beyond rotation and translation, requires careful consideration and justification. In practice, unnecessary difficulties may arise due to the additional degrees of freedom, for example a larger search space in a numerical optimization scheme.

2.2.3 Global and Local Models

There has not been a rigorous distinction between a global and local deformation model. For instance, some kernel-based nonrigid transformation models have so-called compact support, i.e. a movement at one spatial position affects only small region of neighbor locations. They may be considered local but, in theory, it is more global than certain articulated transformation, where each individual joint can move independently. A rather consensual view is that the free-form based and other nonparametric (such as fluid) deformation are more global than a single kernel-based (such as TPS) deformation, whilst the rigid and affine transformations are global. The former usually can only be driven by more data using methods such as the intensity based registration schemes unless heavily regularized, in which case, the local properties are inevitably penalized.

2.2.4 Heuristic, Biophysical and Statistical Deformation Models

There is an engineering approach and there is a physical approach. Convenience in sampling, mathematical representation, or intuitive smoothness constraints motivate the first type of models. We discuss these heuristic *general-purpose* models in details in Section 2.4. *Biophysics-based* deformation models, which only allow physically plausible deformation of biological tissues, are described in Section 2.5. The boundary between these two classes can be overlapping at times as one can add more constraints so a general-purpose deformation model behaves more realistically, or marginalize some assumption in certain biophysics-based model so it becomes more generally applicable, in particular, using a statistical approach.

Statistical models, on the other hand, have capability to describe, summarize or predict certain properties from a set of sample data. In the context of this chapter, the data are deformations that can take either form of general-purpose, biophysics-based or, recursively, nested statistical deformation models.

These two types of deformation models are somewhat related to the concept of diffeomorphism. Conceptually, diffeomorphism is the property of a transformation that is differentiable and has a differentiable inverse. The transformation is therefore smooth, invertible, and has one-to-one mapping. Most biophysical models describing physical motion, for instance, are diffeomorphic, while models that intentionally produce topological change or other types of physical change are not. Whether the intended deformation needs to be diffeomorphic requires careful consideration and depends on the particular biological process and the tolerance of the approximation. When it is required, numerical approaches are possible to maintain this property for general-purpose deformation models. An example for locally affine transformation can be found in [25, 26], where regularization is triggered once the determinant of the Jacobian matrix is smaller than certain pre-defined threshold value.

Some parts of the contents in this chapter are reorganized and extended based on the two PhD theses of the authors [10,22]. Where deemed appropriate, references are given mainly for the purpose for more detailed treatment of the mentioned methods. A comprehensive survey style listing of papers is intentionally dismissed.

2.3 STATISTICAL APPROACHES

From a parametric perspective, the parameter estimation problem for deformation modeling has soon been found a too big problem in many applications. When the underlying deformation model is of high dimensionality, statistical approaches can serve as a dimensionality reduction technique, such as principal component analysis (PCA), arguably, the most popular technique in statistical deformation modeling. When the underlying deformation mode is nonlinear, some statistical analysis can also "re-parameterise" the deformation so the resulting parameters are linear, that has many computationally convenient properties, in particular, for subsequent statistical analysis of the deformation. PCA also can "linearise" the deformation data, but we demonstrate this concept using multiple regression.

In this section, we use two examples of statistical approaches to construct the statistical deformation models based on principal component analysis and multiple linear regression. There are numerous good tutorials about these two techniques. They are simply formalized here to enable implementation without too much in-depth discussion. Interested readers are referred to relevant materials for detailed motivation, derivation and variants.

2.3.1 Principal Component Analysis

Assume a set of N sample data representing certain deformation of interest $\mathrm{T}(\mathbf{x}_n)$, $n = 1, \ldots, N$, in terms of parameter vectors $\mathbf{x}_n$, each having a length of Q. When the deformation model is DDF or point cloud which has M sampled locations, $Q = 3M$, where 3 displacement components for each location are the parameters of the deformation. The PCA can be performed by decomposing the covariance matrix of the sample data, using Eigen-decomposition, as it may be an intuitive way to explain that the *principal modes* are the new basis vectors (eigenvectors) corresponding to the largest P ($P \leq Q$) component variances (eigenvalues). We, however, use an equivalent SVD to solve this problem efficiently by decomposing the data matrix:

$$\frac{1}{\sqrt{N-1}}(\mathbf{X} - \bar{\mathbf{x}} \cdot \mathbf{1}_N^T) = \mathbf{U}\mathbf{\Sigma}\mathbf{V}^T \tag{2.16}$$

where $\bar{\mathbf{x}} = \frac{1}{N}\sum_{n=1}^{N} \mathbf{x}_n$ is the mean data and $\mathbf{1}_N$ is a vector of N ones. By taking only the first P right-singular vectors $\mathbf{P}$, we can approximate the original data retaining most of variance observed in the data. It can be shown that:

$$\mathbf{x}_n = \bar{\mathbf{x}} + \mathbf{V}\mathbf{a}_n \approx \bar{\mathbf{x}} + \mathbf{P}\mathbf{b}_n \tag{2.17}$$

Hence, the deformation becomes a set of linearly combined P new basis:

$$\mathbf{x} = \bar{\mathbf{x}} + \mathbf{P}\mathbf{b} \tag{2.18}$$

PCA has a very wide applicability. One can directly apply this on the parameters of the deformation, although its linearization would be dependent on the underlying deformation model. Alternatively, one can sample DDF or point cloud then apply the PCA directly on the displacement, so the resulting model is linear to the displacement at the sampled locations. The PCA-based statistical deformation models, derived in this way, look like an algebraic process. It is the approximation in the definition equation, which we did not elaborate, that reflects its statistical properties. It is equivalent to $Pr(\mathbf{X}) \approx Pr(\mathbf{B})$.

2.3.2 Multiple Regression Analysis

With the same data, we can use regression analysis to build a statistical deformation model. As one of many supervised learning techniques, it applies on labeled data. For instance, the *predictor* may be an independent signal indicating respiratory cycle, which can be reasonably assumed to govern the repeating motion pattern caused by breathing. Predicting the motion of a liver, for instance, may be achieved by regression between this predictor and the parameter of the underlying deformation parameters. The deformation, sampled at N time points can be "labelled" by the corresponding respiratory signals sampled at the same phase of the respiratory cycle.

Taking[5] a vector respiratory signal $\mathbf{y}_n$ at time point n, we can assume a vector function relationship between $\mathbf{y}$ and $\mathbf{x}$:

$$\mathbf{x} = \mathbf{f}(\mathbf{y}) \tag{2.19}$$

For each x_q, $q = 1, \ldots, Q$, N equations can be listed using the labeled samples:

$$\begin{bmatrix} x_{q,1} = f_q(\mathbf{y}_1) \\ \vdots \\ x_{q,N} = f_q(\mathbf{y}_N) \end{bmatrix} \tag{2.20}$$

Depending on the application, f_q, $q = 1, \ldots, Q$ need not to be in the same form and, computationally, these can be Q different multiple regression analysis, i.e. multivariate multiple regression. Finding the optimal functional form is probably a central problem to statistical learning research, here we only discuss *linear regression*, which can be efficiently solved by least-squares. However, it does not necessary mean only linear relationship between predictor and response (i.e. deformation in this context) can be modeled. Linear regression is referred to as a class of functions, which have a linear combination of the *(regression) parameters*. Nonlinear functions of each predictor and/or of a subset of multiple predictors, when appropriate, can be applied. Examples include various functions: any degree polynomial function, logistic function and any kernel-based function. Note that high degree of freedom functions need more data to have a least-squares solution and may require extra regularization to avoid over-fitting. Now, re-write the set of N

---

[5] This is a different notation from conventional regression analysis, in which $\mathbf{x}$ usually represents predictor whilst $\mathbf{y}$ represents the response. It is, however, for maintaining consistency throughout the chapter.

in terms of regression parameters $\mathbf{r}_q$:

$$x_{q,n} = f_q(\mathbf{y}_n) = \boldsymbol{\varphi}_{q,\mathbf{y}_n} \mathbf{r} \tag{2.21}$$

where $\mathbf{f}_{q,\mathbf{y}_n}^*$ is a set of linear terms in a row vector, functions of $\mathbf{y}_n$, weighted by $\mathbf{r}$. In the matrix notation:

$$\mathbf{x}_q = \boldsymbol{\Phi}_q \mathbf{r}_q \tag{2.22}$$

where $\boldsymbol{\Phi}_q = [\boldsymbol{\varphi}_{q,\mathbf{y}_n}]_n$ is also known as design matrix. Therefore, the regression parameters can be obtained as follows:

$$\mathbf{r}_q = \left(\boldsymbol{\Phi}_q^T \boldsymbol{\Phi}_q\right)^{-1} \boldsymbol{\Phi}_q^T \mathbf{x}_q \tag{2.23}$$

In practice, the regularization parameter λ is set to a small constant to avoid over-fitting while maintaining acceptable residuals. The solution becomes:

$$\mathbf{r}_q = \left(\boldsymbol{\Phi}_q^T \boldsymbol{\Phi}_q + \lambda \mathbf{I}\right)^{-1} \boldsymbol{\Phi}_q^T \mathbf{x}_q \tag{2.24}$$

Inference made with the results from regression analysis on a statistical deformation model can also be considered as conditional distribution of the original deformation distribution on the predictors. It becomes: $Pr(\mathbf{X} \mid \mathbf{Y}) = Pr(\mathbf{R})$. In other words:

$$\boldsymbol{\varphi}_{q,\mathbf{y}_{new}} = E[\boldsymbol{\Theta} \mid \mathbf{y}_{new}] = f_q(\mathbf{y}_{new}) \tag{2.25}$$

Furthermore, it is usually also useful to take into account the shape and size of the organ for an individual patient. The predictor vector can include those and the same approach described above then is adopted.

A problem remaining unsolved is what these linearly combined terms should be for a particular application. Unfortunately, there is no easy answer to this. We describe one special case, where the *kernel function* is used to describe the relationship between the predictors and responses:

$$\boldsymbol{\varphi}_{q,\mathbf{y}_n} = K(\mathbf{y}, \mathbf{y}_{i0}) r_{q,n} \tag{2.26}$$

$$\boldsymbol{\Phi}_\mathbf{m} = \begin{bmatrix} K(\mathbf{y}_{10}, \mathbf{y}_{10}) - \overline{\varphi}_1 & \cdots & K(\mathbf{y}_{10}, \mathbf{y}_{I0}) - \overline{\varphi}_I \\ \vdots & \ddots & \vdots \\ K(\mathbf{y}_{I0}, \mathbf{y}_{10}) - \overline{\varphi}_1 & \cdots & K(\mathbf{y}_{I0}, \mathbf{y}_{I0}) - \overline{\varphi}_I \end{bmatrix} \tag{2.27}$$

where $\overline{\varphi}_k = \frac{1}{I} \sum_{i=1}^{I} K(\mathbf{x}_{k0}, \mathbf{x}_{i0})$, kernel function $K(\mathbf{x}, \mathbf{x}')$ can take many forms, a popular choice is a Gaussian kernel with kernel parameter h, $K(\mathbf{x}, \mathbf{x}') = \exp(-\|\mathbf{x} - \mathbf{x}'\|^2 / 2h^2)$. This parameter can be determined by a nested cross validation. In the next section, we describe a kernel regression method for deformations and the efficient computation on the nonlinear deformation manifold.

These are two examples, how to construct a statistical model with an underlying deformation model. There have been alternatives combining statistical models and others, which are often also known as statistical deformation models. For instance, Monte-Carlo simulations can be augmented on to any existing deformation model to increase its variability. Instead of obtaining a deterministic solution, a statistical method can be directly considered in solving some physical equations to obtain a stochastic solution, such as the proper generalized decomposition (PGD) [3]. Statistical methods can also be used to optimize the computation based on what can be summarized (conditioned) from the application of interest, such as reduced order modeling technique [20]. These, however, are out of the scope of this short chapter.

2.3.3 Kernel Regression of Deformations

In statistical deformation modeling, kernel regression is a non-parametric technique to estimate the conditional expectation of a variable, such as a shape defined by a deformation to the reference, relative to a random variable, such as an indicator of pathological conditions,

$$E(T_c \mid c) = m(c)$$

where $m(\cdot)$ is the function to estimate the expectation relative to the pathological condition descriptor c. Regression of deformations requires the computation on the manifold of deformation. In the following two subsections, we first describe the computation on deformation manifold and then introduce the kernel regression to estimate the expectation of shape given the pathological condition related to a specific subject.

2.3.3.1 Computation on Deformation Manifold

Computation on the manifold of deformations is useful for population-based studies such as regression modeling and manifold learning. To better represent the natural variability of shapes, many works proposed to employ the computation via a Riemannian manifold, instead of the Euclidean vector space. Similarly, the diffeomorphic deformation manifold of medical images can be studied in a subset of a Riemannian manifold, where the distance metric can be naturally defined as the geodesic distance in the manifold and computed via velocity vectors on the Euclidean vector space [27]. In this deformation manifold, the transformation between two coordinates is mapped to a velocity vector field, and the distance of the transformation can be defined as the Euclidean distance of the velocity vectors. In this setting, one can parameterize the shape of an image using a diffeomorphic coordinate transformation between the studied image and the reference image, which is defined as the origin of the coordinates. This provides the mechanism to study the variations of shapes via the computation on diffeomorphic deformation fields.

Given a set of deformations defined on a manifold, $\{T_i | T_i \in \mathbb{T}\}$, the general weighted Fréchet mean is given by

$$m(\{T_i\}) = \underset{\overline{T}}{\operatorname{argmin}} \sum_{i=1...N} w_i \operatorname{dist}(T_i, \overline{T})^2 \qquad (2.28)$$

where $\operatorname{dist}(T_i, \overline{T})$ is the distance between two deformations and $w_i = \frac{1}{N}$ if the same weight is used for all observations. For example, by assuming $\mathbb{T}$ is the Euclidean vector space, the distance metric is then defined as the Euclidean distance, $\operatorname{dist}(T_i, \overline{T}) = |T_i - \overline{T}|_2$. However, diffeomorphism is more representative for image and shape modeling, and therefore $\mathbb{T}$ should be studied in a Riemannian manifold.

The Lagrangian ordinary differential equation (ODE) can be used to estimate a solution for a diffeomorphic transformation between the coordinates on which the studied images or shapes lie, $\frac{d}{ds}\phi_s(\mathbf{x}) = v_s(\mathbf{x})$, where $s \in [0, 1]$ is the time parameter, T_s and v_s are the displacement and velocity vectors respectively, and $T_s(\mathbf{x}) = \mathbf{x} + \phi_s(\mathbf{x})$.

When the constant velocity (flow) field is assumed for a subset of a diffeomorphism manifold, one can estimate the corresponding constant velocity using the Log-Euclidean framework, $v(\mathbf{x}) = \log \phi(\mathbf{x}) = \log(T - \operatorname{Id})(\mathbf{x})$. Similarly, the transformation can be recovered from the constant velocity $T(\mathbf{x}) = \mathbf{x} + \exp v(\mathbf{x})$.

The advantage of using the constant velocity field representation for deformations is that it provides a well-defined distance metric via a Euclidean norm, $\operatorname{dist}(T_1, T_2) = |\log \phi_1 - \log \phi_2|_2 = |v_1 - v_2|_2$. Note that this computation does not guarantee a geodesic distance, but only on the local tangent space of the deformation manifold near the identity.

With the computation of distance, one can compute the Fréchet mean of a set of deformations,

$$m(\{T_i\})(\mathbf{x}) \equiv \mathbf{x} + \exp \overline{v}(\mathbf{x}) = \mathbf{x} + \exp\left(\sum_{i=1}^{N} w_i v_i(\mathbf{x})\right) \qquad (2.29)$$

The velocity vectors are on the Euclidean vector space, hence the Fréchet mean of them can be directly computed. Fig. 2.1 provides the weighted Fréchet mean of two deformations to the reference image, with $w_1 \in [0, 1]$ for the one from the hypertrophic cardiomyopathy and $w_2 = 1 - w_1$ for the other from the right ventricle hypertrophy. The images are generated by applying the deformations to the reference image.

Computing the geodesic distance via velocity fields also provides a mechanism for fast computation of Fréchet mean or expectation of deformations, which can potentially lead to an easy implementation of the Nadaraya–Watson kernel regression function for regression modeling and manifold learning studies.

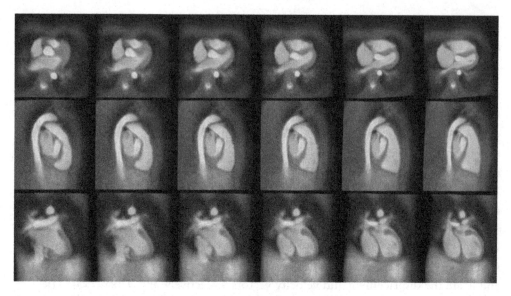

Figure 2.1 Three orthogonal views of the 6 intermediate shapes between the hearts of hypertrophic cardiomyopathy (left ventricle hypertrophy) and right ventricle hypertrophy.

2.3.3.2 Regression of Subject-Specific Pathological Condition

The Nadaraya–Watson kernel regression estimates the expectation using a locally weighted average of the training samples. In the manifold of deformations, the regression is expressed as follows,

$$m(c) \equiv \underset{T \in \mathbb{T}}{\operatorname{argmin}} \frac{\sum_{i=1}^{N} K_h(\operatorname{dist}(c, c_i)) \operatorname{dist}(T, T_i)^2}{\sum_{j=1}^{N} K_h(\operatorname{dist}(c, c_j))}, \tag{2.30}$$

which is the weighted Fréchet mean, where $w_i = \frac{K_h(\operatorname{dist}(c, c_i))}{\sum_{j=1}^{N} K_h(\operatorname{dist}(c, c_j))}$ and $K_h(\cdot)$ is the kernel function for estimating the distribution.

Let I_r be the reference image, such as a known model or atlas image, and $\{I_i\}$ be a set of existing training instances. The shape of a training image is defined by the deformations $\{T_i\}$ computed from an accurate registration between the training image and the reference. For a target image I_u and the pathological condition c, the expected shape of the pathological condition can be estimated using the regression model described above, given the distance metric of pathological conditions is well defined. In [27], the NMI (normalized mutual information) is used to estimate the distance and the kernel function is defined as follows,

$$K_h\big(\operatorname{dist}(c, c_i)\big) = \begin{cases} a(\operatorname{NMI}(I_i, I_u) - b), & \text{if } \operatorname{NMI}(I_i, I_u) - b > \epsilon \\ 0, & \text{otherwise} \end{cases} \tag{2.31}$$

where a will disappear when plugging this formula into the regression equation, b can be estimated using the minimal NMI value between the target image and all the training data. This can be regarded as a regress to the mean shape of a pathological condition represented by the target image I_u.

2.3.4 Common Space and Model Generalization

Raw data in medical image problems come in different forms, such as contours drawn by clinicians, image voxels, vector fields, anatomical landmarks, segmented surfaces. The deformation models we wish to apply statistical analysis on are also different. For some model parameters, it may be a straightforward application as described in the previous sections. However, for the other ones, there is no obvious reason that, for instance, the first voxel – at the top left corner of the first slice – should be in the same dimensionality across different data. This is the so-called *correspondence* problem in statistical shape modeling. This problem is usually considered as a registration problem that finds the corresponding locations between two images, and the deformation is inevitably defined on spatial locations. However, we would like to note that the fundamental goal is to "re-arrange" the deformation data so they can be parameterized in the common high dimensional space for statistical analysis. This should make sense to statistical analysis and does not necessarily need to be spatially correspondent. The choice of the common space is arguably the most important aspect in building a statistical deformation model and it is closely related to the generalization ability the resulting model exhibits.

Moreover, we can only have limited number of data, in particular, in medical image analysis where well protocoled image data are costly, to denote and to organize. Even in the situations where synthetic data such as those from simulations are used, the limitation then is not so much from the size of data set but rises from approximations within the simulation itself, which would never capture the true deformation without deviation. We will elaborate on this point further in the next section.

For the general-purpose models, the *generalizability* and *specificity* can either or both pose a challenge. For instance, a pure elastic model can warp a banana shape to an apple, i.e. a case of lack of specificity. On the other hand, a rigid transformation model would not be able to transform one apple into another due to the lack of *generalizability*. Similarly in medical images, a biomechanical model describing the biomechanical motion of an organ may not be flexible enough to describe a pathological change due to the growth of a tumor.

Statistical approaches provide a means to further constrain the underlying model by imposing a bound on each of the parameters of the model. Monte-Carlo type methods can enhance a very constrained model to artificially improve the model generalization ability.

In practice, fitting the model to new unseen data is a way to measure its generalization ability, and cross validation can be used for obtaining a summary result.

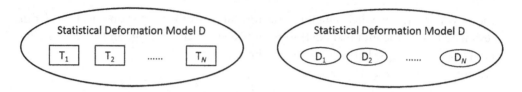

Figure 2.2 Illustration of the population of deformation with underlying deformation representation.

2.3.5 A Note on Population

As indicated in the left part of Fig. 2.2, we interpret the concept of statistical deformation model as: A model describing a deformation population, built by applying statistical approaches on a set of training data of deformation, represented by an underlying parametric deformation model.

Once the statistical model is built, it is ready to make *inference*, i.e. making conclusions about the population. For example, the PCA-based methods re-parameterize the deformation population with fewer and linear parameters; while the regression-based methods classify the original population into sub-populations based on predictors. It is important to emphasize that, modelers should be clear about what the population they are trying to summarize and to what population the inference can be made. For instance, if the model is built using static MR images of female patients, it is unlikely to generalize to prostate cancer cases.

In particular, when the underlying deformation model itself is a statistical model, different types of populations and sub-populations can be involved. For instance, if each underlying deformation model, as indicated in the right part of Fig. 2.2, describing the subject-specific deformation for individual patient and the training data are from different patients. The overall statistical model is no longer a subject-specific model, which might have poor generalization ability if used to model an individual, subject-specific deformation. One solution to this is, again, using statistical approach, such as supervised regression methods, to condition on other available information, so the subject-specific model is inferred.

2.4 GENERAL-PURPOSE DEFORMATION MODELS

2.4.1 A Kernel-Based Approach

We describe a class of global nonlinear transformation models using kernel-based approaches. Kernel-based transformations are related to statistical kernel density estimation and kernel methods in machine learning. For representing a deformation, a kernel function $K(\mathbf{x}, \mathbf{x}_i)$, $i = 1, \ldots, M$ can be considered as a nonlinear scalar distance function between a set of M *control points* $\mathbf{x}_i$ and the location of interest $\mathbf{x}$. In 3D, the displacement $\mathbf{u}(\mathbf{x}) = [u_1(\mathbf{x}), u_2(\mathbf{x}), u_3(\mathbf{x})]^T$, so that $T(\mathbf{x}) = \mathbf{u}(\mathbf{x}) + \mathbf{x}$, at this location then is weighted by a set of coefficients α_i:

$$u_d(\mathbf{x}) = \sum_{i=1}^{M} \alpha_{i,d} K(\mathbf{x}, \mathbf{x}_i), \quad \text{for } d = 1, 2, 3 \tag{2.32}$$

It is this property, having an influence by all the control points, for which transformation models of this type are usually considered as global. It depends on the form of K, and the resulting transformation may or may not be solvable, when estimating $\alpha_i = [\alpha_{i,1}, \alpha_{i,2}, \alpha_{i,3}]^T$ from a set of control points with known displacement. Therefore, it is necessary to add a set of polynomial terms as follows:

$$u_d(\mathbf{x}) = \sum_{i=1}^{M} \alpha_{i,d} K(\mathbf{x}, \mathbf{x}_i) + \sum_{j=1}^{J} \beta_{j,d} P_j(\mathbf{x}) \tag{2.33}$$

where $\boldsymbol{\beta}_j = [\beta_{j,1}, \beta_{j,2}, \beta_{j,3}]^T$ are sets of coefficients in the polynomials. $P_j(\mathbf{x})$ is the J polynomial term up to certain degree. For instance, when first degree polynomials are used, J equals to the number of dimensions of the data plus a constant term, i.e. $J = 4$ for constructing a 3D transformation. On the other hand, if solvability can be ensured by the kernel functions alone, $J = 0$. We shall summarize the corresponding complexity of polynomials and different forms of the kernel functions in the next section. Similar to the kernel functions described earlier, they are also known as the basis functions of the transformation.

The mathematical convenience lies at the fact that, despite of the nonlinear basis functions, the coefficients are linearly combined. Given M control points $\mathbf{x}_m$ with known displacement $\mathbf{u}(\mathbf{x}_m)$, we have, in vector function form:

$$\mathbf{u}(\mathbf{x}_m) = \sum_{i=1}^{M} \alpha_i K(\mathbf{x}_m, \mathbf{x}_i) + \sum_{j=1}^{J} \boldsymbol{\beta}_j P_j(\mathbf{x}_m) \tag{2.34}$$

with the following additional constraints so α_i and $\boldsymbol{\beta}_j$ can be combined in one single linear system:

$$\sum_{i=1}^{M} \alpha_i P_j(\mathbf{x}_i) = \mathbf{0}, \quad \text{for } j = 1, \ldots, J \tag{2.35}$$

Now, α_i and $\boldsymbol{\beta}_j$ can be solved by linear least-squares, using a numerical computation package. And,

$$\mathbf{D} \begin{bmatrix} \mathbf{A} \\ \mathbf{B} \end{bmatrix} = \begin{bmatrix} [\mathbf{T}(\mathbf{x}_m)]_{m=1,\ldots,M}^T \\ \mathbf{0} \end{bmatrix} \tag{2.36}$$

where

$$\mathbf{D} = \begin{bmatrix} \mathbf{K} & \mathbf{P} \\ \mathbf{P}^T & \mathbf{0} \end{bmatrix},$$

and $\mathbf{K} = [K(\mathbf{x}_m, \mathbf{x}_i)]_{m=1,\dots,M, i=1,\dots,M}$ and $\mathbf{P} = [P_j(\mathbf{x}_m)]_{m=1,\dots,M, i=1,\dots,M}$ are $M \times M$ and $M \times J$ submatrix of $\mathbf{D}$ respectively;

$$\mathbf{A} = \begin{bmatrix} \alpha_{1,1} & \alpha_{1,2} & \alpha_{1,3} \\ \vdots & \vdots & \vdots \\ \alpha_{M,1} & \alpha_{M,2} & \alpha_{M,3} \end{bmatrix} \quad \text{and} \quad \mathbf{B} = \begin{bmatrix} \beta_{1,1} & \beta_{1,2} & \beta_{1,3} \\ \vdots & \vdots & \vdots \\ \beta_{J,1} & \beta_{J,2} & \beta_{J,3} \end{bmatrix}$$

are the coefficients to solve; $[\mathrm{T}(\mathbf{x}_m)]_{m=1,\dots,M}^T$ is a $M \times 3$ matrix containing the transformed control points. There are three separate linear systems that can be solved by many least squares solutions. One of them is to use a general solution with SVD:

$$\begin{bmatrix} \mathbf{A} \\ \mathbf{B} \end{bmatrix} = \mathbf{U}\boldsymbol{\Sigma}^{-1}\mathbf{V}^T\mathbf{Z} \tag{2.37}$$

where, $\mathbf{D} = \mathbf{U}\boldsymbol{\Sigma}\mathbf{V}^T$ denotes SVD solution.

An advantage of these kernel-based deformation models is that the solution is straightforward in estimating the deformation parameters, given the correspondence features, i.e. paired control point locations. For applications where correspondence is unknown, many algorithms have been proposed to iteratively find the fitted deformation parameters and the best correspondence, e.g. [5,11,15]. In each iteration, the optimal deformation can be efficiently estimated. A detailed discussion of these methods is beyond the scope of this chapter, but an example will be given in the Application chapter in utilizing a statistical motion model for feature-based registration problems.

2.4.1.1 Smoothing and Approximation

Similar to the rigid regression, additional smoothness can be imposed on the kernel submatrix $\mathbf{K}$ to penalize the norm. It has an interpretation as the landmark localization error (or equivalently, fiducial localization error). The algorithm can be modified to include an additional term when evaluating the kernel matrix:

$$\mathbf{D}^* = \begin{bmatrix} \mathbf{K} + \lambda \mathbf{I}\omega & \mathbf{P} \\ \mathbf{P}^T & \mathbf{0} \end{bmatrix} \tag{2.38}$$

where λ is the scalar parameter to weight vector $\boldsymbol{\omega}$ containing point localization errors in terms of their isotropic inhomogeneous variance, $\boldsymbol{\omega} = [\frac{1}{\sigma_1^2}, \frac{1}{\sigma_2^2}, \dots, \frac{1}{\sigma_M^2}]$. The anisotropic error for individual control points can also be modeled as a covariance matrix and incorporated. It can also be considered readily in the elastic body spline described in later section, where the kernel function is applied on each individual coordinate instead of control point, i.e. $\mathbf{K}$ is $3M \times 3M$.

This smoothing technique has an advantage to potentially avoid the over-fitting problem. It also means if the fitted transformation applies back to the control points, exact displacement would not be obtained. Therefore, it is also referred to as approximating transformation, which arguably reflects the uncertainty and errors often encountered in identifying the control points in real-world applications.

2.4.1.2 Thine-Plate Spline and Multi-Quadrics Splines

Many popular kernels come from a derivation of deforming or bending a thin plate. This leads to a form which is used widely in medical image analysis community:

$$K(\mathbf{x}, \mathbf{x}_i) = \begin{cases} \|\mathbf{x} - \mathbf{x}_i\|_2^2 \log(\|\mathbf{x} - \mathbf{x}_i\|_2), & \text{2D case} \\ -\|\mathbf{x} - \mathbf{x}_i\|_2, & \text{3D case} \end{cases} \tag{2.39}$$

The thin-plate spline (TPS) requires polynomials of degree of 1 for both 2D and 3D cases and has no additional parameter.

More heuristic kernels are based on the multi-quadrics radial basis function:

$$K(\mathbf{x}, \mathbf{x}_i) = \left(\|\mathbf{x} - \mathbf{x}_i\|_2^2 + c^2\right)^{\mu} \tag{2.40}$$

where c and μ are two real-valued parameters. It requires a minimal degree of $\lceil \mu \rceil - 1$, in which $\lceil \mu \rceil$ denotes the smallest integer $\geq \mu$. The resulting transformation can be global as it places influence for points relatively far away, partly contributed by the polynomial part.

2.4.1.3 Gaussian Kernels and Other Locally Supported Models

The Gaussian form of the kernel function is as follows:

$$K(\mathbf{x}, \mathbf{x}_i) = \exp\left(\frac{-\|\mathbf{x} - \mathbf{x}_i\|_2^2}{2\sigma^2}\right) \tag{2.41}$$

where σ^2 is a parameter to control the width of the kernel. Previous research [15] has provided an interpretation for using the Gaussian kernel to describe deformation, which is based on the motion coherence theory. Due to the fact that $\mathbf{K}$ is guaranteed to be positive definite with a Gaussian form, $\mathbf{D}^* = \mathbf{K} + \lambda \mathbf{I} \omega$. The kernel should have a more compact support when σ^2 is small, comparing to TPS. This is a useful property in modeling local deformation, in particular, given sufficiently large data so that less locations need to be predicted without local support from control points.

Other kernels proposed to provide local support include ψ-functions of Wendland. Details can be found in [6] and summarized in Table 2.1.

Table 2.1 Summary of forms of the kernel functions and their corresponding polynomials

Name	Kernel form	Minimum polynomial degree	Parameters
TPS	$K(\mathbf{x}, \mathbf{x}_i) = \begin{cases} \|\mathbf{x} - \mathbf{x}_i\|_2^2 \log(\|\mathbf{x} - \mathbf{x}_i\|_2), & \text{2D} \\ -\|\mathbf{x} - \mathbf{x}_i\|_2, & \text{3D} \end{cases}$	1	none
Multi-quadrics	$K(\mathbf{x}, \mathbf{x}_i) = (\|\mathbf{x} - \mathbf{x}_i\|_2^2 + c^2)^\mu$	$\lceil \mu \rceil - 1$	c and μ
Gaussian	$K(\mathbf{x}, \mathbf{x}_i) = \exp(\frac{-\|\mathbf{x} - \mathbf{x}_i\|_2^2}{2\sigma^2})$	0	σ^2

Table 2.2 Summary of forms of the kernel functions for elastic body splines

Type of force	Kernel form	Affine part
$\mathbf{f}(\mathbf{x}) = \boldsymbol{\alpha}(\mathbf{x} - \mathbf{x}_i)$	$\boldsymbol{\Gamma}(\mathbf{x}, \mathbf{x}_i) = r[(12(1 - \nu) - 1)r^2 \mathbf{I} - 3\boldsymbol{\Upsilon}^2]$	yes
$\mathbf{f}(\mathbf{x}) = \boldsymbol{\alpha}/(\mathbf{x} - \mathbf{x}_i)$	$\boldsymbol{\Gamma}(\mathbf{x}, \mathbf{x}_i) = (8(1 - \nu) - 1)r\mathbf{I} - \boldsymbol{\Upsilon}^2/r$	yes
Gaussian	$\boldsymbol{\Gamma}(\mathbf{x}, \mathbf{x}_i) = [(4(1 - \nu) - 1)c_1 - c_2 + \sigma^2 c_1/r^2]\mathbf{I}$ $+ (c_1/c_2 - 3c_2/r^2 - 3\sigma^2 c_1/r^4)\boldsymbol{\Upsilon}^2$	no

2.4.1.4 Elastic Body Spline

So far, we have described the kernel acting like a function of the Euclidean distance between pairs of control points. The kennel can also be applied on the distance between pairs of each coordinate. This leads to a $3M \times 3M\mathbf{K}$ in 3D case:

$$\mathbf{K} = \begin{bmatrix} \boldsymbol{\Gamma}(\mathbf{x}_1, \mathbf{x}_1) & \cdots & \boldsymbol{\Gamma}(\mathbf{x}_1, \mathbf{x}_M) \\ \vdots & \ddots & \vdots \\ \boldsymbol{\Gamma}(\mathbf{x}_M, \mathbf{x}_1) & \cdots & \boldsymbol{\Gamma}(\mathbf{x}_M, \mathbf{x}_M) \end{bmatrix} \tag{2.42}$$

where $\boldsymbol{\Gamma}(\mathbf{x}, \mathbf{x}_i)$ is similar to a kernel function but outputs a 3×3 matrix determined by the form of the kernel. Therefore, the elastic body splines sometime are also known as 3D splines. Without further explanation, we give three of the most commonly used forms of the kernel in Table 2.2. These are based on different types of the assumed radial symmetric force applied.

In Table 2.2, $r^2 = (\mathbf{x} - \mathbf{x}_i)^T(\mathbf{x} - \mathbf{x}_i)$ and $\boldsymbol{\Upsilon}^2 = (\mathbf{x} - \mathbf{x}_i)(\mathbf{x} - \mathbf{x}_i)^T$; $c_1 = \frac{\text{erf}(r/\sqrt{2}\sigma)}{r}$, $c_2 = \frac{\sigma\sqrt{2/\pi}\exp(-r^2/2\sigma^2)}{r^2}$, erf$(\cdot)$ denotes the error function, $\nu \in [0, 0.5]$ and σ^2 are respectively the Poisson ratio and Gaussian parameter; $\boldsymbol{\alpha}$ is the weighting coefficient vector.

The original function needs to be expanded to accommodate the enlarged size of the kernel. One needs $3M$ coefficients to weight the new kernel. The Gaussian kernel has an advantage in not requiring additional polynomial of degree 1, i.e. equivalent to the affine transformation. But the linear systems still can be solved by very the similar approach using least-squares.

The approximation scheme described earlier still applies. Intuitive generalizations also include 3D-thin-plate splines and volume splines, although few studies have demonstrated the added values from these splines.

2.4.2 Local Deformation Models

As discussed in Section 2.2.3, local deformation models provide more flexibility to describe changes in finer scale but need more densely sampled control points. This could also mean only local information extracted from neighboring data would influence the estimation of the deformation, thus certain locally supported constraints would be necessary to avoid over-fitting.

In this section, we describe two specific local deformation models which have been successfully applied to model deformable organs in medical image computing field.

These models can directly serve as representation of training data for the framework in building a statistical deformation model. Associated registration methods, inherently related to this problem can be used to find the correspondence, or common space in statistical analysis. Therefore, we also included the sections describing image registration algorithms to utilize these local deformation models. On the other hand, the registration can also be useful in providing a common space allowing voting for a best segmentation outcome in a multiple atlas segmentation framework – this also learns implicitly the statistical properties from training data. We also provide an application in this area in the Application chapter.

2.4.2.1 Free Form Deformation (FFD)

The FFD model employs a B-spline mesh to estimate the deformation field. The mesh is controlled by the grid points, referred to as control points, which are scattered in a regular spacing grid. Each control point is associated with a kernel function, which defines the local deformations induced by the displacement of this control point. A popular kernel function chosen is the cubic B-spline kernel function, which is defined as follows:

$$\beta^{(3)}(a) = \begin{cases} \frac{1}{6}(4 - 6a^2 + 3|a|^3), & 0 \le |a| < 1 \\ \frac{1}{6}(2 - |a|)^3, & 1 \le |a| < 2 \\ 0, & 2 \le |a| \end{cases} \tag{2.43}$$

and a FFD transformation T of $\mathbf{x} = [x, y, z]$ in 3D is given by,

$$T(\mathbf{x}) = \mathbf{x} + \sum_{i,j,k} c_{ijk}\beta^{(3)}\left(\frac{x - \phi_{ijk}^x}{l_x}\right)\beta^{(3)}\left(\frac{y - \phi_{ijk}^y}{l_y}\right)\beta^{(3)}\left(\frac{z - \phi_{ijk}^z}{l_z}\right) \tag{2.44}$$

where c_{ijk} and $[\phi^x_{ijk}, \phi^y_{ijk}, \phi^z_{ijk}]$ are respectively the displacement vector and coordinate of the [ijk]th control point ϕ_{ijk}; $[l_x, l_y, l_z]$ is the vector of the grid spacing in three dimensions of the FFD model.

2.4.2.2 Nonrigid Registration

Nonrigid registration based on the FFD model, referred to as FFD registration, is one of the most successful technologies in medical image computing [18,23]. It has the advantages of easy implementation and efficiency in both modeling and computation and has been widely used since its introduction by [18]. This method employs one or a series of B-spline meshes to estimate a deformation field, mapping a reference image to a floating one. To maintain a smooth deformation field, a regularization term penalizing the bending energy of the FFD mesh(es) is commonly adopted. The FFD model does not naturally guarantee a folding-free deformation field [4]. To achieve a diffeomorphic registration, [16] extended the work by constraining the displacements of the control points in a FFD model within a fraction of the spacing of the FFD mesh, which results in a small magnitude of the deformation fields. To model large deformations, a series of such constrained FFD transformations are used and concatenated. The concatenated result can generate large deformations, and also be diffeomorphic.

Rueckert et al.'s FFD registration algorithm employs the normalized mutual information (NMI) as the image similarity measure and is applicable to medical images from different modalities. The registration method is further extended using a spatially encoded mutual information (SEMI) as the similarity metric, which is applicable to dynamic contrast enhanced images or images with severe intensity non–uniformity [23,24].

2.4.2.3 Locally Affine Transformation (LAT)

The LAT is a special type of transformation models, whose degree-of-freedom is between the linear transformations and the fully deformable transformations [25,26]. A LAT model consists of a set of rigid or affine transformations, each of which is only defined to be effective in a local region, such as a sub-volume of the image, a landmark or a boundary. For elsewhere without associated LATs, an interpolation scheme is used to calculate the deformation fields. Fig. 2.3 provides an example of LAT.

An advantage of the LAT model is that it has both the properties of global nonlinearity and local linearity. The global nonlinearity can provide more degree-of-freedom and thus better modeling capability compared to a global affine transformation; and the local linearity can preserve the local shape of substructures which cannot be deformed in certain applications, such as the bone structures.

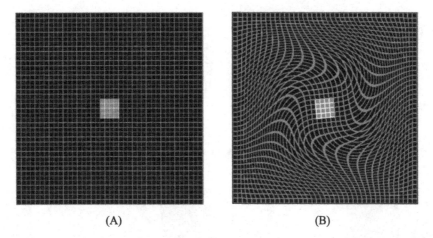

(A)   **(B)**

Figure 2.3 A locally affine transformation model which has a rigid square box in the middle and fixed boundaries of the region-of-field (A). The rigid box can be rotated to generate a locally affine transformation field (B). For more details please refer to [22].

The LAT model is formulated as follows:

$$T(\mathbf{x}) = \begin{cases} G_i(\mathbf{x}), & \mathbf{x} \in V_i \mid_{i=1\ldots n} \\ \text{Interpolate}\{G_i(\mathbf{x}) \mid_{i=1\ldots n}\}, & \mathbf{x} \notin \bigcup_i^n V_i \end{cases} \qquad (2.45)$$

where $\{G_i\}$ are the local affinities, assigned to the corresponding local regions $\{V_i\}$. The interpolation can be computed using linearly weighted sum of the local affinities, similar to (2.4),

$$\text{Interpolate}\{G_i(\mathbf{x}) \mid_{i=1\ldots n}\} = \sum_{i=1}^{n} w_i(\mathbf{x}) G_i(\mathbf{x}) \qquad (2.46)$$

Here $w_i(\mathbf{x})$ is a normalized weighting factor related to the inverse distance between point $\mathbf{x}$ and local regions: $w_i(\mathbf{x}) = d_i(\mathbf{x})^{-e} / \sum_i d_i(\mathbf{x})^{-e}$. In this model, a global affine transformation can also be assigned to the boundary of the region of interest of the subject, such as the example in Fig. 2.3; and e can be set to 2 for computational efficiency.

The method described above provides a global deformation field, which however does not guarantee to be smooth and diffeomorphic. There are two situations which can cause folding. First, the local regions can overlap each other after the individual local affinity. To avoid this overlapping, a correction of the local regions is applied before the interpolation,

$$V_i \xleftarrow{\text{correction}} G_i^{-1}\big(G_i(V_i) - \bigoplus R_{ij}\big) \qquad (2.47)$$

where $R_{ij} = \bigcup_{i \neq j}(G_j(V_i))$ is the volume of other local regions that V_i should not overlap after the local affinities, $\bigoplus$ is the morphological dilation to leave certain initial distance.

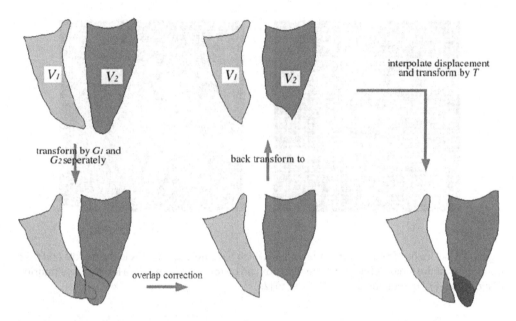

Figure 2.4 The diagram demonstrating the overlap correction procedure of two local regions, V_1 and V_2, after the application of two local affinities.

The length of dilation can be set, for example, to 10 mm in the cardiac image registration. Fig. 2.4 demonstrates the procedure of overlap correction.

The other situation happens when the displacements of the local affinities are too large, the interpolation may produce folding. Therefore, to model a large deformation one can concatenate a series of diffeomorphic LATs, as follows:

$$T_{\text{final}} = T_1 \circ T_2 \cdots T_M \tag{2.48}$$

where each transformation in $\{T_m|_{m=1...M}\}$ is a LAT containing only small deformations. The concatenated deformation is diffeomorphic if each of them is.

2.4.2.4 Locally Affine Registration Method Using Mutual Information

Registration using locally affine transformations, namely the locally affine registration method (LARM), is an attractive registration alternative for applications where a single global affine transformation cannot provide enough accuracy, while a nonrigid registration may incorrectly affect the local topology of the subject due to large shape variations. LARM assigns a local affinity to each substructure, such as the four chambers and great vessels in the application of cardiac image registration, to further initialize them (the application to the particularly challenging cardiac images will be introduced in the Application chapter). This transformation model globally deforms the image, while locally still preserves the shapes of the pre-defined local regions, i.e. the substructures. LARM is normally applied, as a bridge following a global affine registration but before the full deformation registration, to provide a good initialization of the substructures, which is crucial for the deformable registration to achieve a success.

In LARM, the diffeomorphism of the resulting deformation field is guaranteed by monitoring the determinant of the Jacobian matrix of current LAT, $\det(J_T)$, such that when $\det(J_T)$ drops below a threshold (e.g. $\det(J_T) < 0:5$), a new image-pair will be generated from the original one by applying the current LAT, and all the local regions are correspondingly transformed. The Jacobian matrix is computed as follows,

$$J_T = \sum_{i=1}^{n} \left(\frac{\partial w_i}{\partial \mathbf{x}} G_i + w_i \frac{\partial G_i}{\partial \mathbf{x}} \right) = \nabla \mathbf{W} \mathbf{G}^* + \nabla \mathbf{G} \mathbf{W}^* \tag{2.49}$$

where $\mathbf{G}^*$ indicates the transpose of matrix $\mathbf{G}$. Afterwards, the LAT model is reset to identity and a new registration is started with the new local regions and image pair. The final deformation field is the concatenation of the series of LATs.

The image similarity of LARM is based on MI or the normalized measures which are applicable to multi-modality images. The derivative of MI can be obtained by differentiating the entropies, which can be computed from the derivative of the probability distribution function (PDF). Let (l, k) be a joint histogram bin from the target image I_t and source image I_s, $\omega(\cdot)$ be the kernel density estimation function, the derivative of joint PDF with respect to a transformation parameter of local affinity G_i is given by

$$\frac{\partial p(l, k)}{\partial \theta_i} = \sum_x \omega\big(I_t(\mathbf{x})\big) \nabla \omega \nabla I_s(\mathbf{y}|_{\mathbf{y}=T(\mathbf{x})}) \frac{\partial G_i(\mathbf{x})}{\partial \theta_i} w_i(\mathbf{x}) = \sum_x W_{\theta_i}(\mathbf{x}), \tag{2.50}$$

where $p(l, k) = P_\Omega = \sum_{\mathbf{x} \in \Omega} \omega(I_t(\mathbf{x})) \omega(I_s(T(\mathbf{x})))$. Given a point $\mathbf{x}$ having a small value for $d_j(\mathbf{x})$, such that $d_j(\mathbf{x})/d_i(\mathbf{x}) \ll 1$, $w_i(\mathbf{x})$ tends to zero, making $W_{\theta_i}(\mathbf{x})$ tend to zero and negligible. For a dilated local volume Ω_i, associated with local affine T_i, one can rewrite the joint PDF as $p(l, k) = P_{\Omega_i} + P_{\overline{\Omega_i}}$. Therefore, the driving force related to the derivative of the entropy becomes

$$\mathcal{F}(\theta_i) = \frac{\partial H}{\partial \theta_i} = -\sum_{l,k} (1 + \log p(l, k)) \left(\frac{\partial P_{\Omega_i}}{\partial \theta_i} + \frac{\partial P_{\overline{\Omega_i}}}{\partial \theta_i} \right) = \mathcal{F}_{\Omega_i}(\theta_i) + \mathcal{F}_{\overline{\Omega_i}}(\theta_i). \tag{2.51}$$

When the driving force from the complementary of the local volume is negligible due to the relatively small values or unimportance for image registration such as artefacts from background, one can solely compute the driving force of a parameter within the local volume,

$$\mathcal{F}(\theta_i) \equiv \mathcal{F}_{\Omega_i}(\theta_i) = -\sum_{l,k} (1 + \log p(l, k)) \frac{\partial P_{\Omega_i}}{\partial \theta_i}. \tag{2.52}$$

The term $(1 + \log p(l, k))$ preserves the global relation of the intensity information between the images, thus guarantees the robustness of registration. It is also constant for all transformation parameters, and therefore the computation complexity of $\mathcal{F}(\theta_i)$ is mainly determined by $\frac{\partial P_{\Omega_i}}{\partial \theta_i}$, which is only $O(|\Omega_i|)$. This driving force not only speeds up the computation, but also blocks the forces coming from the background which may otherwise affect the robustness of the locally affine registration.

2.5 BIOPHYSICS-BASED DEFORMATION

Although the kernel-based splines have been introduced based on certain physical principles to model elastic deformation, real soft tissues are known to have more complex behavior that are nonlinear, inhomogeneous and anisotropic. General-purpose models do not generally distinguish the detailed cause of the deformation. The resulting method is using a smoothing interpolation and/or extrapolation to best use of the given data, therefore can lack predictive ability. If the main cause of the deformation is known, biophysics-based methods may be employed to deal with specific applications to provide potentially more accurate and informative modeling.

2.5.1 Biomechanical Modeling with Finite Element Analysis

In medical image computing, biomechanical modeling usually refers to applying the principles of classical mechanics to model interactions between internally and externally applied forces and the deformation of organs. Medical image provides a rich source of information from which material properties, forces or displacement may be estimated. It models biological tissue using its estimated mechanical properties so when force or displacement (loading) is applied, the tissue should respond to achieve equilibrium again. Finite element analysis (FEA) is a well-established numerical framework for solving both traditional mechanical problems and emerging biomechanical problems that arise from the study of complex biological systems. When material properties and loadings are estimated given a specific mesh (finite number of elements) representing the geometry of the tissue, the FEA solution may be obtained to contain a complete (dense) displacement field across the model. Other mechanical properties, such as strain or stress field also can be computed, although in the field of medical image computing the usefulness of estimating these parameters is still under investigation. The displacement field, on the other hand, is directly related to the spatial transformation which is of most interest in this chapter.

We summarize the basic steps involved in FEA-based biomechanical modeling as follows:

1. Meshing – Set up a geometrical representation of target organ and surrounding tissues of interest;
2. Material properties – Assign estimated material properties for the tissues of interest;

3. Boundary conditions – Set the boundary conditions for a particular scenario, such as the displacement of an organ surface; and

4. Numerical solution – Solve a system of equations for unknowns of interest, such as displacements, numerically.

In this section, we aim to describe a typical procedure to implement a FEA for biomechanical modeling of an organ or tissue, using established software packages and previously proposed algorithms. Commercially available software packages include ABAQUS (ABAQUS Inc., Rhode Island, USA) and ANSYS (ANSYS Europe Ltd., Oxfordshire, UK). For research purposes, open source codes are also available in MATLAB, C++ and CUDA. Recent developments in fast computational techniques for FE analysis utilize graphical processing units (GPUs), e.g. [12,19], and make FEA methods more practical for simulating soft-tissue motion during surgical procedures.

2.5.1.1 Meshing

The FEA methods share one essential characteristic: the discretization of a continuous, complex domain into a large, finite number of simple geometric elements, e.g. triangles/rectangles in 2D, or tetrahedrons/hexahedrons for 3D domains. Discretization in this way allows mechanical principles to be applied on each element, resulting in a large scale system of partial differential equations and integral equations, which, for real-world biomedical applications, are also defined over a geometrically complex domain that is captured by finite number of these elements.

The first step in FEA is meshing, which some practitioners might argue is the most crucial task in the analysis. In 3D, there are relatively robust methods to automatically generate a tetrahedral mesh, such as Delaunay triangulation employed in ANSYS. Hexahedral elements are more difficult to generate for irregular domains, which are common for biological structures. One reason for this is that a region of interest usually has complex topology so that a structured (or mapped) hexahedral mesh cannot be directly mapped to it. On the other hand, linear tetrahedral elements are vulnerable to so-called 'locking' when the material is assumed to be almost incompressible. Locking refers to an excessive stiffness of the mesh that results in smaller interpolated displacements than that would actually occur. Numerical methods have been proposed to overcome this problem. More complex element shapes include higher order elements in which additional nodes are located at positions other than vertices. These may be considered to trade computational expense for accuracy. A review of the finite element mesh generation can be found in the paper by Ho-Le [8].

In practice, volumetric meshes are generated typically from a geometric representation of a surface, such as a distance function or a surface mesh. The latter provides a simple, discrete representation, which can be converted to most other representations. Therefore, the conversion of triangulated meshes is a popular approach and is supported by most solid meshing algorithms. For instance, to represent a closed surface, spherical harmonic surface representation can be sampled into a triangular surface mesh [21]. Alternative methods exist, which have been designed specifically for surface meshing, e.g. the CGAL Project.[6]

2.5.1.1.1 Surface Meshing

Comparing to other potential methods in shape representation, spherical harmonic (SH) provides a compact parametric form that is well-suited to represent a smooth, start-shaped organ.

A practical alternative is to use well known iso-surface algorithm, which extracts surface from an integer-valued image (a bindery image for only one item of interest) – a 3D labeled volume indicates which voxels are representing the region(s) of interest. However, the resulting surface mesh is simply a separation between adjacent voxel blocks – it may contain a huge number of surface triangles and it is not smoothed with a blocky effect at the size of the original voxels. Furthermore, depending on isotropy of the original voxel dimension, the surface triangles can lead to low quality in the subsequent volumetric mesh. In general, a triangles or tetrahedrons with equal sides are desirable for FEA, which is not guaranteed by the iso-surface algorithm. Quality of the mesh and the number of elements are two important properties in finite element meshing. Therefore, mesh smoothing, simplification and, to a lesser extent, local refinement (so the size of the mesh is not uniformly distributed but with an adaptive density) are usually required after initial iso-surface algorithm.

A simple algorithm for mesh smoothing is based on a Laplacian operator. Its numerical implementation can be found in many numerical packages for geometric modeling and in computer graphics. Mesh reduction using algorithms, such as vertex clustering to remove vertices least affecting properties of the original mesh, can be dangerous as they may not necessarily guarantee preserving topology. A simple work-around is to alternate the smoothing and split-and-merge algorithms so the topology is preserved whilst the mesh is being smoothed. Topology preserving smoothing and mesh reduction is still an active research area where numerous algorithms have been proposed aiming for efficient solutions. Many commercial finite element analysis tools have built-in functions in its geometric modeling capacity. Unfortunately, this is less common for open-source FEA software.

In a 3D FE analysis, surface meshing is not necessary but useful for the subsequent solid meshing as a surface mesh provides a simple, discrete and unambiguous representation of regions of interest, especially comparing to other representations, such as splines, point clouds and binary volumes.

---

[6] An open source project at www.cgal.org with an independent Matlab wrapper at iso2mesh.sourceforge.net.

2.5.1.1.2 Solid Meshing

Again, most commercially available FEA software packages have the capacity of solid meshing. Users usually have choices in geometry representation, such as in spline-based form, surface mesh, regular geometry identities (such as block, sphere, prism, cylinder) and methods to combine these (through so-called Boolean operation such as union and intersection). For medical image applications, because the target objects to be modeled usually have irregular shapes, surface mesh is by far the most used geometric representation as an input.

Implementing a solid meshing tool is not a trivial task, although it has almost always been overlooked by the state-of-the-art research FEA computing software. There are other meshing tools available to generate the solid tetrahedron mesh in the field. For instance, Tetgen is an open source tool to generate quality 3D meshes from surface meshes. On the other hand, a general purpose meshing method for structural meshes, such as four node hexahedrons, is still under active development. This practical consideration is the main motivation of adopting tetrahedron elements from many research applications.

Implementing of a set of simple algorithms is usually required to work with finite element mesh, including finding surface mesh from a solid mesh, finding tetrahedral elements at certain location. Additionally, an interpolation function for the choice of element, shape function, is also useful to obtain numerical estimates of properties (e.g. displacement) at given location, from those at vertex locations. For triangle or tetrahedral elements, one can use barycentric coordinates to represent the weights from vertices.

For regions of interest having a common shape and simple topology, an interesting approach for meshing is to use a general reference mesh, which can be pre-meshed and carefully optimized. For individual segmentation, the reference mesh can then be warped to the target segmentation using surface registration. This may then be refined by simple numerical algorithms to optimize the element quality.

There has not been a general purpose meshing technique available at the time of writing this chapter. It is unlikely such a problem will be solved in near future as not only the geometry encountered in biomedical applications is complex and variable, but also the numerical solution for FEA has been improved continuously which may require different mesh quality. One practical consideration is always to check the resulting solid mesh with the original segmentation to ensure the error caused by surface meshing, solid meshing and their processing is acceptable, for example, by computing distance from surface of the solid mesh to the nearest point in original segmentation representation.

2.5.1.2 Material Properties

The mechanical behavior of biological tissue under load is complicated, so certain simplifications and assumptions have to be made when assigning tissue material properties, which govern the relationship between stress and strain. An interesting argument is that for applications in image-guided interventions, computational speed is more important than the accuracy of the simulations. The general rule is that the more complex the analysis approach and the more detailed model, the greater the computational expense, but the more accurately the real-world biomechanical behaviors can be modeled. A relevant debate in the field is whether accurate material properties (in general, accurate constitute models) are required for modeling of human tissue deformation. Some suggest that material properties should be estimated accurately. Whilst recent developments in computational hardware and parallel computing techniques have significantly reduced the computational burden of FEA, an accurate and complex (therefore realistic) material model is believed generally desirable. The level of the complexity and accuracy, which depends on the application, is one of the active research topics.

In the literature on modeling biomechanical tissue motion, the tissue mechanical properties are typically assigned fixed values based on the results of *ex vivo* experiments. A few studies have attempted to determine the mechanical properties for real soft tissue. *Ex vivo* properties are often poorly representative of the corresponding properties *in vivo*, and accurately measuring *in vivo* mechanical properties is extremely difficult in practice. Furthermore, it is well-known that material properties can have high variation between subjects (patients), particularly when diseased tissue is involved.

However, given knowledge in significantly differed material properties, such as differences between stiffness of fat and muscle, compressibility between airways and solid organ, general guidance is that effort should be made to distinguish those. For any FEA software packages, this is done by simply assigning fixed estimates of values to corresponding groups of elements.

2.5.1.3 Boundary Conditions

Boundary conditions (BCs) – which are also termed "loadings" in some fields of engineering – can be specified in forms of body displacements, velocities, accelerations, and forces and pressures (externally applied and/or due to gravity). However, tissue displacement is the type mostly applied for medical applications, largely because it is usually directly measurable from image data. Gravity is also considered in cases where it applies a significant load on organs that causes deformation [2], such as in change of patient position.

Nodal forces, among other loadings, generally are difficult to estimate from medical images. One attempt is to use the derivative of a similarity measure as a surrogate for applied force [14]. It however is not ideal as the intuitive estimation lacks of physical interpretation or proven efficacy.

Therefore, in the context of medical image registration, most interesting boundary conditions are currently nodal displacement, applied in several forms: by measuring a displacement between two corresponding points, usually on the surfaces of regions of interest and estimated from a surface or a point-based matching algorithm; by using known displacement, such as movement of a rigid organ (e.g. bony structures), and a surgical instrument that can be tracked or accurately estimated; constraints on certain position, by assigning zero displacement where the reference is, such as the body, skull, ribs and pelvic for brain, upper- and lower abdominal areas. One has to analyze the problem at hand, in particular for biomedical problems, how to simplify specifying the loadings. Modeling real interactions between organs and soft tissues can be too complicated to implement for some applications. Some of the interactions need more advanced model altogether, such as contact modeling, fluid dynamic. Although they have been focus of active research, displacement-driven framework is by far a viable choice in applications requiring near real-time computing, such as those in image-guided intervention. We provide an example in the next chapter to describe a commercialized approach to such a problem in a prostate cancer related application.

2.5.1.4 Statistical Modeling of Finite Element Analysis

One interesting approach is to use statistical method to reduce the size of linear system by projecting the full system response onto a subspace with lower dimensionality, in order to accelerate finite element solution. For an explicate analysis with nonlinear system, time integration may be performed in a reduced basis with imposed inhomogeneous essential boundary conditions [20].

Another active research area is to look at principled methods to incorporate the stochastic approaches both in constitutive model (material properties) and boundary conditions, so, instead to solve for a deterministic FEA solution, a space of solution can be estimated with uncertainties in inputs. This is sometimes termed proper generalized decomposition [3].

Such methods are beyond the scope of this chapter but have great potential to accelerate FEA simulations and accommodate greater range of biomedical applications. It also coined some motives behind the statistical deformation models described in this chapter – that is, conceptually, to pre-compute the computationally expensive solution space then find a set of parameter values to instantiate it quickly when needed.

2.5.2 Modeling Intervention-Induced Deformation and Diffusion-Based Modeling of Biological Changes

Instrument interaction and tissue resection are two examples that require more advanced modeling technically beyond simple solid modeling. These processes involve contact between different materials, change of mesh topology, which have not been as well developed as solid modeling for biological tissues. Furthermore, depending on the application, other biological changes, such as interaction induced heat generation and deformation due to temperature and/or loss of blood and other body fluids, may also play a role.

Although development of these biophysics-based modeling techniques has been promising, none has been applied widely in medical image analysis applications. One reason is the availability of the implementation, and more pragmatically, with some approximation, static modeling of solids or fluids with contact modeling [13] may be sufficient for a number of applications, such as surgical simulations.

For instance, swelling due to intervention is a complex biological process involving possible changes in finer cellular scale, it may require not only non-mechanical models but also formidable computational resource. In practice, although it is still very much a research approach, the swelling may be modeled by solving a diffusion problem.

Disease progression is another research area which may cause significant deformation. Solid structure modeling usually is applied to compensate any change caused by motion then a diffusion type model is used to predict the further morphological changes. Unfortunately, to the best of the authors' knowledge, there has not been any revolutionary development in this area. Most applications are, instead, looking at statistical or general-purpose modeling approaches.

REFERENCES

[1] M.A. Audette, F.P. Ferrie, T.M. Peters, Med. Image Anal. 4 (2000) 201–217.
[2] T.J. Carter, M. Sermesant, D.M. Cash, D.C. Barratt, C. Tanner, D.J. Hawkes, Application of soft tissue modelling to image-guided surgery, Med. Eng. Phys. 27 (10) (2005) 893–909.
[3] F. Chinesta, P. Ladeveze, E. Cueto, Arch. Comput. Methods Eng. 18 (2011) 395–404.
[4] Y. Choi, S. Lee, Graph. Models 62 (2000) 411–427.
[5] H. Chui, A. Rangarajan, Comput. Vis. Image Underst. 89 (2003) 114–141.
[6] M. Fornefett, K. Rohr, H.S. Stiehl, Image Vis. Comput. 19 (2001) 87–96.
[7] D.L. Hill, P.G. Batchelor, M. Holden, D.J. Hawkes, Phys. Med. Biol. 46 (2001) R1–R45.
[8] K. Ho-Le, Comput. Aided Des. 20 (1988) 27–38.
[9] B.K.P. Horn, J. Opt. Soc. Am. A 4 (1987) 629.
[10] Y. Hu, Registration of Magnetic Resonance and Ultrasound Images for Guiding Prostate Cancer Interventions, UCL (University College London), 2013.
[11] B. Jian, I.C. Society, B.C. Vemuri, IEEE Trans. Pattern Anal. Mach. Intell. 33 (2011) 1633–1645.
[12] S.F. Johnsen, Z.A. Taylor, M.J. Clarkson, J. Hipwell, M. Modat, B. Eiben, L. Han, Y. Hu, T. Mertzanidou, D.J. Hawkes, et al., Int. J. Comput. Assisted Radiol. Surg. (2014) 1–19.
[13] S.F. Johnsen, Z.A. Taylor, L. Han, Y. Hu, M.J. Clarkson, D.J. Hawkes, S. Ourselin, Int. J. Comput. Assisted Radiol. Surg. (2015) 1–19.
[14] M. Modat, Z.A. Taylor, G.R. Ridgway, J. Barnes, E.J. Wild, D.J. Hawkes, N.C. Fox, S. Ourselin, Nonlinear elastic spline registration: evaluation with longitudinal Huntington's disease data, in: B. Fischer, B.M. Dawant, C. Lorenz (Eds.), Biomedical Image Registration: 4th International Workshop, Proceedings, WBIR 2010, Lübeck, Germany, July 11–13, 2010, Springer, Berlin, Heidelberg, 2010, pp. 128–139.
[15] A. Myronenko, X. Song, IEEE Trans. Pattern Anal. Mach. Intell. 32 (2010) 2262–2275.
[16] D. Rueckert, P. Aljabar, R.A. Heckemann, J.V. Hajnal, A. Hammers, Diffeomorphic registration using B-splines, in: MICCAI, 2006, pp. 702–709.
[17] D. Rueckert, A.F. Frangi, J.A. Schnabel, in: Lect. Notes Comput. Sci., vol. 2208, 2001, pp. 77–84.
[18] D. Rueckert, L.I. Sonoda, C. Hayes, D.L. Hill, M.O. Leach, D.J. Hawkes, IEEE Trans. Med. Imaging 18 (1999) 712–721.

[19] Z.A. Taylor, M. Cheng, S. Ourselin, IEEE Trans. Med. Imaging 27 (2008) 650–663.

[20] Z.A. Taylor, S. Crozier, S. Ourselin, IEEE Trans. Med. Imaging 30 (2011) 1713–1721.

[21] A. Zacharopoulos, S. Arridge, O. Dorn, V. Kolehmainen, J. Sikora, J. Electromagn. Waves Appl. 20 (2006) 1827–1836.

[22] X. Zhuang, Automatic Whole Heart Segmentation Based on Image Registration, University College London, 2010.

[23] X. Zhuang, S. Arridge, D.J. Hawkes, S. Ourselin, A nonrigid registration framework using spatially encoded mutual information and free-form deformations, IEEE Trans. Med. Imaging 30 (2011) 1819–1828.

[24] X. Zhuang, D.J. Hawkes, S. Ourselin, Unifying encoding of spatial information in mutual information for nonrigid registration, in: Proc.: Information Processing in Medical Imaging (IPMI), 2009, pp. 491–502.

[25] X. Zhuang, K. Rhode, S. Arridge, R. Razavi, D. Hill, D. Hawkes, S. Ourselin, An atlas-based segmentation propagation framework using locally affine registration – application to automatic whole heart segmentation, in: Proc. Medical Image Computing and Computer Assisted Intervention (MICCAI), 2008, pp. 425–433.

[26] X. Zhuang, K.S. Rhode, R.S. Razavi, D.J. Hawkes, S. Ourselin, A registration-based propagation framework for automatic whole heart segmentation of cardiac MRI, IEEE Trans. Med. Imaging 29 (2010) 1612–1625.

[27] X. Zhuang, W. Shi, H. Wang, D. Rueckert, S. Ourselin, Computation on shape manifold for atlas generation: application to whole heart segmentation of cardiac MRI, Proc. SPIE (2013) 866941.

CHAPTER 3

Correspondence Establishment in Statistical Shape Modeling: Optimization and Evaluation

Song Wang*, Brent Munsell†, Theodor Richardson‡
*University of South Carolina, Department of Computer Science and Engineering, Columbia, SC 29208
†College of Charleston, Department of Computer Science, Charleston, SC 29401
‡South University, College of Business, Savannah, GA 31405

Contents

3.1	Introduction	67
3.2	PDM and Shape Correspondence	69
3.3	Landmark Sliding for Shape Correspondence	69
	3.3.1 Thin-Plate Splines to Measure Shape Difference	71
	3.3.2 Landmark-Based Shape Representation Error	72
	3.3.3 Topology Preservation and Landmark-Sliding Algorithm	73
	3.3.4 Open-Shape Correspondence	75
3.4	Groupwise Shape Correspondence	75
	3.4.1 Pre-Organizing Shape Instances using Minimum Spanning Tree	76
	3.4.2 Root-Node Selection and Groupwise Shape Correspondence	77
3.5	Performance Evaluation of Shape Correspondence	79
	3.5.1 Generating Shape Instances	80
	3.5.2 Difference Between Two PDMs	81
3.6	Experiments	82
3.7	Conclusions and 3D Shape Correspondence	85
References		86

3.1 INTRODUCTION

Shape plays a critical role in medical imaging and image analysis. Many anatomic structures (organs and tissues) bear specific shapes and the accurate shape modeling can help segment these structures of interest from noisy medical images, which is a fundamental step for further medical image analysis and understanding. In addition, abnormal shape deformation may reflect pathological changes of a structure, which calls for accurate modeling of the normal shape variation of a specific structure. To model the shape and possible shape variation of a structure, statistical shape modeling [4,14] has been attracting considerable research interest in the past decades.

As in traditional statistics, where a statistical distribution is estimated to describe a set of data samples, in statistical shape modeling, a multi-dimensional statistical distribution

Statistical Shape and Deformation Analysis
DOI: 10.1016/B978-0-12-810493-4.00004-3
67

is estimated from a set of shape instances of the structure of interest to describe its shape and possible shape variation. Point distribution model (PDM) [4] is one of the most widely used statistical shape models, where a set of landmark points is identified from each shape instance and concatenating the coordinates of all the landmarks on each shape instance leads to a *shape vector*. Finally all the shape vectors are fitted to a multi-dimensional Gaussian distribution as the desired statistical shape model.

The main challenge in building a PDM is *shape correspondence*, which identifies the same number of landmarks on each shape instance such that landmarks across different shape instances are well corresponded [2,16,27,10]. Furthermore, the identified landmarks on each shape instance must represent the underlying shape very well. As a result, they must show certain level of spatial density and may not be located at anatomically meaningful points. In practice, shape of interest may be of different topology. In particular, the shape of a structure can be closed or open – in 2D it can be in the form of a closed curve or an open curve, while in 3D it can be in the form of a closed surface or an open surface. It is expected that a shape correspondence algorithm can handle both closed and open shapes.

In addition, we need to simultaneously correspond a set of shape instances instead of corresponding a pair of shape instances as in many shape-matching tasks. Typically, this is achieved by corresponding all the shape instances to a common template in a pairwise fashion. This way, shape correspondence performance and the accuracy of PDM construction are highly dependent on the selection of a good template [17,20]. Finally, the performance evaluation of shape correspondence is a challenging problem, since there is usually no unique ground-truth result for shape correspondence. People have used certain statistical properties of the resulting PDM to estimate the performance of shape correspondence, which is indirect and not fully objective [18,19].

In this chapter, we introduce a landmark-sliding based method for shape correspondence that can handle both closed and open shapes. Based on this, we study the groupwise shape correspondence that is not dependent on a common template. Finally, we introduce a more objective performance evaluation method for shape correspondence. For simplicity, we focus on discussing the 2D shape correspondence, with a brief discussion on its 3D extension at the end of this chapter. The models and methods introduced in this chapter are mainly drawn from a sequence of our past publications [26, 18,19,17,20].

The remainder of the chapter is organized as follows. Section 3.2 introduces the point distribution model (PDM) and the problem of shape correspondence. Section 3.3 introduces the landmark-sliding algorithm for pairwise shape correspondence. Section 3.4 discusses the groupwise shape correspondence by pre-organizing shape instances using a rooted tree. Section 3.5 discusses the performance evaluation of the shape correspondence. Section 3.6 reports some experiment results, followed by a brief discussion of the 3D extension and conclusions in Section 3.7.

3.2 PDM AND SHAPE CORRESPONDENCE

For simplicity, we focus on the 2D shape correspondence in this chapter. Given n sample shape instances (*shape contours* in the 2D case) S_i, $i = 1, 2, \ldots, n$ of the structure of interest, PDM [4] is usually constructed by three sequential steps: shape correspondence, shape normalization, and statistical modeling.

As discussed above, shape correspondence aims to identify corresponded landmarks from a set of continuous sample shape instances. More specifically, after shape correspondence we obtain n corresponded landmark sets V_i, $i = 1, 2, \ldots, n$, where $V_i = \{\mathbf{v}_{i1}, \mathbf{v}_{i2}, \ldots, \mathbf{v}_{im}\}$ are m landmarks identified from shape contour S_i and $\mathbf{v}_{ij} = (x_{ij}, y_{ij})$ is the jth landmark identified on S_i. Landmark correspondence means that $\mathbf{v}_{ij}$, $i = 1, 2, \ldots, n$, i.e., the jth landmark in each shape contour, are corresponded, for any $j = 1, 2, \ldots, m$.

In medical imaging, structural shape is usually assumed to be invariant to the transformations of any (uniform) scaling, rotation, and translations. In shape normalization, such transformations are removed among the given n shape contours by normalizing each of the n identified corresponded landmark sets V_i to $\hat{V}_i = \{\hat{\mathbf{v}}_{i1}, \hat{\mathbf{v}}_{i2}, \ldots, \hat{\mathbf{v}}_{im}\}$, $i = 1, 2, \ldots, n$. After the shape normalization, the absolute coordinates of the corresponded landmarks, i.e., $\hat{\mathbf{v}}_{ij} = (\hat{x}_{ij}, \hat{y}_{ij})$, $i = 1, 2, \ldots, n$ are directly comparable. Procrustes analysis [12] is the most widely used tool for shape normalization.

Finally, we construct the statistical shape model by fitting the normalized landmarks sets $\hat{V}_i = \{\hat{\mathbf{v}}_{i1}, \hat{\mathbf{v}}_{i2}, \ldots, \hat{\mathbf{v}}_{im}\}$, $i = 1, 2, \ldots, n$ to a multivariate Gaussian distribution. Specifically, we columnize m landmarks in $\hat{V}_i$ into a $2m$-dimensional vector $\hat{\mathbf{v}}_i = (\hat{x}_{i1}, \hat{y}_{i1}, \hat{x}_{i2}, \hat{y}_{i2}, \ldots, \hat{x}_{im}, \hat{y}_{im})^T$ and call it a *shape vector* of the shape instance $\hat{V}_i$. This way, the (landmark-based) mean shape $\bar{\mathbf{v}}$ and the covariance matrix $\mathbf{D}$ can be calculated by

$$\bar{\mathbf{v}} = \frac{1}{n} \sum_{i=1}^{n} \hat{\mathbf{v}}_i$$

$$\mathbf{D} = \frac{1}{n-1} \sum_{i=1}^{n} (\hat{\mathbf{v}}_i - \bar{\mathbf{v}})(\hat{\mathbf{v}}_i - \bar{\mathbf{v}})^T. \tag{3.1}$$

The Gaussian distribution $\mathcal{N}(\bar{\mathbf{v}}, \mathbf{D})$ is the resulting PDM that models the possible shape-deformation space of the considered structure. Clearly, the accuracy of the PDM is largely dependent on the performance of shape correspondence, i.e., the accuracy in identifying the corresponded landmarks V_i, $i = 1, 2, \ldots, n$.

3.3 LANDMARK SLIDING FOR SHAPE CORRESPONDENCE

In this section, we focus on pairwise shape correspondence – corresponding one shape instance, say S_p, referred to as the target, to another shape instance, say S_q, referred to

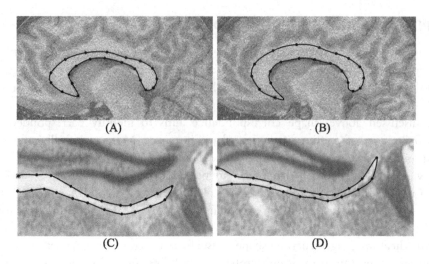

Figure 3.1 An illustrations of (A–B) closed-curve shape correspondence, and (C–D) open-curve shape correspondence. (A–B) are two Callosum shape instances and (C–D) are two stria shape instances in the human brain.

as the template. Extending pairwise correspondence to groupwise correspondence will be discussed in detail in the later Section 3.4.

For simplicity, we first consider the shape in the form of a closed curve, as illustrated in Fig. 3.1A–B, with known parametric representation. Furthermore, we assume that the landmarks in the template S_q are given and our task is to identify a set of corresponding landmarks in the target S_p. Later in the section, we will introduce the extension to open-shape correspondence, as illustrated in Fig. 3.1C–D.

Denote the parametric representation of the template S_q as $\mathbf{v}_q(t) = (x_q(t), y_q(t))^T$, $0 \leq t \leq L_q$, as shown in Fig. 3.1A, and the target S_p as $\mathbf{v}_p(s) = (x_p(s), y_p(s))^T$, $0 \leq s \leq L_p$, as shown in Fig. 3.1B, where L_q and L_p are the total curve lengths of the template and the target, respectively. With the assumption of closed shape, we have $\mathbf{v}_q(0) = \mathbf{v}_q(L_q)$ and $\mathbf{v}_p(0) = \mathbf{v}_p(L_p)$. In this parametrization, $t|L_q$ represents the traversed curve length from $\mathbf{v}_q(0)$ to $\mathbf{v}_q(t)$ and $s|L_p$ represents the traversed curve length from $\mathbf{v}_p(0)$ to $\mathbf{v}_p(s)$, where $a|b$ is the modulus operation.

Let $\mathbf{t} = \{t_0, t_1, \ldots, t_{m-1}\}$ be a set of parameters which generates m sequentially-sampled landmarks $V_q = \{\mathbf{v}_q(t_0), \mathbf{v}_q(t_1), \ldots, \mathbf{v}_q(t_{m-1})\}$. We assume that those landmarks represent the template shape contour S_q well. The task of shape correspondence is to find along the target shape contour S_p the corresponding m parameters $\mathbf{s} = \{s_0, s_1, \ldots, s_{m-1}\}$ such that the landmarks $V_p = \{\mathbf{v}_p(s_0), \mathbf{v}_p(s_1), \ldots, \mathbf{v}_p(s_{m-1})\}$ best match the landmarks V_q in the template. As a landmark representation, V_p must also well represent the underlying continuous contour of the target S_p.

Given the template landmarks, shape correspondence can be formulated as seeking the optimal parameters **s** to minimize the cost function

$$\phi(\mathbf{s}) = d(V_q, V_p) + \lambda R(\mathbf{s}),$$

where $d(V_q, V_p)$ measures the (landmark-based) shape difference and therefore reflects the global shape deformation between the template and target shape contours. $R(\mathbf{s})$ is the representation error when using the m landmarks V_p to represent the underlying target shape contour. We desire small representation error $R(\mathbf{s})$ such that landmark-based shape difference $d(V_q, V_p)$ accurately reflects the difference between the underlying continuous shape contours S_q and S_p. One additional constraint to be considered in this formulation is the preservation of the shape topology, i.e., the resulting landmarks $\mathbf{v}_p(s_0), \mathbf{v}_p(s_1), \ldots, \mathbf{v}_p(s_{m-1})$ should be sequentially located along the target shape contour as landmarks V_q located along the template shape contour.

3.3.1 Thin-Plate Splines to Measure Shape Difference

Following the notations in Section 3.2, we can write $\mathbf{v}_q(t_i)$ as $\mathbf{v}_{qi} = (x_{qi}, y_{qi})^T$ and $\mathbf{v}_p(s_i)$ as $\mathbf{v}_{pi} = (x_{pi}, y_{pi})^T$, $i = 0, 1, \ldots, m-1$. 2D thin-plate splines (TPS) has been widely used to measure the nonrigid deformation between two sets of the landmarks [1]. Specifically, 2D TPS defines a mapping $\mathbf{h} = (f, g)^T : \mathbb{R}^2 \rightarrow \mathbb{R}^2$ that maps the template landmarks V_q to the target landmarks V_p, i.e., $\mathbf{v}_{pi} = \mathbf{h}(\mathbf{v}_{qi}) = (f(\mathbf{v}_{qi}), g(\mathbf{v}_{qi})), i = 0, 1, \ldots, m-1$ by minimizing the TPS bending energy [1]

$$\phi(\mathbf{h}) = \iint_{-\infty}^{\infty} (L(f) + L(g)) dx dy, \tag{3.2}$$

where $L(\cdot) = (\frac{\partial^2}{\partial x^2})^2 + 2(\frac{\partial^2}{\partial x \partial y})^2 + (\frac{\partial^2}{\partial y^2})^2$. The TPS bending energy can also be written as

$$\phi(\mathbf{h}) = \frac{1}{8\pi} (\mathbf{x}_p^T \mathbf{L} \mathbf{x}_p + \mathbf{y}_p^T \mathbf{L} \mathbf{y}_p),$$

where **L** is the $m \times m$ upper left submatrix of

$$\begin{pmatrix} \mathbf{K} & \mathbf{P} \\ \mathbf{P}^T & \mathbf{0} \end{pmatrix}^{-1}. \tag{3.3}$$

Here **K** is an $m \times m$ matrix with $k_{ij} = \|\mathbf{v}_{qi} - \mathbf{v}_{qj}\|^2 \log \|\mathbf{v}_{qi} - \mathbf{v}_{qj}\|$, $i, j = 0, 1, \ldots, m-1$. $\mathbf{P} = (\mathbf{1}, \mathbf{x}_q, \mathbf{y}_q)$ with $\mathbf{x}_q = (x_{q0}, x_{q1}, \ldots, x_{q,m-1})^T$, and $\mathbf{y}_q = (y_{q0}, y_{q1}, \ldots, y_{q,m-1})^T$. Similarly, $\mathbf{x}_p = (x_{p0}, x_{p1}, \ldots, x_{p,m-1})^T$, $\mathbf{y}_p = (y_{p0}, y_{p1}, \ldots, y_{p,m-1})^T$.

Note that **L** is a positive semidefinite matrix and the thin-plate bending energy is invariant to affine transforms, including scaling, rotation and translation, which is desired for shape correspondence in medical-imaging applications. We can use this TPS bending energy to define the landmark-based shape difference, i.e.,

$$d(V_q, V_p) = \frac{1}{8\pi}(\mathbf{x}_p^T \mathbf{L} \mathbf{x}_p + \mathbf{y}_p^T \mathbf{L} \mathbf{y}_p). \tag{3.4}$$

3.3.2 Landmark-Based Shape Representation Error

Let l_{pi}, $i = 0, 1, \ldots m - 1$ be the traversed curve length between two neighboring landmarks $\mathbf{v}_{pi}$ and $\mathbf{v}_{p,i+1|m}$ along the target shape contour. We have

$$l_{pi} = (s_{i+1|m} - s_i)|L_p.$$

Similarly, we have the corresponding curve length between neighboring landmarks in the template shape contour as

$$l_{qi} = (t_{i+1|m} - t_i)|L_q.$$

Not allowing two landmarks to coincide, we have

$$0 < l_{pi} < L_p, 0 < l_{qi} < L_q, \quad i = 0, 1, \ldots, m - 1.$$

Given that $\mathbf{v}_{q0}, \mathbf{v}_{q1}, \ldots, \mathbf{v}_{q,m-1}$ are consecutively sampled landmarks along the closed template shape contour S_q, we have

$$\sum_{i=0}^{m-1} l_{qi} = L_q. \tag{3.5}$$

Given that the landmarks V_q well represent the underlying continuous contour S_q with sufficient spatial sampling density, we can require the landmarks in the target, i.e., V_p, to show a similar spatial distribution to well represent the continuous contour S_p, i.e.,

$$\frac{l_{pi}}{l_{p,i+1|m}} \approx \frac{l_{qi}}{l_{q,i+1|m}}, \quad i = 0, 1, \ldots, m - 1. \tag{3.6}$$

Based on this, we can define the landmark representation error in the target shape contour as

$$R(\mathbf{s}) = \sum_{i=0}^{m-1} (l_{pi} l_{q,i+1|m} - l_{p,i+1|m} l_{qi})^2.$$

3.3.3 Topology Preservation and Landmark-Sliding Algorithm

To preserve the shape topology, we require the resulting landmarks $\mathbf{v}_{p0}, \mathbf{v}_{p1}, \ldots, \mathbf{v}_{p,m-1}$ to be consecutively located along the target shape contour as their corresponding landmarks in the template. Similar to (3.5), this can be reflected by the following constraint to the shape correspondence cost function,

$$\sum_{i=0}^{m-1} l_{pi} = L_p.$$

This way, the landmark-based shape-correspondence problem can be formulated as

$$\min_{\mathbf{s}} \left\{ (\mathbf{x}_p^T \mathbf{L} \mathbf{x}_p + \mathbf{y}_p^T \mathbf{L} \mathbf{y}_p) + \lambda \sum_{i=0}^{m-1} (l_{pi} l_{q,i+1|m} - l_{p,i+1|m} l_{qi})^2 \right\}, \tag{3.7}$$

subject to

$$0 < s_i < L_p$$
$$l_{pi} = (s_{i+1|m} - s_i)|L_p, \quad i = 0, 1, \ldots, m-1$$
$$\sum_{i=0}^{m-1} l_{pi} = L_p,$$

where $\lambda > 0$ is a regularization factor that balances the contribution from the shape difference and the shape representation error.

Finding the global optimal solution to this constrained optimization problem is difficult because of the nonlinearity inherent in the parametric representation of the shape contours. Instead, we develop an iterative algorithm to get a local optimal solution. Specifically, we initialize an $\mathbf{s}$ by sampling a sequence of m landmarks along the target shape contour S_p with the same spatial density and distribution as their corresponding landmarks along the template shape contour, i.e.,

$$\frac{l_{pi}}{L_p} = \frac{l_{qi}}{L_q}.$$

We then iteratively slide these initialized landmarks along the target shape contour to minimize the cost function (3.7) subject to the topology-preserving constraint.

Specifically, each iteration of the landmark sliding consists of two steps: (a) sliding all the landmarks along their respective tangent directions, and (b) projecting the updated landmarks back to the target shape contour. Let $\mathbf{s} = \{s_0, s_1, \ldots, s_{m-1}\}$ be the parameters for the currently estimated landmarks $V_p = \{\mathbf{v}_{p0}, \mathbf{v}_{p1}, \ldots, \mathbf{v}_{p,m-1}\}$. Each step of landmark sliding actually seeks an improved parameters $\mathbf{s}' = \{s_0', s_1', \ldots, s_{m-1}'\}$ for updated landmarks $V_p' = \{\mathbf{v}_{p0}', \mathbf{v}_{p1}', \ldots, \mathbf{v}_{p,m-1}'\}$.

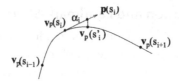

Figure 3.2 An illustration of the landmark sliding algorithm – each iteration consisting of a step of landmark sliding along the tangent direction, followed by a step of projection back to the shape contour.

Let $\mathbf{p}(s) = (p_x(s), p_y(s))$ be the (normalized) tangent direction of the target shape contour at $\mathbf{v}_p(s)$. As illustrated in Fig. 3.2, let α_i be the desired sliding distance of $\mathbf{v}_{pi}$ along the tangent direction. The step (a) moves the landmark $\mathbf{v}_{pi}$ to $\mathbf{v}_{pi} + \alpha_i \mathbf{p}(s_i)$ and the step (b) further projects it back onto the shape contour at

$$\mathbf{v}'_{pi} = \mathbf{v}_p(s_i + \alpha_i),$$

for all $i = 0, 1, \ldots, m-1$.

In step (a), we seek optimal sliding distance $\boldsymbol{\alpha} = (\alpha_0, \alpha_1, \ldots, \alpha_{m-1})^T$ to minimize the cost function (3.7) which can be rewritten as

$$(\mathbf{x}_p + \mathbf{P}_x\boldsymbol{\alpha})^T \mathbf{L}(\mathbf{x}_p + \mathbf{P}_x\boldsymbol{\alpha}) + (\mathbf{y}_p + \mathbf{P}_y\boldsymbol{\alpha})^T \mathbf{L}(\mathbf{y}_p + \mathbf{P}_y\boldsymbol{\alpha})$$
$$+ \lambda \sum_{i=0}^{m-1} \left\{ (l_{pi} + \alpha_i)l_{q,i+1|m} - (l_{p,i+1|m} + \alpha_{i+1|m})l_{qi})^2 \right\}, \tag{3.8}$$

where matrices $\mathbf{P}_x = \mathrm{diag}(p_x(s_0), p_x(s_1), \ldots, p_x(s_{m-1}))$ and $\mathbf{P}_y = \mathrm{diag}(p_y(s_0), p_y(s_1), \ldots, p_y(s_{m-1}))$. We can calculate the updated parameters $\mathbf{s}'$ by

$$s'_i = (s_i + \alpha_i)|L_p,$$

and the updated curve-segment length l'_{pi} by

$$l'_{pi} = (s'_{i+1} - s'_i)|L_p = (l_{pi} - \alpha_i + \alpha_{i+1})|L_p.$$

The shape-topology preservation constraint can be rewritten as

$$l_{pi} - \alpha_i + \alpha_{i+1} > 0, \quad i = 0, 1, \ldots, m-1, \tag{3.9}$$

i.e., no landmark can move beyond its neighbors during the sliding/projection.

It can be seen that minimizing the cost function (3.8) subject to constraints (3.9) is a simple quadratic programming problem, whose optima can be efficiently calculated. We can iteratively run steps (a) and (b) until convergence or a preset maximum number of iterations has reached.

3.3.4 Open-Shape Correspondence

Open shapes may occur when the structure of interest is occluded or cropped by image perimeter, as shown in Fig. 3.1C–D. The above landmark-sliding algorithm can be extended to handle the open-shape correspondence. Specifically, in the cost function (3.7), the shape-difference term will stay unchanged because the TPS bending energy is defined by only the landmarks. But the representation error term $R(\mathbf{s})$ and the additional constraints need to be updated to reflect open-curve topology.

Considering two open-shape instances S_p (target) and S_q (template), each of which is an open curve, the simplest case is that the endpoints are known to be corresponded between these two shape instances, as shown in Fig. 3.1C–D. This case can be addressed by making the following two changes to the above landmark-sliding algorithm: (a) the first landmark $\mathbf{v}_{p0}$ and the last landmark $\mathbf{v}_{p,m-1}$, i.e., the two endpoints of the curve, should be fixed in the landmark sliding. This can be enforced by additional linear constraints $\alpha_0 = 0$ and $\alpha_{m-1} = 0$; (b) In defining the representation error $R(\mathbf{s})$ and topology-preservation constraint, we remove the term corresponding to the curve segment between $\mathbf{v}_{p0}$ and $\mathbf{v}_{p,m-1}$.

A more complex case is that the endpoints do not correspond across the two open-shape instances. In this case, one shape instance may only be matched to a portion of the other shape instance. To adapt the landmark-sliding algorithm to handle this case, we need to allow the endpoints ($\mathbf{v}_{p0}$ and $\mathbf{v}_{p,m-1}$) to move along the shape curve, but limited to a preset maximum sliding distance in case all the landmarks on the target are aggregated together. More details on this case can be found from [21,22].

3.4 GROUPWISE SHAPE CORRESPONDENCE

In the last section, we introduce the landmark-sliding algorithm to correspond a pair of shape instances (either open or closed). In practice, we need to correspond a group of n shape instances for constructing a PDM, as discussed in Section 3.2. One simple strategy to achieve this goal [2] is to perform multiple rounds of the pairwise shape correspondence – we can select one shape instance as a template and then take each of the other $n-1$ shape instances as a target and correspond each target to the template independently using a pairwise correspondence algorithm as described in the last section. To further reduce the dependence on the initially selected template, we can compute the PDM, take the mean shape $\bar{\mathbf{v}}$ as the new template and correspond each of the n shape instance to this new template using the same pairwise correspondence algorithm. This process can be repeated in multiple iterations for a better shape correspondence and PDM. Since this is a direct application of the pairwise correspondence to handle a group of shape instances, we still call it a *pairwise* method in this chapter. While the pairwise correspondence is usually computational efficient, it may produce poor results when the shape instances to be corresponded show large variance [19].

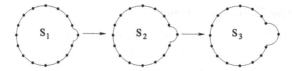

Figure 3.3 An illustration of the groupwise shape correspondence.

Another category of the methods for corresponding a group of n shape instances is to formulate a global cost function involving all the shape instances and then find landmarks on all the shape instances by minimizing the cost function. For example, Minimum Description Length (MDL) [10] method has been widely used for shape correspondence where the global cost function is defined as the total number of bits to represent all the shape instances and the constructed PDM. Another global shape correspondence method is Spherical Harmonics (SPHARM) [11], which is developed specifically for 3D closed shape by mapping each shape instance to a sphere and parameterizing the sphere. *Global* methods, such as MDL and SPHARM, are usually computationally intensive, especially in 3D case, because of the high complexity and nonlinearity in the optimized cost function.

In this section we introduce a *groupwise* correspondence method to address the limitations of global and pairwise methods. This method pre-organizes all the shape instances by constructing a minimum spanning tree (MST), where each node in the MST represents a shape instance and each edge in the MST connects two similar shape instances. A root node that represents the starting template shape instance is selected. All the shape instances that are the children of the root are taken as the targets and corresponded to the root using a pairwise method. Each of the child node is then taken as the template and corresponded to its own children using the same pairwise method. Recursively this will propagate the landmarks from the root to the leaf nodes to achieve the correspondence of all the shape instances. On one hand, this groupwise method is computationally efficient, because it performs a sequence of pairwise correspondences by following the MST. On the other hand, it can be more accurate and reliable than the pairwise method because in this groupwise method, the pairwise correspondence is only performed between similar shape instances. As shown in Fig. 3.3, following the MST, this groupwise method may choose to correspond S_2 to S_1 first, and then S_3 to S_2. Both shape pairs show good similarity, which may lead to higher correspondence accuracy than a pairwise method, e.g., corresponding both S_2 and S_3 to a common template S_1, since S_1 and S_3 show large difference.

3.4.1 Pre-Organizing Shape Instances using Minimum Spanning Tree

The basic goal of pre-organizing the shape instances is to identify shape pairs with high similarity. As in Section 3.2, let's consider a set of n shape instances S_i, $i = 1, 2, \ldots, n$ in the form of continuous shape contours. We can independently correspond each pair of shape instances S_i and S_j by identifying m landmarks on each of them, indicated by V_i and V_j respectively. Using Procrustes analysis [12], we can normalize the landmarks V_i and V_j to $\hat{V}_i$ and $\hat{V}_j$, respectively. Applying the same normalization to the original shape contours S_i and S_j, we can obtain normalized shape contours $\hat{S}_i$ and $\hat{S}_j$, respectively. We define the dissimilarity between S_i and S_j as

$$w_{ij} = w(S_i, S_j) = \Delta(\hat{S}_i, \hat{S}_j) \times \Omega(V_i, V_j) \tag{3.10}$$

for each pair of (i, j), $i \neq j$, $i, j = 1, 2, \ldots, n$.

In Eq. (3.10), the first term $\Delta(\hat{S}_i, \hat{S}_j)$ is the Jaccard coefficient

$$\Delta(\hat{S}_i, \hat{S}_j) = 1 - \frac{|\mathcal{R}(\hat{S}_i) \cap \mathcal{R}(\hat{S}_j)|}{|\mathcal{R}(\hat{S}_i) \cup \mathcal{R}(\hat{S}_j)|}, \tag{3.11}$$

where $\mathcal{R}(S)$ is the region enclosed by contour S, and $|\mathcal{R}|$ computes the area of the region $\mathcal{R}$. For open shapes, the first and last landmarks are connected to enclose the region. The Jaccard coefficient takes a value between [0, 1] with zero indicating that the two shape instances are exactly coincident with each other.

The second term $\Omega(V_i, V_j)$ measures the landmark-based nonrigid shape difference between S_i and S_j. We can use the TPS bending energy as defined in Section 3.3. More specifically, we define

$$\Omega(V_i, V_j) = \frac{d(V_i, V_j) + d(V_j, V_i)}{2},$$

where $d(V_i, V_j)$ and $d(V_j, V_i)$ are the TPS bending energies [1] as defined in Eq. (3.4).

Given n shape instances, we can then build a fully connected graph with n nodes, with each node representing a shape instance. Between each pair of nodes, say S_i and S_j, we connect them by an edge, with edge weight w_{ij} as defined in Eq. (3.10). By applying an MST algorithm, such as Prim's or Kruskal's algorithms [5], we can obtain an MST which connects all the n nodes using $n - 1$ edges with minimum total edge weight, as shown in Fig. 3.4. In the MST, only pair of nodes with relative small edge weight, i.e., only pair of shape instances with high similarity, are directly connected by an edge.

3.4.2 Root-Node Selection and Groupwise Shape Correspondence

In the MST, we select a root node as the initial template to start the correspondence (to the child nodes) and then recursively propagate the pairwise correspondence layer-by-layer until getting to the leaf nodes. Clearly, any shape-correspondence errors in the early recursions (upper layers) may be propagated to the later recursions (lower layers). Therefore, we need to select a root node such that in the MST, the edges with smaller depths, i.e., closer to the root, have smaller edge weights.

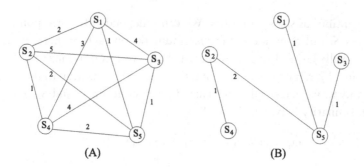

Figure 3.4 An illustration of a fully-connected graph (A) and its MST (B). Number on each edge is its weight.

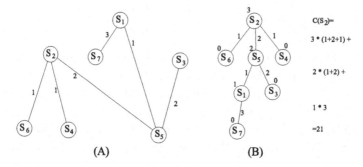

Figure 3.5 An illustration of the rooted MST and the root node selection. (A) An MST derived from the graph. (B) Selecting S_2 as root leads to a root cost of 21. Number on each edge is its weight and number on each node is its height.

To achieve this goal, we define the following root cost function $C(S_k)$ for each node (shape instance) S_k, $k = 1, 2, \ldots, n$. By selecting S_k as the root, we can draw the rooted MST, denoted as $MST(S_k)$, layer by layer as shown in Fig. 3.5. Denote all the edges in $MST(S_k)$ as (S_i, S_j) such that S_i is the parent node of S_j. For each node S_i in $MST(S_k)$ we further define its height $h(S_i)$ to be the number of edges along the longest path from this node S_i to a leaf node. As shown in Fig. 3.5, we define the root cost function

$$C(S_k) = \sum_{(S_i, S_j) \in MST(S_k)} w_{ij} \cdot h(S_i).$$

Then we select the shape instance (node)

$$\arg\min_{S_k} C(S_k)$$

as the root note of the MST for groupwise shape correspondence.

Based on the rooted MST, we can recursively propagate the landmarks from the root to each leaf and achieve the groupwise shape correspondence. Specifically, using the landmark-sliding algorithm described in Section 3.3, the groupwise shape correspondence algorithm consists of the following steps:

1. Sample landmarks along the root shape instance as the template.
2. Take each child node of the root in the MST as the target and correspond each target to the template using the landmark-sliding algorithm.
3. All the target shape instances in Step 2, with their identified landmarks, are then treated as the new templates. For each new template, find its children in the rooted MST as the new targets and perform landmark-sliding algorithm for shape correspondence, i.e., identifying corresponded landmarks on the new targets.
4. Repeat Step 3 to propagate the landmarks layer by layer in the rooted MST until all the leaf nodes are processed.

3.5 PERFORMANCE EVALUATION OF SHAPE CORRESPONDENCE

Objective and quantitative evaluation of shape correspondence is a challenging problem, because of the lack of the ground truth and the non-uniqueness in selecting the corresponded landmarks. Davies et al. [23] suggested the use of three general metrics – compactness, specificity, and generality – for evaluating shape-correspondence performance in terms of the PDM construction. These three metrics describe the properties of the PDM constructed from the identified landmarks – it follows the steps in Section 3.2 to construct a PDM $\mathcal{N}(\bar{\mathbf{v}}, \mathbf{D})$, and then defines the three metrics as the function of the number of considered principal eigenvectors of $\mathbf{D}$. However, it has been shown that the same shape space may show different values of compactness, generality, and specificity when represented by different sets of corresponded landmarks [19].

Using synthesized shape contours from a given PDM can well address these issues and achieve more objective performance evaluation. Specifically, we start from a given ground-truth PDM, from which we can randomly generate a set of synthetic shape contours. The shape-correspondence algorithm to be evaluated is applied to identify corresponded landmarks from these shape contours and leads to a PDM by following the steps in Section 3.2. The better the resulting PDM reflects the ground-truth PDM, the better the shape-correspondence performance.

As shown in Fig. 3.6, the performance evaluation consists of the following five components: (C1) specifying a PDM $\mathcal{N}(\bar{\mathbf{v}}^t, \mathbf{D}^t)$ as the ground truth, (C2) using this PDM to randomly generate a set of shape contours $S_1, S_2, \dots, S_n$, (C3) running the test shape-correspondence algorithm on these shape contours to identify corresponded landmarks, (C4) deriving a PDM $\mathcal{N}(\bar{\mathbf{v}}, \mathbf{D})$ from the identified landmarks using Eq. (3.1), and (C5) comparing the derived PDM $\mathcal{N}(\bar{\mathbf{v}}, \mathbf{D})$ to the ground-truth PDM $\mathcal{N}(\bar{\mathbf{v}}^t, \mathbf{D}^t)$

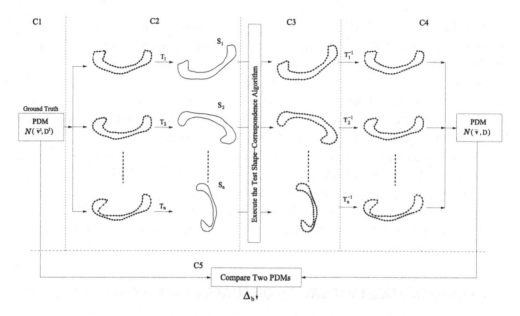

Figure 3.6 An illustration of the proposed shape-correspondence evaluation method.

and using their difference to measure the performance of the test shape-correspondence algorithm.

In essence, this approach evaluates the shape-correspondence algorithm's capability to recover the underlying statistical shape model from a set of sampled shape instances. This reflects the role of the shape correspondence in statistical shape modeling. In these five components, (C3) and (C4) are for shape correspondence and PDM construction, which have been discussed in detail in previous sections. The task of Component (C1) is to specify a mean shape vector $\bar{\mathbf{v}}^t$ and a covariance matrix $\mathbf{D}^t$. Ideally, they can take any values only if $\mathbf{D}^t$ is positive definite. In practice, we can pick them to resemble some real structures. In this section, we focus on developing algorithms for Components (C2) and (C5).

3.5.1 Generating Shape Instances

Given the ground-truth PDM $\mathcal{N}(\bar{\mathbf{v}}^t, \mathbf{D}^t)$ with k landmarks (k might be different from m, the number of landmarks identified by shape correspondence in Component (C3)), we can randomly generate as many sample shape vectors $\hat{\mathbf{v}}_i^t$, $i = 1, 2, \ldots, n$ as possible. More specifically, with $\hat{\mathbf{p}}_j^t$ and $\hat{\lambda}_j^t$, $j = 1, 2, \ldots, 2k$ being the eigenvectors and eigenvalues of $\mathbf{D}^t$, we can generate shape instances in the form of

$$\hat{\mathbf{v}}^t = \bar{\mathbf{v}}^t + \sum_{j=1}^{2k} \hat{b}_j^t \hat{\mathbf{p}}_j^t, \tag{3.12}$$

where $\hat{b}_j^t$ is independently and randomly sampled from the 1D Gaussian distribution $\mathcal{N}(0, \hat{\lambda}_j^t), j = 1, 2, \ldots, 2k$.

Each shape vector $\hat{\mathbf{v}}_i^t$, $i = 1, 2, \ldots, n$ in fact defines k landmarks $\{\hat{\mathbf{v}}_{i1}^t, \hat{\mathbf{v}}_{i2}^t, \ldots, \hat{\mathbf{v}}_{ik}^t\}$. By assuming that these k landmarks are sequentially sampled from a continuous shape contour, we can estimate this continuous contour $\hat{S}_i$ by landmark interpolation. For constructing a closed shape contour, we interpolate the portion between the last landmark $\hat{\mathbf{v}}_{ik}^t$ and the first landmark $\hat{\mathbf{v}}_{i1}^t$. For constructing an open shape contour, we do not interpolate the portion between $\hat{\mathbf{v}}_{ik}^t$ and $\hat{\mathbf{v}}_{i1}^t$. We use the Catmull–Rom cubic spline [3] for interpolating these landmarks into a contour.

For each synthetic shape contour $\hat{S}_i$, we then apply a random affine transformation T_i, consisting of a random rotation, a random (uniform) scaling and a random translation to simulate the real cases in practice, to obtain shape contour S_i. We record the affine transformations T_i, $i = 1, 2, \ldots, n$ and then pass $S_1, S_2, \ldots, S_n$ to the test shape-correspondence algorithm. Note that the recorded affine transformations T_i, $i = 1, 2, \ldots, n$ are not passed to the test shape-correspondence algorithm (Component (C3)), because in real applications, these transformations are not priorly known and it is expected that the shape correspondence algorithm can handle them. If the test shape-correspondence algorithm introduces further transformations, such as Procrustes analysis, in Component (C3), we record and undo these transformations before outputting the shape-correspondence result. This ensures the corresponded landmarks identified by the test shape-correspondence algorithm are placed directly back onto the input shape contours $S_1, S_2, \ldots, S_n$. Then in Component (C4), we directly apply the inverse transform T_i^{-1}, $i = 1, 2, \ldots, n$, to the landmarks identified on S_i. This guarantees the correct removal of the random affine transformation T_i before PDM construction in Component (C4).

3.5.2 Difference Between Two PDMs

The PDM $\mathcal{N}(\bar{\mathbf{v}}, \mathbf{D})$ derived in Component (C4) and the ground-truth PDM $\mathcal{N}(\bar{\mathbf{v}}^t, \mathbf{D}^t)$ each represents a shape space. The closer these two shape spaces, the better the shape correspondence that is used to construct PDM $\mathcal{N}(\bar{\mathbf{v}}, \mathbf{D})$. However, directly computing a distance metric between these two PDMs, e.g., a Kullback–Leibler distance [15], is not applicable in this task, because the landmarks in $\bar{\mathbf{v}}$ and $\bar{\mathbf{v}}^t$ are not corresponded. Actually, $\bar{\mathbf{v}}$ and $\bar{\mathbf{v}}^t$ may be of different dimensions because they may consist of different number of landmarks, i.e., $\bar{\mathbf{v}} \in \mathbb{R}^{2m}$, $\bar{\mathbf{v}}^t \in \mathbb{R}^{2k}$ and $m \neq k$, where m and k are the number of landmarks along each shape contour in these two PDMs.

To compare the underlying shape spaces, we compare the two PDMs in the continuous shape space instead of using the sampled landmarks. Specifically, we use a random-simulation strategy: randomly generating a large set of N shape vectors from each PDM using Eq. (3.12), interpolating these landmarks defined by these shape vectors into continuous shape contours using Catmull–Rom cubic spline, and then measuring the similarity between these two sets of shape contours. We denote the N continuous shape contours generated from PDM $\mathcal{N}(\bar{\mathbf{v}}, \mathbf{D})$ to be $\hat{S}_1^c, \hat{S}_2^c, \ldots, \hat{S}_N^c$ and the N continuous shape contours generated from the ground-truth PDM $\mathcal{N}(\bar{\mathbf{v}}^t, \mathbf{D}^t)$ to be $\hat{S}_1^t, \hat{S}_2^t, \ldots, \hat{S}_N^t$. When N is sufficiently large, the difference between these two sets of continuous shape contours can well reflect the difference of the shape spaces underlying these two PDMs.

Given a pair of continuous shape contours $\hat{S}_i^c$ and $\hat{S}_j^t$, generated from two PDMs respectively, we can define their difference by Jaccard coefficient $\Delta(\hat{S}_i^c, \hat{S}_j^t)$ using Eq. (3.11), because the possible affine transformation between these two shape contours have been removed as described in Section 3.5.1. Based on this, we can estimate the difference between two shape-contour sets $\{S_i^c\}_{i=1}^N$ and $\{S_j^t\}_{j=1}^N$ using the bipartite-matching algorithm. We run bipartite matching between these two shape-contour sets so that the total matching cost, which is defined as the total difference between the matched shape contours, is minimal. Then we define a difference measure between these two PDMs as

$$\Delta_b \triangleq \frac{\sum_{i=1}^N \Delta(\hat{S}_i^c, \hat{S}_{b(i)}^t)}{N}, \tag{3.13}$$

where $\hat{S}_i^c$ and $\hat{S}_{b(i)}^t$ are the matched pair of shape contours in the bipartite matching. In this difference measure, we introduce a normalization over N so that Δ_b takes values in the range of $[0, 1]$. Using the bipartite-matching algorithm, the measure Δ_b not only assesses whether the two shape spaces (defined by the two PDMs) contain similar shape contours, but also examines whether a shape contour shows a similar probability density in these two shape spaces. The smaller the difference Δ_b, the better the shape-correspondence performance.

3.6 EXPERIMENTS

In this section, we evaluate the performance of the shape-correspondence algorithm based on landmark sliding, as described in Section 3.3 and Section 3.4, using the measure Δ_b as described in Section 3.5. The landmark-sliding algorithm was implemented in C++ and the balance factor λ in Eq. (3.7), and the maximum number of iterations used by the quadratic programming solver for these experiments, are set to 2 and 30 respectively. For comparison, we also evaluate the performance of several other shape correspondence algorithms. CPU time reported in this section for each evaluated shape correspondence algorithm is obtained from Linux workstations running Intel Xeon 3.4 GHz processor with 4 GB of RAM.

Specifically, we evaluate six algorithms in the experiments. *SDI*: the pairwise shape correspondence algorithm where all the shape instances are corresponded to a common

Figure 3.7 From left to right, ground-truth mean shapes that resemble the human hand, corpus callosum, and femur, and human face silhouette shape structures.

template by pairwise landmark–sliding algorithm. Initial template is a randomly selected shape instance and in later iterations the template is updated to be the mean shape computed from the previous iteration as described at the beginning of Section 3.4. *MST*: the minimum spanning tree (MST) based groupwise correspondence algorithm as described in Section 3.4. *T-MDL*: Thodberg's implementation of MDL with approximations [25,24]. *E-MDL*: Ericsson and Karlsson's implementation of the MDL [13]. *E-MDL+CUR*: Ericsson and Karlsson's implementation of the MDL method with curvature distance minimization [13]. *EUC*: Ericsson and Karlsson's implementation of the parameterization method by minimizing Euclidean distance [13].

As described in Section 3.5, we need to specify a ground-truth PDM to compute Δ_b for evaluating the shape correspondence performance. Four ground-truth PDMs were specified that resemble the hand, corpus callosum (callosum for short), femur, and human face silhouette (face for short) for the performance evaluation. These four ground-truth PDMs $\mathcal{N}(\bar{\mathbf{v}}^t, \mathbf{D}^t)$ are defined by a mean shape vector $\bar{\mathbf{v}}^t$ with dimension 256 and a covariance matrix $\mathbf{D}^t$ with dimension 256×256, i.e. each shape instance consists of $k = 128$ landmarks. The mean shapes for the four ground-truth PDMs are shown in Fig. 3.7. For each of the four ground-truth PDMs $n = 800$ synthetic shape contours were generated by following the steps described in Section 3.5. These 800 shape contours are used to test the performance of the six shape correspondence algorithms. All the test shape correspondence algorithms are set to identify $m = 64$ landmarks along each shape contour. $N = 2000$ synthetic shape contours are generated from both the PDM $\mathcal{N}(\bar{\mathbf{v}}^t, \mathbf{D}^t)$ and the PDM $\mathcal{N}(\bar{\mathbf{v}}, \mathbf{D})$ for estimating the difference Δ_b by following the steps described in Section 3.5. To check the stability of the Δ_b measure, this random simulation process of generating shape instances is repeated for 50 rounds.

For each shape-correspondence algorithm and each ground-truth PDM, the Δ_b values from the 50 rounds of random simulations are shown in Fig. 3.8. Their average Δ_b values over 50 rounds of simulations are shown in Table 3.1. We can see that the proposed groupwise method, i.e., MST, demonstrates the best performance for the ground-truth PDMs that resembles the callosum, and femur shape structures, has comparable performance to E-MDL for the ground-truth PDM that resembles the face, and has comparable performance to E-MDL and E-MDL+CUR for the ground-truth PDM that resembles the hand.

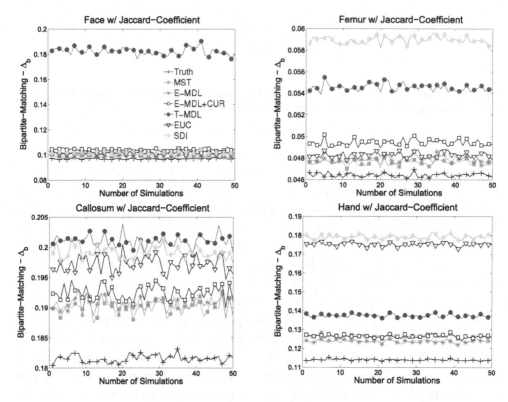

Figure 3.8 Δ_b values for the four ground-truth PDMs resulting from the six test shape correspondence methods using the Jaccard coefficient. The x-axis indicates the round of the random simulation. The curves with the "+" symbols are the values of Δ_b between each ground-truth PDM and itself.

Table 3.1 Δ_b values resulting from the six test shape correspondence algorithms. Each Δ_b value is the average value over the 50 random simulations

	Hand	Callosum	Femur	Face
MST	0.12656	0.19021	0.04762	0.10026
T-MDL	0.13733	0.20105	0.05455	0.18226
E-MDL	0.12381	0.19053	0.04788	0.09847
E-MDL+CUR	0.12646	0.19248	0.04943	0.10432
EUC	0.17520	0.19727	0.04830	0.10148
SDI	0.17897	0.19916	0.05893	0.10189

Table 3.2 reports the CPU time (in seconds) taken by the six test shape correspondence algorithms to correspond the $n = 800$ continuous shape contours generated by each of the four ground-truth PDMs. We can see that the proposed groupwise shape correspondence algorithm, e.g., MST, is much faster than the four global shape correspondence algorithms, i.e., T-MDL, E-MDL, E-MDL+CUR, and EUC. The pairwise

Table 3.2 CPU time (in seconds) taken by the six test shape correspondence methods

	MST	T-MDL	E-MDL	E-MDL+CUR	EUC	SDI
Hand	2927	50784	107317	304504	29572	739
Callosum	2318	44732	107506	278832	28420	703
Femur	1757	59663	109875	261093	28538	740
Face	2417	50710	103822	259551	28286	745

SDI algorithm is about 4 times faster than the proposed MST algorithm. However, from Table 3.1 we can see that SDI demonstrates the worst Δ_b performance for the ground-truth PDMs that resemble the hand and femur shape structures, and demonstrates the second worst performance for the ground-truth PDM that resembles the callosum shape structure.

3.7 CONCLUSIONS AND 3D SHAPE CORRESPONDENCE

Shape correspondence, i.e., identifying a set of landmark points across a population of the shape instances, is one of the most challenging steps in constructing statistical shape models. In this chapter, we focused on discussing about the 2D shape correspondence and its performance evaluation. Specifically, we introduced a landmark-sliding algorithm that can move the landmarks along the shape contours to get better correspondence. We further introduced a groupwise correspondence algorithm that pre-organizes all the shape instances into a rooted minimum spanning tree and the landmarks are propagated from the root shape instance to the leaf nodes layer-by-layer using pairwise correspondence. Finally, we introduced more objective method to evaluate shape correspondence by comparing the shape spaces underlying the ground-truth PDM and the PDM constructed by the shape correspondence. Comparison experiments against several other well known shape correspondence algorithms were reported using this performance evaluation method.

With 3D medical imaging, such as MRI and CT, we can get 3D shape instances of many organs and structures. This leads to 3D statistical shape modeling and 3D shape correspondence. Just like 2D shape, 3D shape can be closed or open, as illustrated in Fig. 3.9. The increased geometric complexity and data size in 3D further increase the computational burden of shape correspondence. For example, global methods, such as MDL and SPHARM, have been used for 3D shape correspondence, but taking extensive CPU time. The landmark-sliding algorithm introduced in Section 3.3 can be easily extended to handle 3D shape correspondence to achieve a good balance between accuracy and running time. Please refer to [7,6,9] for more details on 3D landmark-sliding algorithms. The basic ideas underlying the groupwise correspondence algorithm based on MST in Section 3.4 and the performance evaluation method in Section 3.5 are also applicable to 3D shape [8].

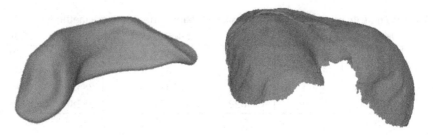

Figure 3.9 3D shape of the hippocampus (left) and diaphragm (right) structures. Note that hippocampus has a closed shape while diaphragm has an open shape.

REFERENCES

[1] F. Bookstein, Principal warps: thin-plate splines and the decomposition of deformations, IEEE Trans. Pattern Anal. Mach. Intell. 11 (6) (1989) 567–585.

[2] F. Bookstein, Landmark methods for forms without landmarks: morphometrics of group differences in outline shape, Med. Image Anal. 1 (3) (1997) 225–243.

[3] E. Catmull, R. Rom, A class of local interpolating splines, Comput. Aided Geom. Des. (1974) 317–326.

[4] T. Cootes, et al., Active shape models – their training and application, Comput. Vis. Image Underst. 61 (1) (1995) 38–59.

[5] T. Cormen, et al., Introduction to Algorithms, MIT Press, Cambridge, MA, 1990.

[6] P. Dalal, et al., 3D open-surface shape correspondence for statistical shape modeling: identifying topologically consistent landmarks, in: International Conference on Computer Vision, 2009.

[7] P. Dalal, et al., A fast 3D correspondence method for statistical shape modelling, in: IEEE Conference on Computer Vision and Pattern Recognition, IEEE, 2007.

[8] P. Dalal, et al., Multiple cortical surface correspondence using pairwise shape similarity, in: International Conference on Medical Image Computing and Computer Assisted Intervention, vol. 1, 2010, pp. 349–356.

[9] P. Dalal, S. Wang, Landmark Sliding for 3D Shape Correspondence, IGI Global, 2012.

[10] R. Davies, et al., A minimum description length approach to statistical shape modeling, IEEE Trans. Med. Imaging 21 (5) (2002) 525–537.

[11] G. Gerig, et al., Shape analysis of brain ventricles using SPHARM, in: IEEE Workshop on Mathematical Methods in Biomedical Image Analysis, 2001, pp. 171–178.

[12] J. Gower, G. Dijksterhuis, Procrustes Problems, Oxford University Press, 2004.

[13] J. Karlsson, A. Ericsson, A ground truth correspondence measure for benchmarking, in: Int. Conf. Pattern Recognit., vol. 3, 2006, pp. 568–573.

[14] D. Kendall, et al., Shape and Shape Theory, John Wiley & Sons, Ltd., 1999.

[15] S. Kullback, R.A. Leibler, On information and sufficiency, Ann. Math. Stat. 22 (1) (1951) 79–86.

[16] C. Lorenz, N. Krahnstover, Generation of point-based 3D statistical shape models for anatomical objects, Comput. Vis. Image Underst. 77 (2000) 175–191.

[17] B. Munsell, et al., Fast multiple shape correspondence by pre-organizing shape instances, in: IEEE Conference on Computer Vision and Pattern Recognition, 2009, pp. 840–847.

[18] B.C. Munsell, et al., A new benchmark for shape correspondence evaluation, in: International Conference on Medical Image Computing and Computer Assisted Intervention, 2007, pp. 507–514.

[19] B.C. Munsell, et al., Evaluating shape correspondence for statistical shape analysis: a benchmark study, IEEE Trans. Pattern Anal. Mach. Intell. 30 (11) (2008) 2023–2039.

[20] B.C. Munsell, et al., Pre-organizing shape instances for landmark-based shape correspondence, Int. J. Comput. Vis. 97 (2012) 210–228.

[21] T. Richardson, S. Wang, Nonrigid shape correspondence using landmark sliding, insertion and deletion, in: International Conference on Medical Image Computing and Computer Assisted Intervention, Part II, 2005, pp. 435–442.

[22] T. Richardson, S. Wang, Open-curve shape correspondence without endpoint correspondence, in: International Conference on Medical Image Computing and Computer Assisted Intervention, Part I, 2006, pp. 17–24.

[23] M. Styner, et al., Evaluation of 3D correspondence methods for model building, in: Information Processing in Medical Imaging Conference, 2003.

[24] H. Thodberg, Adding curvature to minimum description length shape models, in: British Machine Vision Conference, vol. 2, 2003, pp. 251–260.

[25] H. Thodberg, Minimum description length shape and appearance models, in: Information Processing in Medical Imaging Conference, 2003, pp. 51–62.

[26] S. Wang, et al., Shape correspondence through landmark sliding, Part I, in: IEEE Conference on Computer Vision and Pattern Recognition, 2004, pp. 143–150.

[27] J. Xie, P. Heng, Shape modeling using automatic landmarking, Part II, in: International Conference on Medical Image Computing and Computer Assisted Intervention, 2005, pp. 709–716.

CHAPTER 4

Landmark-Based Statistical Shape Representations

Bulat Ibragimov*, Tomaž Vrtovec[†]
*Stanford University School of Medicine, Department of Radiation Oncology, USA
[†]University of Ljubljana, Faculty of Electrical Engineering, Slovenia

Contents

4.1 Introduction	89
4.2 Landmark-Based Shape Representation	91
4.2.1 Initial Shape Alignment	92
4.2.2 Complete Graph	93
4.2.3 Contour and Mesh	94
4.2.4 k-Fan Graph	95
4.2.5 Graphical Lasso	96
4.2.6 Transportation-Based Graphs	97
4.2.6.1 Optimal Transportation-Based Graph	98
4.2.6.2 Optimal Transportation-Based Graph With Landmark Clustering	99
4.2.6.3 Optimal Assignment-Based Graph With Landmark Clustering	100
4.3 Shape-Based Landmark Detection	101
4.3.1 Principal Component Analysis	101
4.3.2 Sparse Linear Combination for Shape Modeling	103
4.3.3 Minimal Cost Path	104
4.3.4 Markov Random Fields	107
4.3.5 Game-Theoretic Optimization	109
4.3.6 Machine Learning-Based Optimization	110
4.4 Conclusion	111
References	112

4.1 INTRODUCTION

Landmarking as the process of detecting intrinsic salient points, i.e. landmarks, has an essential role in modern image analysis because of its wide use for localization, quantification and visualization of objects of interest. In medical image analysis, for example, it can be applied to both normal and pathological anatomical structures, therefore providing assistance towards a better diagnosis, treatment planning and execution, and monitoring of treatment results and disease progression. A landmark is a unique point in an image that can be anatomically or mathematically defined. Anatomical landmarks mark important points of the observed object and are often identified for quantitative morphometric image measurements, while mathematical landmarks are located at the

Statistical Shape and Deformation Analysis
DOI: 10.1016/B978-0-12-810493-4.00005-5

89

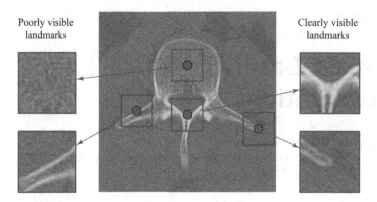

Figure 4.1 An example of clearly visible landmarks located at tips and corners of the observed object, and poorly visible landmarks that have a relatively similar appearance as their neighborhood. (The observed object is a vertebra shown in an axially reconstructed computed tomography cross-section.)

extrema of object boundaries, terminal points and edge intersections. A third type of landmarks, the so-called pseudo-landmarks, is sometimes introduced to fill the regions of the observed object that are poorly represented by landmarks [1,2]. In general, any point that can be uniquely defined in images of a similar (anatomical) region can be considered a landmark [3]. However, some landmarks, for example those located in the center of the observed object, may have anatomically clearly or well-defined positions, but are more difficult to be visually detected than other landmarks, for example those located at tips of the observed object parts (Fig. 4.1). As a result, the selection of landmarks can considerably influence the outcome of their detection, i.e. determine whether it is successful or not. Although a single landmark can hardly provide any information about the observed object, a set of spatially dependent landmarks provides a more information-rich description because the numerical properties of spatial relationships among landmarks define the shape of the observed object. Optimal extraction and description of spatial relationships among landmarks can considerably improve the accuracy of landmark detection, and simplify further qualitative and quantitative analysis of the observed objects. In the field of medical image analysis, for example, the observed objects represent specific anatomical structures [4,5].

There are two main considerations in landmark-based image analysis that are related to the shape of the observed object, namely the selection of landmarks for which pairwise connections are established and corresponding spatial relationships are evaluated [6], and the optimization method that combines the information obtained from these spatial relationships with the appearance information of individual landmarks to find the optimal positions of the same landmarks in previously unseen images of the observed object [7]. Often, the optimization method determines the success or failure of landmark detection, and may therefore impose strict conditions on the number and properties of landmark connections. However, if the conditions are flexible, a proper selection of landmark connections can positively affect the quality of landmark detection. In this chapter, we will describe the existing approaches for selecting adequate landmark connections that correspond to information-rich spatial relationships among landmarks, methods for extracting statistical properties of such spatial relationships, and techniques for finding the optimal solutions of corresponding landmark detection problems. In most cases, the formulation of the shape-based landmark detection can be used for analyzing images of arbitrary dimensionality. However, as image analysis is mainly focused on two-dimensional (2D) and three-dimensional (3D) images, we will use a *2D formulation* everywhere throughout this chapter, except where specific methodological adjustments are required due to a different dimensionality. As landmarks are often positioned on the boundary of the observed object, which can be represented by its contour in 2D images or surface in 3D images, the term *adjacent* or *neighboring landmarks* describes landmarks that are connected by a contour segment in 2D or a surface patch in 3D, and the term *landmark normal* describes the vector at the selected landmark position that is oriented outwards the observed object and is orthogonal to the object contour in 2D or surface in 3D. However, landmarks can be also positioned at specific clearly distinguishable points inside or outside the observed object, therefore the aforementioned terms can be theoretically extended to landmarks not located on its boundary or surface. Finally, the observed object may be represented by different disconnected or nested objects (e.g. lung fields, or external and internal heart ventricle walls), and landmarks from such multi-object systems can be usually detected using approaches for single-object systems with eventual incremental methodological modifications. We will therefore use the term *object of interest* for any configuration of landmarks that describes a single-object or a multi-object system of anatomical structures.

4.2 LANDMARK-BASED SHAPE REPRESENTATION

Let the object of interest be described by a set of landmarks $\mathcal{P}$ with known spatial coordinates for each landmark $p \in \mathcal{P}$, let its shape representation be augmented by pairwise connections among landmarks that capture their spatial relationships and therefore describe distinctive shape properties of the object of interest, and let a set E contain connections only for selected pairs of landmarks $p, q \in \mathcal{P}$. There are three main conditions that have to be taken into account for defining set E. The first condition requires that the set E satisfies specific graph properties, for example, that the number of connections attributed to each landmark $p \in \mathcal{P}$ is constant. This condition does not require any information except the total number of landmarks $|\mathcal{P}|$. The second condition requires that the set E satisfies specific geometrical properties, for example, that only connections belonging to the boundary of the object of interest described by landmarks are selected. This condition therefore requires at least one sample of spatial positioning for

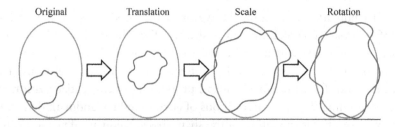

Figure 4.2 A schematic illustration of shape alignment that consists of normalizing translation, scale and rotation of an object sample (blue contour) to the reference object (red contour). (For interpretation of the references to color in this figure, the reader is referred to the web version of this chapter.)

all landmarks in $\mathcal{P}$. The third condition requires that the set E satisfies specific statistical properties, for example, that only connections that are statistically highly dependent on each other are selected. This condition therefore requires multiple samples of spatial positioning for all landmarks in $\mathcal{P}$, which form the so-called training set of samples. Different landmark-based shape representations rely on only one condition, e.g. the k-fan graph (Section 4.2.4), or on two/three conditions, e.g. the optimal transportation–based graph (Section 4.2.6). This section presents various approaches for defining the set E of established connections among landmarks.

4.2.1 Initial Shape Alignment

Landmark-based image analysis always starts with a training set of landmark samples (usually defined in a training set of images), from which we can study the appearance of individual landmarks and evaluate spatial relationships among groups of landmarks. Let matrix $\mathbf{V}$ of size $N \times |\mathcal{P}|$ contain N samples of landmarks $\mathcal{P}$, where i-th row corresponds to the spatial positioning of i-th sample, and suppose that considerable translation, scaling and rotation differences exist among these landmark samples (Fig. 4.2). If we compare individual positions for each landmark $p \in \mathcal{P}$, and distances and angles for pairs of landmarks $p, q \in \mathcal{P}$ from such samples, a considerable variability of the obtained values will be observed despite the fact that these landmark samples form a relatively similar shape. This variability that originates from the similarity transformation (i.e. translation, scaling and rotation of the object of interest) will hamper the correct understanding of the underlying object shape, and therefore it must be minimized by unifying all samples from matrix $\mathbf{V}$ so that only properties related to describing the object shape are retained. This unification is performed using the Procrustes analysis, where translation, scaling and rotation parameters are first extracted from all samples, then normalized across samples and finally applied back to each sample. The translation component is removed from a sample $\mathbf{v} = \{(x_p, y_p)\}$ of size $1 \times |\mathcal{P}|$; $\mathbf{v} \in \mathbf{V}$, by aligning its centroid $(\overline{x}, \overline{y})$ with the origin of the coordinate system:

$$(\overline{x}, \overline{y}) = \left(\frac{1}{|\mathcal{P}|} \sum_{p \in \mathcal{P}} x_p, \frac{1}{|\mathcal{P}|} \sum_{p \in \mathcal{P}} y_p \right); \quad (x_p, y_p) \leftarrow (x_p - \overline{x}, y_p - \overline{y}). \quad (4.1)$$

Uniform scaling is then performed so that the root mean square distance from landmarks $\mathcal{P}$ to the centroid of $\mathbf{v}$ (i.e. to $(0,0)$, as the translation component was removed) is normalized to 1:

$$\delta = \sqrt{\frac{1}{|\mathcal{P}|} \sum_{p \in \mathcal{P}} (x_p^2 + y_p^2)}; \quad (x_p, y_p) \leftarrow \left(\frac{x_p}{\delta}, \frac{y_p}{\delta} \right). \quad (4.2)$$

Finally, to remove rotation, samples are slightly rotated against each other until the sum of the squared distances among corresponding landmarks from these samples is minimized (Fig. 4.2). The obtained matrix $\mathbf{P}$ of size $N \times |\mathcal{P}|$ with N aligned samples of landmarks $\mathcal{P}$ can be used for studying the properties of a previously unseen object of a similar shape, usually called the target object (observed in a previously unseen target image). The described initial shape alignment is often used prior to modeling of the object shape, otherwise shape modeling would be very noisy and could potentially hamper the search for consistent shape patterns among landmark samples.

4.2.2 Complete Graph

The most straightforward approach for generating a landmark-based shape representation is to establish connections for every pair of landmarks $p, q \in \mathcal{P}$, which results in the *complete graph* (CG) with the set of connections among landmarks equal to $E = \{(p,q)\}: \forall p \neq q; \ p, q \in \mathcal{P}$ (Fig. 4.3A). The total number of connections in CG is therefore $|\mathcal{P}|(|\mathcal{P}| - 1)/2$. The inclusion of all possible connections among landmarks into E guarantees that all landmarks from $\mathcal{P}$ equally contribute to the resulting shape representation, which is relatively rigid, meaning that it prevents non-plausible shape deformations and considerable landmark misdetections. At the same time, CG limits the elasticity of the shape representation, and therefore landmarking of objects with highly variable or pathological shape may become challenging. Another disadvantage of CG is the complexity of the resulting shape representation, as a linear growth of the number of landmarks is reflected in a quadratic growth of landmark connections, and evaluating statistical properties of every connection may therefore result in considerably long computation times while taking up the available memory, which is usually not tolerated by practical applications. This issue may not be an obstacle in 2D image analysis, where the number of landmarks is usually relatively low (i.e. in practice, it rarely exceeds a hundred), but can considerably limit 3D image analysis, where incomparably more landmarks are required for an accurate

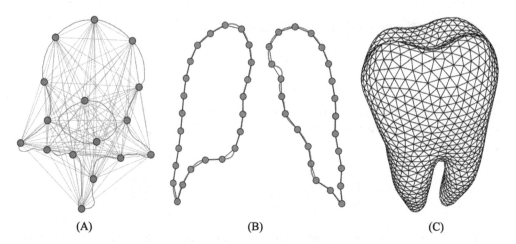

(A)   (B)   (C)

Figure 4.3 (A) The complete graph (CG) of connections (shown for an axial cross-section of a vertebra). (B) A contour-based shape representation (shown for an outline of the left and right lung fields). (C) A mesh-based shape representation (shown for a 3D model of a tooth).

description of the object of interest (i.e. in practice, often more than a hundred thousand).

4.2.3 Contour and Mesh

In applications such as image segmentation, landmarks $\mathcal{P}$ are often positioned on the boundary of the object of interest, therefore its shape can be defined by connections that pass through adjacent landmarks. In the case the boundary is a 2D *contour*, each landmark is connected with exactly two adjacent landmarks (Fig. 4.3B), whereas in the case the boundary is a 3D surface or *mesh*, each landmark is connected with several adjacent landmarks (Fig. 4.3C). The simplicity of contours enables the use of highly efficient graph-based optimization methods for landmark detection, such as dynamic programming and the max-flow-min-cut algorithm, while meshes can compactly describe objects with a large number of landmarks, which makes them the most common 3D object shape representation in computer graphics. As the shape representation in the form of contour or mesh is, from the perspective of graph theory, weaker than CG (Section 4.2.2), the occurrence of non-plausible shape deformations cannot be prevented. On the other hand, contours and meshes are at the same time very elastic, and therefore suitable for analyzing objects with a highly variable shape. Moreover, when landmarks are densely distributed on the object boundary, the connections between adjacent landmarks that form contours and meshes also align with the object boundary, and can therefore be easily detected as they pass through strong image intensity gradients.

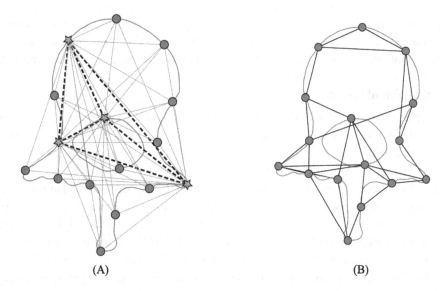

(A)   (B)

Figure 4.4 (A) A shape representation in the form of a 4-fan graph. Landmarks marked by stars correspond to a 4-clique, dashed lines correspond to intra-clique connections, and thin lines correspond to connections among landmarks from the clique and other landmarks. (B) A schematic example of a shape representation in the form of a graphical lasso (GL).

4.2.4 k-Fan Graph

When the shape representation of the object of interest is defined by CG (Section 4.2.2), a set of connection lengths uniquely defines the shape of the object of interest, while sets of connection lengths of a contour or mesh (Section 4.2.3) can correspond to objects with different shape (e.g. the mesh of a sphere can be compressed into a disk without modifying the length of individual mesh edges). It is therefore important to estimate the number of connections that can be removed until the uniqueness of the shape representation is lost. The graph theory introduces the *k-fan graph* that has very few connections but still satisfies the above mentioned condition.

- *globally rigid graph*: A graph $\mathcal{G} = (\mathcal{P}, E)$ is globally rigid if the length of connections from E uniquely determines the distance between any two landmarks $p, q \in \mathcal{P}$;
- *k-clique*: A fully-connected subset of k landmarks from $\mathcal{P}$ is a k-clique on graph $\mathcal{G}$;
- *k-fan graph*: If graph $\mathcal{G}$ has a k-clique and all other landmarks are connected with each landmark from the k-clique, $\mathcal{G}$ is called a k-fan graph (Fig. 4.4A).

Any k-fan graph with $k > m$ was shown to be globally rigid on domain $\mathbb{R}^m$ [8], therefore respectively for 2D ($m = 2$) and 3D ($m = 3$) landmark sets, 3-fan and 4-fan graphs are already globally rigid. The total number of connections in the k-fan graph of landmarks $\mathcal{P}$ is $k||\mathcal{P}| - k| + (k-1)(k-2)/2$, which is for a large number of landmarks much smaller than the total number of connections in CG. The main disadvantage of k-fan graphs is that landmarks from k-cliques are randomly selected without considering the statistical properties of the object shape. As landmarks from k-cliques are connected with all other landmarks, their contribution to the shape representation can be considerably information-rich and therefore they should be carefully selected [8,9].

4.2.5 Graphical Lasso

Shape representations in the form of CG (Section 4.2.2), contour or mesh (Section 4.2.3), and k-fan graph (Section 4.2.4) rely solely on graph properties of the observed landmark configuration, and therefore can be generated from a single sample of landmarks $\mathcal{P}$. As some landmarks are usually more dependent on each other, it would be beneficial to preserve connections among highly dependent landmarks and neglect connections among relatively independent landmarks. For a given matrix $\mathbf{P}$ of size $N \times |\mathcal{P}|$ with N aligned samples of landmarks $\mathcal{P}$, the *graphical lasso* (GL) is an algorithm for estimating the covariance matrix $\mathbf{\Sigma}$ of $\mathbf{P}$ under the assumption that its inverse matrix $\mathbf{\Theta} = \mathbf{\Sigma}^{-1}$ is sparse, which is important when the number of samples N is not very large. To find the optimal $\mathbf{\Theta}$, GL minimizes the following l_1-regularized negative log-likelihood:

$$\min_{\mathbf{\Theta} \succ 0}\bigl(-\log\det(\mathbf{\Theta}) + \operatorname{tr}(\mathbf{S}\mathbf{\Theta}) + \lambda\|\mathbf{\Theta}\|_1\bigr), \qquad (4.3)$$

where $\mathbf{S}$ is the sample covariance matrix computed directly from $\mathbf{P}$, $\operatorname{tr}(\cdot)$ denotes the matrix trace (i.e. the sum of matrix diagonal elements), $\|\mathbf{\Theta}\|_1$ is the l_1 norm of $\mathbf{\Theta}$ (i.e. the sum of matrix absolute values), and λ is a weighting factor. The minimum of Eq. (4.3) is obtained where its derivative equals zero:

$$-\mathbf{\Theta}^{-1} + \mathbf{S} + \lambda\mathbf{\Gamma}_{\mathbf{\Theta}} = 0, \qquad (4.4)$$

where $\mathbf{\Gamma}_{\mathbf{\Theta}}$ is the sign matrix of $\mathbf{\Theta}$. By defining matrix $\mathbf{W}$ as the estimate of $\mathbf{\Sigma}$, $\mathbf{W} = \mathbf{\Theta}^{-1}$, Eq. (4.4) can be solved by iterating through the columns of $\mathbf{W}$ and considering all variables in other columns as fixed. For p-th column (Eq. (4.4)), we obtain:

$$\mathbf{W}_{\bar{p},\bar{p}}\frac{\boldsymbol{\theta}_{\bar{p}}}{\mathbf{\Theta}_{p,p}} + \mathbf{s}_{\bar{p},p} + \lambda\boldsymbol{\gamma}_{\bar{p},p} = 0, \qquad (4.5)$$

where matrix $\mathbf{W}_{\bar{p},\bar{p}}$ corresponds to $\mathbf{W}$ without p-th row and p-th column, $\boldsymbol{\theta}_{\bar{p}}$ is p-th column of matrix $\mathbf{\Theta}$ without element $\mathbf{\Theta}_{p,p}$, $\mathbf{s}_{\bar{p},p}$ is the element from $\mathbf{S}$, and $\boldsymbol{\gamma}_{\bar{p},p}$ is the element from $\mathbf{\Gamma}_{\mathbf{\Theta}}$. This problem (Eqs. (4.4), (4.5)) can be solved using the element-wise coordinate descent optimization algorithm, and the obtained results update $\mathbf{W}$ and $\mathbf{\Theta}$ (additional details on the GL optimization can be found in [10,11]). The solution of GL indicates landmark pairs that are highly dependent on each other (Fig. 4.4B), and therefore the corresponding connections that should be retained. At the same time, parameter λ is inversely proportional to the number of selected connections so that the resulting graph can be made more rigid or more flexible.

4.2.6 Transportation-Based Graphs

Despite GL (Section 4.2.5) selects the most dependent pairs of landmarks, the resulting set E can contain weakly connected landmarks, i.e. landmarks with a relatively low number of attributed connections in comparison to other landmarks. The application of such shape representations, which lack rigidity at some parts, may result in non-plausible shapes. In the non-discriminative case, when the number of connections is equal for every landmark, and, at the same time, relatively large, each landmark contributes an equal amount of geometric information to the resulting shape representation.

A non-discriminative shape representation, where the same number of connections is attributed to every landmark, can be defined as:

$$1^* = \arg\min_{\{1_{p,q}\}} \sum_{p \in \mathcal{P}} \sum_{q \in \mathcal{P} \setminus \{p\}} 1_{p,q} \cdot u_{p,q},$$
$$\forall p \in \mathcal{P}: \sum_{q \in \mathcal{P} \setminus \{p\}} 1_{p,q} = n, \tag{4.6}$$

where $u_{p,q}$ is a measure of the cost that is made to the shape representation by establishing the connection between landmarks p and q, and which sum is being minimized. The indicator function $1_{p,q} = \{0, 1\}$ equals one if the connection between landmarks p and q is established, and zero otherwise, and the set $1^* = \{1_{p,q}^*\}$ represents the optimal selection of connections for all pairs of landmarks. The constant $n \leq |\mathcal{P}| - 1$ defines the number of connections that are attributed to every landmark (the equality is valid for CG (Section 4.2.2), where every landmark has exactly $|\mathcal{P}| - 1$ connections). Because the number of connections n is equal for all landmarks, the resulting shape representation is a regular graph. If landmark p is connected with landmark q, then landmark q is also connected with landmark p, therefore $1_{p,q} = 1_{q,p}$, and therefore the resulting shape representation is a non-oriented graph. Consequently, the cost of the connection between landmarks p and q is also reciprocal, i.e. $u_{p,q} = u_{q,p}$.

The problem of finding optimal connections with minimal cost (i.e. maximal benefit) can be observed from the perspective of transportation theory [12], which studies optimal transportation and allocation of goods, and provides globally optimal and computationally efficient solutions for a variety of applications based on graph theory. Let set $\mathcal{P}$ be observed as two separate but equal sets $\mathcal{P}_1$ and $\mathcal{P}_2$; $\mathcal{P}_1 = \mathcal{P}_2 = \mathcal{P}$. Each element $p \in \mathcal{P}_1$ acts as a source and has n amount of goods, and each element $q \in \mathcal{P}_2$ acts as a destination and requires n amount of goods. If $p \neq q$, then source p is connected with destination q by a road, and the cost of transporting one unit of goods from p to q via this road equals $u_{p,q}$ (Eq. (4.6)). Moreover, the capacity of the road is one, meaning that at most one unit of goods can be transported via this road. The goal is to transport the available goods from sources $\mathcal{P}_1$ to destinations $\mathcal{P}_2$ by minimizing the cost of such transportation. The formulated optimization is called the transportation problem:

$$c^* = \arg\min_{\{c_{p,q}\}} \sum_{p\in\mathcal{P}_1} \sum_{q\in\mathcal{P}_2\setminus\{p\}} c_{p,q} \cdot u_{p,q},$$

$$\forall p \in \mathcal{P}_1: \quad \sum_{q\in\mathcal{P}_2\setminus\{p\}} c_{p,q} = n$$

$$\forall q \in \mathcal{P}_2: \quad \sum_{p\in\mathcal{P}_1\setminus\{q\}} c_{p,q} = n \qquad (4.7)$$

$$\forall p \in \mathcal{P}_1, \; q \in \mathcal{P}_2: \quad 0 \le c_{p,q} \le 1,$$

where $c_{p,q}$ is the varying amount of goods, and $c_{p,q}^*$ is the optimal amount of goods transported via the road connecting source p and destination q. The optimal transportation plan is represented by set $c^* = \{c_{p,q}^*\}$, which comprises of the selected optimal amounts of goods that are transported via the corresponding roads to obtain the minimal transportation cost (i.e. maximal benefit). The cost paid for using the road between source p and destination q is defined by $u_{p,q}$. The solution of this transportation problem (Eq. (4.7)) is globally optimal and can be obtained by computationally efficient algorithms.

The described transportation problem can be transferred to the landmark-based shape representation by considering landmarks as sources and destinations, all possible landmark connections as roads between these sources and destinations, and established landmark connections as goods that are transported via these roads. According to this formulation, cost $u_{p,q}$ is based on statistical properties between landmarks p and q computed from matrix $\mathbf{P}$ of size $N \times |\mathcal{P}|$ with N aligned samples of landmarks $\mathcal{P}$. By the presented principles of transportation theory, three types of sets E can be derived: the optimal transportation-based graph (Section 4.2.6.1), the optimal transportation-based graph with landmark clustering (Section 4.2.6.2), and the optimal assignment-based graph with landmark clustering (Section 4.2.6.3).

4.2.6.1 Optimal Transportation-Based Graph

The solution of the transportation problem (Eq. (4.7)) indicates whether a road should be used for transporting goods from the perspective of minimizing transportation costs and, equivalently, if a connection between landmarks p and q should be included into the shape representation. For natural values of the number of connections n attributed to every landmark, which means that n units of goods should pass through this landmark, the optimal amount of goods $c_{p,q}^*$ will be zero or one, and $c_{p,q}^*$ will therefore act as the indicator function $1_{p,q}^* = \{0, 1\}$ (Eq. (4.6)). The total number of connections is therefore $\frac{1}{2}|\mathcal{P}|n$, as each landmark is connected with exactly n other landmarks (Fig. 4.5A). We refer to such landmark-based shape representation as the *optimal transportation-based graph* (OTG). However, because every landmark represents a source and, at the same time, a destination, additional restrictions have to be applied. First, a regular graph, where each landmark $p \in \mathcal{P}$ has the same number of attributed connections, does not exist for

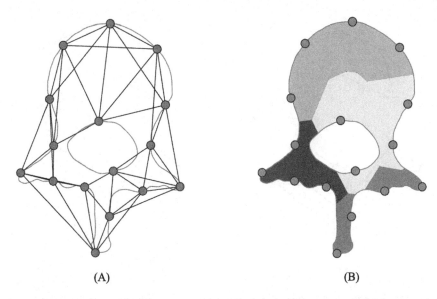

(A)   (B)

Figure 4.5 (A) The optimal transportation-based graph (OTG), where each landmark is connected with exactly five other landmarks so that the total length of connections among landmarks is minimized. (B) An example of the separation of landmarks into four clusters.

every $n \leq |\mathcal{P}| - 1$. The required condition for a regular graph to exist is that $|\mathcal{P}|n$ is an even number. Second, a non–oriented graph, where $c_{p,q}^* = c_{q,p}^*$, is not always guaranteed. In the case $c_{p,q}^* \neq c_{q,p}^*$, the corresponding solution is removed from further consideration, and the next optimal solution is selected [13–15].

4.2.6.2 Optimal Transportation-Based Graph With Landmark Clustering

The presented OTG (Section 4.2.6.1) guarantees that the selected connections have the minimal total cost according to $u_{p,q}$. However, the selection of $u_{p,q}$ may cause separation of landmarks into weakly connected subsets (i.e. partitioning by low-ratio sparse graph cuts). To avoid such unsupervised clustering, we first separate landmarks into L clusters $\mathcal{C} = \{\mathcal{C}_i\}$, each with the same number of landmarks $|\mathcal{C}_i| = |\mathcal{C}_j|$; $\forall i, j$, so that $\mathcal{C}_1 \cup \mathcal{C}_2 \cup \cdots \cup \mathcal{C}_L = \mathcal{P}$ (Fig. 4.5B). For landmarks within each cluster $\mathcal{C}_i$, we find $n = n_1$ optimal intra-cluster connections by solving the transportation problem (Eq. (4.7)). Moreover, for each unordered pair of clusters $\mathcal{C}_i$ and $\mathcal{C}_j$; $i \neq j$, we consider landmarks from cluster $\mathcal{C}_i$ as sources and landmarks from cluster $\mathcal{C}_j$ as destinations, and find $n = n_2$ optimal inter-cluster connections, again by solving the transportation problem (Eq. (4.7)). As a result, every landmark from cluster $\mathcal{C}_i$ is connected with n_1 landmarks from the same cluster $\mathcal{C}_i$ and with n_2 landmarks from every other cluster $\mathcal{C}_j$. The total number of connections is therefore $\frac{1}{2}|\mathcal{P}|(n_1 + n_2(L-1))$ (Fig. 4.6A) [13–15]. We refer to such landmark-based shape representation as the *optimal transportation-based*

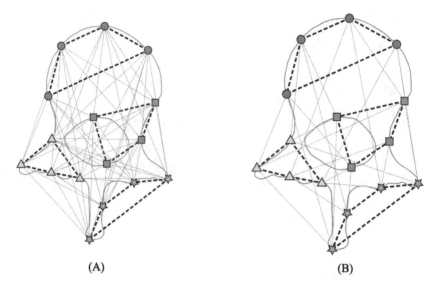

Figure 4.6 (A) The optimal transportation-based graph with landmark clustering (OTG-C). (B) The assignment-based graph with landmark clustering (OAG-C). Landmarks are first separated into clusters (Fig. 4.5B, shown here in different shape and color), and then landmarks from individual clusters are connected by solving the transportation problem, while landmarks from different clusters are connected by solving again the transportation problem in the case of OTG-C, or the assignment problem in the case of OAG-C. (For interpretation of the colors in this figure, the reader is referred to the web version of this chapter.)

graph with landmark clustering (OTG-C). However, searching for inter–cluster connections by solving the transportation problem is possible only if every cluster has the same number of landmarks, i.e. $|\mathcal{C}_i| = |\mathcal{C}_j|$; $\forall i, j$. In the case the same cardinality cannot be achieved for every cluster, a supplementing number of "dummy" landmarks must be added to corresponding clusters. Each dummy landmark $\tilde{p}$ is an artificial entity without specific spatial position, but its cost measured by $u_{\tilde{p},q}$ is equal against every other landmark q; $u_{\tilde{p},q} = const.$; $\forall q \in \mathcal{P}$. As a result, dummy landmarks do not interfere with the optimization process but ensure its regularity.

4.2.6.3 Optimal Assignment-Based Graph With Landmark Clustering

In OTG-C (Section 4.2.6.2), the number of inter–cluster connections attributed to every landmark is n_2 $(L - 1)$. The selection of n_2 therefore directly affects the complexity of the resulting shape representation. In terms of transportation theory, $n_2 = 1$ means that exactly one unit of goods is stored at every source and is, at the same time, required at every destination. The optimal transportation plan can be therefore formulated as the optimal assignment of sources to destinations in a one-to-one manner. Such a special case of the transportation problem is called the assignment problem and is used in a wide range of applications, for example, in assigning workers to tasks [12]. Moreover, the assignment problem has several computationally efficient solutions, e.g. the Hungarian algorithm [16]. As the application of optimal assignment to intra-cluster connections would result in a weak shape representation, we find $n = n_1$ optimal intra-cluster connections for landmarks from each cluster C_i by solving the transportation problem (Eq. (4.7)), i.e. the same as in the case of OTG-C. On the other hand, for each unordered pair of clusters C_i and C_j; $i \neq j$, we consider landmarks from cluster C_i as sources and landmarks from cluster C_j as destinations, and find $n_2 = 1$ optimal inter-cluster connections by solving the assignment problem [13–15]. As a result, every landmark from cluster C_i is connected to n_1 landmarks from the same cluster C_i and exactly to one landmark from every other cluster C_j. The total number of connections is therefore $\frac{1}{2}|\mathcal{P}|(n_1 + L - 1)$ (Fig. 4.6B). We refer to such shape representation as the *optimal assignment-based graph with landmark clustering* (OAG-C).

4.3 SHAPE-BASED LANDMARK DETECTION

The graph, geometrical and statistical properties of matrix **P** of size $N \times |\mathcal{P}|$ with N aligned samples of landmarks $\mathcal{P}$ are used for selecting the optimal connections among landmarks. At the same time, **P** contains essential information about the quality of the spatial configuration of landmarks $\mathcal{P}$ in terms of the underlying shape representation. If the corresponding properties are analyzed and consistent patterns that describe the shape are extracted, we can estimate the similarity between a new configuration of landmarks $\mathcal{P}$ and samples from **P**. However, the process of finding optimal landmark positions in a previously unseen target image is a non-trivial task. For integrating the shape representation, its consistent patterns and appearance models of individual landmarks into a landmark detection framework, most of the existing approaches make use of the numerical analysis of the object appearance and shape, e.g. principal component analysis (Section 4.3.1), sparse linear combination for shape modeling (Section 4.3.2), minimal cost paths (Section 4.3.3), Markov random fields (Section 4.3.4), game-theoretical approaches (Section 4.3.5) and machine learning approaches (Section 4.3.6).

4.3.1 Principal Component Analysis

Given **P**, the goal is to numerically estimate the probability that a previously unseen sample of landmarks $\mathcal{P}$ corresponds to the shape of the object of interest, and to devise instruments for modifying a sample of landmarks $\mathcal{P}$ that does not towards a sample that does represent the shape of the object of interest. To mathematically define such properties, the covariance matrix **cov(P)** of **P** is first computed:

$$\mathbf{cov}(\mathbf{P}) = \frac{1}{N-1} \sum_{\mathbf{x} \in \mathbf{P}} (\mathbf{x} - \bar{\mathbf{x}})(\mathbf{x} - \bar{\mathbf{x}})^T, \tag{4.8}$$

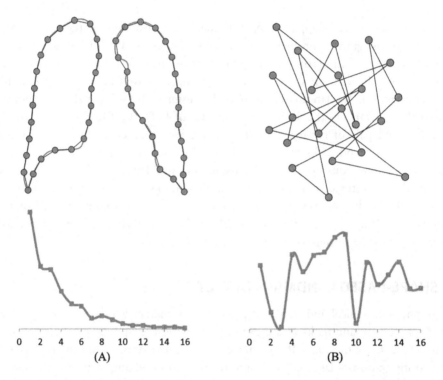

Figure 4.7 (A) Landmark-based lung field shape representation (top) and the corresponding linear combination coefficients that are in agreement with the eigenvalues (bottom). (B) A sample of randomly positioned landmarks (top) with the corresponding linear combination coefficients that are randomly distributed (bottom).

where $\bar{\mathbf{x}}$ is the mean configuration of landmarks $\mathcal{P}$ computed from $\mathbf{P}$. From $\mathbf{cov}(\mathbf{P})$, a matrix of eigenvectors $\mathbf{A}$ is computed with the corresponding eigenvalues λ. The eigenvectors are put in the descending order according to λ, and those eigenvectors that correspond to the largest eigenvalues represent the most considerable changes among samples from $\mathbf{P}$. As eigenvectors are orthogonal, their linear combination can describe any possible combination of $|\mathcal{P}|$ points. However, the linear combination coefficients are unpredictable when describing a random set of $|\mathcal{P}|$ points and, in general, in agreement with the eigenvalues when describing a set of $|\mathcal{P}|$ points that corresponds to the shape of the object of interest. The eigenvalues do not only indicate the corresponding eigenvectors that contribute most to the shape of the object of interest, but also represent the variance of the corresponding eigenvectors. By limiting the linear combination coefficients to $\pm 3\sqrt{\lambda}$, we ensure that eigenvectors generate plausible samples of the shape of landmarks $\mathcal{P}$. Using eigenvectors in $\mathbf{A}$ and eigenvalues λ, we can numerically estimate the similarity between a previously unseen sample of landmarks $\mathcal{P}$ and the shape of the object of interest (Fig. 4.7).

The procedure of describing object samples by eigenvectors and eigenvalues is called the *principal component analysis* (PCA) and is widely used in observing shape through the application of active shape models (ASM) [17,18]. According to ASM, the mean configuration $\overline{\mathbf{x}}$ of landmarks $\mathcal{P}$ is first positioned in a new target image, and then, for each candidate point s_p of landmark $p \in \mathcal{P}$, a more adequate candidate point s_p^* is searched along the boundary normal vector at s_p. The candidate points are then modified towards points $\mathcal{S}^* = \{s_p^*\}$ while being restricted by principal components $\overline{\mathbf{A}}$, i.e. the eigenvectors in $\mathbf{A}$ that correspond to the largest eigenvalues λ:

$$\mathbf{x} \leftarrow \min_{\Delta, \delta, \theta, \mathbf{b}} \left| \mathcal{S}^* - T_{\Delta, \delta, \theta}(\overline{\mathbf{x}} - \overline{\mathbf{A}}\mathbf{b}) \right|, \tag{4.9}$$

where $T_{\Delta, \delta, \theta}(\cdot)$ is a similarity transformation operator that translates (for shift Δ), scales (for factor δ) and rotates (for angle θ) points $\overline{\mathbf{x}}$, and $\mathbf{b}$ is a set of linear coefficients for eigenvectors in $\overline{\mathbf{A}}$. The procedure in Eq. (4.9) searches for the optimal positioning of the shape that describes the object of interest according to candidate points $\mathcal{S}^*$, and is, until convergence, repeated with the updated candidate points $\mathcal{S}^*$.

A strong methodological foundation, its simplicity and universality made PCA/ASM one of the most popular approaches in the field of landmark-based image analysis [19]. The main advantage of ASM is that a solution with a plausible shape is guaranteed and that the morphometry of the observed object can be explicitly computed using eigenvector coefficients $\mathbf{b}$. The main disadvantage of ASM is the need for initialization, and if the mean configuration $\overline{\mathbf{x}}$ of landmarks $\mathcal{P}$ is initially positioned far away from the observed object of interest, ASM may fail. Moreover, the search for new candidate points is performed along the normal vectors of existing candidate points s_p, which considerably restricts ASM.

4.3.2 Sparse Linear Combination for Shape Modeling

When a representative number of landmark training samples of the object of interest is available in set $\mathbf{P}$, the underlying shape can be described as a linear combination of these samples:

$$\mathbf{x} \leftarrow \min_{\Delta, \delta, \theta, \mathbf{b}} \left| \mathbf{P}\mathbf{b} - T_{\Delta, \delta, \theta}(\mathcal{S}^*) \right|^2, \tag{4.10}$$

where, similarly to PCA (Section 4.3.1), $T_{\Delta, \delta, \theta}(\mathcal{S}^*)$ is a similarity transformation of landmark candidate points $\mathcal{S}^*$ and $\mathbf{b}$ is a set of coefficients that makes the corresponding linear combination of training samples $\mathbf{P}$ most similar to $\mathcal{S}^*$. Clearly, the more training samples we have, the better we can approximate landmark candidate points $\mathcal{S}^*$. If the number of samples N is much larger than $|\mathcal{P}|$ and there is a $|\mathcal{P}| \times |\mathcal{P}|$ basis in $\mathbf{P}$, i.e. some $|\mathcal{P}|$ training samples form a linearly independent set of vectors, we can always find a set of coefficients $\mathbf{b}$ that will result in an exact description of $\mathcal{S}^*$. However, if we can obtain an exact description of any set of points $\mathcal{S}^*$, Eq. (4.10) has no predicative power and it is therefore impossible to distinguish whether points $\mathcal{S}^*$ describe the object of interest or are randomly distributed in space. Moreover, even if there is no $|\mathcal{P}| \times |\mathcal{P}|$ basis in $\mathbf{P}$, several optimal sets of coefficients $\mathbf{b}$ may exist and there is a need to select the most adequate one. As it is better to use only a few samples from $\mathbf{P}$ that are most similar to $\mathcal{S}^*$, Eq. (4.10) can be constrained to prefer sparse sets $\mathbf{b}$:

$$\mathbf{x} \leftarrow \min_{\Delta,s,\theta,\mathbf{b}} \left|\mathbf{P}\mathbf{b} - T_{\Delta,s,\theta}(\mathcal{S}^*)\right|^2; \quad \|\mathbf{b}\|_0 \leq \bar{b}, \tag{4.11}$$

where $\|\mathbf{b}\|_0$ is the l_0 norm of $\mathbf{b}$, and $\|\mathbf{b}\|_0 \leq \bar{b}$ ensures that the total number of non-zero coefficients in $\mathbf{b}$ is lower than a predefined constant $\bar{b}$. The obtained problem (Eq. (4.11)) can no longer be optimized in polynomial time due to the nonconvexity of the l_0 norm operation. However, it was shown that such problems with sparse constrains can be approximated using the l_1 norm operation [20,21]:

$$\mathbf{x} \leftarrow \min_{\Delta,s,\theta,\mathbf{b}} \left(\left|\mathbf{P}\mathbf{b} - T_{\Delta,s,\theta}(\mathcal{S}^*)\right|^2 + \lambda_b\|\mathbf{b}\|_1\right). \tag{4.12}$$

The solution of Eq. (4.12) allows to verify the similarity between candidate points $\mathcal{S}^*$ and the shape of the object of interest, and provides information about which points should be changed to make $\mathcal{S}^*$ more similar to the shape of the object of interest [22, 23]. The main challenge of using the described *sparse linear combinations for shape modeling* (SLC) is that the procedure may require a lot of training samples to model elastic or complex objects. This issue is intensified when a large number of landmarks $|\mathcal{P}|$ is used to define the shape of the object of interest, and therefore represents a considerable limitation in 3D image analysis that is usually associated with a large number of landmarks that correspond to 3D mesh vertices.

4.3.3 Minimal Cost Path

To reduce the dependence on initialization, a number of approaches perform landmark detection by first identifying landmark candidate points $\mathcal{S} = \{\mathcal{S}_p\}$ that can be searched in whole images, and then selecting the optimal candidate points using a predefined shape model of the object of interest. The set of candidate points $\mathcal{S}_p = \{s_p\}$ for landmark $p \in \mathcal{P}$ is usually represented by locations of local maxima of a selected appearance likelihood map f_p for landmark p, trained on N annotated images. The total number of all possible combinations of candidate points $\mathcal{S}$ to represent landmarks $\mathcal{P}$ equals to $|\mathcal{S}_1| \cdot |\mathcal{S}_2| \cdot \ldots \cdot |\mathcal{S}_p| \cdot \ldots \cdot |\mathcal{S}_{|\mathcal{P}|}|$, which is in practice impossible to analyze in real time even if the number of candidate points $|\mathcal{S}_p|$ for each landmark p is relatively small. This issue limits the applicability of holistic models such as PCA (Section 4.3.1), and requires focusing on pairwise spatial relationships among landmarks as well as on computationally fast and efficient optimization approaches.

The *minimal intensity and shape cost path* (MISCP) algorithm computes pairwise spatial relationships between adjacent landmarks on the boundary of the object of interest. Let the connection between two adjacent landmarks p and p_+ be of length ε_{p,p_+}, and let $\mu_{\varepsilon_{p,p_+}}$ be the mean length and $\sigma_{\varepsilon_{p,p_+}}$ the corresponding standard deviation of the length of the same connection in the training set of N landmark samples $\mathbf{P}$. The likelihood that the positioning of candidate points s_p and s_{p_+} for landmarks p and p_+, respectively, is in agreement with the positioning of the same two landmarks in $\mathbf{P}$, i.e. the so-called shape likelihood map, is computed using the Mahalanobis distance:

$$g_{p,p_+}(s_p, s_{p_+}) = 1 - (\varepsilon_{s_p,s_{p_+}} - \mu_{\varepsilon_{p,p_+}})^T \sigma_{\varepsilon_{p,p_+}}^{-1} (\varepsilon_{s_p,s_{p_+}} - \mu_{\varepsilon_{p,p_+}}), \qquad (4.13)$$

where $\varepsilon_{s_p,s_{p_+}}$ is the length of the connection between s_p and s_{p_+}. Although the defined shape likelihood map g_{p,p_+} is not invariant to rotation and scaling of the object of interest, these transformations can be incorporated by considering the rotation and size of the previously analyzed boundary segment of length $\varepsilon_{s_{p_-},s_p}$, represented by the connection between candidate points s_{p_-} and s_p for adjacent landmarks p_- and p, respectively. For this aim, we introduce a $\varepsilon_{s_{p_-},s_p}$-based coordinate system $(\mathbf{e_x}, \mathbf{e_y})$:

$$\mathbf{e_x} = \frac{\varepsilon_{s_{p_-},s_p}}{\|\varepsilon_{s_{p_-},s_p}\|}, \qquad \mathbf{e_y} = \begin{bmatrix} 0 & -1 \\ 1 & 0 \end{bmatrix} \mathbf{e_x}, \qquad (4.14)$$

so that the normalized values of $\varepsilon_{s_p,s_{p_+}}$ can be expressed as:

$$\overline{\varepsilon}_{s_p,s_{p_+}} = \frac{\|\varepsilon_{s_{p_-},s_p}\|}{\|\mu_{\varepsilon_{p_-,p}}\|} \overline{\varepsilon}_{s_p,s_{p_+}}^T [\mathbf{e_x}, \mathbf{e_y}], \qquad (4.15)$$

and that the resulting shape likelihood map g_{p,p_+} equals to:

$$g_{p,p_+}(s_p, s_{p_+}, \varepsilon_{s_{p_-},s_p}) = 1 - (\overline{\varepsilon}_{s_p,s_{p_+}} - \mu_{\overline{\varepsilon}_{p,p_+}})^T \sigma_{\overline{\varepsilon}_{p,p_+}}^{-1} (\overline{\varepsilon}_{s_p,s_{p_+}} - \mu_{\overline{\varepsilon}_{p,p_+}}). \qquad (4.16)$$

By combining the appearance likelihood map f_p and shape likelihood map g_{p,p_+} for landmark candidate points $\mathcal{S}$, the optimal candidate points $\mathcal{S}^*$ can be selected. As each landmark is connected with exactly two neighboring landmarks, optimization is reformulated into finding the optimal path with the minimal cost:

$$\mathcal{S}^* = \arg\max_{\mathcal{S}} \sum_{p \in \mathcal{P}} (f_p(s_p) + \gamma g_{p,p_+}(s_p, s_{p_+})), \qquad (4.17)$$

if the first shape likelihood map is selected (Eq. (4.13)), and:

$$\mathcal{S}^* = \arg\max_{\mathcal{S}} \sum_{p \in \mathcal{P}} (f_p(s_p) + \gamma g_{p,p_+}(s_p, s_{p_+}, \varepsilon_{s_{p_-},s_p})), \qquad (4.18)$$

if the second shape likelihood map (Eq. (4.16)) is selected. For both approaches, parameter γ balances the contribution of intensity and shape likelihood maps. The choice for the first or second shape likelihood map depends on the properties of the object of interest, namely if we can anticipate that the object of interest is considerably rotated or scaled, it is better to select the more complex second shape likelihood map. The globally optimal solution of Eqs. (4.17) and (4.18) can be efficiently found using dynamic programming, however, as the optimization forms a cycle through all landmarks $\mathcal{P}$ so that the first landmark in the path is connected with the last one, an extended version of dynamic programming should be used [24].

The described MISCP finds globally optimal landmark positions in the target image in linear time. However, it relies on dynamic programming, which requires each landmark to be connected with no more than two other landmarks. If landmarks are positioned on the boundary of a 2D object, this condition is naturally fulfilled by connecting adjacent landmarks, i.e. defining the object contour. In 3D, however, the object boundary is described as a surface mesh $\{\mathcal{P}, \mathcal{E}, \mathcal{F}\}$, where $\mathcal{P}$, $\mathcal{E}$ and $\mathcal{F}$ define sets of vertices (i.e. landmarks), edges between vertices and faces, respectively. Each landmark is, on average, connected with 5–6 other landmarks so that it is not entirely clear how a path through all landmarks can be established. The solution is to not analyze the complete set of landmarks simultaneously but work with random paths through connected landmarks. Assuming that $\forall p, q \in \mathcal{P}: |\mathcal{S}_p| = |\mathcal{S}_q|$, let us initialize matrix $\mathfrak{P} = \mathbf{0}$ of size $\mathcal{P} \times |\mathcal{S}_p|$ that will store probabilities for each candidate point s_p to represent the corresponding landmark p, and let subset $\rho(p_1, p_2, p_3, \ldots, p_t)$ contain t landmarks p_1, p_2, $p_3, \ldots, p_t$ that are adjacent on mesh $\{\mathcal{P}, \mathcal{E}, \mathcal{F}\}$, i.e. $\forall i = 2, \ldots, t: (p_{i-1}, p_i) \in \mathcal{E}$. The optimal path through these landmarks can be found using the above described MISCP algorithm (Eq. (4.18)). Candidate points s_p of the obtained path indicate points that most likely represent landmarks from ρ, and the elements of probability matrix $\mathfrak{P}$ at cell locations (p, s_p) are accordingly increased by one. The procedure is repeated for new subsets of adjacent landmarks until:

$$\sum_{s_p \in \mathcal{S}_p} \mathfrak{P}(p, s_p) \geq \mathfrak{p}, \tag{4.19}$$

where $\mathfrak{p}$ is a predefined minimal number of analyzed landmark subsets that contain landmark p. The optimal candidate point s_p^* for landmark p is then determined by:

$$s_p^* = \underset{s_p \in \mathcal{S}_p}{\arg\max} \, \mathfrak{P}(p, s_p). \tag{4.20}$$

In contrast to the 2D minimal cost path, landmark subset ρ does not pass through all landmarks from $\mathcal{P}$, and landmarks p_1 and p_t are not necessarily connected so that landmarks from ρ do not form a cycle of edges. In such a case, each landmark p_1 or p_t

would be spatially restricted by only individual landmarks, i.e. p_2 or p_{t-1}, respectively. This difference can be taken into account by not modifying probabilities for p_1 and p_t in $\mathfrak{P}$, or by modifying them only with a weighting factor [25].

4.3.4 Markov Random Fields

The main limitation of minimal cost path approaches (Section 4.3.3) is that optimization is performed on landmark paths, i.e. each landmark depends on two other adjacent landmarks. To overcome this, a more universal approach has to be considered, where each landmark can be connected with an arbitrary number of landmarks, and the optimal candidate points can be detected considering all of these pairwise connections. In such a system, the position of landmark p depends only on its neighbors, whereas for example in PCA (Section 4.3.1), we need to know the positions of all other landmarks $\mathcal{P} \backslash p$ to estimate the correctness of landmark p positioning. This condition is called the Markov property, and a set of landmarks with connections that function according to the Markov property is called a *Markov random field* (MRF) [26,27]. For a graph $\mathcal{G} = \{\mathcal{P}, \mathcal{E}\}$ with landmarks $\mathcal{P}$ and connections among landmarks $\mathcal{E}$, and for candidate points $\mathcal{S} = \{\mathcal{S}_p\}$ for landmarks $\mathcal{P}$, the optimal candidate points $\mathcal{S}^* = \{s_p^*, s_q^*, \ldots\}$ can be obtained by solving (Fig. 4.8A):

$$\mathcal{S}^* = \arg\min_{\mathcal{S}} \left(\sum_{p \in \mathcal{P}} \bar{f}_p(s_p) - \gamma \sum_{(p,q) \in \mathcal{E}} \bar{g}_{p,q}(s_p, s_q) \right), \tag{4.21}$$

where the appearance likelihood map $\bar{f}_p = 1 - f_p$ is called the unary potential, the shape likelihood map $\bar{g}_{p,q} = 1 - g_{p,q}$ is called the pairwise potential, and γ is a weighting factor. If the main advantage of minimal cost path approaches (Section 4.3.3) was that globally optimal paths can be obtained in linear time, the computational complexity of the MRF problem is higher, as it can be proved that the optimal solution of Eq. (4.21) cannot be found in polynomial time. Let us consider the problem of coloring graph vertices, i.e. landmarks $\mathcal{P}$, in such a way that vertices connected by an edge do not share the same color. Clearly, if the number of colors is higher or equal to $\mathcal{P}$, graph coloring is trivial and each vertex receives a unique color, but finding the smallest number of colors that can color graph $\mathcal{G}$ is a well-known non–deterministic polynomial-time (NP)-hard problem that cannot be solved in polynomial time. However, we can observe that graph coloring is a special case of the MRF problem. Let us define the unary potential as $\bar{f}_p = 0$, the weighting factor as $\gamma = 1$, and binary pairwise potentials that are equal for all vertices as:

$$\forall p, q \in \mathcal{P}: \bar{g}_{p,q}(s_p, s_q) = \begin{cases} 0 & s_p \neq s_q \\ 1 & s_p = s_q \end{cases}. \tag{4.22}$$

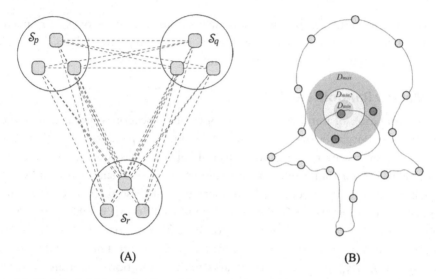

(A)   (B)

Figure 4.8 (A) Sets S_p, S_q and S_r of candidate points for landmarks p, q and r, respectively, form a graph $\mathcal{G} = \{\mathcal{P}, \mathcal{E}\}$, where $\mathcal{P}$ are landmarks ($p, q, r \in \mathcal{P}$) and $\mathcal{E}$ are connections among landmarks. (B) Examples of positive (green) and negative (red) landmark positioning defined by circles with radii D_{min}, D_{min2} and D_{max}. (For interpretation of the references to color in this figure, the reader is referred to the web version of this chapter.)

If the globally optimal solution of the MRF problem (Eqs. (4.21), (4.22)) has the minimal total cost above zero, the coloring of graph $\mathcal{G}$ is impossible considering the number of available colors $|S_p|$, but by increasing the number of available colors and solving the MRF problem, we can find the minimal number of colors suitable for coloring graph $\mathcal{G}$. We therefore observe that the graph coloring problem can be solved using MRF, and, because graph coloring is NP-hard, MRF also cannot be solved in polynomial time. However, an approximate solution of the MRF problem can be found in polynomial time by reformulating MRF as an integer programming problem:

$$\min_{S} \sum_{p \in \mathcal{P}} \sum_{s_p \in S_p} \bar{f}_p(s_p) x_p(s_p) + \sum_{(p,q) \in \mathcal{E}} \sum_{s_p \in S_p} \sum_{s_q \in S_q} \bar{g}_{p,q}(s_p, s_q) x_{pq}(s_p, s_q) \qquad (4.23)$$

$$\sum_{s_p \in S_p} x_p(s_p) = 1$$

$$\sum_{s_p \in S_p} x_{p,q}(s_p, s_q) = 1$$

$$\sum_{s_q \in S_q} x_{p,q}(s_p, s_q) = 1$$

$$x_p(s_p), x_{p,q}(s_p, s_q) \in \{0, 1\}.$$

Without the last restriction that variables $x_p(s_p)$ and $x_{p,q}(s_p, s_q)$ must be integer, the problem can be solved in linear time. In the current form, however, its solution can be approximated by using the duality principle numeric optimization, which is also built in a publicly available MRF implementation [28]. According to the duality principle, there is a unique dual problem to the primal problem, i.e. to Eq. (4.23). Moreover, any solution that does not violate the conditions of the primal problem represents an upper bound for the globally optimal solution, whereas any solution of the dual problem represents a lower bound for the globally optimal solution. As the dual problem is a regular linear optimization problem, we can first solve the dual problem and then find an approximated solution for the primal problem. If the difference between two solutions is large, a next approximation is generated, while if it is small, the obtained approximation is close to the globally optimal solution. There are no theoretical restrictions on the number of connections among landmarks, however, it is unwise to use only connections between adjacent landmarks, because in such a case minimal cost path approaches (Section 4.3.3) are more efficient. On the other hand, including too many connections can considerably slow down the MRF optimization process.

4.3.5 Game-Theoretic Optimization

The selection of optimal candidate points $\mathcal{S}^*$ can be observed from a *game-theoretic* (GT) perspective, so that landmarks are players, landmark candidate points are strategies, and likelihoods that each candidate point represents a landmark are payoffs. For each pair of landmarks p and $q \in \mathcal{P}$, which can, respectively, be represented by landmark candidate points $s_p \in \mathcal{S}_p$ and $s_q \in \mathcal{S}_q$ from the corresponding sets $\mathcal{S}_p$ and $\mathcal{S}_q$, a matrix $W_{p,q}$ of partial payoffs of size $|\mathcal{S}_p| \times |\mathcal{S}_q|$ is generated. The element of $W_{p,q}$ at location (s_p, s_q) represents the partial payoff of landmark p if landmarks p and q are represented by, respectively, candidate points s_p and s_q. Similarly as in other approaches, the partial payoffs are calculated as a linear combination of the appearance likelihood map f_p and shape likelihood map $g_{p,q}$:

$$W_{p,q}(s_p, s_q) = f_p(s_p) + \gamma g_{p,q}(d_{s_p,s_q}, \varphi_{s_p,s_q}, \theta_{s_p,s_q}, \Delta, \Upsilon, \Theta), \qquad (4.24)$$

where d_{s_p,s_q}, φ_{s_p,s_q} and θ_{s_p,s_q} are, respectively, the distance, azimuth angle and polar angle between candidate point s_p for landmark p and candidate point s_q for landmark q in the target image. Parameters Δ, Υ and Θ compensate for additional scaling and rotation of a landmark configuration, if the scale and/or orientation of the object of interest in the target image is not in agreement with the average scale and/or orientation of the object of interest in images from the training set (note that θ_{s_p,s_q} and Θ are used only for 3D images). The payoff of landmark p is then the sum of its partial payoffs:

$$w_p = \sum_{q \in \mathcal{P} \setminus \{p\}} 1_{p,q}^* \cdot W_{p,q}(s_p, s_q), \qquad (4.25)$$

where $1^*_{p,q} = \{0, 1\}$ is the indicator function, computed from images in the training set, that determines whether the connection between landmarks p and q in the target image is included into the shape representation, e.g. CG, GL, OTG (Section 4.2). By assuming that all landmarks simultaneously participate in maximizing their cumulative likelihoods, we can consider landmark detection as a cooperative game with the grand coalition [21], and the combination of optimal candidate points $\mathcal{S}^* = \{s^*_p, s^*_q, \ldots\}$ for corresponding landmarks $\{p, q, \ldots\}$ therefore corresponds to the maximal joint payoff ϑ^*:

$$\vartheta^*(\mathcal{P}, \mathcal{S}, \mathcal{W}) = \arg\max_{\omega} \left(\sum_{p \in \mathcal{P}} w_p \right), \qquad (4.26)$$

where $\omega = \{w_p, w_q, \ldots\}$ is an admissible combination of payoffs, and $\mathcal{W}$ is the set of partial payoff matrices for all pairs of landmarks.

By maximizing the joint payoff of candidate points in sets $\mathcal{S}_p$; $\forall p \in \mathcal{P}$, we find the initial combination of optimal candidate points $\mathcal{S}^* = \{s^*_p, s^*_q, \ldots\}$. The final combination of optimal candidate points that represents landmarks in the target image is found by re-evaluating the position of each candidate point s^*_p for landmark p while considering the position of the remaining candidate points $\mathcal{S}^* \backslash \{s^*_p\}$ as fixed. The re-evaluation is performed for every candidate point $s^*_p \in \mathcal{S}^*$ within a spatial domain $\mathcal{Q}_p$ in the target image that encompasses all candidate points from $\mathcal{S}_p$, and consecutive re-evaluation of all candidate points forms a single re-evaluation iteration, which is performed until no considerable changes in candidate point positions are detected [29].

4.3.6 Machine Learning-Based Optimization

In the above mentioned approaches, the shape likelihood maps that describe the spatial relationships among landmarks are based on statistical properties of the complete configuration of landmarks, e.g. PCA (Section 4.3.1), or on individual connections between landmarks, e.g. MRF (Section 4.3.4). An alternative approach is to use *machine learning* approaches for modeling the difference between positive samples, i.e. correct landmark configurations, and negative samples, i.e. cases when specific landmarks are mis-positioned and the resulting shape does not correspond to the shape of the object of interest. A straightforward way of generating positive samples is to use landmarks from N training samples, however, modern machine learning approaches often require a much larger number of training samples. To increase the number of training samples, a set $\boldsymbol{p}^+_I = \{p^+_I\}$ of positive landmark positions can be artificially generated for each landmark $p \in \mathcal{P}$ and each training image I by (Fig. 4.8B):

$$p^+_I = p_I + \mathrm{rand}(D_{min}), \qquad (4.27)$$

where p_I is the reference position of landmark p in training image I, and $\mathrm{rand}(D_{min})$ is a random displacement with the maximal magnitude D_{min}. Although these artificial samples are similar to the original landmark configuration $\mathcal{P}_I$ from image I, they can enrich the set of positive training samples. On the other hand, there are several ways to generate negative samples p_I^-, for example, by generating a randomly positioned set of points with cardinality $|\mathcal{P}|$. A machine learning descriptor trained on such negative and positive sets of samples would be able to distinguish between the object of interest and random point configurations. However, knowing that sets S_p of landmark candidate points do not contain randomly displaced points, there is no need to automatically recognize random point configurations. Moreover, configurations of landmark candidate point with the highest response in the intensity likelihood map of individual points $f_p(s_p)$ are usually very similar to the shape of the object of interest, and therefore they most probably contain a number of optimal candidate points. As a result, a more information-rich set $p_I^- = \{p_I^-\}$ of negative samples for each landmark $p \in \mathcal{P}$ and each training image I can be generated by (Fig. 4.8B):

$$p_I^- = p_I + \text{rand}(D_{max}); \quad |p_I^-, p_I^+| > D_{min2}, \tag{4.28}$$

where D_{max} is the maximal displacement magnitude, required because candidate points S_p for landmark p are detected in the neighborhood of the correct landmark position, and therefore very distant points are not considered. Points located in interval $(D_{min}; D_{min2})$ are too far from the reference landmark position p_I to represent adequate positive samples, but, at the same time, not far enough to represent adequate negative samples. A machine learning classifier $\mathcal{M}_{\mathcal{P}}$ (e.g. random forests, neural networks) is then trained to distinguish between positive and negative samples of landmark positions. For a selected point set $\{s_p, s_q, \dots\}$, the classification result of $\mathcal{M}_{\mathcal{P}}$ is a probability vector $\zeta = \{\zeta_0, \dots, \zeta_p, \zeta_q, \dots\}$, where element ζ_p stores the probability that candidate point s_p is misdetected, and element ζ_0 stores the probability that all landmarks are detected correctly. This classifier is then used to detect optimal candidate points in set S_p computed from the target image [30].

4.4 CONCLUSION

Algorithms that combine landmark detection and shape analysis have an important role in modern image analysis. The rapid development of machine learning approaches for modeling the appearance of individual landmarks provides considerable support, however, without capturing, modeling and quantifying of the object shape, the applicability of landmark detection and quality of the obtained results would be very limited.

In this chapter, we summarized some of the existing algorithms for landmark-based shape representation and shape-based landmark detection. Shape representation is based on measuring the numerical properties of the object of interest, and focusing on those properties that best describe the spatial relationships of its parts and components, where pairwise connections among landmarks are mainly used for this aim. Selecting the optimal connections is a problem of finding a balance between the elasticity, rigidity and representativeness of the shape representation, and the computational complexity of the subsequent optimization. Optimization algorithms search for the agreement between the appearance of individual landmarks and spatial relationships among landmarks that are represented by landmark connections, and some optimization algorithms may only be suitable for a certain type of shape representations. The algorithms in this field often show a superior performance, for example in medical image computational challenges, and demonstrate exceptional performance in image segmentation, registration, pathology detection and computer-assisted diagnosis.

REFERENCES

[1] D.G. Kendall, A survey of the statistical theory of shape, Stat. Sci. 4 (2) (May 1989) 87–99.
[2] K. Rohr, Landmark-Based Image Analysis, vol. 21, Springer Netherlands, Dordrecht, 2001.
[3] K. Mikolajczyk, C. Schmid, A performance evaluation of local descriptors, IEEE Trans. Pattern Anal. Mach. Intell. 27 (10) (Oct. 2005) 1615–1630.
[4] C.R. Rao, S. Suryawanshi, Statistical analysis of shape of objects based on landmark data, Proc. Natl. Acad. Sci. USA 93 (22) (Oct. 1996) 12132–12136.
[5] T. Heimann, H.-P. Meinzer, Statistical shape models for 3D medical image segmentation: a review, Med. Image Anal. 13 (4) (Aug. 2009) 543–563.
[6] S. Hanaoka, A. Shimizu, M. Nemoto, Y. Nomura, S. Miki, T. Yoshikawa, N. Hayashi, K. Ohtomo, Y. Masutani, Automatic detection of over 100 anatomical landmarks in medical CT images: a framework with independent detectors and combinatorial optimization, Med. Image Anal. 35 (Jan. 2017) 192–214.
[7] B. Ibragimov, B. Likar, F. Pernuš, T. Vrtovec, A game-theoretic framework for landmark-based image segmentation, IEEE Trans. Med. Imaging 31 (9) (Sep. 2012) 1761–1776.
[8] T.S. Caetano, T. Caelli, D. Schuurmans, D.A.C. Barone, Graphical models and point pattern matching, IEEE Trans. Pattern Anal. Mach. Intell. 28 (10) (Oct. 2006) 1646–1663.
[9] A. Besbes, N. Paragios, Landmark-based segmentation of lungs while handling partial correspondences using sparse graph-based priors, in: 2011 IEEE International Symposium on Biomedical Imaging: From Nano to Macro, 2011, pp. 989–995.
[10] J. Friedman, T. Hastie, R. Tibshirani, Sparse inverse covariance estimation with the graphical lasso, Biostatistics 9 (3) (Jul. 2008) 432–441.
[11] R. Mazumder, T. Hastie, The graphical lasso: new insights and alternatives, Electron. J. Stat. 6 (2012) 2125–2149.
[12] A. Schrijver, Combinatorial Optimization, 1st edition, Springer, Berlin, New York, 2003.
[13] B. Ibragimov, B. Likar, F. Pernuš, T. Vrtovec, Statistical shape representation with landmark clustering by solving the assignment problem, in: SPIE Medical Imaging, 2013, 86690E.
[14] B. Ibragimov, B. Likar, F. Pernuš, T. Vrtovec, Segmentation of vertebrae from 3D spine images by applying concepts from transportation and game theories, in: Computational Methods and Clinical Applications for Spine Imaging, Springer International Publishing, 2014, pp. 3–14.
[15] B. Ibragimov, B. Likar, F. Pernuš, T. Vrtovec, Shape representation for efficient landmark-based segmentation in 3-d, IEEE Trans. Med. Imaging 33 (4) (Apr. 2014) 861–874.
[16] H.W. Kuhn, The Hungarian method for the assignment problem, Nav. Res. Logist. Q. 2 (1–2) (Mar. 1955) 83–97.

[17] T.F. Cootes, A. Hill, C.J. Taylor, J. Haslam, Use of active shape models for locating structures in medical images, Image Vis. Comput. 12 (6) (1994) 355–365.

[18] T.F. Cootes, C.J. Taylor, D.H. Cooper, J. Graham, et al., Active shape models-their training and application, Comput. Vis. Image Underst. 61 (1) (1995) 38–59.

[19] C. Lindner, P.A. Bromiley, M.C. Ionita, T.F. Cootes, Robust and accurate shape model matching using random forest regression-voting, IEEE Trans. Pattern Anal. Mach. Intell. 37 (9) (Sep. 2015) 1862–1874.

[20] D.L. Donoho, For most large underdetermined systems of equations, the minimal l1-norm near-solution approximates the sparsest near-solution, Commun. Pure Appl. Math. 59 (6) (Jun. 2006) 797–829.

[21] J.L. Starck, M. Elad, D.L. Donoho, Image decomposition via the combination of sparse representations and a variational approach, IEEE Trans. Image Process. 14 (10) (Oct. 2005) 1570–1582.

[22] S. Zhang, Y. Zhan, M. Dewan, J. Huang, D.N. Metaxas, X.S. Zhou, Towards robust and effective shape modeling: sparse shape composition, Med. Image Anal. 16 (1) (Jan. 2012) 265–277.

[23] C. Chen, W. Xie, J. Franke, P.A. Grutzner, L.-P. Nolte, G. Zheng, Automatic X-ray landmark detection and shape segmentation via data-driven joint estimation of image displacements, Med. Image Anal. 18 (3) (Apr. 2014) 487–499.

[24] D. Seghers, D. Loeckx, F. Maes, D. Vandermeulen, P. Suetens, Minimal shape and intensity cost path segmentation, IEEE Trans. Med. Imaging 26 (8) (Aug. 2007) 1115–1129.

[25] D. Seghers, P. Slagmolen, Y. Lambelin, J. Hermans, D. Loeckx, F. Maes, P. Suetens, Landmark based liver segmentation using local shape and local intensity models, in: Proc. Workshop of the 10th Int. Conf. on MICCAI, Workshop on 3D Segmentation in the Clinic: A Grand Challenge, 2007, pp. 135–142.

[26] R. Donner, G. Langs, B. Mičušik, H. Bischof, Generalized sparse MRF appearance models, Image Vis. Comput. 28 (6) (Jun. 2010) 1031–1038.

[27] R. Donner, B.H. Menze, H. Bischof, G. Langs, Global localization of 3D anatomical structures by pre-filtered Hough Forests and discrete optimization, Med. Image Anal. 17 (8) (Dec. 2013) 1304–1314.

[28] N. Komodakis, G. Tziritas, Approximate labeling via graph cuts based on linear programming, IEEE Trans. Pattern Anal. Mach. Intell. 29 (8) (Aug. 2007) 1436–1453.

[29] B. Ibragimov, J.L. Prince, E.Z. Murano, J. Woo, M. Stone, B. Likar, F. Pernuš, T. Vrtovec, Segmentation of tongue muscles from super-resolution magnetic resonance images, Med. Image Anal. 20 (1) (Feb. 2015) 198–207.

[30] B. Ibragimov, B. Likar, F. Pernuš, T. Vrtovec, Accurate landmark-based segmentation by incorporating landmark misdetections, in: 2016 IEEE 13th International Symposium on Biomedical Imaging (ISBI), 2016, pp. 1072–1075.

CHAPTER 5

Probabilistic Morphable Models

Bernhard Egger, Sandro Schönborn, Clemens Blumer, Thomas Vetter
Department of Mathematics and Computer Science, University of Basel, Switzerland

Contents

5.1	Introduction	115
	5.1.1 Related Work	119
5.2	Methods	120
	5.2.1 Probabilistic Morphable Model	120
	5.2.2 MCMC Sampling for Inference	122
	5.2.3 Integration of Bottom-Up Cues by Filtering	123
	5.2.4 Explicit Background Modeling	125
	5.2.5 Occlusion-Aware Morphable Model	126
5.3	Applications and Results	129
5.4	Conclusion	131
References		133

5.1 INTRODUCTION

In this chapter, we present Probabilistic Morphable Models – a fully probabilistic framework to interpret face images. We reconstruct 3D faces from 2D images with a statistical model in an Analysis-by-Synthesis setting. A given target face is represented by model instances which are similar to the target image. So far, most Morphable Model adaptation techniques relied on manual initialization and were prone to outliers and occlusions. We urge to use a fully probabilistic framework to obtain an automated and occlusion-aware system.

Statistical models have been applied for segmentation in CT, MRI or 2D photographs. Usually the model adaptation is initialized manually and solved by optimization techniques. This approach is sensitive to initialization and prone to occlusions and outliers. The optimization often leads to local minima. In our probabilistic setting, we do not aim for a single best solution through optimization, but we search for the posterior probability distribution of possible model explanations of the input image. During the model adaptation process we only have uncertain correspondence between the target image and the face model. The aim of model adaptation is to find the location and orientation of the face (pose), statistical model parameters and the illumination condition. The likelihood functions are highly non–convex as the dependence on the input image renders the adaptation rough and highly nonlinear. Occlusions make it even

Statistical Shape and Deformation Analysis
DOI: 10.1016/B978-0-12-810493-4.00006-7

harder to target this problem with standard optimization techniques. In order to handle occlusions and to include uncertain information our framework is fully probabilistic.

We present two challenges to highlight the benefit of using a fully probabilistic framework. First, our model adaptation process includes uncertain detection results for feature points, such as eye or mouth corners. The localization of such feature points is still a challenge in computer vision and produces unreliable results. In a classical feed-forward optimization procedure, the uncertainty is ignored. This leads to pipelines which take early and possibly wrong decisions that can not be reconsidered in later steps. Our probabilistic approach enables us to integrate this uncertain information source and guide the overall adaptation process. Second, faces are often occluded by various objects like glasses, hands or microphones. Our probabilistic setting makes possible to build an occlusion-aware adaptation framework. We detect occlusions using our strong appearance prior of the statistical model, combined with knowledge arising from classical image segmentation. The segmentation enforces smoothness of the labels assigned to neighboring pixels and is not a simple pixel-wise thresholding. The face model is adapted to the image and its uncertainty guides the segmentation of occlusion. The segmentation then drives the model adaptation to explain contiguous regions and guides it to explain as much as possible by the face model.

We use a 3D Morphable Model (3DMM) [5] as appearance prior for faces. The 3DMM is a Parametric Appearance Model (PAM). PAMs are able to generate images controlled by parameters θ. PAMs are widely used in generative setups, especially in the field of face image analysis. In medical image analysis, related models, namely Active Shape Models [7] are prominent. Instead of modeling appearance, they define gradients or higher level image features. An application to face photographs brings some additional challenges compared to medical data. Whilst medical data are most often recorded in a controlled setting, facial photography is highly unconstrained. Besides the shape, we also need to estimate facial color, pose and illumination from the single input image. Together with a camera model, 3DMMs can synthesize new face images. Due to pose and shape variations, facial parts can be invisible by self-occlusion effects. Especially the effects of a color model and illumination add additional challenges as presented in Fig. 5.1.

We reinterpret the 3DMM to build a fully probabilistic framework. The model consists of a shape and a color model. We build both models in the Gaussian Process framework proposed by Lüthi et al. [20]. It is the basis for our statistical prior on face shape and color appearance. We use a multi-linear face model to handle expressions.

Model adaptation is the most challenging part of face image analysis. Through the model adaptation process, we search model instances which match the input image. This process is called "fitting". It is solved with different methods which we summarize in the section on related work. We infer the posterior distribution of possible image explanations by our face model. Our approach is a Data Driven Markov Chain Monte Carlo

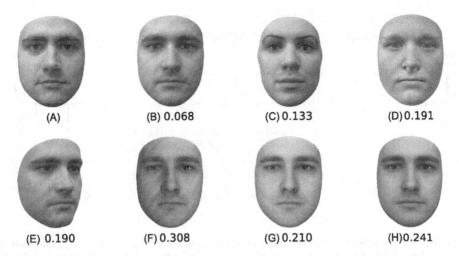

(A)   (B) 0.068   (C) 0.133   (D) 0.191

(E) 0.190   (F) 0.308   (G) 0.210   (H) 0.241

Figure 5.1 We illustrate the dominance of illumination effects on facial appearance. We show the target image (A) and its best fitting model instance (B). We manipulated a block of parameters to obtain the other images (C–H). We indicate the RMS-distance to the target image (A) in color space for each rendered image. We inverted shape (C) and color (D) parameters and also changed the yaw angle to 45 degree (E). All those changes significantly influence the facial appearance. However, the illumination changes to an illumination from the side (F), the front (G) or a real world illumination from another image (H) have a higher influence on the RMS-distance. When adapting the model parameters and searching for the best instance (b), the illumination is therefore dominant. (For interpretation of the references to color in this figure, the reader is referred to the web version of this chapter.)

(DDMCMC) sampling technique. In contrast to other 3DMM adaptation techniques, it does not aim for a single model instance as output but approximates the posterior distribution of possible solutions. The distribution carries information about the certainty of a fit.

As this posterior distribution cannot be computed analytically, we use the Metropolis –Hastings algorithm to generate samples from the posterior distribution. The algorithm consists of proposal and verification steps. It is based on a proposal distribution to generate samples and enforces consistency to the observed data and the model in the verification step. This partitioning is a key feature of the algorithm and represents a propose-and-verify architecture. All proposals are evaluated in the verification steps, therefore they can be explorative and do not have to always improve the result. The verification step accepts and rejects proposals based on their likelihood. In the verification steps, we use a filtering strategy. Different evaluation criteria, like the L_2 image difference or detection responses, are integrated by cascading Metropolis–Hastings acceptance stages with the respective likelihoods, see Fig. 5.2. For example, a proposal is evaluated against a feature point likelihood in an early stage where bad proposals can be filtered out quickly. Filtering allows us to focus computing time on promising regions which are more expensive to evaluate, like the image difference.

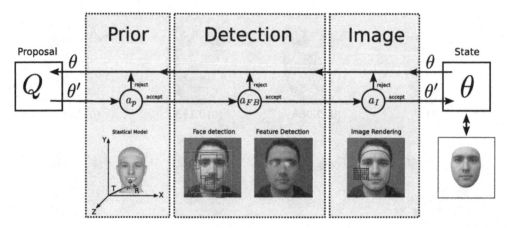

Figure 5.2 Our sampling framework is built on a Metropolis–Hastings algorithm. A parameter update θ' is drawn from the proposal generator Q and evaluated through three filters. At each filtering stage a_x a sample can be accepted or rejected by its corresponding likelihood function following the Metropolis–Hasting acceptance rule. If a sample is accepted from the prior, detection and image filtering stage, it builds the new state θ of the Markov Chain. The sequence of states in the Markov chain builds the posterior distribution over θ.

We explicitly distinguish between face and non-face regions in the target image to handle occlusions. The image is segmented into regions which can be explained by the face model and regions which are explained by a simple background color model. This segmentation is defined on the 2D image plane and integrated into the model likelihood.

During model adaptation, we have to find face and non-face regions simultaneously. The strong appearance prior from the statistical face model is used to decide whether a pixel is considered face or background. The likelihood of a pixel being part of the face region changes during fitting. Therefore, we constantly reestimate the segmentation in an Expectation-Maximization (EM) procedure during the whole model adaptation process.

The illumination conditions in unconstrained face photographs vary heavily. In the beginning of the model adaptation, the distance between the current estimation and the target image is large and dominated by illumination mismatch, see Fig. 5.1. Measuring only color distance, e.g. shadowed regions can differ stronger than occlusions. Especially under occlusion, illumination has to be estimated in a robust way. Occlusions would mislead the illumination estimation strongly. We exploit the illumination by using a RANSAC-based robust illumination estimation technique. This gives us a proper initialization of the illumination conditions and a first guess of occluded pixels in the image.

The following paragraphs are organized as follows: First we give a short overview over related work in the field of 3DMMs in Section 5.1.1. We then present our probabilistic face model (Section 5.2.1). In Section 5.2.2 we present how inference can be performed given a target image and integrate detection information (Section 5.2.3). Finally we extend the framework to handle background (Section 5.2.4) and become aware of occlusions (Section 5.2.5). We include most experiments directly in the parts on corresponding methods and show the performance of the full framework in Section 5.3.

5.1.1 Related Work

PAMs are common in computer vision. The first successful PAM was the Eigenfaces approach [15,29]. Principal Component Analysis (PCA) was performed on pixel intensity values of roughly aligned face images. This led to a parametric and generative representation of face images. The models were successful in a strongly constrained face recognition task. The next step in parametric face modeling has been Active Appearance Models [8], which combine Active Shape Models [7] and the idea of the Eigenfaces approach. Whilst the Eigenfaces approach assumes the images to have pixel-wise correspondence, AAMs add a shape model to handle different face shapes. Shape deformations are modeled separately from the appearance. The shape model is learned from a set of 2D correspondences while the appearance model is restricted to shape-normalized images. Both models use PCA to find an efficient parameterization. Active Appearance Models became successful by the availability of specialized fast fitting algorithms [21,3]. However, those models are defined in 2D and therefore cannot handle strong pose variation with self-occlusion. The next development for parametric modeling of faces are 3DMMs. The 3DMM models a face as a 3D object. The image formation process is explicitly modeled using a pinhole camera and a Phong reflectance model. The model is built on 3D face scans which are in dense correspondence and combines separate color and shape models. Contrary to the AAM, the shape and color models only describe face variation while the rendering part is handled by the explicit camera and illumination models. 3DMMs can therefore handle self-occlusion and head pose in a very natural way.

There are different approaches for adaptation of Morphable Models to images. The original approach by Blanz and Vetter [5] used a stochastic gradient descent method. Romdhani et al. [25,24] presented a multiple-features fitting approach. Aldrian et al. presented a fast model adaptation method based on inverse rendering [2]. Recently, we presented an approach based on sampling which is able to include unreliable information sources in a probabilistic way [27]. Current developments in machine learning also investigate Supervised Descent Methods [30,14] and probabilistic methods combined with deep learning [17]. Those learning methods achieve promising results for shape model adaptation with limited pose. However, they do not include a color or illumination model and are therefore not fully generative. All those model adaptation techniques rely on good initialization and are characterized by standard optimization techniques which are prone to local minima.

Occlusions can severely mislead the model adaptation process. Only few generative model adaptation approaches are able to handle occlusions. There is previous work, for all kind of PAMs, on how to handle occlusions using robust error measures or robust strategies like RANSAC [12]. In contrast, for 3DMM adaptation only few robust approaches exist. Most of them rely on manual labeling of occlusions or knowledge about how much of the face is occluded. Other approaches implement robust error measures. As appearance is dominated by illumination, robust error measures work only for almost ambient illumination settings. Those approaches tend to exclude facial parts which can be explained by complex illumination. Another main source of error are regions which are difficult to explain by the face model [25,9,23]. Examples for such regions are the eye, eyebrow, nose and mouth region, they vary much stronger in color appearance than e.g. the cheek. The pixels in those regions are harder to fit by the model but crucial for representing facial characteristics. Note that previous works on occlusion handling using a 3DMM focused on databases with artificial and homogeneous, frontal illumination settings. Our approach includes a robust illumination estimation which allows us to adapt the model in presence of occlusions even under complex illumination settings.

5.2 METHODS

We give an overview over all components of our Analysis-by-Synthesis approach. We present our generative probabilistic face model and explain the image formation process. The most challenging part of the overall process is the inference for model adaptation to a given target image. We show how DDMCMC sampling can be used for this inference task and how it allows us to integrate various sources of unreliable information into the model adaptation process. The face model is only defined in the face region, to explain the whole image, we need a model for the background. We present our approach of background modeling and explain why it is important for generative models. Finally we want to be able to adapt our model to unconstrained face images. Occlusions are a challenge in those settings and can mislead the Analysis-by-Synthesis process. Therefore we extend the model and its adaptation to simultaneously segment the target image into face and non-face regions and become occlusion-aware.

5.2.1 Probabilistic Morphable Model

Our generative 3D Morphable Face Model is a variant of the Basel Face Model (BFM) [22], built from face scans of 200 people taken with a structured light 3D scanner. We extend the original face model to a multi-linear model to handle expressions as described in [4]. The multi-linear statistical model consists of two independent Probabilistic PCA (PPCA) models for face shape and face color as well as an additional model for the deformations by facial expression. Facial expression is modeled as a difference to the neutral face shape. The PPCA models are learned on the aligned vertex positions $\vec{s}$ and color $\vec{c}$. Each face mesh consists of 21,662 vertices. This leads to a distribution over facial color $P(\vec{c} \mid \vec{\theta})$ and shape $P(\vec{s} \mid \vec{\theta})$ and also handles observation noise in the training data:

$$P(\vec{s} \mid \vec{\theta}) = \mathcal{N}\left(\vec{s} \mid \vec{\mu}_S + U_S D_S \vec{\theta}_S + \vec{\mu}_E + U_E D_E \vec{\theta}_E, \sigma_S^2 I\right) \tag{5.1}$$

$$P(\vec{c} \mid \vec{\theta}) = \mathcal{N}\left(\vec{c} \mid \vec{\mu}_C + U_C D_C \vec{\theta}_C, \sigma_C^2 I\right) \tag{5.2}$$

All the models (subscripts S for shape, C for color and E for expression) consist of a mean $\vec{\mu}$, the principal components in matrix U and the variances along each component in diagonal matrix D. The additive Gaussian noise is only added for shape and for color. The parameters $\vec{\theta}$ follow a standard normal distribution in latent space:

$$P(\theta_x) = \mathcal{N}\left(\vec{\theta}_x \mid \vec{0}, I\right) \tag{5.3}$$

Previous ad hoc probabilistic interpretations of the 3D Morphable Model have been introduced for shape reconstruction in [6,1,18]. We resort to the recently introduced Gaussian Process Morphable Model of Lüthi et al. [20] as a consistent and clean framework to understand and create probabilistic shape models. The PPCA prior can be understood as a Gaussian Process with a covariance function consisting of a statistical kernel of N samples $\vec{s}_i(\vec{x})$ and independent Gaussian noise with variance σ^2 (only shown for shape):

$$K(\vec{x}, \vec{y}) = \frac{1}{N} \sum_{i=1}^{N} (\vec{s}_i(\vec{x}) - \vec{\mu}(\vec{x})) \left(\vec{s}_i(\vec{y}) - \vec{\mu}(\vec{y})\right)^T + \sigma^2 \delta(\vec{x}, \vec{y}) I_3 \tag{5.4}$$

Additionally to the shape process, we add an expression deformation process and also define a color appearance process analogously. The Gaussian Process model is discretized on the vertices of our reference mesh and parametrized through a low-rank expansion of the statistical part of the kernel (corresponds to PCA). The independent Gaussian kernel is handled without approximation (corresponds to PPCA). These Gaussian Process models are identical to the PPCA models described above.

The 3DMM also defines a rendering process $\mathfrak{R}$ to generate synthetic face images I. We use a pinhole camera model and a spherical harmonics illumination model [28]:

$$I(\theta) = \mathfrak{R}\big(M(\theta_S, \theta_C, \theta_E); \theta_P, \theta_L\big) \tag{5.5}$$

The shape θ_S, color θ_C and expression θ_E parameters are coupled to the principal components of the statistical PPCA prior, the camera θ_P and illumination parameters θ_L are handled separately. We assume a multivariate Gaussian distribution as a prior for the spherical harmonics expansion parameters and a uniform distribution on the camera parameters. The distribution of the illumination parameters is empirically estimated from successful fittings of the AFLW [16] database. This leads to a full face model prior distribution, including shape, color, expression, camera and illumination.

5.2.2 MCMC Sampling for Inference

Standard optimization for adapting the model to a given target image leads to a single local optimal parameter set. In our probabilistic setting, we perform probabilistic inference instead of optimization. The result of inference is not a single maximum, but a posterior distribution of the model parameters θ conditioned on the target image $\tilde{I}$:

$$P(\theta \mid \tilde{I}) = \frac{P(\tilde{I} \mid \theta)P(\theta)}{\int P(\tilde{I} \mid \theta)P(\theta)\mathrm{d}\theta} \quad (5.6)$$

The posterior is intractable and can only be evaluated point-wise with respect to an unknown multiplicative constant. The involvement of the target image in the likelihood with unknown correspondence leads to a highly non-convex distribution without a closed-form representation. A single evaluation involves rendering an image and comparing it to the target.

We use the Metropolis–Hastings algorithm for inference. The algorithm defines a Markov Chain which delivers samples approximately from the posterior distribution $P(\theta \mid \tilde{I})$. The algorithm achieves this by stochastic acceptance of random samples from a simpler probability distribution, called the proposal distribution $Q(\theta' \mid \theta)$. The probability of accepting a new sample is given by

$$a = \min\left\{\frac{P(\theta' \mid \tilde{I})}{P(\theta \mid \tilde{I})} \frac{Q(\theta \mid \theta')}{Q(\theta' \mid \theta)}, 1\right\}. \quad (5.7)$$

To calculate the acceptance probability, it is sufficient to evaluate posterior ratios which allow for an unnormalized evaluation. If a proposal is rejected the last sample in the chain remains as the state of the chain. The algorithm defines a formal propose-and-verify structure. It decouples finding solutions from validating them which allows us to also use uncertain and unreliable proposals. Verification with the model likelihood and the prior ensures a consistent sample.

The proposal distribution has to be designed carefully. On one hand, it has to be easy to sample from, on the other hand, it has to be complex enough to explore the high dimensional parameter space. In our case, we use a mixture of a set of different proposals. We group logical parts together and only adapt a single block at once. A proposal consists of a shape, color, illumination or camera parameter update. The proposal can be unreliable – the verification step enforces consistency by evaluating the proposal against a likelihood function and the model assumptions. Most proposals are simple normally distributed random walks of fine to coarse scale.

During the verification steps, the proposals are evaluated according to their likelihood e.g. the likelihood given the target image. When adapting to an image, our main likelihood measures the color distance between the rendering of the current model estimate $I_i(\theta)$ at each pixel i and the raw pixels of the target image $\tilde{I}_i$:

$$\ell_{\text{face}}(\theta; \tilde{I}_i) = \frac{1}{N} \exp\left(-\frac{1}{2\sigma^2} \left\| \tilde{I}_i - I_i(\theta) \right\|^2\right) \qquad (5.8)$$

In practice, the verification step also includes information about face and feature point detections and is split into different filtering stages described in Section 5.2.3.

We draw samples from our Markov Chain to obtain a posterior distribution. The Markov Chain needs to be initialized by a specific set of parameters. In the beginning, the chain depends on the initialization. After a so-called burn-in phase, we draw samples from the Markov Chain which are independent of the initialization. Those samples are used to approximate the posterior distribution.

Some applications need not a posterior distribution but a single optimal parameter estimate. Using our Metropolis–Hastings algorithm, the best observed sample can be used as MAP estimate. The longer the sampling runs, the better the approximation gets. It can be stopped when the desired precision is obtained.

The proposed Metropolis–Hastings algorithm has some additional advantages: Due to the verification step it can easily integrate various sources of information. It is extendable to proposals which e.g. include gradients or parameter updates from learning techniques (e.g. cascaded regression techniques). The proposals are allowed to be imperfect as they are validated in the verification step. Last but not least, the algorithm is simple and easy to implement.

5.2.3 Integration of Bottom-Up Cues by Filtering

The ability to integrate Bottom-Up cues is a key strength of the proposed MCMC inference method. It is open to include information from arbitrary, unreliable Bottom-Up methods, even from different sources at the same time. We give an overview on how those Bottom-Up cues can be integrated in the Metropolis–Hasting algorithm and then concrete examples including face and facial feature point detectors.

In non-probabilistic algorithms, Bottom-Up cues, e.g. detections, are usually included in a feed-forward manner. This leads to early and possibly wrong decisions which cannot be reviewed in later steps. In our fully probabilistic setting, we can use unreliable input of Bottom-Up methods with uncertainty and include this input directly in the inference process.

The integration of detection methods is simple and takes part in the evaluation step of the Metropolis–Hastings algorithm. Thus, we need a likelihood function for measur-

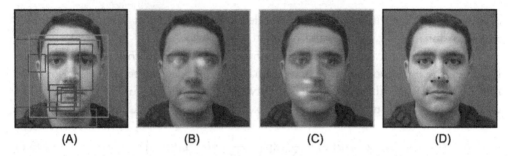

(A)   (B)   (C)   (D)

Figure 5.3 Face and feature point detections: In (A) the 10 strongest face detections are visualized. The probability of the detections is indicated by light red color. In the middle, we show the detection probability of the left inner eye corner (B) and for the right lip corner (C). We can observe that the lip corner detection is much more distributed over the image whilst the eye corner is more precisely located. However for both points we also get strong wrong detections at other facial features locations. In (D) we show all the nine feature points we trained detectors for. (For interpretation of the references to color in this figure, the reader is referred to the web version of this chapter.)

ing how likely the observed data are under the current model estimate. The different likelihoods of Bottom–Up cues are considered in a step-by-step manner. We cascade multiple stochastic acceptance steps of the Metropolis–Hastings algorithm to integrate one likelihood after the other, see Fig. 5.2. Only a proposal which is compatible with the respective likelihood at each stage is accepted. This process is called filtering. Filtering is very flexible and allows us to combine various information sources in a single algorithm efficiently. The cascade prevents a computationally expensive evaluation with respect to all likelihood functions if an early stage does not agree with the sample. This leads to early rejection of bad samples and improves the performance of the fitting procedure.

We include face and facial feature point detection into our model adaptation process to demonstrate the filtering chains. We search for the 10 best face detections in the target image and run facial feature detection for prominent landmark points like the eye or mouth corners, see Fig. 5.3. Instead of using the strongest detection, we integrate the detection evidence as a likelihood model as described above. The exact procedure of detection integration is described in [28].

The likelihood of a face detection candidate considers landmarks detection maps $\mathcal{D}_l$ and the face box $\mathcal{B}$ with location $\vec{p}$ and scale s. The face box likelihood ℓ_B (Eq. (5.9)) consists of a Gaussian likelihood with respect to its center position and a log–normal distribution for face scale. The landmark map likelihood ℓ_LM (Eq. (5.10)) combines a noisy Gaussian landmarks observation model with the detection map.

$$\ell_\mathrm{B}(\theta; \mathcal{B}_i) = \mathcal{LN}\left(s(\theta) \mid \mathcal{B}_i.s, \sigma_\mathrm{bs}\right) \mathcal{N}\left(p(\theta) \mid \mathcal{B}_i.\vec{p}, \sigma_\mathrm{bp}\right) \tag{5.9}$$

$$\ell_\mathrm{LM}(\vec{x}; \mathcal{D}) = \max_{\vec{t}} \mathcal{N}\left(\vec{t} \mid \vec{x}, \sigma_\mathrm{LM}^2\right) \mathcal{D}\left(\vec{t}\right) \tag{5.10}$$

We perform a max convolution to find the best possible combination of detection and distance at each point in the image (as described in [27]). We combine the corresponding face and feature point detections:

$$\ell_i(\theta; \mathcal{B}_i, \mathcal{D}_i) = \ell_{\mathrm{B}}(\theta; \mathcal{B}_i) \prod_l \ell_{\mathrm{LM}}(\vec{x}(\theta); \mathcal{D}_l) \qquad (5.11)$$

Our assumption is that the image contains exactly one face, therefore we optimize for the best candidate i of face and feature point detection:

$$\ell_{\mathrm{FB}}(\theta; \mathcal{B}, \mathcal{D}) = \max_i \ell_i(\theta; \mathcal{B}_i, \mathcal{D}_i) \qquad (5.12)$$

We roughly initialize the 3D pose of the face by evaluating only with the detection or landmark likelihoods in the beginning. In later model adaptation steps, the consistency to the detections through filtering guides the adaptation procedure, see Fig. 5.2. A proposal is first evaluated by its likelihood given the detections ℓ_{FB} and then also against the image ℓ_{I}:

$$P(\theta) \xrightarrow{\ell_{\mathrm{FB}}(\theta; \mathcal{B}, \mathcal{D})} P(\theta \mid \mathcal{B}, \mathcal{D}) \xrightarrow{\ell_{\mathrm{I}}(\theta; \tilde{I})} P(\theta \mid \mathcal{B}, \mathcal{D}, \tilde{I}) \qquad (5.13)$$

5.2.4 Explicit Background Modeling

Most PAMs do not model the full image but solely the object of interest. Therefore, those generative models synthesize the foreground $\mathcal{F}$ of the image containing the face. The parts that cannot be modeled are referred to as background and are ignored by most adaptation methods. Ignoring the background leads to problems during model adaptation. The most prominent problem is shrinking of the face during the fitting process. Shrinking can be overcome by evaluating in normalized model space of 2D models rather than on the image. This solution does not work for 2D analysis with 3D models. Parts of the face are not always visible, depending on the 3D pose, due to self-occlusion. In a frontal view, different parts of the face are visible differently than in a profile view. If change in visibility is ignored, the pose cannot be adapted properly during the model adaptation. Earlier approaches handled this by assuming a fixed and predetermined visibility for each point on the face. This requires an accurate initialization of pose and strongly restricts the flexibility of the model with respect to pose and shape.

The effects of ignoring background in generative modeling are independent of the concrete optimization mechanism and should be handled on a model level to retain the full model flexibility. Our completely probabilistic approach allows us to include a background model directly into the likelihood function

$$\ell(\theta; \tilde{I}) = \prod_{i \in \mathcal{F}} \ell_{\mathrm{face}}(\theta; \tilde{I}_i) \prod_{i' \in \mathcal{B}} b(\tilde{I}_{i'}). \qquad (5.14)$$

Classical Model   Occlusion-Aware Face Model

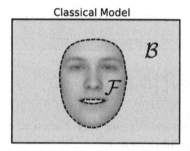 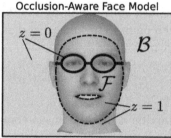

Figure 5.4 On the left, we show the classical model presented in Eq. (5.14). The classical model distinguishes pixels in the foreground $\mathcal{F}$, where the face model is defined, from pixels in the background $\mathcal{B}$. The occlusion-aware model adds an additional variable z which is a label for face ($z = 1$) or non-face ($z = 0$), compare Eq. (5.15). This additional label allows to exclude pixels in the foreground $\mathcal{F}$ from the face model explanation.

The idea is to distinguish between foreground pixels $\mathcal{F}$ which are covered by the rendered face and background pixels $\mathcal{B}$, compare Fig. 5.4. The pixels in the background are then covered by an explicit background model with likelihood b. In some settings, it makes sense to build a complex background model which is able to adapt to the expected background appearance. In medical image analysis, the surroundings of objects in the body (e.g. organs or bones) are often similar and can be modeled. In the setting of face image analysis, background is unconstrained and a concrete background model is infeasible. In order to counteract the negative effects of background, it is already sufficient to use a simple background model. We resort to simple global color models. Such models are either general, e.g. a uniform color distribution, or adapted to the concrete target, e.g. a histogram. In Schönborn et al. [26] we discuss the problem and different background models in detail. However, the actual choice of background model is not very critical.

5.2.5 Occlusion-Aware Morphable Model

Images of faces often contain occlusions of the face by other objects. The most common sources are not only unrelated objects in front of the face but also glasses, beards and hair. In an Analysis-by-Synthesis setting, occlusions act as outliers which render the fitting problem even harder. The appearance prior of our model and the ability to generate images is successfully applied to handle occlusions. Unhandled occlusions can disrupt the fitting process due to their unpredictable color and location, see Fig. 5.5. The strong appearance prior is used to detect and segment occlusions in the image as those parts which cannot be explained by the model. Our probabilistic approach allows us to build an occlusion-aware model which deals with uncertainty of occlusion segmentation and the adapted face model.

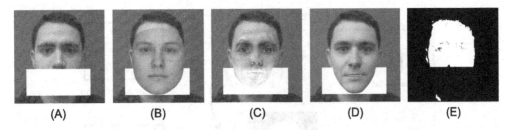

(A)   (B)   (C)   (D)   (E)

Figure 5.5 We use a partially occluded face as target image (A). Even the model adaptation is properly initialized manually (B), when using the likelihood from Eq. (5.14), the adaptation is strongly misled by the occluded part (C). The model is able to deform color appearance strongly. With our occlusion-aware likelihood from Eq. (5.15) and the segmentation approach, we can properly fit the target face (D) and segment face versus non-face (E). The fitting of the not-occluded target image is contained in Fig. 5.2. (For interpretation of the references to color in this figure, the reader is referred to the web version of this chapter.)

We propose to segment the image into face and non-face parts. The image likelihood is therefore extended with an additional binary label z to mark face pixels (compare Fig. 5.4):

$$\ell\left(\theta; \tilde{I}, z\right) = \prod_i \ell'_{\text{face}}\left(\theta; \tilde{I}_i\right)^z \cdot \ell_{\text{non-face}}\left(\theta; \tilde{I}_i\right)^{1-z} \tag{5.15}$$

The general idea is very similar to background modeling. Contrary to the background $\mathcal{B}$, occlusion can also appear within the foreground region $\mathcal{F}$. Occluded parts are excluded from model explanation and are not evaluated with the standard foreground image likelihood. The occlusion segmentation becomes part of the model parameters and needs to be included in model adaptation.

We choose the variational framework suggested by Chan and Vese as a basis for segmentation. We replace the plain average color models by our more complicated face model likelihoods for foreground and background to adapt it to our problem and model. The segmentation algorithm then correctly evaluates with respect to the likelihood of model fit rather than using simply its mean color. The original Chan–Vese algorithm is formulated with an energy term E. We reformulate it as a posterior to obtain the label z for a given parameter set θ and the target image $\tilde{I}$:

$$-\log p\left(z|\tilde{I}, \theta\right) = E = \Psi + \int_\Omega z(x) \log \ell'_{\text{face}}(\theta; \tilde{I}(x)) + (1 - z(x)) \log \ell_{\text{non-face}}(\theta; \tilde{I}(x))\, \mathrm{d}x \tag{5.16}$$

Ψ is the length term of the classical Chan–Vese formulation and regularizes the boundary of the segmentation.

The segmentation relies on the model adaptation and vice versa. Therefore both parts have to be optimized together. We choose an EM-like procedure to optimize

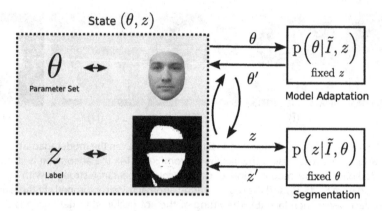

Figure 5.6 The occlusion handling is integrated into the probabilistic framework. A state consists of both model parameters θ and the segmentation label z. During the model adaptation we assume a fixed segmentation and during segmentation we assume fixed model parameters. Inference of the parameter set and the segmentation is performed in an EM-style manner in alternation.

them in close alternation. During segmentation, we assume fixed model parameters and during fitting we assume a given segmentation, see Fig. 5.6. However, in both steps we assume an uncertainty of the other estimate. By considering also the neighboring pixels N in the segmentation likelihood, we assume the fit not to be perfect yet:

$$
\ell'_{\text{face}}\left(\theta; \tilde{I}_i\right) = \begin{cases} \frac{1}{N} \exp\left(-\frac{1}{2\sigma^2} \min_{n \in N(i)} \left\| \tilde{I}_i - I_{i,n}(\theta) \right\|^2\right) & \text{if } i \in \mathcal{F} \\ b_{\mathcal{F}, z=1}\left(\tilde{I}_i\right) & \text{if } i \in \mathcal{B} \end{cases} \tag{5.17}
$$

Taking the neighborhood into the likelihood function considers small misfits of the position on the whole face. Since the likelihood has to be defined on the whole image, we extend the likelihood to cover also pixels in the background. Therefore, we use a color model learned on foreground pixels labeled as face. We respect the uncertainty of segmentation by allowing the face model to include pixels which it can explain better than the background model:

$$
\ell_{\text{non-face}}\left(\theta, \tilde{I}_i\right) = \begin{cases} \max\left(\ell_{\text{face}}\left(\theta, \tilde{I}_i\right), b\left(\tilde{I}_i\right)\right) & \text{if } i \in \mathcal{F} \\ b\left(\tilde{I}_i\right) & \text{if } i \in \mathcal{B} \end{cases} \tag{5.18}
$$

As presented in Fig. 5.1 the appearance of faces is dominated by illumination. Therefore we initialize the segmentation and fitting, by a robust illumination estimation with a RANSAC-like procedure as described in [10]. This gives us a first guess of the illumination conditions and the pixels which probably belong to the face region see Fig. 5.7.

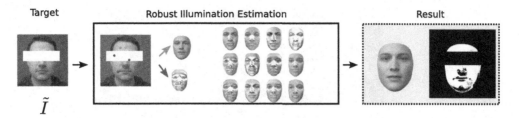

Figure 5.7 We randomly select points in the foreground $\mathcal{F}$ for robust illumination estimation. Those points are used to estimate the spherical harmonics illumination. Depending on the selection of points we get different illumination estimations. We added exemplary estimations on the three points which are suited for illumination estimation (green), three points on occluded parts (red) and some more on other random points. The iterative RANSAC-like algorithm chooses the most probable illumination setting. The results are illumination parameters θ_L and a mask of pixels which can be explained by this illumination setting. The mask is used as starting point for the label z in our occlusion-aware model adaptation. (For interpretation of the references to color in this figure, the reader is referred to the web version of this chapter.)

5.3 APPLICATIONS AND RESULTS

In this section, we present experiments and practical application of our framework. We do not strive for a complete evaluation because all proposed elements of the probabilistic Morphable Model image analysis framework, including model fitting, are already thoroughly validated elsewhere. Instead, we highlight the power of the method and quality of our results with a few rather different example applications.

In the following we make use of the elements described above. Technical details, such as the final proposal distribution, feature point and face detectors and the histogram background model are described in [27,10,28,11]. The software implementation is based on the Statismo [19] and Scalismo[1] software frameworks. Parts of our implemented Probabilistic Morphable Model framework are contributed to both frameworks and therefore available open source.

In our first experiment, we explore the posterior distribution under occlusion. The posterior distribution is the result of our image analysis process conditioned on detections and the target image. The posterior distribution can be visualized to analyze the uncertainty of the model adaptation. After a burn-in phase of 50,000 samples we draw further 50,000 samples to estimate the posterior distribution. We use a collective likelihood for this experiment as described in [27]. With the posterior distribution, we investigate the remaining shape flexibility under occlusion, see Fig. 5.8. We observe that parts of the model can be recovered with high certainty whilst other parts are more flexible and therefore can only be recovered with low certainty.

[1] Scalismo – A Scalable Image Analysis and Shape Modeling Software Framework available as Open Source under https://github.com/unibas-gravis/scalismo.

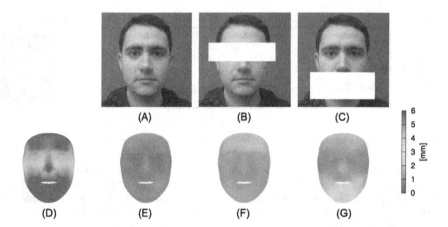

Figure 5.8 Posterior variability of the face shape for different target images. The shape standard deviation of the model prior is shown in (D). Conditioning on a target image without occlusion (A) we get a small posterior variance (E) for the whole face region. With occlusion in the eye region (B), the model gets less certain about the shape, but can narrow down the variability by the strong appearance prior of the statistical model (F). Under occlusion of the chin (C) the variation in the chin region increases strongly (G), this uncertainty is based on the strong influence of expressions which lead to high variability in this region.

In a second experiment our model adaptation recovers from wrong initialization as seen in Fig. 5.9. We therefore show an example where the strongest face detection is the wrong one. Our cascaded filtering presented in Eq. (5.13) and Fig. 5.2 succeeds to find the solution with the highest probability which is in this case the correct face. We draw 10,000 samples for the model adaptation conditioned on the image and the detections.

In our third experiment we perform a qualitative evaluation on the Labeled Faces in the Wild database (LFW) [13]. It contains occlusions, expressions and real world illumination settings. We again draw 10,000 samples and present the best fitting result. The LFW database is challenging for generative models. The presented images depict the variability of the database. Pose, illumination, expressions and occlusions are challenging variations. By using a Morphable Model including facial expressions, we add an additional challenge for the background model. The 3DMM does not contain the inner mouth region which is explained by the background model. We use feature point detections at the mouth corner (compare Fig. 5.3). The expression, respectively the opening of the mouth is solely inferred in the Analysis-by-Synthesis loop from the image. With our explicit background model we succeed to open and close the mouth matching the target image. The most prominent occlusion in the LFW database are glasses. We present results for a variety of faces (neutral, with expression, glasses, sunglasses, occluding hair and other occlusion). Selected results are shown in Fig. 5.10.

Figure 5.9 We show the 10 strongest face detection results for this real world target image. The yellow facebox is the strongest face detection. Our fully probabilistic framework succeeds to recover from this wrong strongest detection and adapts to the most consistent detection. On the right we show the best fitting result and the label map. Image: KEYSTONE/AP/Richard Drew. (For interpretation of the references to color in this figure, the reader is referred to the web version of this chapter.)

In a fourth experiment we present a straightforward application. The fit of our model to an image infers correspondence information of every face pixel to a vertex point in the model space. The correspondence information can be used for face image manipulation, see Fig. 5.11. We manipulate parameters and therefore facial appearance in model space and render the changes back into the face image. The transfer from the manipulation in 3D is rendered in a 2D warp field on the image. The result is a photo-realistic face image manipulation. The presented manipulation is achieved by removing the expression mean $\vec{\mu}_E$ and the expression deformation $\vec{\theta}_E$ from the face model to neutralize facial expressions from the target image.

All parts of the proposed framework have been published and evaluated individually. Additional qualitative experiments on different tasks are presented in [27,28,11,26]. Those publications contain extensive experimental evaluation on pose-invariant face recognition, pose estimation, attribute description, eye gaze estimation, shape reconstruction and also experiments to investigate characteristics of the DDMCMC sampling strategy.

5.4 CONCLUSION

We present an unprecedented completely automatic fitting framework for 3DMM adaptation including detection and occlusions. It is generic and based on a fully probabilistic approach. Our probabilistic setting is capable of robust and fully automatic face image

Figure 5.10 Exemplary fitting results of the fully automatic probabilistic framework on the LFW database. The target image is shown on the left, the best fit (MAP estimate) in the middle and the segmentation label *z* on the right.

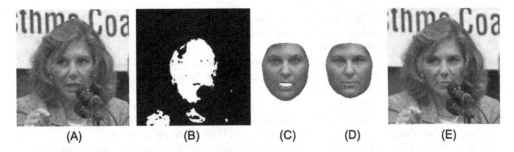

Figure 5.11 Photo-realistic image manipulation becomes possible through the high-quality results of our probabilistic analysis. The target image (A) is fitted using our fully automatic and robust model adaptation framework. The occlusion label (B) and fitting result (C) hold correspondence information from the face model to the target image. The expression manipulation is performed in model space by removing the expression components (D). We obtain the final manipulation result (E) by rendering the 3D manipulation back into a warp-field for the target image.

analysis. We demonstrated the flexibility of our approach by integrating discriminative Bottom-Up cues and segmentation of occlusion. The Bottom-Up integration is superior to feed-forward methods and can recover from unreliable detection results. We build an occlusion-aware model by exploiting the uncertainty of the face model to segment occlusions which cannot be explained otherwise. Both, the occlusion awareness and the Bottom-Up integration can be naturally integrated in the likelihood functions of the probabilistic framework. We extended our 3DMM with facial expressions in order to adapt it to unconstrained face photographs. We highlight the importance of background modeling for Analysis-by-Synthesis settings. The Metropolis–Hastings algorithm builds the core of our probabilistic framework and makes future extensions easy. Our approach is promising for image analysis and scene understanding. The resulting framework is generic and flexible. The proposed approach is not specific for faces but can also be applied for occlusion handling or initialization of other statistical shape models. The showcased integration of Bottom-Up cues and segmentation are only two examples of extensions. The model adaptation process would highly profit from more Bottom-Up cues like edge or pose information. Due to its propose-and-verify architecture, our framework is open to include arbitrary uncertain information sources.

REFERENCES

[1] Thomas Albrecht, Reinhard Knothe, Thomas Vetter, Modeling the remaining flexibility of partially fixed statistical shape models, in: 2nd MICCAI Workshop on Mathematical Foundations of Computational Anatomy, 2008, pp. 160–169.

[2] Oswald Aldrian, William A.P. Smith, Inverse rendering of faces with a 3d morphable model, IEEE Trans. Pattern Anal. Mach. Intell. 35 (5) (May 2013) 1080–1093.

[3] Brian Amberg, Andrew Blake, Thomas Vetter, On compositional image alignment, with an application to active appearance models, in: IEEE Conference on Computer Vision and Pattern Recognition, 2009, CVPR 2009, IEEE, 2009, pp. 1714–1721.

[4] Brian Amberg, Reinhard Knothe, Thomas Vetter, Expression invariant 3d face recognition with a morphable model, in: 8th IEEE International Conference on Automatic Face & Gesture Recognition, 2008, FG'08, IEEE, 2008, pp. 1–6.

[5] Volker Blanz, Thomas Vetter, A morphable model for the synthesis of 3d faces, in: Proceedings of the 26th Annual Conference on Computer Graphics and Interactive Techniques, SIGGRAPH '99, ACM Press/Addison-Wesley Publishing Co., New York, NY, USA, 1999, pp. 187–194.

[6] Volker Blanz, Thomas Vetter, Reconstructing the complete 3d shape of faces from partial information (rekonstruktion der dreidimensionalen form von gesichtern aus partieller information), IT, Inf. Technol. 44 (6/2002) (2002) 295.

[7] Tim F. Cootes, Chris J. Taylor, D.H. Cooper, James Graham, Active shape models-their training and application, Comput. Vis. Image Underst. 61 (1) (January 1995) 38–59.

[8] Tim F. Cootes, G.J. Edwards, Chris J. Taylor, Active appearance models, IEEE Trans. Pattern Anal. Mach. Intell. 23 (6) (June 2001) 681–685.

[9] Michael De Smet, Rik Fransens, Luc Van Gool, A generalized em approach for 3d model based face recognition under occlusions, in: 2006 IEEE Computer Society Conference on Computer Vision and Pattern Recognition, vol. 2, IEEE, 2006, pp. 1423–1430.

[10] Bernhard Egger, Andreas Schneider, Clemens Blumer, Andreas Morel-Forster, Sandro Schönborn, Thomas Vetter, Occlusion-aware 3d morphable face models, in: Proceedings of the British Machine Vision Conference (BMVC), September 2016.

[11] Bernhard Egger, Sandro Schönborn, Andreas Forster, Thomas Vetter, Pose normalization for eye gaze estimation and facial attribute description from still images, in: German Conference on Pattern Recognition, Springer, 2014, pp. 317–327.

[12] Martin A. Fischler, Robert C. Bolles, Random sample consensus: a paradigm for model fitting with applications to image analysis and automated cartography, Commun. ACM 24 (6) (1981) 381–395.

[13] Gary B. Huang, Marwan Mattar, Tamara Berg, Eric Learned-Miller, Labeled faces in the wild: a database for studying face recognition in unconstrained environments, 2008.

[14] Patrik Huber, Zhen-Hua Feng, William Christmas, Josef Kittler, Matthias Rätsch, Fitting 3d morphable face models using local features, in: 2015 IEEE International Conference on Image Processing (ICIP), IEEE, 2015, pp. 1195–1199.

[15] Michael Kirby, Lawrence Sirovich, Application of the Karhunen–Loeve procedure for the characterization of human faces, IEEE Trans. Pattern Anal. Mach. Intell. 12 (1) (1990) 103–108.

[16] Martin Köstinger, Paul Wohlhart, Peter M. Roth, Horst Bischof, Annotated facial landmarks in the wild: a large-scale, real-world database for facial landmark localization, in: 2011 IEEE International Conference on Computer Vision Workshops (ICCV Workshops), 2011, pp. 2144–2151.

[17] Tejas D. Kulkarni, Pushmeet Kohli, Joshua B. Tenenbaum, Vikash Mansinghka, Picture: a probabilistic programming language for scene perception, in: Proceedings of the IEEE Conference on Computer Vision and Pattern Recognition, 2015, pp. 4390–4399.

[18] Marcel Lüthi, Thomas Albrecht, Thomas Vetter, Probabilistic modeling and visualization of the flexibility in morphable models, in: IMA International Conference on Mathematics of Surfaces, Springer, 2009, pp. 251–264.

[19] Marcel Lüthi, Remi Blanc, Thomas Albrecht, Tobias Gass, Orcun Goksel, Philippe Buchler, Michael Kistler, Habib Bousleiman, Mauricio Reyes, Philippe C. Cattin, et al., Statismo-a framework for PCA based statistical models, Insight J. (2012) 1–18.

[20] Marcel Lüthi, Christoph Jud, Thomas Gerig, Thomas Vetter, Gaussian process morphable models, CoRR, abs/1603.07254, 2016.

[21] Iain Matthews, Simon Baker, Active appearance models revisited, Int. J. Comput. Vis. 60 (2) (2004) 135–164.

[22] Paysan Paysan, Reinhard Knothe, Brian Amberg, Sami Romdhani, Thomas Vetter, A 3d face model for pose and illumination invariant face recognition, in: 2009 Advanced Video and Signal Based Surveillance, 2009, pp. 296–301.

[23] Jean-Sébastien Pierrard, Thomas Vetter, Skin detail analysis for face recognition, in: IEEE Conference on Computer Vision and Pattern Recognition, 2007, CVPR'07, IEEE, 2007, pp. 1–8.

[24] Sami Romdhani, Thomas Vetter, Estimating 3d shape and texture using pixel intensity, edges, specular highlights, texture constraints and a prior, in: IEEE Computer Society Conference on Computer Vision and Pattern Recognition, 2005, CVPR 2005, vol. 2, June 2005, pp. 986–993.

[25] Sami Romdhani, Thomas Vetter, Efficient, robust and accurate fitting of a 3d morphable model, in: Ninth IEEE International Conference on Computer Vision, 2003, Proceedings, 2003, pp. 59–66.

[26] Sandro Schönborn, Bernhard Egger, Andreas Forster, Thomas Vetter, Background modeling for generative image models, Comput. Vis. Image Underst. 136 (July 2015) 117–127.

[27] Sandro Schönborn, Bernhard Egger, Andreas Morel-Forster, Thomas Vetter, Markov chain Monte Carlo for automated face image analysis, Int. J. Comput. Vis. (2016) 1–24.

[28] Sandro Schönborn, Andreas Forster, Bernhard Egger, Thomas Vetter, A Monte Carlo strategy to integrate detection and model-based face analysis, in: Pattern Recognition, in: Lect. Notes Comput. Sci., vol. 8142, Springer, Berlin, Heidelberg, 2013, pp. 101–110.

[29] Matthew Turk, Alex Pentland, Eigenfaces for recognition, J. Cogn. Neurosci. 3 (1) (1991) 71–86.

[30] Xuehan Xiong, F. De La Torre, Supervised descent method and its applications to face alignment, in: 2013 IEEE Conference on Computer Vision and Pattern Recognition (CVPR), June 2013, pp. 532–539.

CHAPTER 6

Object Statistics on Curved Manifolds

Stephen M. Pizer*, J.S. Marron[†]

*University of North Carolina at Chapel Hill, Department of Computer Science
[†]University of North Carolina at Chapel Hill, Department of Statistics and Operations Research

Contents

6.1	Objectives of Object Statistics	138
6.2	Objects Live on Curved Manifolds	139
	6.2.1 Point Distribution Models	139
	6.2.2 Spherical Harmonic Models	140
	6.2.3 Normals Distribution Models	140
	6.2.4 Deformation Models	141
	6.2.5 Skeletal Models	142
	6.2.5.1 Medial Models	*143*
	6.2.5.2 S-reps	*143*
6.3	Statistical Analysis Background	144
	6.3.1 Statistical Analysis in Euclidean Space	144
	6.3.2 Fully Extrinsic Analysis on Manifolds	146
	6.3.3 Distance-Based Statistical Analysis Methods	147
	6.3.4 Tangent Plane Statistical Analysis Methods	147
6.4	Advanced Statistical Methods for Manifold Data	148
	6.4.1 Principal Nested Spheres	148
	6.4.2 Composite Principal Nested Spheres	149
	6.4.3 Polysphere PCA	150
	6.4.4 Barycentric Subspace Analysis	151
	6.4.5 Bayes Methods for Manifold Data	151
	6.4.6 Manifold Techniques Based on Brownian Motion	152
	6.4.7 Manifold Classification	152
6.5	Correspondence	153
	6.5.1 Correspondence via Reparameterization-Insensitive Metrics	153
	6.5.2 Correspondence via Entropy Minimization	154
6.6	How to Compare Representations and Statistical Methods	155
	6.6.1 Classification Accuracy	156
	6.6.2 Hypothesis Testing Power	156
	6.6.3 Specificity, Generalization, Compactness	156
	6.6.4 Compression Into Few Modes of Variation	157
	6.6.5 Quality in Application, Esp. Segmentation	157
6.7	Results of Classification, Hypothesis Testing, and Probability Distribution Estimation	157
	6.7.1 Generalized Rotations	158
	6.7.2 Classification of Schizophrenia via Hippocampus S-reps	158
	6.7.3 Hypothesis Testing via S-reps	158

Statistical Shape and Deformation Analysis
DOI: 10.1016/B978-0-12-810493-4.00007-9

      6.7.4 Shape Change Statistics   158

      6.7.5 Tu Correspondence Evaluation   161

6.8 Conclusions   161

Acknowledgements   162

References   162

6.1 OBJECTIVES OF OBJECT STATISTICS

We understand an object to be a region of space, typically with a smooth, or at least piecewise smooth, boundary (Fig. 6.1). Examples are anatomic entities such as the kidney or the brain structure called the hippocampus or an everyday object such as a table. While objects can have holes, we will concentrate on objects with no holes, i.e., with spherical topology in 3D or circular topology in 2D.

The statistical objectives on training samples from populations of such objects include probability distribution estimation, classification, and hypothesis testing. For classification we will restrict ourselves to discrimination between two classes. Likewise, in hypothesis testing we will restrict ourselves to finding object features that have statistically significant differences between two classes. In this chapter we will be concerned with not just whether there *are* differences but with where on the object they are and what geometric type they have, e.g., bulging, bending, or twisting.

The standard statistical methods for reaching these objectives assume that the feature tuples describing these entities lie in a Euclidean space, i.e., a "flat manifold", where the Pythagorean Theorem applies. This is the case for the main method of Gaussian probability distribution estimation, Principal Component Analysis (PCA); the main methods for classification, including Support Vector Machines (SVM) and Fisher linear discriminants; and the main methods for hypothesis testing, including the t-test.

Figure 6.1 An example of a smooth object, seen from two points of view (by permission of T. Fletcher).

6.2 OBJECTS LIVE ON CURVED MANIFOLDS

But as explained below, geometric object properties (GOPs) do not lie on an abstract flat manifold. Rather, that manifold is best understood as curved, so the standard statistical methods do not strictly apply. The following discusses some of the most common GOPs and explains why they live in curved spaces.

6.2.1 Point Distribution Models

The simplest GOPs are tuples of locations, most commonly spatial sampling the boundary of the object (Fig. 6.2). These are called "point distribution models (PDMs)" [7,6]. We will call PDMs 0th order GOPs. Let $\mathbf{z} = (\underline{x}_1, \underline{x}_2, \ldots \underline{x}_m)$ be such a tuple; it has dimension $\mathcal{D}m$ for $\mathcal{D}$-dimensional objects. They initially appear to have Euclidean properties. However, they typically need centering: let us center the tuple by subtracting its center of mass (average of the $\underline{x}_i$) from each location. Then let us compute $\gamma =$ the square root of the sum of squares of the distances of the centered locations to the origin, taking γ as a GOP. Finally, let us normalize each centered location by dividing it by γ to produce a GOP formed by the tuple of normalized, centered locations: $((x''_{11}, x''_{12}, \ldots, x''_{1\mathcal{D}}), (x''_{21}, x''_{22}, \ldots, x''_{2\mathcal{D}}), \ldots, (x''_{m1}, x''_{m2}, \ldots, x''_{m\mathcal{D}}))$, where the first subscript indexes the points and in the second subscript "1" refers to the x dimension, "2" refers to the y dimension, $\ldots$, and "$\mathcal{D}$" refers to the $\mathcal{D}$th spatial dimension. This GOP satisfies $\sum_{i=1}^{m} \sum_{j=1}^{D} (x_{ij})^2 = 1$, which is the equation for a point on a unit sphere of dimension $\mathcal{D}m - 1$ (a circle is a sphere of dimension 1). Strictly, this sphere is of dimension $\mathcal{D}m - \mathcal{D} - 1$: $S^{\mathcal{D}m-\mathcal{D}-1}$, since $\mathcal{D}$ dimensions have been taken up by the centering. Thus a population of that GOP consists of points on a unit sphere, a curved manifold.

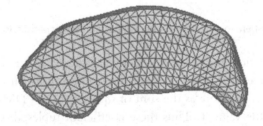

Figure 6.2 An object boundary in 3D represented by the PDM formed by the vertices of the triangular tiles. The inter-vertex connections are strictly not part of the PDM.

Moreover, the scale parameter γ also is not Euclidean because it cannot take on negative values. Different scales are related to each other by a multiplicative relation, not an additive one. We will produce a "Euclideanized" version of γ by taking its logarithm, and then we will statistically center this value $\log \gamma$, by subtracting its mean over the training cases. This yields a Euclideanized GOP $\overline{\gamma}\log(\gamma/\overline{\gamma})$, where $\overline{\gamma}$ is the geometric mean of the training cases' γs.

We conclude that even PDMs, which initially appear to live on a flat space, can be better understood as living on the Cartesian product of a sphere and a 1D flat space, $\mathcal{R}^1$, once the scale parameter is Euclideanized using the logarithm.

6.2.2 Spherical Harmonic Models

A PDM on a normalized, centered object with spherical topology can be alternatively represented by orthonormal basis functions mapping the unit sphere to the boundary of the object (Fig. 6.3). In 2D these functions are the Fourier (sinusoidal) basis functions, and in 3D they are the spherical harmonics [25]. Then a particular object's boundary is described by a weighted sum of these basis functions; that is, it is represented by the $(\mathcal{D}m - \mathcal{D} - 1)$-dimensional tuple of the weights, the coefficients of the basis functions. One should realize that these mappings vary according to the parameterization of the object surface.

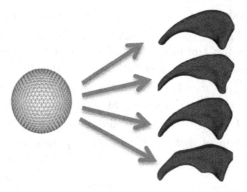

Figure 6.3 Mapping a sphere to objects via spherical harmonics (by permission of M. Styner).

While these coefficient tuples are normally taken to lie in a Euclidean space, Parseval's theorem guarantees that when they are derived from normalized, centered objects, their sum of squares is the same as the sum of squares of the PDM-points' distances to the origin, i.e., has the value 1. Thus these coefficient tuples also live on a sphere of dimension $\mathcal{D}m - \mathcal{D} - 1$: $S^{\mathcal{D}m-\mathcal{D}-1}$.

6.2.3 Normals Distribution Models

Derivatives of 0th order boundary properties yield a normal vector $\mathbf{u}$ at each boundary point: 1st order properties. Intuition suggests that the tuple of normal directions $\underline{\mathbf{z}} = (\mathbf{u}_1, \mathbf{u}_2, \ldots, \mathbf{u}_m)$ better captures the shape (by encoding surface information from a small neighborhood) than the positions themselves.

But each normal is a direction vector; it lives on a unit sphere of dimension $\mathcal{D} - 1$: $S^{\mathcal{D}-1}$. Thus a tuple of normals lives on a Cartesian product of unit $\mathcal{D} - 1$-dimensional spheres, a curved space.

A view on objects given by Srivastava et al. [44] also yields a representation by boundary normals. Their objective was to consider every object's boundary as an equivalence class over the spherical parameterizations described in section 6.2.4. They wished to build a distance function between pairs of such equivalence classes as a distance between the corresponding objects. Section 6.3.3 discusses the use of this non-Euclidean distance function for object statistics. The use of this non-Euclidean metric is equivalent to regarding a model as living on a curved manifold.

Srivastava et al. first studied such distances modulo parameterization on objects in 2D [26], and then they extended the method to objects in 3D [28]. In both cases they showed that a natural distance measure with these properties could be built upon Euclidean comparisons across the boundary if it was represented by a dense sampling of the boundary normal, as follows:

$$\| [q_1], [q_2] \| = \min_{O \in SO^n, \; \gamma \in \Gamma} d_c(q_1, O(q_2 \circ \gamma))/\sqrt{\dot{\gamma}},$$

where $[q]$ refers to the equivalence class of object representations that are reparameterizations of q, γ is a reparameterization of the object boundary, Γ is the set of all reparameterizations, SO^n is the set of rotations, and d_c is the L^2 norm on the boundary normals distribution.

Models of spatial derivative order higher than 1, e.g., those based on curvatures, are also possible, but they are not common and will not be discussed here.

6.2.4 Deformation Models

Another representation of an object is the deformation of Euclidean space from an atlas for the object, i.e., a reference object formed by a collection of voxels (pixels in 2D) assigned to values 1 if they are interior to the reference object and 0 if they are exterior to that object. This deformation is commonly understood as a displacement vector **d** at each voxel, so the object representation consists of an M-tuple of **d** values, where M is the number of voxels. That is, the representation is of dimension $\mathcal{D}M$. Alternatively, the deformation of each pixel is represented by a curved path formed by series of short "velocity" vectors over pseudo-time divided into m intervals, whereby the object representation is of dimension $\mathcal{D}mM$.

Usually these deformation models are understood to live in Euclidean space. Theoretically, this is justified by arguing that deformations are geodesics on a curved space and geodesics from the reference model can be understood as an initializing direction on that curved space, which can be understood as a tangent space, i.e., a flat space [1].

However, if we look at the component displacement or velocity vectors, it is more geometrically intuitive to see each of them as a direction in $\mathcal{D}$ dimensions and a length. The direction lives on a $\mathcal{D}-1$-dimensional unit sphere: $\mathbf{S}^{\mathcal{D}-1}$. Each length l, like the aforementioned scale factors γ, requires the application of the logarithm to produce $\bar{l}\log(l/\bar{l})$ before it lives in a flat space ($\bar{l}$ is the geometric mean of the lengths at that voxel). This formulation understands an object as living in a space $(\mathbf{S}^{\mathcal{D}-1})^M \times \mathcal{R}^M$ (after Euclideanization) or of $(\mathbf{S}^{\mathcal{D}-1})^{mM} \times \mathcal{R}^{mM}$, depending on whether the deformation is represented by displacements or velocities.

6.2.5 Skeletal Models

An object has an interior, not just a boundary, and the connections from one boundary position to the other "opposite" it across the object are intuitively important. The length of these connections captures the notion of local object width. Skeletal models (Fig. 6.4) [42] capture these relations by having the following two components:

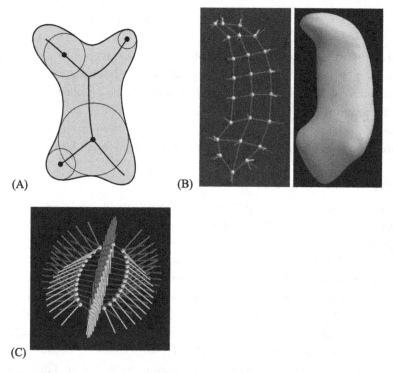

(A)   (B)

(C)

Figure 6.4 A) An object boundary in 2D and its skeletal (here, medial) model; the medial model is built from the bitangent circles; B) left: a skeletal surface for a hippocampus with selected spokes – the curve along the exterior of the mesh forms the curve at which the skeletal surface is folded; right: the boundary implied by it. C) An s-rep for an ellipsoid in 3D (by permission of J. Vicory).

1. A skeleton: a curve in 2D and a surface in 3D ideally halfway between the connected boundary points. This skeleton is a sort of collapsed version of the boundary and thus has the same topology as the boundary; the fold curve (points in 3D) of that surface divides one side of the boundary from the other.
2. Non-crossing vectors that we call "spokes" extending from each skeletal point to the boundary. The lengths of these spokes capture the object half-widths. The mathematics of such models, algorithms for computing them from the object boundary, and applications of these are covered at length in [42].

6.2.5.1 Medial Models

The earliest forms of such skeletal models were called "medial models". The most common form of medial model was due to Blum [4]. The Blum medial axis consists of a skeleton which is a doubling (two-sided, folded version) of the locus of spheres bitangent to the object boundary and interior to the object, together with spokes formed by the radii connecting the sphere center to the points of boundary bitangency. These spokes are orthogonal to the boundary. The spokes proceeding from the fold curve touch the boundary at crest points (a type of curvature extremum [27]).

The Blum medial model for a typical object is highly branching, with different branching patterns for different objects in a population. This makes statistical analysis of these models problematic. Pruning the branches to form a representation with a fixed branching pattern over a population usually makes the boundary implied by the locus of spoke vector ends too far from the true object boundary to be useful.

Two forms of medial model computer representations have been developed: one due to Pizer et al. produced by sampling the medial locus and associated spokes, and another due to Yushkevich based on parameterizing splines. Both are described in Ch. 8 of Siddiqi and Pizer [42]. While some statistical analysis has been done using each, the s-reps below have been shown to be more successful for statistical analysis.

6.2.5.2 S-reps

S-reps are skeletal models that are designed for statistical analysis by fixing the object branching pattern but still having the skeletally implied boundary close to the actual object boundary. They do this by still having the structure of a folded surface but relaxing the conditions of having the skeletal surface be precisely centered from the boundary and of having the spokes precisely orthogonal to the object boundary; instead they fit a reference model [36,47] according to an objective function that rewards fit to the object boundary and penalizes deviations from mediality and spoke-to-boundary orthogonality. A computer representation is formed by sampling the skeleton and associated spokes (Fig. 6.4B). There are spokes on the skeletal manifold as well as on the skeletal fold curve; the latter proceed to crest points on the implied boundary. Thus this representation consists of a tuple of n spokes, each consisting of a skeletal point $\mathbf{p}$, a spoke direction (unit vector) U, and a spoke length r. In implementation the skeletal locus is sampled, assuring that spokes on the fold are included. Such representations live on the manifold formed by the Cartesian product of the PDM manifold of the n skeletal points: $S^{3n-4} \times \mathcal{R}^1$, the n 2-spheres S^2 on which the spoke directions live, and $\mathcal{R}^n$, containing the Euclideanized spoke length values $\bar{r} \log(r/\bar{r})$, where $\bar{r}$ is the geometric mean of the lengths of *that* spoke over the population.

6.3 STATISTICAL ANALYSIS BACKGROUND

This section reviews the basic statistical methods used in the shape analysis contexts of this chapter. Section 6.3.1 considers the case of conventional Euclidean data. Section 6.3.2 describes fully intrinsic approaches. Tangent plane methods are described in Section 6.3.3.

6.3.1 Statistical Analysis in Euclidean Space

Several statistical tasks are routinely addressed in shape analysis, i.e. summarizing and modeling populations of shapes, usually based on a sample from that population. The first task is understanding *centerpoint*, which is usually calculated as the sample mean. After the center is understood, the next task is to consider the variation about that center. Because shape statistics are typically High Dimension, Low Sample Size problems (using terminology from [18]) in nature, standard quantifications of variability such as the *covariance matrix* are hard to estimate. Thus it is natural to analyze variability using *Principal Component Analysis* (PCA, Fig. 6.5). See [23] for a good introduction and discussion of many important aspects of PCA.

The third task, useful for many purposes, for example incorporating external information such as anatomical structure into an analysis through Bayes-like methods, is *probability distribution modeling*. Gaussian distributions are far and away the most commonly used, mostly because many natural distributions are approximately Gaussian due to the Central Limit Theorem. A second reason that Gaussian distributions commonly appear in shape analysis is their Bayes conjugacy properties; that is, Gaussian priors and likelihoods lead to Gaussian posteriors. This is important as it allows closed form calculations instead of the complicated simulation-based approaches used in most modern Bayes analyses.

A fourth task is statistical *classification* (also called *discrimination*). For this task training data, with known class labels, is given and is used to develop a classification rule for assigning new data to one of the classes. For Euclidean data objects, there are many methods available; see [10] for a good overview.

The most common methods for classification are based on Euclidean spaces. Particularly widely used in these days is the method called support vector machines (SVM); see, for example, [38] for detailed discussion. SVM is based on optimizing the gap in

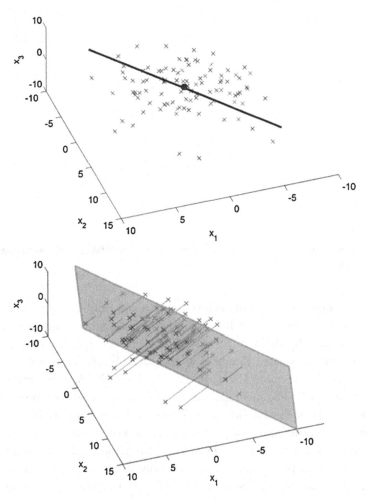

Figure 6.5 The mean, best fitting line, and best fitting plane derived by PCA for a set of observations in $\mathcal{R}^3$.

feature space between the training cases in the two classes. A more statistically efficient method called Distance-Weighted Discrimination (DWD), still in a Euclidean feature space was developed by Marron et al. [31]; its efficiency derives from using all of the training data, not just those near the gap. Both SVM and DWD yield a direction in feature space that optimally separates the classes (Fig. 6.6). Classification then involves projecting the feature tuple onto the separation direction and deriving the class or the class probability from the resulting scalar value.

The above tools are useful in situations where the full set of GOPs is available, and they are preferable in that case. But in some situations only pairwise distances between the data objects, i.e., the *distance matrix*, are available. Some analysis can still be done in

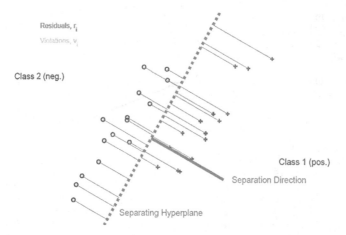

Figure 6.6 The idea of the separation direction determined from SVM or DWD (each method will determine a somewhat different separation direction).

that case. For example, a notion of centerpoint that can still be computed is the *Fréchet mean*, proposed by Fréchet [15]. It has a number of synonyms including *barycenter*. The Fréchet mean is actually defined, for an arbitrary metric space, as the point which minimizes the sum of squared distances to the data objects (or points, since the arg minimum may not be unique). There also is an analog of PCA, called *multi-dimensional scaling*, see [45] and [17]. The main idea here is to solve an optimization problem whose solution is a set of points in Euclidean space whose distance matrix best approximates the given one. When the input distance matrix is Euclidean, multi-dimensional scaling results in scores that are the same as PC scores. Multi-dimensional scaling scores can also provide a surrogate population for further analysis, but an important issue that will be central to the following discussion is that for non-Euclidean (e.g., manifold-based shapes) data, this represents an approximation. In some situations the approximation is adequate. Situations where that approximation is inadequate and alternate approaches give better results are the focus of Section 6.4.

6.3.2 Fully Extrinsic Analysis on Manifolds

The challenge of statistical analysis of populations lying on manifolds has been addressed in a number of different ways. The diversity of these can be seen already in the simple but nontrivial case where the data objects lie on the unit circle. That topic is often called *directional data* and naturally arises in contexts where the data objects are angles such as wind or magnetic field directions. As seen in the monographs of Mardia [30] and Fisher [12], there is a large literature on the statistical analysis of directional data. In that context, even for the simplest statistical summary of mean or center point of a data set, there are divergent reasonable choices, which have been characterized as extrinsic and intrinsic (see [2,3,32] for an overview of these issues). In general, extrinsic summaries are first computed in the ambient space (the Euclidean space in which the manifold is embedded) and then projected back to the nearest point on the manifold. Thus the extrinsic mean for directional data is the vector average in $\mathcal{R}^2$ of the data vectors, which is projected to the circle by using the angular part of its polar coordinate representation. Intrinsic summaries strive to work more within the manifold.

Extrinsic versions of PCA are not widely used in shape analysis, perhaps because there is potential for very strong distortions while projecting the resulting PC scores back to the manifold. Hotz [21] did a detailed comparison of the intrinsic and extrinsic mean for data that consists of points in the unit circle. A number of criteria were considered, and each form of mean had its relative strengths and weaknesses.

6.3.3 Distance-Based Statistical Analysis Methods

For data lying on a manifold, a natural distance is the geodesic distance (the length of the shortest path along the manifold). Computing the matrix of pairwise distances between the collection of data points leads to approximate analyses as described in Section 6.3.1. This includes the Fréchet mean, which in this case is also called the geodesic mean.

An important example of the non-uniqueness of the Fréchet mean is a data set that is distributed widely around the equator of S^2, the 2D sphere, where both the north and south poles are minimizers of the Fréchet sum of squares. Neither of these means is close to the data in that example.

Another distance-based analysis approach, which is a variation of the Euclideaniza-tion idea discussed above, is to represent the data using the multi-dimensional scaling scores of each object to each of the objects and then to proceed with standard Euclidean analysis methods.

6.3.4 Tangent Plane Statistical Analysis Methods

The first generation of intrinsic analogs of PCA for the analysis of manifold data are based on the definition of a manifold, which is a surface in the ambient space being smooth in the sense of having an approximating hyperplane (in the sense of shrinking neighborhoods) at every point. In this spirit, [13] proposed *Principal Geodesic Analysis* (PGA) (Fig. 6.7). This is based on the plane that is tangent at the Fréchet mean. The data on the surface of the manifold are represented as points in the tangent plane using the *Log map*. PCA is then performed there, and the resulting eigenvectors and summarized data are mapped back into the manifold using the *Exponential map*. The corresponding scores give another type of Euclideanization.

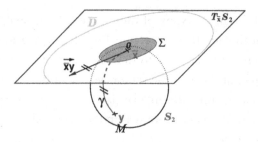

Figure 6.7 Tangent plane statistical analysis (by permission from X. Pennec).

6.4 ADVANCED STATISTICAL METHODS FOR MANIFOLD DATA

While much useful shape analysis has been done using the statistical methods described in Section 6.3, for many data sets large gains in statistical efficiency have been realized through the development of more sophisticated methods. The S^2 example given in Section 6.3.3 above, where the data lies very near the equator of a sphere, illustrates an important aspect of this challenge. While the data are essentially *one dimensional* in nature (since they just follow a single geodesic), the PGA requires *two* components (where the projections follow a circle) to appropriately describe the variation in the data. This type of consideration has motivated a search for more statistically efficient approaches to statistical analysis of data lying on a manifold. The first of these is *Principal Nested Spheres* (PNS), motivated and described in Section 6.4.1. Extension of this to more complicated manifolds, such as the polyspheres central to s-rep shape representations is given in Section 6.4.2. Section 6.4.3 discusses a yet more efficient approach to polysphere analysis involving a high-dimensional spherical approximation followed by a PNS analysis.

6.4.1 Principal Nested Spheres

In the case of data lying in a high dimensional sphere S^k embedded in $\mathcal{R}^{k+1}$, a useful intrinsic version of PCA is *Principal Nested Spheres* (PNS), proposed by Jung et al. [24]. The central idea is to iteratively find a nested (through dimension) series of subspheres, each of which provides an optimal fit to the data (Fig. 6.8). In particular, at each step the dimension of the approximation is reduced by 1, finding the subsphere which best fits the data in the sense of minimum sum of squared residuals, measured using arc length along the surface of the sphere. The signed residuals are also saved as PNS scores for that component. The concatenation of these scores, over the dimensions, becomes the PNS Euclideanization of each data point. The advantages of this approach are tractability, since each lower dimensional manifold is determined by the imposition of a single (usually easy to find) constraint, and statistical efficiency.

One reason that PNS was an important statistical landmark is that it motivated the more general idea of *Backwards PCA* as a general paradigm for finding principal com-

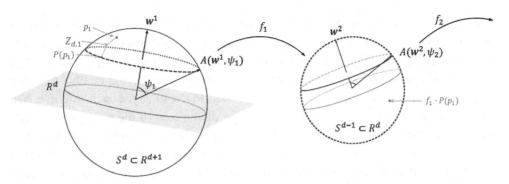

Figure 6.8 An optimal small subsphere, geodesic distances ($Z_{d,1}$) forming scores, and projections onto the subsphere (by permission from S. Jung).

ponents in non-Euclidean data contexts. The full generality of this idea can be found in [8]. A key concept is that the general utility of backwards PCA follows from thinking of PCA in terms of a nested series of constraints. Backwards tends to be easier to work with because from that viewpoint, the constraints can be found sequentially, instead of needing to know the full set and then sequentially relaxing them. This idea is seen to generate (or have the potential to generate) useful analogs of PCA in a variety of other non-Euclidean settings such as Nonnegative Matrix Factorization, Manifold Learning, and on other manifolds.

The S^2 example given above, with data widely distributed around the equator, also illustrates a sense in which the Fréchet mean can be a poor choice that is not at all representative of the data. The *backwards mean* is an intrinsic mean that is much more representative of the data in that example than the Fréchet mean. Generally this is computed by taking one more step in PNS. In particular, the backwards mean is the Fréchet mean of the rank 1 circular representation of the data, which can then be viewed as the best backwards rank 0 approximation.

6.4.2 Composite Principal Nested Spheres

Pizer et al. [36] proposed extending PNS to manifolds that involve products of spheres, such as those for various shape representations discussed in Section 6.2 above, using the idea of Composite Principal Nested Spheres (CPNS). The idea here is to first develop the PNS representation for each spherical component, and then concatenate these, together with Euclidean components, into a large Euclidean representation, followed by PCA on the result.

The commensuration between the components, before the PCA is done, is important. In preparation for the comparisons between using original features of PDMs of hippocampi and the alternative Euclideanized features in [20], an experimental paradigm was designed to determine the most reasonable commensuration on the Euclideanized features, more precisely, on the scale factor and the sphere-resident features derived from the centered, normalized PDMs. First, the features were transformed to be in the same units: in our case the log-transformed version of the scale factor γ (i.e., $\overline{\gamma} \log(\gamma/\overline{\gamma})$) had units of millimeters, and the Euclideanized shape features derived by PNS from the high-dimensional unit sphere, on which the scaled PDM tuple for each case live, had units of radians (unitless values θ_i for each of the dimensions of the unit sphere). Thus we multiplied each PNS-derived feature by $\overline{\gamma}$ to put them into units of distance. The problem then is to determine the scalar factor to commensurate the feature capturing scale with the PNS-derived features. We determined that scalar factor by creating a new population that would have a non-varying shape consistent with those in the original population and a scale variation that was the same as those in the original population. To do this, we formed the new population by applying the measured log-transformed γ values for each case to the hippocampus of median scale scaled to have $\gamma = 1$. By comparing the total variances of the original and created populations, respectively, one could determine the correct commensuration factor between $\overline{\gamma} \log(\gamma/\overline{\gamma}))$ and the Euclideanized features from PNS, namely, $\overline{\gamma}\theta_i$. The experiment concluded that the correct commensuration factor was 1.0, up to sample variation. This idea of separately treating scale can be used for problems of commensuration of other types of variation.

6.4.3 Polysphere PCA

Data spaces that are products of spheres, such as the skeletal model spaces of Section 6.2.5, have been called polyspheres by [11]. That paper goes on to propose a new method, *Polysphere PCA*. This is a potentially large improvement over CPNS which allows a more flexible modeling of the dependence between features than is done by the Gaussian PCA on the Euclideanized data in CPNS. This is achieved by approximating the polysphere manifold with a higher dimensional tangent ellipsoid, projecting the data onto that, then squashing the ellipsoid into a sphere, and using PNS on that (Fig. 6.9).

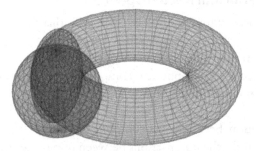

Figure 6.9 The ellipsoid tangent to a torus, which is a polysphere $S^1 \times S^1$, together with the sphere into which the ellipsoid is squashed (by permission of B. Eltzner).

On skeletal shape data, this approach has been shown to give a lower dimensional probability distribution representation than is available from CPNS.

6.4.4 Barycentric Subspace Analysis

An important new general view to PCA analogs for manifold data has been provided by Pennec [33]. A fundamental observation motivating that work is that both the backwards methods (e.g., PNS) that were explicitly described in [8], and forward methods such as PGA and Geodesic PCA [22] rely upon greedy sequential searches (in opposite directions in some sense). This new work proposes the appealing idea of a simultaneous search. The goal of such a search is an appropriate analog of a properly nested sequence of subspaces that is found by a complete (in the sense of simultaneity over subspaces of all ranks) PCA in Euclidean space. This analog is called a *flag*, which is a sequence of *nested* sub-manifolds (where sub-manifolds of lower dimension are contained in those of higher dimension, as in Euclidean PCA). Appropriate components for the flag subspaces in manifolds are constructed as *barycentric subspaces* which are affine combinations (a generalization of Fréchet, Karcher or exponential weighted averages where negative weights allow appropriate extrapolation) of a set of *reference points*. Nesting of the manifolds is elegantly achieved through an appropriate nesting of the control points. In this mechanism geodesics and thus inter-object distances are nicely handled by restriction to regions bounded by *cut loci* associated with each of the reference points and determined by the original object's topology. The computational implications of this property are still a matter of research.

6.4.5 Bayes Methods for Manifold Data

Motivated by the desire to develop Bayes statistical methods on manifolds, Fletcher [14] has noted that the statistical approaches based on Gaussians in Euclidean spaces follow three criteria:

1. that the probability is *shift-equivariant* (this is the mathematical term; engineers call this "shift-invariant"). That is, for all vectors $\mathbf{v}$,

$$p(\underline{x} \mid \underline{\mu}, \sigma) = p(\underline{x} + \mathbf{v} \mid \underline{\mu} + \mathbf{v}, \sigma).$$

2. that the probability distribution is *scale-equivariant* (for engineers: "scale-invariant"). That is, for all scale factors s, $p(\underline{x} \mid \underline{\mu}, \sigma) = (1/s^d)p(s\underline{x} \mid s\underline{\mu}, s\sigma)$, where d is the dimension of the space.

3. that the Fréchet mean is the mode of the probability distribution. That is, $\arg\min_{\underline{\mu} \in \mathcal{R}^d} \sum_{i=1}^{N} |\underline{x}_i - \underline{\mu}|^2 = \arg\max_{\underline{\mu} \in \mathcal{R}^d} \prod_{i=1}^{N} p(\underline{x}_i \mid \underline{\mu}, \Sigma)$, where d is the dimension of the variables $\underline{x}$ and N is the number of points.

He gives requirements for the same properties to be followed for probability distributions on a curved manifold. When these requirements are satisfied, operations such as probability distribution estimation, regression, classification, and hypothesis testing can be accomplished by means very similar to those used in Euclidean spaces.

6.4.6 Manifold Techniques Based on Brownian Motion

Sommer [43] understands Gaussian distributions as formed by Brownian motion, with a covariance that is stationary in space. He derives a curved-manifold counterpart to Brownian motion in Euclidean space by augmenting the positions on the surface with a fitted frame (coordinate system) that is transported from the origin of the motion along a curve on the surface by parallel transport. This allows the definition of stationary covariance that is dependent on the curve, and this in turn allows the definition of a Brownian-produced probability that is dependent on the curve. Based on this careful mathematics, an implementation calculating the most probable curve from the origin to any other point on the manifold, given the covariance matrix, has been completed (Fig. 6.10), and an implementation calculating the regression curve to a set of data points, given the covariance matrix, is in progress.

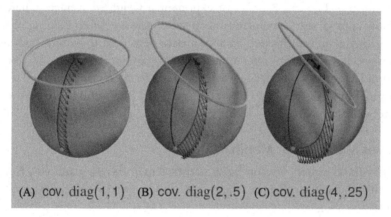

(A) cov. $\mathrm{diag}(1,1)$ (B) cov. $\mathrm{diag}(2,.5)$ (C) cov. $\mathrm{diag}(4,.25)$

Figure 6.10 Most probable paths on a sphere with different Brownian covariances. In each subfigure the curve without arrows is the geodesic path between the points (by permission of S. Sommer).

6.4.7 Manifold Classification

Use of DWD with object data is often done by Euclideanizing the object features and then applying DWD to the result. As shown in [20], this can improve classification accuracy over using the object features directly in DWD.

A large open research problem is the development of intrinsic classification methods on manifolds. The first attempt, using SVM ideas, can be found in [41] (Fig. 6.11). Given the major improvements in PCA from using intrinsic ideas discussed in Sections 6.4.1–6.4.3, we believe that large gains in classification error rates can be made by creative work in this direction.

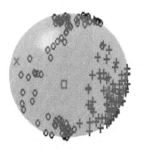

Figure 6.11 Classification on the manifold by optimizing two control points, shown as "×", such that the separation direction is the geodesic between the control points [41].

6.5 CORRESPONDENCE

Statistics on objects depend on a good correspondence between the primitives on each object in the training or target set with the others. For example, it would not do to have the position or normal direction at a fingertip on one hand object in the set being considered as the corresponding GOP to that at the knuckle in another hand object in the set. Putting things in correspondence can be understood as a reparameterization of the object representation. But how can this correspondence be determined? Roughly, it can be done in a geometric way or a statistical way.

The geometric way of producing correspondence works by finding the reparameterization on objects that brings them closest to each other in ambient space. This requires a metric between a pair of objects. For example, this could be the L^2 norm of the GOP differences between corresponding positions, with the integration across the object (e.g., its boundary or its skeletal surface). This metric is best if it recognizes that the objects lie on a curved manifold. With such a metric, one can find the correspondence between two objects by reparameterizing the second to minimize the metric between the two. This approach can be applied between all pairs of objects in the training set, or it can be applied between a reference object and each of the objects in the training set. Such a method is discussed in section 6.5.1 below.

The statistical means of producing correspondence is based on properties of the whole set of training objects, in particular on the probability distribution estimated from those objects. The concept is that miscorrespondence widens that distribution, so correspondence optimization should involve reparameterizing each object in the set so as to optimally narrow the distribution, while each object's GOP is still a good descriptor of the whole object (Fig. 6.12). Such methods are discussed in section 6.5.2 below.

6.5.1 Correspondence via Reparameterization-Insensitive Metrics

As discussed earlier in section 6.2.3, [26] produced a method for objects in 2D that allowed a metrics between equivalence classes of objects over reparameterizations. The

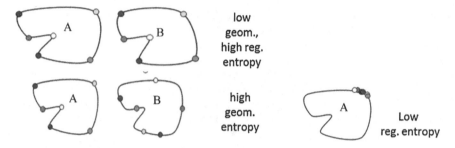

Figure 6.12 Correspondence: good (low geometric entropy) vs. bad (high geometric entropy); irregular sampling (low regularity entropy) vs. regular sampling (high regularity entropy).

mathematics required that the comparison be over derivatives of the object boundary. It follows that Kurtek's [28] generalization to 3D had to be over normal directions on the boundary. These metrics are seen to be invariant with respect to reparameterization, which allows them to be easily adapted to the quotient space under reparameterizations. That gives a notion of shape as orbits, i.e., equivalence classes with respect to reparameterization.

The method for computing the metrics of the difference between two objects described in section 6.2.3 involves picking a particular parametrization of one of the objects and finding the parametrization of the other yielding the closest field of surface normals (together with a global rotation). It thereby yields a correspondence between the locations on the surfaces of the object pair.

Finding the Fréchet mean equivalence class, and a central representer of the class gives a template mean representative. Their method allows a distance to be calculated between a reference object, e.g., the template mean, and each object in the training set. As just described, it allows a distance to be calculated between objects, so the multidimensional scaling approach can be used.

6.5.2 Correspondence via Entropy Minimization

The tightness of a probability distribution can be measured in many ways, but the ones that have turned out to be the most effective are based on information theory. Taylor and his team [9] pioneered a form based on minimum description length (MDL), and an almost equivalent form was based on entropy, as developed in Whitaker's laboratory [5]. The basic idea is to fit the population of GOPs by a Gaussian distribution and minimize its entropy over all object reparameterizations.

The entropy of an n-dimensional Gaussian distribution is given by $n[1/2\ln(2\pi+1)+$ the mean of the principal variances]. The difficulty is that when the idea is applied to a probability distribution estimated from data by a PCA or PCA-like method, the successive sorted principal variances get successively smaller and eventually are dominated

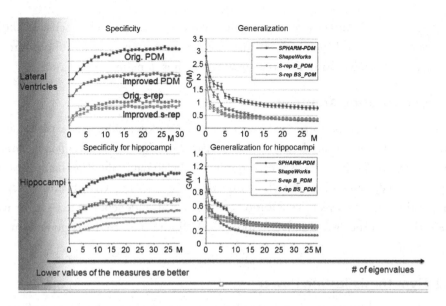

Figure 6.13 Generalization and specificity improvements due to s-rep-based and PDM-based correspondence optimization.

by noise in the data, and each negative logarithm of these small principal variances is very large, so these noise-induced principal variances dominate the population principal variances in the entropy calculation. Cates et al. [5] dealt with this problem by adding a small constant to each principal variance. Tu et al. [48] dealt with it by cutting off the series when the ratio of the principal variance to the total variance fell below a certain threshold.

Cates et al. [5] showed that entropy minimization for GOPs that were a boundary PDM could yield improvements in hypothesis testing significances. In their work boundary points were slid along the boundary to minimize entropy. Tu et al. [48] developed the same idea for s-reps, whereby the skeletal spokes were slid along the skeletal surface to minimize entropy. They showed (Fig. 6.13) the surprising result that, according to the measures described in section 6.6.3 on a training set of hippocampi, when the s-rep-based correspondence method was applied, the boundary points at the ends of the spokes had superior statistical properties (specificity and generalization, see section 6.6.3) than those produced by the Cates boundary point shifting method.

6.6 HOW TO COMPARE REPRESENTATIONS AND STATISTICAL METHODS

Given all of the different shape representations discussed in Section 6.2, and the analytic approaches discussed in Sections 6.3 and 6.4, an important issue is comparison of them.

One basis for this is knowledge of how they work, which can lead to sensible choices in the wide variety of shape problems that one may encounter. But it is also interesting to consider quantitative bases for comparison, which is done in this section.

6.6.1 Classification Accuracy

When a data set is naturally grouped into 2 (or perhaps more) subsets, e.g., pathology vs. normal controls, various methods can be compared on the basis of *classification accuracy*. The classification problem starts with a group of labeled data called the *training set*, and the goal is to develop a rule for classifying new observations. Classification accuracy is simply the rate of correct classifications, either for an independent test set, or using some variation of the cross-validation idea. See [10] for access to the large literature on classification.

6.6.2 Hypothesis Testing Power

A related approach to comparing shape methodologies, again based on two well labeled subgroups in the data, is to construct a formal hypothesis test for the difference between the groups. Quantification of the difference then follows from the level of statistical significance, allowing a different type of comparison of the relative merits of various approaches to the analysis. This was done to good effect in [40].

6.6.3 Specificity, Generalization, Compactness

In object statistics work in medicine two common measures of a probability distribution derived from data via its eigenmodes are specificity and generalization [9]. Specificity is a measure of how well the estimated probability distribution represents only valid instances of the object. It is computed as the average distance between random samples in the computed shape space with their nearest members of the data. Generalization is a measure of how close new instances of the object are to the probability distribution estimated from the training cases. It is calculated by computing a shape space, spanned by the eigenmodes, on all but one of the training cases and computing the distance between the last shape and its projection onto this shape space. Thus both specificity and generalization are measured in units of GOP differences, e.g., positional differences when the GOP is a position tuple or normal direction differences when the GOP is a normal direction.

Compactness is a measure of the tightness of a probability distribution. The entropy of the distribution or the determinant of its covariance matrix (total variance) are often used.

6.6.4 Compression Into Few Modes of Variation

The Euclidean PCA decomposition of data into modes of variation is also usefully understood from a signal processing viewpoint. In the case of a low dimensional signal, in the presence of noise (which is high dimensional by its definition of spreading energy across the spectrum), PCA provides a data driven basis (in the sense of linear algebra) that puts as much of the low dimensional signal as possible into a few basis elements with largest variance. Many of the gains in statistical efficiency, such as those discussed in Section 6.4, can be understood in terms providing better signal compression in these terms.

However, this analogy fails in the presence of even moderate noise in Euclidean data, and the problem appears even more strongly in non-Euclidean contexts such as shape analysis. In particular, the standard Euclidean assumption of all of the signal being present on the first few eigenvalues is usually misleading because both nonlinear shape signals and noise typically spread some signal power among many eigenvalues. The result is that very noisy data can be measured to require fewer eigenmodes to achieve a given fraction of total variance than less noisy data. Hence standard *dimension reduction* approaches, based on "total signal power" or "percent of variation explained", are usually inappropriate.

Yet there is still a natural desire to think in terms of *effective dimensionality*, i.e., the concept that the true underlying signal has much less variation than is present in noisy data. Ideas based in *random matrix theory* are promising avenues for research in this direction. See [51] for good discussion of using this powerful theory in the context of Euclidean PCA. A first important part of this theory is the asymptotic probability distribution of the full collection of eigenvalues under a pure noise model, called the Marčenko–Pastur distribution [29]. Second is the corresponding limiting distribution of the largest eigenvalue, called the Tracey–Widom distribution [46].

6.6.5 Quality in Application, Esp. Segmentation

One more approach to comparing shape analysis methods is to study their impact when used for various applications. An important application has been to the *segmentation* problem in medical image analysis, where the goal is to find the region of an image occupied by an object such as a particular organ. A series of successive improvements in shape analysis resulting in improved image segmentation can be found in Pizer et al. [34–36], Fletcher et al. [13], Rao et al. [37], Gorczowski et al. [16], and Vicory et al. [49,50].

6.7 RESULTS OF CLASSIFICATION, HYPOTHESIS TESTING, AND PROBABILITY DISTRIBUTION ESTIMATION

This section reviews some recent work, with a number of specific applications of the above ideas.

6.7.1 Generalized Rotations

Schulz et al. [39] studied statistics on generalized rotations as characterized by boundary normal directions at a small number of boundary locations. The generalized rotations they considered were global rotation, bending, and twisting. They showed that given a collection of cases with one of these rotations but with statistically varying angles of rotation about a fixed axis, they could derive the axis as well as variance of the rotation angle. The axis derivation was based on the realization that the data fell on coaxial small circles on the sphere of directions (Fig. 6.14). Next, they studied compositions of two of these types of generalized rotations, each about its own axis. Finally, they developed hypothesis tests for establishing statistical significance of variation in these directions and used it for new scientific insights into human knee movement.

6.7.2 Classification of Schizophrenia via Hippocampus S-reps

Hong et al. [20], compared PDMs vs. s-reps and Euclideanization vs. direct Euclidean analysis of the ambient space coordinate values in classifying a hippocampus as to whether it was from a typical individual or from a first-episode schizophrenic. They showed that, according to areas under the Receiver Operating Characteristic curve (ROC) [19], Euclideanizing boundary PDMs produced better classification than without Euclideanization and that Euclideanized s-reps produced better classification than either of the boundary-based analyses (Fig. 6.15). They also showed the usefulness of displaying the variation of the object along the vector in the Euclideanized feature space passing through the pooled mean and in the separation direction of the classes (Fig. 6.16).

6.7.3 Hypothesis Testing via S-reps

Schulz et al. [40] demonstrated the ability to test hypotheses on shape variation between two objects using s-rep features. They showed how to do not only global hypothesis tests but also GOP-by-GOP and location-by-location tests. The method involved permutation tests that recognize that means benefit from being backward means and that GOP differences need to use a metric appropriate for curved manifolds. This work reported a new method for compensating for the correlations of these GOPs. With this approach they were able to analyze which locations on hippocampi of first-episode schizophrenics had statistically significant GOP differences from controls, and which GOPs had those statistically significant differences. Thus they found important shape differences between the classes.

6.7.4 Shape Change Statistics

Vicory [50] studied statistics on the change of object shape between two stages. Realizing that statistics on change requires transporting each object pair such that the starting

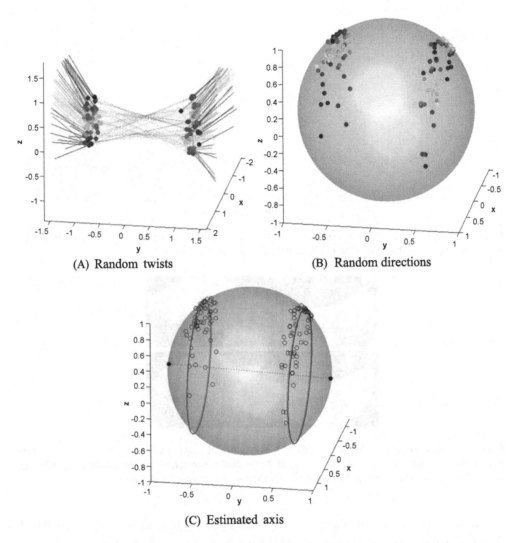

(A) Random twists   (B) Random directions

(C) Estimated axis

Figure 6.14 Random twists applied to two normals, those normal on the 2-sphere, and the twist axis estimated from the co-axial circles.

object was at a fixed place on the manifold of objects, he proposed solving this problem by pooling all $2n$ objects in the n shape pairs in the training set, producing a polar system by PNS on the pool of GOP tuples, Euclideanizing each object according to that polar system, and then producing Euclidean (ordinary) differences between the Euclideanized features of each pair.

Vicory studied this method in two applications. The first was on ellipsoids that were randomly bent and/or twisted. The GOPs in this study were the boundary point co-

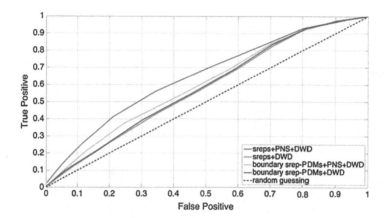

Figure 6.15 ROCs for classifying hippocampi as to schizophrenia vs. normal using s-reps and PDMs, each with original features and their Euclideanized counterparts.

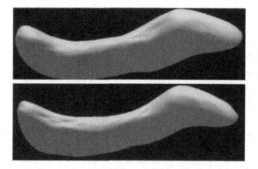

Figure 6.16 Boundaries implied by s-reps of hippocampi at (top) 2 standard deviations from the pooled mean in one direction and (bottom) at 2 standard deviations from the pooled mean in the other direction, for classification between first episode schizophrenics and controls.

ordinates provided by spherical harmonic analysis. He found that when the simulated data was produced from random samples from a single transformation type, analysis either with or without Euclideanization yielded a single correct eigenmode, but the analysis with Euclideanization allowed more accurate estimation of an object's bending or twisting angle and, when the mode of variation was visualized, the object moved more naturally than using PCA alone. He also found that when the simulated data came from random bending cascaded with twisting, both types of analysis yielded two modes of variation but Euclideanization allowed a more correct estimate of the relative variances and mean deformations closer to the expected deformations than their PCA-alone counterparts.

Vicory's second test was on prostates segmented from MRI and the same prostate within 3D transrectal ultrasound; the latter was deformed by the ultrasound transducer.

The shape change eigenmodes were used as a shape space in which to segment the prostate from the ultrasound image, given the prostate shape of the patient derived from his MRI. The GOPs he used were the skeletal points, spoke lengths, and spoke directions from fitted s-reps. He found that using the shape change space resulting from Euclideanization followed by subtraction yielded more accurate segmentations than when a shape space was formed by s-rep feature differences applied to a mean of the prostates in the training MRIs and then CPNS was applied to the resulting objects. For a target segmentation using that second method, the feature difference in prostate from the patient's MRI from the MRI-based training mean was applied to the ultrasound mean, which became the starting point for the segmentation within the scale space.

Recently Hong is applying this shape change Euclideanization approach to two classes of pairs of shapes, at two different ages. Pooling all 4 of his hippocampus objects to yield a Euclideanization is yielding informative shape differences between children at high risk for autism who do not develop autistic symptoms and those at high risk who do develop such symptoms.

6.7.5 Tu Correspondence Evaluation

Tu et al. [48] compared analysis approaches through focusing simultaneously on specificity, generalization and compactness. Measuring according to specificity and generalization, they compared probability distributions derived from boundary PDMs on hippocampi in multiple ways. Briefly, they showed notable improvements in specificity and generalization (Fig. 6.13) when improving PDMs' correspondence derived via spherical harmonics by Cates' entropy minimization method and, as compared to the results of correspondence improvements via PDMs, notable improvements in specificity with little change in generalization when improving s-reps fitted from a common reference s-rep.

6.8 CONCLUSIONS

We have shown that object shapes can be considered to reside on curved manifolds and have presented a number of methods for representing the objects and calculating statistics that recognize the manifold curvature. We have presented methods for comparing these representations and statistical methods and have given evidence that doing the statistics recognizing the manifold curvature frequently provides more effective analysis. We have also showed improvements when the object representation captures aspects of boundary normal directions and object widths. The study of how to do statistics on curved manifolds is an active area of research, and we anticipate numerous advances in the next few years.

ACKNOWLEDGEMENTS

We are grateful for contributions to this paper by Hyo-Young Choi, Benjamin Eltzner, Thomas Fletcher, Junpyo Hong, Thomas Hotz, Sungkyu Jung, Xavier Pennec, Jörn Schulz, Stefan Sommer, Anuj Srivastava, Liyun Tu, Stephan Huckemann, and Jared Vicory.

REFERENCES

[1] M.F. Beg, M.I. Miller, A. Trouvé, L. Younes, Computing large deformation metric mappings via geodesic flows of diffeomorphisms, Int. J. Comput. Vis. 61 (2) (2005) 139–157.

[2] R. Bhattacharya, V. Patrangenaru, Large sample theory of intrinsic and extrinsic sample means on manifolds. I, Ann. Stat. (2003) 1–29.

[3] R. Bhattacharya, V. Patrangenaru, Large sample theory of intrinsic and extrinsic sample means on manifolds: II, Ann. Stat. (2005) 1225–1259.

[4] H. Blum, Biological shape and visual science (Part I), J. Theor. Biol. 38 (2) (1973) 205–287.

[5] J. Cates, P.T. Fletcher, M. Styner, M. Shenton, R. Whitaker, Shape modeling and analysis with entropy-based particle systems, in: Proc. Inf. Proc. Med. Imag., vol. 20, 2007, pp. 333–345.

[6] D.H. Cooper, T.F. Cootes, C.J. Taylor, J. Graham, Active shape models—their training and application, Comput. Vis. Image Underst. 61 (1995) 38–59.

[7] T.F. Cootes, C.J. Taylor, Active shape model search using local grey-level models: a quantitative evaluation, in: J. Illingworth (Ed.), Proc. British Machine Vision Conference, BMVA Press, 1993, pp. 639–648.

[8] J. Damon, J.S. Marron, Backwards principal component analysis and principal nested relations, J. Math. Imaging Vis. 50 (1–2) (2014) 107–114.

[9] R.H. Davies, C.J. Twining, P.D. Allen, T.F. Cootes, C.J. Taylor, Building optimal 2D statistical shape models, Image Vis. Comput. 21 (2003) 1171–1182.

[10] R.O. Duda, P.E. Hart, D.G. Stork, Pattern Classification, John Wiley & Sons, 2012.

[11] B. Eltzner, S. Jung, S. Huckemann, Dimension reduction on polyspheres with application to skeletal representations, in: International Conference on Networked Geometric Science of Information, Springer International Publishing, October 2015, pp. 22–29.

[12] N.I. Fisher, Statistical Analysis of Circular Data, Cambridge University Press, 1995.

[13] P.T. Fletcher, C. Lu, S.M. Pizer, S. Joshi, Principal geodesic analysis for the study of nonlinear statistics of shape, IEEE Trans. Med. Imaging 23 (8) (2004) 995–1005.

[14] P.T. Fletcher, Geodesic regression and the theory of least squares on Riemannian manifolds, Int. J. Comput. Vis. 105 (2) (2013) 171–185.

[15] M. Fréchet, Les éléments aléatoires de nature quelconque dans un espace distancié, Ann. Inst. Henri Poincaré 10 (4) (1948) 215–310.

[16] K. Gorczowski, M. Styner, J.Y. Jeong, J.S. Marron, J. Piven, H.C. Hazlett, S.M. Pizer, G. Gerig, Statistical shape analysis of multi-object complexes, in: 2007 IEEE Conference on Computer Vision and Pattern Recognition, IEEE, June 2007, pp. 1–8.

[17] J.C. Gower, Some distance properties of latent root and vector methods used in multivariate analysis, Biometrika 53 (3–4) (1966) 325–338.

[18] P. Hall, J.S. Marron, A. Neeman, Geometric representation of high dimension, low sample size data, J. R. Stat. Soc., Ser. B, Stat. Methodol. 67 (3) (2005) 427–444.

[19] J.A. Hanley, B.J. McNeil, The meaning and use of the area under a receiver operating characteristic (ROC) curve, Radiology 143 (1) (1982) 29–36.

[20] J. Hong, J. Vicory, J. Schulz, M. Styner, J. Marron, S.M. Pizer, Classification of medically imaged objects via s-rep statistics, Med. Image Anal. 31 (2016) 37–45.

[21] T. Hotz, Extrinsic vs intrinsic means on the circle, in: F. Nielsen, F. Barbaresco (Eds.), Geometric Science of Information, in: Lect. Notes Comput. Sci., vol. 8085, Springer-Verlag, 2013, pp. 433–440.

[22] S. Huckemann, T. Hotz, A. Munk, Intrinsic shape analysis: geodesic principal component analysis for Riemannian manifolds modulo Lie group actions. Discussion paper with rejoinder, Stat. Sin. 20 (2010) 1–100.

[23] I. Jolliffe, Principal Component Analysis, John Wiley & Sons, Ltd., 2002.

[24] S. Jung, I.L. Dryden, J.S. Marron, Analysis of principal nested spheres, Biometrika 99 (3) (2012) 551–568.

[25] M. Kazhdan, T. Funkhouser, S. Rusinkiewicz, Rotation invariant spherical harmonic representation of 3d shape descriptors, in: Symposium on Geometry Processing, vol. 6, June 2003, pp. 156–164.

[26] E. Klassen, A. Srivastava, M. Mio, S.H. Joshi, Analysis of planar shapes using geodesic paths on shape spaces, IEEE Trans. Pattern Anal. Mach. Intell. 26 (3) (2004) 372–383.

[27] J.J. Koenderink, Solid Shape, MIT Press, 1990.

[28] S. Kurtek, A. Srivastava, E. Klassen, H. Laga, Landmark-guided elastic shape analysis of spherically-parameterized surfaces, Comput. Graph. Forum 32 (2, pt. 4) (May 2013) 429–438.

[29] V.A. Marčenko, L.A. Pastur, Distribution of eigenvalues for some sets of random matrices, Math. USSR Sb. 1 (4) (1967) 457.

[30] K.V. Mardia, Statistics of Directional Data, Academic Press, 2014.

[31] J.S. Marron, M.J. Todd, J. Ahn, Distance-weighted discrimination, J. Am. Stat. Assoc. 102 (480) (2007) 1267–1271.

[32] V. Patrangenaru, L. Ellingson, Nonparametric Statistics on Manifolds and Their Applications to Object Data Analysis, 2015.

[33] X. Pennec, Barycentric subspace analysis on manifolds, arXiv preprint arXiv:1607.02833, 2016.

[34] S.M. Pizer, D.S. Fritsch, P.A. Yushkevich, V.E. Johnson, E.L. Chaney, Segmentation, registration, and measurement of shape variation via image object shape, IEEE Trans. Med. Imaging 18 (10) (1999) 851–865.

[35] S.M. Pizer, S. Joshi, P.T. Fletcher, M. Styner, G. Tracton, J.Z. Chen, Segmentation of single-figure objects by deformable M-reps, in: International Conference on Medical Image Computing and Computer-Assisted Intervention, Springer, Berlin, Heidelberg, October 2001, pp. 862–871.

[36] S.M. Pizer, S. Jung, D. Goswami, J. Vicory, X. Zhao, R. Chaudhuri, J.N. Damon, S. Huckemann, J.S. Marron, Nested sphere statistics of skeletal models, in: Innovations for Shape Analysis, Springer, Berlin, Heidelberg, 2013, pp. 93–115.

[37] M. Rao, J. Stough, Y.Y. Chi, K. Muller, G. Tracton, S.M. Pizer, E.L. Chaney, Comparison of human and automatic segmentations of kidneys from CT images, Int. J. Radiat. Oncol. Biol. Phys. 61 (3) (2005) 954–960.

[38] B. Schölkopf, A.J. Smola, Learning with Kernels: Support Vector Machines, Regularization, Optimization, and Beyond, MIT Press, 2002.

[39] J. Schulz, S. Jung, S. Huckemann, M. Pierrynowski, J.S. Marron, S.M. Pizer, Analysis of rotational deformations from directional data, J. Comput. Graph. Stat. 24 (2) (2015) 539–560.

[40] J. Schulz, S.M. Pizer, J.S. Marron, F. Godtliebsen, Non-linear hypothesis testing of geometric object properties of shapes applied to hippocampi, J. Math. Imaging Vis. 54 (1) (2016) 15–34.

[41] S.K. Sen, M. Foskey, J.S. Marron, M.A. Styner, Support vector machine for data on manifolds: an application to image analysis, in: 5th IEEE International Symposium on Biomedical Imaging: From Nano to Macro, 2008, ISBI 2008, IEEE, May 2008, pp. 1195–1198.

[42] K. Siddiqi, S. Pizer (Eds.), Medial Representations: Mathematics, Algorithms and Applications, vol. 37, Springer Science & Business Media, 2008.

[43] S. Sommer, Anisotropic distributions on manifolds: template estimation and most probable paths, in: S. Ourselin, et al. (Eds.), Information Processing in Medical Imaging, in: Lect. Notes Comput. Sci., vol. 9123, Springer, 2015, pp. 193–204.

[44] A. Srivastava, E. Klassen, S.H. Joshi, I.H. Jermyn, Shape analysis of elastic curves in Euclidean spaces, IEEE Trans. Pattern Anal. Mach. Intell. 33 (7) (2011) 1415–1428.

[45] W.S. Torgerson, Multidimensional scaling: I. Theory and method, Psychometrika 17 (4) (1952) 401–419.

[46] C.A. Tracy, H. Widom, Level-spacing distributions and the Airy kernel, Commun. Math. Phys. 159 (1) (1994) 151–174.

[47] L. Tu, D. Yang, J. Vicory, X. Zhang, S.M. Pizer, M. Styner, Fitting skeletal object models using spherical harmonics based template warping, IEEE Signal Process. Lett. 22 (12) (2015) 2269–2273.

[48] L. Tu, M. Styner, J. Vicory, B. Paniagua, J.C. Prieto, D. Yang, S.M. Pizer, Skeletal shape correspondence via entropy minimization, in: SPIE Medical Imaging, International Society for Optics and Photonics, March 2015, pp. 94130U.

[49] J. Vicory, M. Foskey, A. Fenster, A. Ward, S.M. Pizer, Prostate segmentation from 3DUS using regional texture classification and shape differences, in: Symposium on Statistical Shape Models & Applications, June 2014, p. 24.

[50] J. Vicory, Shape Deformation Statistics and Regional Texture-Based Appearance Models for Segmentation, PhD dissertation, University of North Carolina at Chapel Hill, 2016 (on the internet at http://midag.cs.unc.edu, PhD dissertations button).

[51] J. Yao, Z. Bai, S. Zheng, Large Sample Covariance Matrices and High-Dimensional Data Analysis (No. 39), Cambridge University Press, 2015.

CHAPTER 7

Shape Modeling Using Gaussian Process Morphable Models

Marcel Lüthi, Andreas Forster, Thomas Gerig, Thomas Vetter
Department of Mathematics and Computer Science, University of Basel, Switzerland

Contents

7.1	Introduction	165
7.2	Shape Modeling Using Gaussian Processes	166
	7.2.1 Classical Statistical Shape Models Revisited	167
	7.2.2 Gaussian Process Morphable Models	168
	7.2.3 Defining Gaussian Processes	169
	7.2.3.1 Learning Deformations from Data	*169*
	7.2.3.2 Modeling Smooth Deformations	*170*
	7.2.3.3 Combining Kernels	*170*
	7.2.3.4 Modeling Symmetric Variations	*171*
7.3	Non-Rigid Registration Using Gaussian Process Priors	172
	7.3.1 Hybrid Landmark and Surface Registration	173
7.4	Case Study: Building a Statistical Shape Model of the Skull	174
	7.4.1 Data and Experimental Setup	174
	7.4.2 Modeling Shape Priors for the Skull	175
	7.4.2.1 Modeling Smooth Deformation Using a Single Gaussian Kernel	*175*
	7.4.2.2 Modeling Deformations on Multiple Scale Levels	*177*
	7.4.2.3 Incorporating Symmetries into the Prior	*178*
	7.4.2.4 Quantitative Comparison of Different Prior Models	*179*
	7.4.3 Skull-Registration Using Model Fitting	181
	7.4.4 Computing the Statistical Shape Model	182
7.5	Modeling and Analyzing Pathologies	184
	7.5.1 Simultaneous Fitting of Healthy and Pathological Deformations	184
	7.5.2 Building a Model for Pathologies	185
	7.5.3 Experiment	186
7.6	Conclusion	186
	Appendix 7.A	188
	7.A.1 Approximating Eigenfunctions Using the Nyström Method	188
	7.A.2 Obtaining the Ground-Truth	189
	References	190

7.1 INTRODUCTION

Statistical shape models (SSMs) are a well established tool in medical image analysis [9]. The core idea is that a statistical shape model represents the normal shape variation of

Statistical Shape and Deformation Analysis
DOI: 10.1016/B978-0-12-810493-4.00008-0

a class of shapes, which is then used as prior knowledge in an algorithm. The most important examples of statistical shape models are the Active Shape Model [5] and the Morphable model [4], which learn the shape variation from given training examples, and represent the shape variation using the leading principal components. These models are linear, parametric models with mathematically convenient properties. As they can only represent shapes that are in the linear span of the given training examples, they lead to algorithms that are robust towards artifacts and noise. The downside of this specificity is that to learn an expressive model (i.e. one that can express all possible target shapes), a lot of training data is needed. Lüthi et al. have recently introduced a generalization of these classical statistical shape models, called Gaussian Process Morphable Models (GPMM), which use the mathematical framework of Gaussian processes to model shape variations [13]. GPMMs make it possible to construct expressive shape priors using analytically defined covariance functions, even when there are no or only few example shapes available to learn the shape variations from.

The main goal of this chapter is to give an introduction to GPMMs and to show that with this not too complicated concept, we can solve many shape modeling problems in a unified way, which previously required specialized algorithms. As a theoretical contribution we introduce a novel covariance function, which allows us to formulate GPMMs that enforce axial symmetric shape deformations. We will illustrate the use of GPMMs in an extended practical use case, which addresses the problem of building a skull model from a set of noisy skull surfaces. Our intention is that this use case does not only illuminate the concepts, but also serves as a practical guide on how to do shape modeling using GPMMs. In particular we want to illustrate the main considerations that need to be taken into account when we design prior models, how we can visualize and quantitatively asses the model quality and how we can choose the free parameters.

In the final section of this article, we will also introduce a novel application of GPMMs, which is the analysis of pathological shapes using a statistical shape model. The main idea is that we combine a classical statistical shape model, which is learned from example data, with an analytically defined model for the pathology. This combined model is fitted to the pathological shape and subsequently divided into the anatomically normal part, represented by the classical shape model, and the pathological deformation. This approach does not only make it possible to use a statistical shape model to fit pathological shapes, but also provide us with the most likely anatomical normal shape.

7.2 SHAPE MODELING USING GAUSSIAN PROCESSES

Gaussian process morphable models are a generalization of classical (i.e. PCA-based) statistical shape models. We therefore start by revisiting the basic concepts of statistical shape models, before discussing how these models can be interpreted using the formalism of Gaussian processes.

7.2.1 Classical Statistical Shape Models Revisited

PCA-based statistical shape models assume that the space of all possible shape deformations can be learned from a set of typical example shapes $\{\Gamma_1, \ldots, \Gamma_n\}$. Each shape Γ_i is represented as a discrete set of landmark points, i.e.

$$\Gamma_i = \{x_k^i \mid x_k \in \mathbb{R}^3, k = 1, \ldots, N\},$$

where N denotes the number of landmark points. In early approaches, the points typically denoted anatomical landmarks, and N was consequently small (in the tens). Most modern approaches use a dense set of points to represent the shapes. In this case, the number of points is typically in the thousands. The crucial assumption is that the points are in correspondence among the examples. This means that the k-th landmark point x_k^i and x_k^j of two shapes Γ_i and Γ_j represents the same anatomical point of the shape. These corresponding points are either defined manually, or automatically determined using a registration algorithm. To build the model, a shape Γ_i is represented as a vector $\vec{s}_i \in \mathbb{R}^{3N}$, where the x-, y-, z-components of each point are stacked into a large vector:

$$\vec{s}_i = (x_{1x}^i, x_{1y}^i, x_{1z}^i, \ldots, x_{Nx}^i, x_{Ny}^i, x_{Nz}^i).$$

This vectorial representation makes it possible to apply standard multivariate statistics to model a probability distribution over shapes. The usual assumption is that the shape variations can be modeled using a normal distribution

$$\vec{s} \sim \mathcal{N}(\vec{\mu}, \Sigma)$$

where the mean $\vec{\mu}$ and covariance matrix Σ are estimated from the example data:

$$\vec{\mu} = \bar{s} := \frac{1}{n} \sum_{i=1}^{n} \vec{s}_i \tag{7.1}$$

$$\Sigma = S := \frac{1}{n-1} \sum_{i=1}^{n} (\vec{s}_i - \bar{s})(\vec{s}_i - \bar{s})^T. \tag{7.2}$$

As the number of points N is usually large, the covariance matrix Σ cannot be represented explicitly. Fortunately, as it is determined completely by the n example datasets, it has at most rank n and can therefore be represented using at most n basis vectors. The basis vectors are found by performing a Principal Component Analysis (PCA) [11]. In its probabilistic interpretation, PCA leads to a model of the form

$$\vec{s} = \bar{s} + \sum_{i=1}^{n} \alpha_i \sqrt{d_i} \vec{v}_i \, \alpha_i \sim N(0, 1) \tag{7.3}$$

where $(\vec{v}_i, d_i)$, $i = 1, \ldots, n$, are the eigenvectors and eigenvalues of the covariance matrix Σ [23]. It is easy to check that under these assumptions, $\vec{s} \sim \mathcal{N}(\vec{s}, S)$. Thus, we have an efficient, parametric representation of the distribution.

7.2.2 Gaussian Process Morphable Models

The literature of PCA-based statistical shape models usually emphasizes the point that it is the shape that is modeled. Eq. (7.3) however, gives rise to a different interpretation: A statistical shape model is a model of deformations $\vec{\phi} = \sum_{i=1}^{n} \alpha_i \sqrt{d_i} \vec{v}_i \sim \mathcal{N}(0, S)$, with which the mean shape $\vec{s}$ is deformed. The probability distribution is on the deformations. This is the interpretation we use when we generalize these models to define Gaussian Process Morphable Models. We define a probabilistic model directly on the deformations. To stress that we are modeling deformations (i.e. vector fields defined on the reference domain Γ_R), and to become independent of the discretization, we model the deformations as a Gaussian process.

Let $\Gamma_R \subset \mathbb{R}^3$ be a reference shape and denote by $\Omega \subset \mathbb{R}^3$ a domain, such that $\Gamma_R \subseteq \Omega$. We define a Gaussian process $u \in GP(\mu, k)$ with mean function $\mu : \Omega \to \mathbb{R}^3$ and covariance function $k : \Omega \times \Omega \to \mathbb{R}^{3 \times 3}$. Note that any deformation $\hat{u}$ sampled from $GP(\mu, k)$, gives rise to a new shape Γ by warping the reference shape Γ_R:

$$\Gamma = \{x + \hat{u}(x) \mid x \in \Gamma_R\}.$$

Similar to the PCA representation of a statistical shape model used in (Eq. (7.3)), a Gaussian process $GP(\mu, k)$ can be represented in terms of an orthogonal set of basis functions $\{\phi_i\}_{i=1}^{\infty}$

$$u(x) \sim \mu(x) + \sum_{i=1}^{\infty} \alpha_i \sqrt{\lambda_i} \phi_i(x), \quad \alpha_i \in \mathcal{N}(0, 1), \tag{7.4}$$

where (λ_i, ϕ_i) are the eigenvalue/eigenfunction pairs of the integral operator

$$T_k f(\cdot) := \int_{\Omega} k(x, \cdot) f(x) \, d\rho(x),$$

where $\rho(x)$ denotes a measure. The representation (7.4) is known as the Karhunen–Loève expansion of the Gaussian process [3]. Since the random coefficients α_i are uncorrelated, the variance of u is given by the sum of the variances of the individual components. Consequently, the eigenvalue λ_i corresponds to the variance explained by the i-th component. This suggests that if the λ_i decay sufficiently quickly, we can accurately approximate the process using the first r components only:

$$\tilde{u}(x) \sim \mu(x) + \sum_{i=1}^{r} \alpha_i \sqrt{\lambda_i} \phi_i(x). \tag{7.5}$$

The expected error of this approximation is given by the tail sum

$$\sum_{i=r+1}^{\infty} \lambda_i.$$

The resulting model is a finite dimensional, parametric model, similar to a standard statistical shape model. Note, however, that there is no restriction that the covariance function k needs to be the sample covariance matrix. Any valid positive definite covariance function can be used.

The question remains how we can compute the eigenfunction/eigenvalue pairs in (7.5). Although for some kernel functions analytic solutions are available (see e.g. [2,8]) for most interesting models we need to resort to numeric approximations. Fortunately, this problem has been widely studied in the literature (see e.g. Flannery et al. Chapter 18 [6]). A classical method, which we use in our implementation, is the Nyström method [17]. This method is discussed in detail in Appendix 7.A.1.

7.2.3 Defining Gaussian Processes

To define a Gaussian process $\mathrm{GP}(\mu, k)$ for modeling shape deformations, we need to define the mean function $\mu : \Omega \to \mathbb{R}^3$ and a covariance function $k : \Omega \times \Omega \to \mathbb{R}^{3 \times 3}$. In the following, we will discuss two important strategies for choosing these parameters. As we are only interested in modeling the deformations on the reference surface Γ_R, we always let $\Omega = \Gamma_R$.

7.2.3.1 Learning Deformations from Data

Similar to classical statistical shape models, we can learn the mean and covariance structure of the models from example data. Let $\{\Gamma_1, \dots, \Gamma_n\}$ be the example surfaces and $\{u_1, \dots, u_n\}, u_i : \Gamma_R \to \mathbb{R}^d$ denote the corresponding deformation fields, which establish correspondence between the reference Γ_R and the respective surface, i.e.

$$\Gamma_i = \{x + u(x) | x \in \Gamma_R\}.$$

We can define the mean at every point x as the sample mean

$$\mu_{\mathrm{SSM}}(x) = \frac{1}{n} \sum_{i=1}^{n} u_i(x) \tag{7.6}$$

and the covariance function at the points x and x' by the sample covariance

$$k_{\mathrm{SSM}}(x, x') = \frac{1}{n-1} \sum_{i=1}^{n} (u_i(x) - \mu_{\mathrm{SSM}}(x))(u_i(x') - \mu_{\mathrm{SSM}}(x'))^T. \tag{7.7}$$

We refer to this kernel k_{SSM} as the *sample covariance kernel* or *empirical kernel*.

7.2.3.2 Modeling Smooth Deformations

We can also define a Gaussian process model when we have no example shapes. In this case, we usually assume that the chosen reference shape is close to the (hypothetical) average shape of the modeled class of shapes and let the mean deformation be zero

$$\mu(x) = (0, 0, 0)^T.$$

In absence of other prior knowledge, the most basic assumption when modeling shape deformations is that shape deformations vary smoothly. A well known covariance function that enforces smooth functions is the scalar-valued *Gaussian kernel*, defined by

$$g(x, x') = \exp(-\|x - x'\|^2/\sigma^2),$$

where σ^2 defines the range over which the function values at the points x and x' are correlated. The larger the values of σ, the more smoothly varying the resulting deformations fields will be. In order to use this scalar-valued kernel for modeling deformations, we can define a matrix-valued Gaussian kernel as

$$k(x, x') = s \cdot \text{diag}(g) := s \begin{pmatrix} g(x, x') & 0 & 0 \\ 0 & g(x, x') & 0 \\ 0 & 0 & g(x, x') \end{pmatrix}.$$

The diagonal structure of the resulting 3×3 matrix means that the x, y, z components of the modeled vector field are independent. The parameter $s \in \mathbb{R}$ determines the variance (i.e. scale) of a deformation vector. This construction can be generalized by defining the matrix-valued kernel k as

$$k(x, x') = A^T g(x, x')A, \quad A \in \mathbb{R}^{3 \times 3}, \tag{7.8}$$

which would allow us to introduce anisotropic scaling and correlations between the components [14].

7.2.3.3 Combining Kernels

From a mathematical point of view, the only requirement that is needed to define a valid Gaussian process is that the covariance function is a symmetric and positive semi-definite kernel [1]. It is well known that the Gaussian kernel satisfies this property [17]. While it is in general rather difficult to prove that a symmetric function is positive semi-definite, the following rules can be used to construct kernels that have this property [21,14]:

Theorem 1. *Let $g, h : \Omega \times \Omega \to \mathbb{R}$ be two symmetric positive semi-definite kernels and $f : \Omega \to \mathbb{R}$ an arbitrary function. Then the following rules can be used to generate new positive semi-definite kernels:*

1. $k(x, x') = g(x, x') + h(x, x')$
2. $k(x, x') = \alpha g(x, x'), \alpha \in \mathbb{R}_+$
3. $k(x, x') = g(x, x') h(x, x')$
4. $k(x, x') = f(x) f(x')$
5. $B^T h(x, x') B, \ B \in R^{r \times n}$
6. $k(x, x') = k_3(\phi(x), \phi(x')) \ k_3 : \mathbb{R}^n \times \mathbb{R}^n \to \mathbb{R}, \ \phi : \Omega \to \mathbb{R}^n$

Using these simple rules we can easily prove that the covariance functions that we defined above are positive semi-definite.[1] More importantly, however, they allow us to combine existing kernels to build more complicated models. We will make extensive use of these rules in Sections 7.4 and 7.5, in order to define application specific shape priors.

7.2.3.4 Modeling Symmetric Variations

Many biological shapes are mirror symmetrical. It turns out that we can also encode this property directly into a covariance function, and thus obtain shape priors that enforces such symmetries. Without loss of generality, we assume that the mirror plane is orthogonal to the first dimension of the domain over which the Gaussian process is defined and that this plane is positioned at the origin. The mirrored position of a given point can then be expressed by flipping the sign of the first coordinate. We can define a symmetry-kernel over $\mathbb{R}^3$ from a valid scalar-valued kernel $k(\cdot, \cdot)$ as

$$k_{sym}(x, x') = I k(x, x') + \bar{I} k(x, \bar{x}') , \tag{7.9}$$

where I is the 3×3 identity matrix and

$$\bar{I} = \left\{ \begin{matrix} -1 & 0 & 0 \\ 0 & 1 & 0 \\ 0 & 0 & 1 \end{matrix} \right\}, \quad \bar{x}' = \left\{ \begin{matrix} -x'_1 \\ x'_2 \\ x'_3 \end{matrix} \right\} . \tag{7.10}$$

The flipped sign of $\bar{x}'$ reflects the point at the above mentioned mirror plane.[2] The negative sign for a specific diagonal entry in $\bar{I}$ favors opposed oriented deformations for symmetric points along this dimension. A proof that this kernel is positive semi-definite is given in [15].

It turns out that defining a GPMM using the symmetric kernel k_{sym} leads to a model with increased variance compared to the original covariance function k. First, the total

---

[1] Combine rules 4, 1 and 2 for deriving the empirical covariance kernel, and rule 5 for the matrix-valued Gaussian kernel.

[2] This corresponds to the sum over the orbit method for constructing a kernel following this mirror invariance (see [7]).

variance changes for points close to the symmetry plane. Second, the variance parallel to the symmetry plane increases while it is reduced in the perpendicular direction. Exactly on the symmetry plane the perpendicular variance is zero. While the change in perpendicular direction is expected, we correct the increased variance using a correction function $S(x, x')$, which reduces the variance depending on the distance to the symmetry plane:

$$S(x, x') = \begin{Bmatrix} 1 & 0 & 0 \\ 0 & f(x)f(x') & 0 \\ 0 & 0 & f(x)f(x') \end{Bmatrix}, \quad f(x) = \sqrt{\frac{1.0}{2.0 - k(x, x) + k(x, \bar{x})}}. \quad (7.11)$$

The final symmetric kernel is given as:

$$k_{sym}(x, x') = k(x, x') + S(x, x')\bar{I}k(x, \bar{x}'). \quad (7.12)$$

7.3 NON-RIGID REGISTRATION USING GAUSSIAN PROCESS PRIORS

In this section we show how we can use GPMMs as prior models for non-rigid surface registration. The idea is that we define a GPMM of possible shape deformations $u \sim GP(\mu, k)$ to model the shape variability with respect to a reference shape Γ_R. Our main assumption is that we can identify each point $x_R \in \Gamma_R$ to the corresponding point $x_T \in \Gamma_T$ of the target surface Γ_T. The goal of the registration problem is to recover the deformation field u that relates the two surfaces.

To this end, we formulate the problem as a MAP estimate:

$$\arg\max_u p(u|\Gamma_T, \Gamma_R) = \arg\max_u p(u)p(\Gamma_T|\Gamma_R, u), \quad (7.13)$$

where $p(u) \sim GP(\mu, k)$ is a Gaussian process prior over deformation fields and the likelihood function is given by $p(\Gamma_T|\Gamma_R, u) = \frac{1}{Z} \exp(-\eta^{-1}\mathcal{D}[\Gamma_T, \Gamma_R, u])$, where $\mathcal{D}$ is a metric that measures the similarity of two surfaces, $\eta \in \mathbb{R}$ is a weighting parameter and Z is a normalization constant.

In order to find the MAP solution, we reformulate the registration problem as an energy minimization problem. Taking logarithms in (7.13) we arrive at the equivalent minimization problem

$$\arg\min_u \mathcal{D}[\Gamma_R, \Gamma_T, u] - \eta \ln p(u) \quad (7.14)$$

Using the low-rank approximation (Eq. (7.5)) we can restate the problem in the parametric form

$$\arg\min_\alpha \mathcal{D}[\Gamma_R, \Gamma_T, \mu + \sum_{i=1}^{r} \alpha_i \sqrt{\lambda_i}\phi_i] + \eta'\|\alpha\|^2. \quad (7.15)$$

Here we used that the coefficients α in (7.5) are independent and hence $p(u) \propto \exp(-\|\alpha\|^2)$. Denoting the model by

$$\mathcal{M}[\alpha](x) = \mu(x) + \sum_{i=1}^{r} \alpha_i \sqrt{\lambda_i} \phi_i(x)$$

we can write this in the simpler form:

$$\arg\min_{\alpha} \mathcal{D}[\mathcal{M}[\alpha], \Gamma_T] + \eta' \|\alpha\|^2. \tag{7.16}$$

The final registration formulation (7.16) is a parametric optimization problem, which can be approached using standard optimization algorithms. This formulation also makes it clear that the registration problem is really a problem of model fitting, and that all the prior assumptions about possible deformations are represented in the GPMM $\mathcal{M}[\alpha]$.

7.3.1 Hybrid Landmark and Surface Registration

Using Gaussian processes as a prior for surface registration also gives rise to a simple way to include landmark constraints. Let $L_R = \{l_R^1, \ldots, l_R^n)\}$ and $L_T = \{l_T^1, \ldots, l_T^n)\}$ be two sets of corresponding landmarks. Assume that a user has defined a number of landmark points $L_R = \{l_R^1, \ldots, l_R^n)\}$ on a reference shape together with the matching points $L_T = \{l_T^1, \ldots, l_T^n)\}$ on a target surface. These landmarks provide us with known deformations at the matching points, i.e.

$$L = \{(l_R^1, l_T^1 - l_R^1), \ldots, (l_R^n, l_T^n - l_R^n)\}$$
$$=: \{(l_R^1, \hat{u}^1), \ldots, (l_R^n, \hat{u}^n)\}.$$

If we choose the likelihood function as

$$p(L|u) = \prod_{i=1}^{N} \mathcal{N}(u(l_R^{i})|\hat{u}^i, \sigma^2 \mathcal{I}_d),$$

which asserts that the inaccuracies of the landmarks can be modeled as independent Gaussian noise, then the problem is an instance of Gaussian process regression and the posterior distribution $p(u|L)$ is known in closed form [17]. It turns out that the distribution $p(u|L)$ is again a Gaussian process $GP(\mu_p, k_p)$, and its mean μ_p and covariance k_p are given by

$$\mu_p(x) = \mu(x) + K_X(x)^T (K_{XX} + \sigma^2 \mathcal{I})^{-1} \hat{U} \tag{7.17}$$
$$k_p(x, x') = k(x, x') - K_X(x)^T (K_{XX} + \sigma^2 \mathcal{I})^{-1} K_X(x'). \tag{7.18}$$

Here, we defined

$$K_X(x) = (k(x, x_i))_{i=1}^{n} \in \mathbb{R}^{3n \times 3},$$
$$K_{XX} = \left(k(x_i, x_j) \right)_{i,j=1}^{n} \in \mathbb{R}^{3n \times 3n}$$

and

$$\hat{U} = (\hat{u}^1 - \mu(x), \ldots, \hat{u}^n - \mu(x))^T \in \mathbb{R}^{3n}.$$

This implies that the posterior $p(u|L)$ is again a valid GPMM (that is, a shape model), which already includes information about a given target surface. Thus we can use this posterior as the prior in the registration formulation (7.13). This results in a hybrid registration approach for surface registration, where the landmark information is implicitly enforced by the shape prior.

7.4 CASE STUDY: BUILDING A STATISTICAL SHAPE MODEL OF THE SKULL

In this section we show how Gaussian process morphable models can be used in practice. We consider the task of building a statistical shape model of the human skull. There are several aspects that make this task challenging: 1) The skull is a large and complex organ, with many structures that are difficult to model. 2) Due to the limited CT resolution, data segmented from CT usually contains holes and missing data. 3) Most data that we get in practice are images of elderly people, and hence teeth are often missing and dental fillings lead to large metal artifacts. 4) Finally, it is difficult to acquire sufficiently many example shapes to estimate the full shape variability of skulls.

The most important step in building a statistical model of the skull is to establish correspondence between the example shapes, which is performed using surface registration. In order to make this process robust to artifacts in the data, we already need a strong shape prior in this step. We will, of course, use a GPMM for this purpose.

In a first step we build a GPMM and discuss how we specifically tailored the prior to the problem of skull registration, using only analytically known covariance functions. In the second step we use this prior to establish correspondence between surfaces in a non-rigid registration task. In the final step, we learn a statistical shape model from the registered data, and demonstrate how to overcome the problem of not having a sufficiently large dataset to capture all shape variation.

7.4.1 Data and Experimental Setup

As a basis for all our modeling tasks we use a manually segmented, anatomically normal reference skull, which we denote by Γ_R. This reference skull is depicted in Fig. 7.1. An

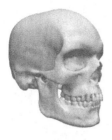 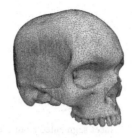 

Figure 7.1 The reference skull used to build the skull models. Cranium and mandible are modeled separately. The cranium mesh has approximately 46,000 vertices and the mandible mesh approximately 26,000.

issue that a good skull model should address is that the mouth can open and close. As this is not conveniently modeled using Gaussian processes, we have separated the cranium from the mandible and built separate models for each part. In this section we will only discuss the model built for the cranium, which is the more complicated of the two structures. The same considerations apply also when building a model of the mandible.

As example data, we use a set of automatically segmented CT images of the skull. Due to the poor quality of the example surfaces, we cannot use them directly to evaluate the quality of our models. In order to generate a ground-truth dataset, we have performed a surface-to-surface registration, which was guided by 16 manually defined landmark points. We then warped the reference skull with the resulting deformation field. A detailed description of how registration was done is given in Appendix 7.A.2. Fig. 7.2 shows typical example surfaces together with our generated ground truth.

For computing the low-rank approximation we choose 700 points on the surface of the reference skull Γ_R, which are used in the Nyström approximation (cf. Appendix 7.A.1). For all the experiments we approximate the original Gaussian process using the 200 leading eigenfunctions.

7.4.2 Modeling Shape Priors for the Skull

The first step is to build a shape prior that can be used in surface registration. In this section we will develop several different prior models, starting from simple smoothness priors to more sophisticated models that also enforce symmetry. For all the models we use a zero mean Gaussian process $GP(\vec{0}, k)$, where $\vec{0} = (0, 0, 0)^T$ is the zero vector and $k : \Gamma_R \times \Gamma_R \to \mathbb{R}^{3 \times 3}$ is a matrix-valued covariance function. As discussed previously, the assumption that the process is zero-mean, implies that we believe that the shape of the reference skull is approximately average.

7.4.2.1 Modeling Smooth Deformation Using a Single Gaussian Kernel

The most basic modeling assumption we can make is that the deformations that relate two shapes of the same shape family are smooth. As discussed in Section 7.2.3.2

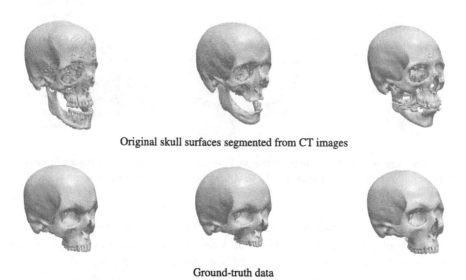

Original skull surfaces segmented from CT images

Ground-truth data

Figure 7.2 Three examples of skull surfaces, which were segmented from CT data, and the corresponding ground-truth shapes, which we use for our evaluation.

smooth deformations can, for example, be modeled using a Gaussian kernel for each component.

$$k_g^{(s,\sigma)}(x, x') = s \operatorname{diag}(\exp(-\frac{\|x - x'\|^2}{\sigma^2})).$$

There are two parameters that we have to define: the smoothness σ and the scale of the deformation s. Both parameters have a natural interpretation and unit (usually mm). The scale directly translates to the variance of the size of the deformation. Hence it is measured in mm^2. Taking into account that with high probability the deformation component is not larger than 3 standard deviations, we have an indication of how to choose the parameter. The parameter σ^2 determines the strength of the correlation between two points x and x'. We know that for the Gaussian kernel, that values farther away than 2σ from the mean are already small, and hence almost uncorrelated. We should therefore choose the parameter such that 2σ correspond to approximately the range of correlations (in mm) that we expect.

An important feature of Gaussian process models is that they are generative, and thus we can visualize samples from the prior to see if our assumptions are reasonable. To visualize the samples, we draw a random deformation field u_i form the Gaussian process model and warp the reference surface with this deformation. More precisely, the new surface is defined by warping all the points of the reference with the deformation field. Fig. 7.3 shows random samples for different parameters of s and σ.

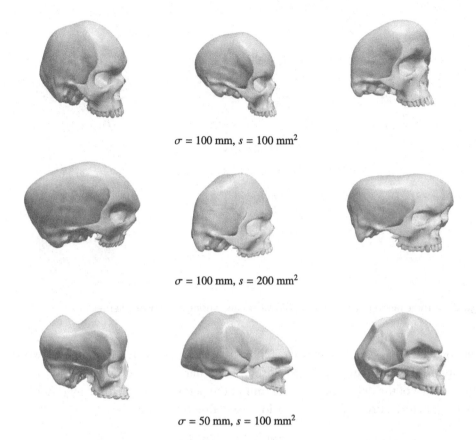

$$\sigma = 100 \text{ mm}, \ s = 100 \text{ mm}^2$$

$$\sigma = 100 \text{ mm}, \ s = 200 \text{ mm}^2$$

$$\sigma = 50 \text{ mm}, \ s = 100 \text{ mm}^2$$

Figure 7.3 Random samples from a GPMMs defined using a Gaussian kernel with different values for scale parameter s and smoothness parameter σ.

From visual inspection, we clearly see that choosing $s = 200$ mm^2 leads to unnaturally big deformations. We therefore prefer to use the value $s = 100$ mm^2 in our model, which seems approximately right. For the smoothness parameter σ^2, the samples with $\sigma = 100$ mm look more natural than those where $\sigma = 50$ mm. On the other hand, such a strong smoothness assumption also implies that the registration algorithm will not be able to explain more detailed shape variations.

7.4.2.2 Modeling Deformations on Multiple Scale Levels

The dilemma that small deformations are needed to explain detailed variations, but that these seem to produce unnaturally looking samples can be resolved by combining smaller and larger deformations in one model. The rules for combining kernels defined in Section 7.2.3 makes it possible to mix properties of different models. In particular, we can sum two covariance functions $g, h : \Gamma \times \Gamma \to \mathbb{R}^{3,3}$ to obtain a new valid covariance

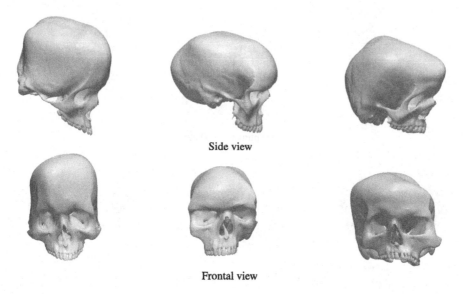

Side view

Frontal view

Figure 7.4 Three random samples from a GPMM defined using a multiscale kernel.

function $k(x, x') = g(x, x') + h(x, x')$. Under this new covariance function k, two deformations $u(x), u(x')$ are correlated, if they are either correlated by $g(x, x')$ or $h(x, x')$. The variance of the deformation is the sum of the deformations defined by g and h. For modeling deformations of our skull prior, we define the kernel

$$k_{\mathrm{ms}}(x, x') = k_g^{(70,100)}(x, x') + k_g^{(30,50)}(x, x').$$

This *multiscale kernel* assumes that large shape deformations are smoothly varying, but still allows for more local, but smaller changes.[3] Note that while the smoothness parameter is a combination of the above defined models, it has the same total variance as the kernels $k_g^{(100,50)}$ and $k_g^{(100,100)}$ defined above. From the random samples in Fig. 7.4 we see that this model produces more natural shape variations, but still allows for flexible deformations.

7.4.2.3 Incorporating Symmetries into the Prior

One important aspect of the skull is not yet represented in our model. Skulls are nearly symmetric about the sagital plane. From the generated samples we see, however, that the models we built so far do not enforce such a symmetry. To incorporate symmetry into the model, we use the method discussed in 7.2.3.4 and define the symmetric kernel

---

[3] This idea of defining multiscale kernels is discussed in more detailed by Opfer et al. [16].

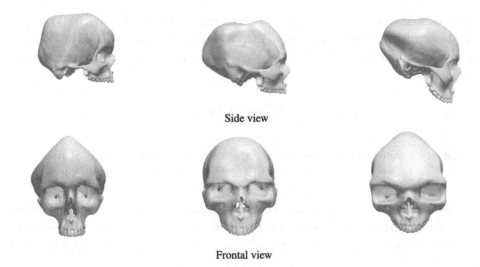

Side view

Frontal view

Figure 7.5 Three random samples from a GPMM defined a symmetric kernel.

k_{psym} from the multiscale kernel k_{ms}[4]:

$$k_{\text{psym}}(x, x') = k_{ms}(x, x') + S(x, x')\bar{I}k_{ms}(x, \bar{x}'),$$

where S and $\bar{x}$ and $\bar{I}$ are defined as in Section 7.2.3.4. The kernel k_{psym} models perfect symmetry, which is usually not given in natural shapes. We therefore also allow for small asymmetric deformations, which we again model using a Gaussian kernel. The final kernel has the form

$$k_{\text{sym}}(x, x') = k_{\text{psym}}(x, x') + \text{diag}(k_g^{(2,100)}(x, x'))$$

Sampling from this model already leads to much more realistically looking skull shapes, as shown in Fig. 7.5.

7.4.2.4 Quantitative Comparison of Different Prior Models

So far we have selected the model using purely qualitative arguments and with the help of the visualized samples. In this section we support our choices by providing a quantitative evaluation of our models. For this purpose we use the standard metrics of generalization, specificity as well as a compactness introduced by Styner et al. [22].

---

[4] To define the symmetry plane we used 3 manually clicked points on the sagital plane. We then use this information to rigidly transform the skull so that the symmetry plane aligns with the y–z plane at the coordinate axis.

The *generalization* ability measures how accurately the model, denoted by $\mathcal{M}[\alpha]$, can fit the datasets from the modeled class of shapes, which is here represented by the set of ground truth datasets $\{\Gamma_1, \ldots, \Gamma_m\}$. More precisely, we define generalization for a model $\mathcal{M}$ as:

$$\text{gen}(M) = \frac{1}{m} \sum_{i=1}^{m} \min_{\alpha \in \mathbb{R}^n} D[\mathcal{M}[\alpha], \Gamma_i].$$

As a distance measure we use the averaged squared Euclidean distance of the corresponding points. As the model is in correspondence with the ground-truth dataset, we can compute the minimization in closed form.[5]

Specificity is a measure of how well randomly generated samples from the model resemble valid instances from this class of shapes. It is evaluated by comparing the distance of k randomly generated sample shapes (determined by the set of model parameters $\{\tilde{\alpha}_1, \ldots, \tilde{\alpha}_k\}$) to the closest shape of a set of valid example shapes

$$\text{spec}(M) = \frac{1}{k} \sum_{j=1}^{k} \min_{i} D[\mathcal{M}[\tilde{\alpha}_j], \Gamma_i].$$

Finally, *compactness* indicates how much variance the model represents when it is represented using only the first r basis functions. This variance is given by the sum of the leading r eigenvalues:

$$\text{var}(M) = \sum_{i=1}^{r} \lambda_i.$$

We report here two measures of compactness: 1) how many basis functions are needed to represent 99% of the total variance of the model and 2) how much of the total variance of the process (given by $\int_{\Gamma_R} k(x, x)\, dx$) is approximated using the first 200 eigenfunctions). Compactness can be seen as a measure of complexity. The more complex the model (i.e. the more flexible it is to represent many different shapes accurately) the more eigenfunctions are needed to accurately represent the model.

Table 7.1 shows the 3 measures applied to the different prior models. We see that the numbers confirm our visual impression that the symmetric model is clearly the best, followed by the multiscale model. For both models we managed to keep the compactness and specificity low by incorporating more knowledge about the true structure of the shape variability. That the symmetric model can at the same time be the most specific and the most general is a consequence of the strong prior assumptions. The leading 200

---

[5] To compute the minimum in practice, we use Gaussian process regression with an anisotropic noise of 0.1 mm, on a 1000 uniformly sampled points.

Table 7.1 Quantitative evaluation of the model quality for different models. The last two columns show the compactness of the model. The second last column shows how many basis functions were required to approximate 99% of the variance, while the last column shows how much of the model's total variance was approximated using 200 basis functions

Model	Generalization	Specificity	99% var.	Approx. var.
$k_g^{(100,100)}$	0.30 mm	14.6 mm	48	0.999
$k_g^{(200,100)}$	0.30 mm	20.6 mm	48	0.999
$k_g^{(100,50)}$	0.28 mm	15.5 mm	183	0.978
Multiscale	0.24 mm	14.7 mm	126	0.994
Symmetric	0.21 mm	12.5 mm	73	0.999

eigenfunctions, which are used to represent the model, capture most of the characteristic variations in the class of skull shapes. In contrast, the leading 200 eigenfunctions of the more generic models include variations that are not occurring in this class.

7.4.3 Skull-Registration Using Model Fitting

Once we have the prior model defined, we can perform the actual surface registration. Recall from Section 7.3 that to perform the registration we minimize the functional

$$\arg\min_{\alpha} \mathcal{D}[\Gamma_T, \mathcal{M}[\alpha]] + \eta' \|\alpha\|^2. \tag{7.19}$$

For the distance measure $\mathcal{D}$, we choose a robust distance measure

$$\mathcal{D}[\Gamma_1, \Gamma_2] = \int_{\Gamma_1} \rho(x - CP_{\Gamma_2}(x)) \tag{7.20}$$

where $CP_\Gamma(x) = \arg\min_{x' \in \Gamma}(\|x - x'\|)$ finds the closest points to x on the surface Γ and ρ is the Huber loss function [10] defined by

$$\phi(x) = \begin{cases} x^2/2 \\ (|x| - k/2) \end{cases} \quad k = 1.345.$$

Fig. 7.6 shows the accuracy of registration results evaluated using the ground-truth data. We see that the results correspond to what we expected from the model metrics. The symmetric kernel clearly yields the best results, followed by the multiscale kernel. Fig. 7.7 shows a typical registration result, where the red surface shows the registration solution and the white (transparent) surface the ground truth. We see that the shape is well matched in general. Upon close inspection, however, we can spot places where there are still larger errors (indicated by the arrow). We could now start to tune the parameters of our model in the hope that we can find values which reduce this error. A more direct and efficient solution is to define landmark points to enforce the correct

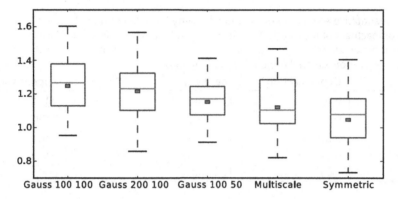

Figure 7.6 Registration results for the different models. The numbers represent the Procrustes distance (in mm) between the registered surface and the ground truth.

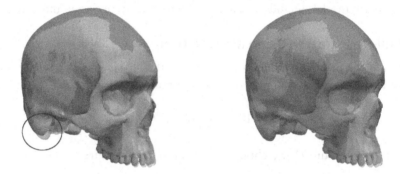

Figure 7.7 Left: A typical registration result. The shape is in generally well matched, but some local features still show some registration error. Right: The same registration result, but when the correct matching is enforced using a landmark that is placed in the erroneous region.

deformations manually and to include these as known observations into the model. As discussed in Section 7.3.1, the resulting posterior model is again a Gaussian process, and thus can itself be used as the prior model in the registration functional (7.19). Fig. 7.7 (right) shows the same registration, but this time with landmarks. We see that the error is greatly reduced when we enforce the deformation using a landmark. That adding landmark points will reduce the error for all models is shown in Fig. 7.8, where we used 6 manually defined landmarks on all the skulls.

7.4.4 Computing the Statistical Shape Model

In the final step we can use the registered data to compute a statistical shape model [5,4]. Let $\{u_1, \ldots, u_n\}$, $u_i : \Gamma_R \to \mathbb{R}^d$ denote the deformation field resulting from the registration, which establishes correspondence between the surfaces Γ_R and Γ_i. As discussed

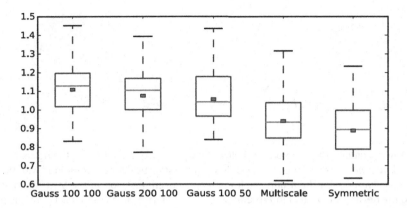

Figure 7.8 Registration results for the different models when 5 landmarks were used to guide the registration. The numbers represent the Procrustes distance (in mm) between the registered surface and the ground truth.

Table 7.2 Quantitative evaluation of the model quality for the classical statistical shape models and the augmented model

Model	Generalization	Specificity	99% var.	Approx. var.
SSM	0.89 mm	3.6 mm	24	1.000
SSM (augmented)	0.23 mm	4.6 mm	23	0.999

above, in the Gaussian process framework, a statistical shape model is simply a Gaussian process model $GP(\mu_{SSM}, k_{SSM})$, where the mean and covariance functions are estimated from the data using the formulas for the sample mean and sample covariance (Eqs. (7.6) and (7.7)). It is interesting to compute the quantitative measures from the previous section (cf. Table 7.1) also for this model.[6] In Table 7.2 we see that the generalization error is comparably large, but the model is much more specific and also more compact than the previously defined models. The good specificity value constitute the big advantage of statistical shape models.

The relatively large generalization error of 0.89 mm is due to the small number of example shapes. It shows that 46 examples are not sufficient to cover the full shape variation of the skull. Fortunately, combining different kernels also yields a simple solution to this problem. The idea is to model the missing variability as a small, but smoothly varying deformation using e.g. a Gaussian kernel $k_g(100, 1)$. Combining this with the statistical shape model leads to the *augmented* model

$$k_{aug}(x, x') = k_{SSM}(x, x') + k_g(100, 1),$$

[6] In this case we perform a leave-one-out procedure to estimate the generalization error.

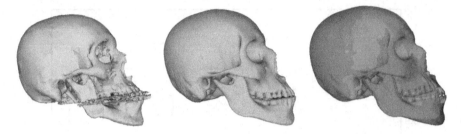

Figure 7.9 Registration of a pathological skull with overbite. On the *left* the pathological target skull is illustrated. The *middle* shows the fitting result using the statistical model combined with the pathological model. On the *right* the fitting result in green as overlay on top of the SSM part only in red, predicting the healthy shape. (For interpretation of the references to color in this figure, the reader is referred to the web version of this chapter.)

where we assumed that the scale of the error is around 1 mm and is smoothly varying. Computing the usual metrics reveals that this final model generalizes much better, while it still provides excellent specificity and compactness (see Table 7.2). Hence, in practical applications this model is likely to be superior to a classical statistical shape model.

7.5 MODELING AND ANALYZING PATHOLOGIES

That GPMMs allow us to increase the variability of a statistical shape model makes it possible to use statistical shape models for fitting pathological shapes. In this section we discuss this important topic on the exemplar case study of fitting a skull model to a CT scan of a patient with an overbite. The idea is to use the statistical skull model in order to devise a surgical plan to attain a desired aesthetic facial profile [20].

Fitting a statistical shape model built from healthy examples only does not work in this case. The model would trade-off the fitting accuracy in the healthy regions against a better explanation of the pathological deformations. To overcome this difficulty, we explicitly model the pathological part using an analytically defined Gaussian process. While we could use exactly the same approach, as for extending the model variation in a statistical skull model (cf. Section 7.4.4), we use here a variant of this approach, which does not compute a low-rank approximation of the full model, but approximates the statistical shape model and the model for the pathological part separately. This allows us to separate the healthy part of the deformation from the combined deformation including also the pathological part after the fitting, as illustrated in Fig. 7.9 (right).

7.5.1 Simultaneous Fitting of Healthy and Pathological Deformations

Let $u_{\mathrm{SSM}} \sim GP(\mu_{\mathrm{SSM}}, k_{\mathrm{SSM}})$ and $u_p \sim GP(0, k_g(x, x'))$ be two GPMMs which explain the healthy and the pathological part of a shape deformation. Assuming that we have the low-rank models computed for both models (cf. Section 7.2.2),

$$\mathcal{M}_{\text{SSM}}[\alpha^{\text{SSM}}] = \mu_{\text{SSM}} + \sum_{i=1}^{r} \alpha_i^{\text{SSM}} \sqrt{\lambda_i^{\text{SSM}}} \, \phi_i^{\text{SSM}} \tag{7.21}$$

$$\mathcal{M}_{\text{p}}[\alpha^{\text{p}}] = \sum_{j=1}^{r'} \alpha_j^p \sqrt{\lambda_j^p} \, \phi_j^p, \tag{7.22}$$

we can write any shape Γ as

$$\Gamma = \{x + \mathcal{M}_{\text{SSM}}[\alpha^{\text{SSM}}](x) + \mathcal{M}_{\text{p}}[\alpha^{\text{p}}](x) | x \in \Gamma_R\}$$

for some sets of parameters α^{SSM} and α^p.

Thus, we can write the combined registration problem as

$$\underset{\alpha^{\text{SSM}}, \alpha^{\text{p}}}{\arg\min} \mathcal{D}\left[\mathcal{M}_{\text{SSM}}[\alpha^{\text{SSM}}] + \mathcal{M}_{\text{p}}[\alpha^p], \Gamma_T\right] + \eta\|\alpha^{\text{SSM}}\|^2 + \eta\|\alpha^{\text{p}}\|^2,$$

where $\mathcal{D}$ is the distance measure defined in (7.20). The optimal parameter-sets α^{SSM} and α^p, represent the explanations of the different deformation models. The anatomically normal skull shape is thus given by the model $\mathcal{M}_{\text{SSM}}$ and the associated model parameters α^{SSM}.

7.5.2 Building a Model for Pathologies

The main modes of deformation of the combined model are not necessarily orthogonal. This means that it might be possible to explain certain shape deformation with either of the models. To reduce this effect, the model explaining the pathological deformations should not account for deformations outside the region where we expect the shape to depict pathological variability. A model can be restricted to a predefined region of the full domain by using a spatially varying kernel $k_{sv}(x, x')$. The spatially varying property of the kernel is defined through a function $a(x)$ that damps the size of the deformations a kernel allows based on the location of the points. We define the spatially-varying kernel as

$$k_{sv}(x, x', k) = a(x)k(x, x')a(x')$$

where k itself is an arbitrarily chosen kernel and $a(x)$ is the function that damps the kernel functions magnitude except in a defined region.

In our example we expect the pathological deformation to appear in the region of the mandible. Therefore we defined the bias model only in the region of the mandible, where we expect the pathology and hence need additional flexibility. We use a function

$$a(x) = \exp\left(-\|x - x_c\|^2/\sigma^2\right)$$

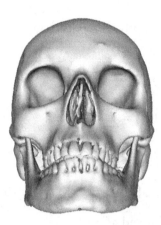

Figure 7.10 Landmark placement on the reference shape. For this example we have placed 8 landmarks in the teeth and mandible region of the skull. The landmarks in the teeth region prevent the model from adapting to the metal artifacts in the teeth region.

defining a region around the mandible with x_c as the closest point to x on the mandible. The parameter σ controls the size of the region around the mandible with function values of 1.0 for points that are close to the mandible and smoothly go to 0.0 for points that are further away.

7.5.3 Experiment

We applied this procedure to fit the skull shown in Fig. 7.9. As there are large metal artifacts in the teeth region we decided to place 8 landmarks to increase the fitting accuracy (see Fig. 7.10). The fitting results are depicted in Fig. 7.9. The green surface (*middle/right*) is our final fitting result while on the *left* the original target skull is depicted. On the *right* the part explained by the statistical shape model only is depicted in red. The visible difference, of the fitting result of the combined model and the statistical shape model only reconstruction, demonstrates the ability to separate the pathological deformation using the combined model approach. The results are also shown in the profile view for 2D slices through the skull in Fig. 7.11. The target is shown in blue, the fitting result in green and the part explained exclusively by the statistical shape model in red. Again, it can be seen that the statistical shape model does not explain the overbite, but the pathological part of the deformation is explained by the added spatially varying model.

7.6 CONCLUSION

In this chapter we have discussed GPMMs, a generalization of PCA-based statistical models. With GPMMs we are not restricted to estimate all the covariance structure

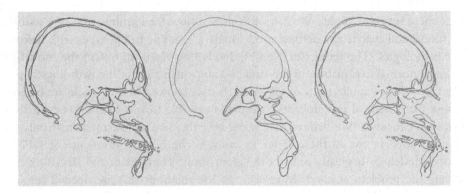

Figure 7.11 Sliced mesh representation of the fitting result. In all the illustrations we can see the registration target in blue, the fitting result with bias model in green and the part explained by the statistical shape model in red. On the *left* we see the fitting result to the blue target. The region of the mandible of the shape is well aligned. The *middle* shows the parts explained by the statistical shape model in red and the additional part from the bias model in green. It can be seen that the bias model allows for additional flexibility in the mandible region, which originates from the spatially varying kernel. On the *right* we see that most of the target skull is explained well by the statistical shape model. However, in the mandible region it is helpful to add more flexibility with the bias model. (For interpretation of the references to color in this figure, the reader is referred to the web version of this chapter.)

from training examples, but can additionally use analytically defined kernels to model our prior assumptions. Typically, these analytically defined kernels encode smoothness assumptions of the deformation. However, also more complex constraints can be formulated, such as that the model should match observed deformations, or that the model shape variation is symmetric around a given axis. The rules for constructing kernels allow us to combine learned and analytically defined kernels, and thus to build expressive shape priors even when there are not sufficiently many example surfaces available to learn the full shape variability.

We have presented an extended use case, where we discussed how GPMMs can be used for the problem of building a model of the skull. We have shown how we can start with simple prior assumptions, and gradually incorporate more knowledge into the model. We have seen that a practical way to assess if our prior is reasonable is by visualizing random samples from the model. Our quantitative evaluation of the different models confirmed that incorporating more prior knowledge into the model does not only lead to a better model as measured by the model metrics, but that this translates directly to better registration results. Our experiments also confirmed the usefulness of combining a learned statistical shape model with an analytically defined prior. In cases when we do not have sufficiently many example surfaces to learn the shape variation, we showed that augmenting it with an analytically defined model greatly enhances the generalization ability, without sacrificing much of the specificity of the model. Indeed, the result suggest that whenever we have training data available, it is always beneficial to include these into the model. We have also shown how the combination of a statistical shape model and analytically defined model make it possible to fit shapes with deformities or pathologies. The result can even be divided up again, to obtain the most likely anatomical normal explanation of the data and an explanation of the pathological part.

All the different applications of GPMMs that we showed in this article are based on the same mathematical principle and are made possible thanks to the great flexibility of Gaussian processes. We believe that being able to use a single, mathematically well established concept for all these different tasks, is the biggest advantage of GPMMs for shape modeling. It greatly reduces the algorithmic complexity and lets us focus on modeling the problem at hand. Moreover, all the methods that are needed for using GPMMs are available as open source, as part of the modeling software Scalismo [19] and Statismo [12].

APPENDIX 7.A

7.A.1 Approximating Eigenfunctions Using the Nyström Method

The goal of the Nyström method is to obtain a numerical estimate for the eigenfunctions/eigenvalues of the integral operator

$$\mathcal{T}_k f(\cdot) := \int_\Omega k(x, \cdot) f(x) d\rho(x). \tag{7.23}$$

The pairs (ϕ_i, λ_i), satisfying the equation

$$\lambda_i \phi_i(x') = \int_\Omega k(x, x') \phi_i(x) \, d\rho(x), \ \forall x' \in \Omega \tag{7.24}$$

are sought. The Nyström method is intended to approximate the integral in (7.24). This can, for example, be achieved by letting $d\rho(x) = p(x) \, dx$ where $p(x)$ is a density function defined on the domain Ω, and to randomly sample points $X = \{x_1, \ldots, x_n\}$ according to p. The samples $(x_l)_{l=1,\ldots,N}$ for x' in (7.24) lead to the matrix eigenvalue problem

$$K u_i = \lambda_i^{mat} u_i, \tag{7.25}$$

where $K_{il} = k(x_i, x_l)$ is the kernel matrix, u_i denotes the i-th eigenvector and λ_i^{mat} is the corresponding eigenvalue. Note, that since the kernel is matrix-valued ($k : \Omega \times \Omega \to \mathbb{R}^{d \times d}$), the matrices K and k_X are block matrices: $K \in \mathbb{R}^{nd \times nd}$ and $k_X \in \mathbb{R}^{nd \times d}$. The eigenvalue λ_i^{mat} approximates λ_i, while the eigenfunction ϕ_i in turn is approximated with

$$\tilde{\phi}_i(x) = \frac{\sqrt{n}}{\lambda_i^{mat}} k_X(x) u_i \approx \phi_i(x), \tag{7.26}$$

where $k_X(x) = (k(x_1, x), \ldots, k(x_n, x))$.

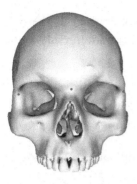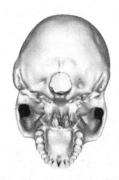

Figure 7.12 16 landmark points were defined on each skull, to enforce the correct correspondences for the ground-truth registrations. Integrating the landmark points improves the registration results especially for cases where there are large artifacts or holes in the data.

Clearly, the quality of this approximation improves with the number of points n, which are sampled. In our applications, we usually find that 500 to 1000 randomly sampled points are sufficient to obtain a good approximation. For a more thorough evaluation of how the number of points affects the approximation quality, we refer to Rosasco et al. [18] for a theoretical treatment, and Lüthi et al. for a practical evaluation for shape modeling [13].

7.A.2 Obtaining the Ground-Truth

Due to the low quality of the available example shapes, which contain missing parts (e.g. missing teeth), segmentation artifacts as well as metal artifacts around the teeth, we decided to perform our evaluations on a clean ground-truth dataset. The idea is to start from a clean reference shape (shown in Fig. 7.12), which is used as an atlas, which is then registered on each data set. In this way it is ensured that our numerical comparison are not distorted by the large amount of artifacts in the data. For registration of the reference skull to the examples, we use the registration approach discussed in Section 7.3. As a shape prior we used a combination of a multiscale kernel, which was symmetrized, plus two additional Gaussian kernel to allow for non-symmetric deformations.

$$k_{sym}(k_g^{50,100}(x, x') + k_g^{20,50}(x, x')) + k_g^{(2,100)}(x, x') + k_g^{(20,5)}(x, x'). \tag{7.27}$$

The last kernel in Eq. (7.27) allows for non-symmetric, local deformations and thus ensures that the model is not bias by the symmetric kernel and can accurately represent the target surface. To also retain this flexibility in the low-rank approximation, we represent the model using the leading 400 basis functions. The registration has been performed using the same setup as discussed in Section 7.4.3. In order to ensure correct correspondences, we have incorporated 16 landmarks, which are shown in Fig. 7.12.

In absence of meaningful registration metrics (which are not available as the original data is too noise) we ensured by visual inspection that the registration result accurately matched its target.

REFERENCES

[1] Petter Abrahamsen, A Review of Gaussian Random Fields and Correlation Functions, Norsk Regnesentral/Norwegian Computing Center, 1997.

[2] Yali Amit, Ulf Grenander, Mauro Piccioni, Structural image restoration through deformable templates, J. Am. Stat. Assoc. 86 (414) (1991) 376–387.

[3] Alain Berlinet, Christine Thomas-Agnan, Reproducing Kernel Hilbert Spaces in Probability and Statistics, vol. 3, Springer, 2004.

[4] Volker Blanz, Thomas Vetter, A morphable model for the synthesis of 3d faces, in: SIGGRAPH '99: Proceedings of the 26th Annual Conference on Computer Graphics and Interactive Techniques, ACM Press, 1999, pp. 187–194.

[5] T.F. Cootes, C.J. Taylor, D.H. Cooper, J. Graham, et al., Active shape models-their training and application, Comput. Vis. Image Underst. 61 (1) (1995).

[6] Brian P. Flannery, Wiliam H. Press, Saul A. Teukolsky, William Vetterling, Numerical Recipes in C, Press Syndicate of the University of Cambridge, New York, 1992.

[7] David Ginsbourger, Xavier Bay, Olivier Roustant, Laurent Carraro, Argumentwise invariant kernels for the approximation of invariant functions, Ann. Fac. Sci. Univ. Toulouse 21 (3) (2012) 501–527.

[8] Ulf Grenander, Michael I. Miller, Computational anatomy: an emerging discipline, Q. Appl. Math. 56 (4) (1998) 617–694.

[9] Tobias Heimann, Bram Van Ginneken, Martin Styner, Yulia Arzhaeva, Volker Aurich, Christian Bauer, Andreas Beck, Christoph Becker, Reinhard Beichel, György Bekes, et al., Comparison and evaluation of methods for liver segmentation from ct datasets, IEEE Trans. Med. Imaging 28 (8) (2009) 1251–1265.

[10] Peter J. Huber, Robust Statistics, Springer, 2011.

[11] Ian Jolliffe, Principal Component Analysis, Wiley Online Library, 2002.

[12] M. Lüthi, R. Blanc, T. Albrecht, T. Gass, O. Goksel, P. Büchler, M. Kistler, H. Bousleiman, M. Reyes, P. Cattin, T. Vetter, Statismo – A Framework for PCA Based Statistical Models, http://hdl.handle.net/10380/3371, 2012, University of Basel.

[13] Marcel Lüthi, Christoph Jud, Thomas Gerig, Thomas Vetter, Gaussian process morphable models, preprint, arXiv:1603.07254, 2016.

[14] C.A. Micchelli, M. Pontil, On learning vector-valued functions, Neural Comput. 17 (1) (2005) 177–204.

[15] Andreas Morel-Forster, Generative Shape and Image Analysis by Combining Gaussian Processes and MCMC Sampling, PhD thesis, Universität Basel, 2016.

[16] R. Opfer, Multiscale kernels, Adv. Comput. Math. 25 (4) (2006) 357–380.

[17] C.E. Rasmussen, C.K.I. Williams, Gaussian Processes for Machine Learning, Springer, 2006.

[18] Lorenzo Rosasco, Mikhail Belkin, Ernesto De Vito, On learning with integral operators, J. Mach. Learn. Res. 11 (2010) 905–934.

[19] Scalismo – scalable image analysis and shape modelling, http://github.com/unibas-gravis/scalismo.

[20] Kamal Shahim, Philipp Jürgens, Philippe C. Cattin, Lutz-P. Nolte, Mauricio Reyes, Prediction of cranio-maxillofacial surgical planning using an inverse soft tissue modelling approach, in: International Conference on Medical Image Computing and Computer-Assisted Intervention, Springer, 2013, pp. 18–25.

[21] John Shawe-Taylor, Nello Cristianini, Kernel Methods for Pattern Analysis, Cambridge University Press, 2004.

[22] Martin A. Styner, Kumar T. Rajamani, Lutz-Peter Nolte, Gabriel Zsemlye, Gábor Székely, Christopher J. Taylor, Rhodri H. Davies, Evaluation of 3d correspondence methods for model building, in: Information Processing in Medical Imaging, Springer, 2003, pp. 63–75.

[23] Michael E. Tipping, Christopher M. Bishop, Probabilistic principal component analysis, J. R. Stat. Soc., Ser. B, Stat. Methodol. 61 (3) (1999) 611–622.

CHAPTER 8

Bayesian Statistics in Computational Anatomy

Christof Seiler
Department of Statistics, Stanford University

Contents

8.1	Introduction	193
8.2	Parametric Bayesian Statistics	194
	8.2.1 Background	195
	8.2.2 Example	196
	8.2.3 Model	197
	8.2.4 Markov Chain Monte Carlo	200
	8.2.5 Hamiltonian Monte Carlo	201
	8.2.6 Software Implementations	202
	8.2.7 Convergence Diagnostics	202
	8.2.8 Related Work	203
8.3	Nonparametric Bayesian Statistics	204
	8.3.1 Background	204
	8.3.2 Example	206
	8.3.3 Model	208
	8.3.4 Gibbs Sampler	210
	8.3.5 Software Implementations	211
	8.3.6 Related Work	211
8.4	Conclusions and Open Problems	212
References		212

8.1 INTRODUCTION

The goal of this book chapter is to showcase two applications of Bayesian statistics we have successfully employed in our own research. The first application is on quantifying uncertainty in image registration using a parametric Bayesian model and an efficient sampling algorithm. The second application is on clustering of deformations into spatially contiguous regions using Bayesian nonparametrics.

Before introducing our statistical work, we would like to define what we understand by computational anatomy. The aim of computational anatomy is to describe the anatomy not by what it looks like but by how it deforms. Finding such deformations is usually referred to as image registration; throughout this text we will use computational anatomy and image registration interchangeably. Statistical estimation and analysis of de-

Statistical Shape and Deformation Analysis
DOI: 10.1016/B978-0-12-810493-4.00009-2

formations of a group of medical images can for instance be used for early diagnosis and prediction of disease. The emerging field of computational anatomy has many facets. It lays at the interface between medicine, geometry, computing and statistics. Ideally, mathematicians define a notion of shape and a deep theory that is rooted in geometry. The computing community takes these notions and theories and implements them on a computer. Statisticians then quantify the uncertainty when running these computer programs on real world data. Finally, medical doctors make treatment decisions based on the calculated statistics.

Going back in time, we can trace the origins of computational anatomy to Riemann and his "Habilitationsschrift" in 1854 where he linked manifolds to shape of solid figures. He conceived the analysis of shapes of a solid figure as an important part of geometry. In 1917, the biologist-mathematician D'Arcy Thompson wrote an entire chapter on the analysis of shapes using deformations in his book on "Growth and form" [30]. His key idea was to study the shapes not by what they look like, but by how they deform relative to each other. He showed how one could relate various species, different kinds of fish, monkeys and humans to each other by stretching, scaling, and, by some more complicated deformations like conformal maps. In 1966, Arnold wrote his groundbreaking paper on employing modern geometry to describe incompressible fluids [4]. From the nineties until today, Grenander, Miller, Trouvé, Younes, Holm [13, 31,39,16], and many others, modeled the human anatomy using fluid dynamics and developed Arnold's ideas to establish the foundations of computational anatomy.

Today, computational anatomy is surrounded by a vibrant community and several workshops during the MICCAI conference have been held over the past years on its mathematical foundations.[1] In 2015, an entire research program on the theoretical foundations of computational anatomy and its applications was organized at the University of Vienna.[2]

8.2 PARAMETRIC BAYESIAN STATISTICS

Image registration is one of the major work horses of medical image analysis. One of the most important applications of image registration is to "normalize" brains to a common brain atlas. This is a crucial "preprocessing" step in most multimodal brain studies: For instance, in structural MRI studies, image registration is used to measure local brain morphology; or in functional and diffusion MRI studies, image registration is used to bring all subjects in the same anatomical space for comparison. Other imaging based communities have extended and adapted registration based method for their own

---

[1] http://www-sop.inria.fr/asclepios/events/MFCA13/.
[2] http://www.mat.univie.ac.at/~shape2015/.

application. For instance, orthopedic research uses image registration to estimate implant designs and evaluate fracture risk of bones.

Image registration builds on assumptions and approximations. From a statistical viewpoint, image registration is the task of estimating the parameters that define non-linear deformations. In the first part of this book chapter, we review how estimation can be done using Bayesian statistics.

8.2.1 Background

Before going into the modeling and computational details of how Bayesian statistics can be used in image registration, we review the general concepts of Bayesian statistics. We start with a model

$$M = \{f(y \mid \theta) : \theta \in \Omega\}$$

and refer to function f as the likelihood and the variable θ as parameters. The parameters can take values in simple one dimensional spaces or high dimensional spaces; they can be discrete or continuous. This model describes the data as random samples from the likelihood function given a fixed parameter

$$Y_1, \ldots, Y_n \sim f(y \mid \theta).$$

A Frequentist would stop here, a Bayesian needs to define a prior distribution on the parameters

$$\pi(\theta)$$

to complete the model. We can think of the prior as a way to incorporate our beliefs from previous experiments into our analysis. Combining everything using Bayes rule will give us the posterior distribution of the parameters after having observed data and given our prior beliefs

$$\pi(\theta \mid y) = \frac{\prod_{i=1}^{n} f(y_i \mid \theta) \pi(\theta)}{m(y)}.$$

The denominator $m(y)$ is called the marginal distribution,

$$m(y) = m(y_1, \ldots, y_n) = \int f(y_1, \ldots, y_n \mid \theta) \pi(\theta) \, d\theta = \int \prod_{i=1}^{n} f(y_i \mid \theta) \pi(\theta) \, d\theta,$$

and can be seen as the measurement of how well our model explains the data on average over all possible parameter values weighted by their prior probability. With the posterior distribution in hand, we can now compute posterior means

$$\bar{\theta} = E(\theta \mid y) = \int \theta \, \pi(\theta \mid y) \, d\theta$$

or 95% credible intervals between

$$0.025 = \int_{-\infty}^{\theta_l} \theta\,\pi(\theta\mid\gamma)\,d\theta \quad \text{and} \quad 0.975 = \int_{\theta_h}^{\infty} \theta\,\pi(\theta\mid\gamma)\,d\theta,$$

where 95% of the posterior mass is between θ_l and θ_h. In most applications, we will not be able to calculate these integrals analytically but we will have to resort to computational approximations. Variational inference and Markov Chain Monte Carlo (MCMC) are the main computational approximations employed in practice. Even when $m(\gamma)$ is unknown, we can use MCMC to draw samples from the posterior. Once can think of sampling as an alternative way of describing a distribution. If one were able to sample infinitely many times from a distribution, one would know everything about the distribution. However, in practice we will resort to finite sample approximations, and thus we will know the distribution up to approximation errors.

Sampling from high dimensional probability distributions is a crucial step towards the Bayesian treatment of computational anatomy. The goal is to setup a statistical model of computational anatomy, this includes constructing a prior for deformations and linking it through the likelihood to fixed and moving images. To infer the deformations parameters, we then sample many times from the posterior distribution that is a combination of the likelihood and prior distribution. We can then build uncertainty estimates form these samples. In computational anatomy, we need efficient samplers that work well in high dimensions. One promising candidate is the Hamiltonian Monte Carlo (HMC) method. It is a promising candidate for two reasons: it is efficient in high dimensions, and it provides geometric structures very similar to that encountered in computational anatomy.

Due to the computational complexity the focus has been mostly on finding efficient algorithms and implementations to obtain point estimates by optimizing an objective function and finding its maximum value. However, computational anatomy is clearly a statistical problem considering that images are noisy and registered brains between subjects and template never match perfectly. With recent advances in computational power and Bayesian computations it has now become possible to add error bars to solutions. This provides the practitioner with additional information of the uncertainty associated to registration results. This is crucial in a clinical setting where it is important to obtain reproducible results.

8.2.2 Example

We start with an example from an ongoing back pain project. Fig. 8.1 shows the uncertainty map when registering two participants in our back pain dataset. On the left, the *posterior mean* shows the expected deformation, overlaid with a contour map that shows 95% credible interval lengths derived from the posterior distribution of deforma-

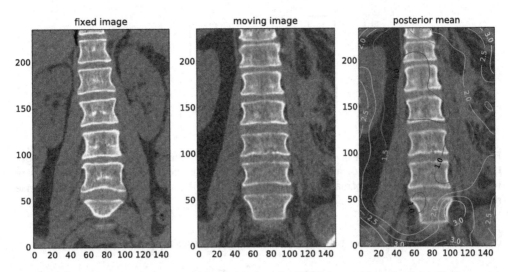

Figure 8.1 The contour plot (most right image) shows credible interval length of the displacement magnitude in millimeters when deforming the moving image (middle image) to the fixed image (most left image).

tion parameters. We see that lumbar spine vertebrae L1 to L4 (L4 is the fifth vertebra from the top) deformations have an uncertainty around 1 mm. In contrast, the lumbar vertebra L5 deformation has an uncertainty up to 3 mm. This may indicate that the registration failed for L5. In three dimensional examples similar problems can occur due to anatomical abnormalities.

This illustrates the uncertainty estimation on the subject level between one fixed and one moving image. We generated this uncertainty map by sampling form a posterior distribution of deformations using Hamiltonian Monte Carlo (HMC). We will now carefully develop the model underlying this example and introduce the HMC sampler, which is a member of the Markov Chain Monte Carlo family.

8.2.3 Model

We analyze geometric differences of the spine anatomy through geometric deformations. Our model describes patient images $\mathbf{I}_k$ as deformations φ_k from a common template image $\mathbf{I}_0$ (Fig. 8.2). We model deformations with cubic multidimensional B-spline polynomial basis functions $\mathcal{B} : \mathbb{R}^3 \to \mathbb{R}^3$ [32]. The model parameters are the weights associated to each basis function. If we pick C control points placed on a uniform grid over the template image $\mathbf{I}_0$, then we need to estimate a total of $q \in \mathbb{R}^{3C}$ weights in three-dimensional volumetric CT images with voxels positions $x_i \in \mathbb{R}^3$. The

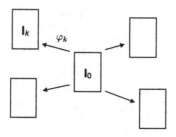

Figure 8.2 The template image I_0 represents a typical anatomy. We model individual patients' images I_k as non-linear deformations φ_k from the template. We will estimate the deformations from moving to fixed images with the goal of mapping all images into the same coordinate system, thus the inverse notation.

estimated deformations ($\mathcal{N}_x$ is the support of the B-splines at spatial position x)

$$\varphi_k(x, q) = x + \sum_{x_i \in \mathcal{N}_x} q_i \, B\,(x - x_i)$$

for each subject k can then be statistically analyzed.

There are two sources of uncertainty when estimating model parameters q from two images I_0 and I_k: image noise and model misspecification. The images are acquired with CT scanners which like other measurement devices introduce noise. The model misspecification is due to the crude assumption that the template can be deformed into a subject anatomy not accounting for shape difference that cannot be capture by such a simple model, e.g. missing part or entire vertebra in a patient. To capture these uncertainties, we assign a prior distribution to the parameters and sample form its posterior to compute posterior mean and posterior errors.

To control the spatial smoothness of the deformations we put a normal prior distribution

$$q \sim N(0, (\lambda_1 \Lambda)^{-1}) = \frac{1}{Z} \exp\left(\lambda_1 \sum_{i=1}^{N} \| \operatorname{Jac} \varphi_k(x_i, q) \|_{\ell_2}^2 \right) = \frac{1}{Z} \left(-\lambda_1 \operatorname{Reg}(q)\right)$$

on the deformation parameters. The term $\operatorname{Reg}(q)$ can be seen as a form of regularization, penalizing large first order derivatives of the deformations; lower values represent more regular deformations. The block precision matrix Λ is defined by the element-wise partial derivatives of the B-spline basis matrix ($\otimes$ denotes the Kronecker product)

$$\Lambda = I_3 \otimes \sum_{i=1}^{3} \left(\frac{\partial B^{\mathsf{T}}}{\partial x^i} \frac{\partial B}{\partial x^i} \right)$$

and the Jac is the Jacobian matrix. For more details see [3].

We then make the link to the imaging data through the likelihood term

$$(\mathbf{I}_k \mid q) \sim \frac{1}{Z}\left(-\sum_{i=1}^{N}(\mathbf{I}_k \circ \varphi_k(x_i, q) - \mathbf{I}_0(x_i))^2\right) = \frac{1}{Z}\exp\left(-\text{Dist}(\mathbf{I}_0, \mathbf{I}_k, q)\right)$$

measuring the dissimilarity $\text{Dist}(\mathbf{I}_0, \mathbf{I}_k, q)$ between template image $\mathbf{I}_0$ and subject image $\mathbf{I}_k$ after applying the deformation φ_k; lower values represent a better match. Combining the prior and likelihood term into the posterior distribution yields

$$(q \mid \mathbf{I}_k) \sim \frac{1}{Z}\exp\left(-\text{Dist}(\mathbf{I}_0, \mathbf{I}_k, q) - \lambda_1 \text{Reg}(q)\right),$$

which completes the Bayesian model for geometric deformations.

To make inference about deformation parameters, we can now compute different functionals of the posterior distribution. For example, the posterior mean

$$\theta = \mathbb{E}(q \mid \mathbf{I}_k) = \int q\,\pi(q \mid \mathbf{I}_k)\,dq.$$

This $\mathbb{R}^{3C}$-dimensional integral is intractable analytically and to solve it one must resort to numerical methods. The first idea of evaluating the posterior at evenly distributed grid points is infeasible due to the large number of grid points. A clever alternative are Markov Chain Monte Carlo (MCMC) sampling algorithms. The main idea for sample-based estimators of integrals is to use the sample mean estimator

$$\widehat{\theta} = \frac{1}{T}\sum_{t=1}^{T} q_t$$

with draws from the posterior distribution

$$q_1, \ldots, q_T \sim (q \mid \mathbf{I}_k).$$

A wide range of MCMC samplers exists. In the next section, we will introduce the basic principle and elaborate on the more advanced Hamiltonian Monte Carlo.

In addition to the sample mean, we can also compute credible intervals from samples by calculating sample quantiles. For Fig. 8.1, we computed element-wise quantiles of the parameter vector

$$q = \left(q^1, \ldots, q^{3C}\right)^{\mathsf{T}}$$

to obtain a 95% confidence interval

$$\hat{\theta}_l = \left(q_{(\alpha)}^1, \ldots, q_{(\alpha)}^{3C}\right)^{\mathsf{T}} \quad \text{and} \quad \hat{\theta}_h = \left(q_{(1-\alpha)}^1, \ldots, q_{(1-\alpha)}^{3C}\right)^{\mathsf{T}}$$

by ordering parameters from smallest to largest indicated by $_{()}$, where $_{(\alpha)}$ is the αth smallest value. We can then compute deformations by plug-in and compute the desired credible interval lengths as the difference between the 2.5% and 97.5% percentiles

$$|\varphi(x, \hat{\theta}_h) - \varphi(x, \hat{\theta}_l)|.$$

8.2.4 Markov Chain Monte Carlo

As the name suggests, MCMC is composed of a Markov chain and a Monte Carlo simulation. For lower dimensional integrals Monte Carlo simulations without Markov chains are possible. This works by drawing independent samples from $\pi(\theta \mid \mathbf{I}_k)$

$$\hat{\theta} = \frac{1}{T} \sum_{i=1}^{T} q_i, \quad q_1, \ldots, q_T \overset{indep.}{\sim} \pi(q \mid \mathbf{I}_k).$$

However Monte Carlo simulation breaks down for problems of three or more dimensions. In this case, we resort to constructing a Markov chain that generates dependent samples

$$\hat{\theta} = \frac{1}{T} \sum_{i=1}^{T} q_i, \quad q_1, \ldots, q_T \overset{dep.}{\sim} \pi(q \mid \mathbf{I}_k).$$

Metropolis algorithm is the simplest MCMC algorithm. Consider a finite sample space. Think of it as a state space, where each outcome corresponds to a state. Metropolis can sample from an unnormalized probability $\pi(x)$ on finite state space $\mathcal{X}$. Define a Markov transition matrix $J(x, y)$ that assigns nonzero probabilities of moving from x to y and y to x. Metropolis changes $J(x, y)$ to a new matrix $K(x, y)$ that corresponds to a possibly unnormalized version of $\pi(x)$.

The algorithm contains the following steps:
- Pick initial point in sample space x_0
- Pick potential next move from $J(x, y)$ with $J(x, y) > 0$ and $J(y, x) > 0$
- Evaluate

$$A(x, y) = \frac{\pi(y)J(y, x)}{\pi(x)J(x, y)}$$

- If $A(x, y) \geq 1$ move to y
- If $A(x, y) < 1$ flip a coing with this success probability
  - and move to y if success
  - otherwise stay at x

We can write this in matrix form

$$K(x, y) = \begin{cases} J(x, y) & \text{if } x \neq y, A(x, y) \geq 1 \\ J(x, y)A(x, y) & \text{if } x \neq y, A(x, y) < 1 \\ J(x, y) + \sum_{z:A(x,z)<1} J(x, z)(1 - A(x, z)) & \text{if } x = y \end{cases}$$

Then we can use the Fundamental Theorem of Markov Chains to prove that

$$K^n(x, y) \to \pi(y) \quad \text{for each } x, y \in \mathcal{X}$$

or in other words, the matrix $K^n = K^1 K^2 \cdots K^n$ converges to a matrix K

$$\pi K = \lambda \pi \quad \leftrightarrow \quad \pi K = \pi$$

with one left eigenvector π and one eigenvalue $\lambda = 1$.

To sample from π, apply J to the left of the current sample position x_t. This will give us the next sample x_{t+1}. A nice introduction to the subject of MCMC is Diaconis [9].

8.2.5 Hamiltonian Monte Carlo

As the name suggest, HMC involves defining a Hamiltonian function $H : \mathbb{R}^d \times \mathbb{R}^d \to \mathbb{R}$ with a potential energy term $V(q)$ which is equal to the $-\log$ transformed posterior distribution and a kinetic energy term $K(p)$ which will have the quadratic form $\frac{1}{2}p^\mathsf{T} Gp$. The sum of both terms

$$H(q, p) = V(q) + K(p)$$

is the Hamiltonian function. It can be shown that the following algorithm produces samples from the distribution of the random variable $(q \mid \mathbf{I}_k)$:

- Fix a starting position q_0
- Draw p_0 from a Gaussian $N(0, G^{-1})$
- Solve the Hamiltonian system

$$\frac{dq}{dt} = \frac{\partial H}{\partial p} \quad \text{and} \quad \frac{dp}{dt} = -\frac{\partial H}{\partial q}$$

for a predefined amount of time and record end point q_1
- Repeat the previous steps $T_0 + T$ times yielding $q_1, \ldots, q_{T_0}, \ldots, q_T$
- Then $q_{T_0}, \ldots, q_T \sim (q \mid \mathbf{I}_k)$ are samples from the posterior. Note that samples before T_0 will be discarded because they could be biased due to a bad starting point.

The MCMC samples are correlated and standard Monte Carlo errors estimates do not apply. It is common to use trace plots to choose the appropriate T, which we found to be $T = 500$ in our spine registration problem. We refer to Seiler et al. [27]; Holmes et al. [17] for a theoretical investigation on this topic. The book chapter by Neal [21] is an excellent source for more background material and illustrative toy examples on HMC.

8.2.6 Software Implementations

The main HMC sampling algorithm is easy to implements and contains only few steps. The tricky part is solving the Hamiltonian system. This is usually done using an Euler numerical scheme. However the fine tuning of parameters is not straightforward and software has been implemented to automize this step [8].

A specific implementation of HMC for medical images can be found on the GitHub repository of the first author.[3]

8.2.7 Convergence Diagnostics

Some statisticians[4] argue that diagnostics for MCMC only finds "obvious, gross, embarrassing problems that jump out of simple plots". One of the concerns has to do with bad starting points. Consider an event B having high probability under the equilibrium distribution, that is, the distribution that is unchanged if we ran the MCMC sampler for a long time. Suppose we are unlucky and start the sampler at a bad starting point and it would take a long time, say longer than the age of the universe to reach B, then the chance of diagnosing this problem will be highly improbable.

Keeping this limitation in mind, convergence diagnostics of MCMC samplers are the only way to quantify the quality of our sample. The following description builds heavily on [25, Chapter 8]. The R package coda offers implementations for the most popular diagnostic tools.

We consider two types of convergence of MCMC methods:

1. Convergence to the stationary distribution: Check that distribution of the chain x_t is the stationary distribution f. In practice this is impossible to test with just one chain. Several chains have to be run to test this and it can actually never be tested exactly. What is tested is how independent the chains are at time t when started at different starting positions.

2. Convergence of averages: Once stationarity is established, we can focus on evaluating the Monte Carlo error. However, in contrast to the usual Monte Carlo error, we have additional problems to take into account: the samples are dependent across time. The stronger this dependence the slower we explore the posterior distribution.

---

[3] Code available: https://christofseiler.github.io/BayesianImageRegistration.
[4] http://users.stat.umn.edu/~geyer/mcmc/diag.html.

To be more concrete, define an estimator

$$\hat{\theta} = \frac{1}{T} \sum_{t=1}^{T} h(x_t)$$

of the parameter

$$\theta = \mathrm{E}(h(X)).$$

The variance of this estimator $\mathrm{Var}(\hat{\theta}_{\mathrm{MC}})$ in case of identically distributed and independent Monte Carlo samples is given by the central limit theorem. The variance of this estimator $\mathrm{Var}(\hat{\theta}_{\mathrm{MCMC}})$ in case of dependent Markov chain samples gets worse by a factor that depends on the amount of correlation between draws. This can be measured by the effective sample size

$$T_{\mathrm{MC}} = \frac{T_{\mathrm{MCMC}}}{\kappa(h)}$$

with the autocorrelation time

$$\kappa(h) = 1 + 2 \sum_{t=1}^{\infty} \mathrm{corr}\left(h(x_0), h(x_t)\right).$$

Intuitively, the larger $\kappa(h)$ the more dependent are draws and we need to increase the sample size to reach the Monte Carlo error. If draws are independent then we have $\kappa(h) = 1$ and the Monte Carlo and MCMC samples are equivalent. Diagnostics tools can now be build on this concept.

For more involved samplers, e.g. the HMC sampler, diagnostics becomes even more complicated. Recently, we have reformulated HMC in the language of Riemannian geometry and applied theorems from Joulin and Ollivier [18] on Markov chain curvature that inspired a new diagnostic tool for HMC [27,17]. The idea of Markov chain curvature is similar to measuring the autocorrelation time of the Markov chain, in fact, the curvature is inverse proportional to the correlation between draws from the Markov chain. Intuitively, a random walk moves faster through low density regions of the space, and slower through high density regions.

8.2.8 Related Work

We will distinguish between two types of deformations: *small deformations* and *large deformations*. To keep this book chapter focused on uncertainty quantification we will not go into the difference between these two types except for saying that one can think of large deformations as compositions of small deformations.

First, we focus on the *small deformations* setting. Allassonnière et al. [1,2] describe a maximum a posterior estimation (MAP) procedure based on an Expectation-Maximization (EM) algorithm to estimate the template from a set of subject images.

Similarly Van Leemput [33] estimated the MAP template using a combination of pseudo-likelihood technique [6], EM algorithm, and Laplace's method. In contrast, Risholm et al. [24,23] sample from the posterior distribution of pairwise registration parameters using Metropolis–Hastings algorithm which is part of the family of MCMC algorithms. In addition to MCMC, Risholm et al. [22] marginalize over hyperparameters (modeling noise in the image intensities) using Laplace's method. Simpson et al. [29] use mean-field variational Bayes to approximate the full posterior distribution. This is an optimization-based alternative to MCMC and much faster in practice. However it is unclear how accurately the posterior distribution is approximated with this method. In contrast, MCMC-type methods enjoy the property that they are sampling exactly from the true posterior distribution given that the sampler is run for long enough. Quantifying running times is however very tricky and usually only possible for simplified toy examples. In Simpson et al. [28] an additional Laplace's method is used for hyperparameters analog to Risholm et al. [22]. Heinrich et al. [14] propose an alternative approach using dynamic programming to find the MAP.

Now, we focus on the *large deformations* setting. Zhang et al. [41] use Hamiltonian Monte Carlo (HMC) to sample diffeomorphic deformations. The HMC algorithm is part of a Monte Carlo EM algorithm to estimate an image template. This is quite computationally expensive and Zhang and Fletcher [40] provide a fast algebraic approximation that is useful in a sampling scheme. Wassermann et al. [36] approximate the full Bayesian posterior distribution by a variational formulation. Their method can be viewed as a combination of the Laplace's method (describing stochastic differential equations by Gaussian processes) and variational Bayes (minimizing the Kullback–Leibler divergence). Yang and Niethammer [38] propose a low rank approximation of the Hessian matrix at the mode of the posterior distribution.

8.3 NONPARAMETRIC BAYESIAN STATISTICS

8.3.1 Background

The following introduction to Bayesian Nonparametrics (BN) builds strongly on the lecture notes by Larry Wasserman.[5] A good reference on the theoretical background are the lecture notes by van der Vaart.[6]

We replace the finite dimensional model from the previous section

$$\{f(y|\theta) : \theta \in \Theta\}$$

---

[5] Lecture notes: http://www.stat.cmu.edu/~larry/=sml/nonparbayes.pdf.
[6] Lecture notes: http://www.math.leidenuniv.nl/~avdvaart/BNP/.

with an infinite dimensional model and constrain the second derivate of possible functions to be finite

$$\mathcal{F} = \left\{ f : \int (f''(\gamma))^2 d\gamma < \infty \right\}.$$

Other constrains are possible but for illustrative purposes we focus on this model. In order to translate parametric Bayesian ideas to the nonparametric setting, we need to address some questions. First, we will need to put a prior π on an infinite dimensional space. For example, suppose we observe

$$X_1, \ldots, X_n \sim F$$

with unknown distribution F with density f. We put a prior π on the set of all distributions $\mathcal{F}$. In many cases, we cannot explicitly write down a formula for π. This follows from technical arguments about the infinite dimensional set $\mathcal{F}$ that we will not cover in this short introduction. How can we describe a distribution π in another way than writing it down? If we know how to draw from π we can get many samples and then even without knowing the formula for π we can plot and summarize it in any way we want. The idea is to find an algorithm to sample from this model

$$F \sim \pi$$
$$X_1, \ldots, X_n \mid F \sim F.$$

The usual frequentist estimate of F is the empirical distribution function

$$F_n(x) = \frac{1}{n} \sum_{i=1}^{n} I(X_i \leq x).$$

To estimate F from a Bayesian perspective we put a prior π on the set of all $\mathcal{F}$. Such a prior was invented by Ferguson [12]. The prior denoted by $\mathrm{DP}(\alpha, F_0)$ has two parameter: F_0 and α. F_0 is a distribution function and should be thought of as a prior guess of F. The number α controls how tightly concentrated the prior is around F_0. The model is

$$F \sim \mathrm{DP}(\alpha, F_0)$$
$$X_1, \ldots, X_n \mid F \sim F.$$

But how to draw samples from this model? First to draw samples from the prior $\mathrm{DP}(\alpha, F_0)$, we follow four steps:

1. Draw $s_1, s_2, \ldots$ independently from F_0.
2. Draw $V_1, V_2, \cdots \sim \mathrm{Beta}(1, \alpha)$.
3. Stick breaking process: Let $w_1 = V_1$ and $w_j = V_j \prod_{i=1}^{j-1}(1 - V_i)$ for $j = 2, 3, \ldots$.

- Imagine a stick of unit length.
- Then w_1 is obtained by breaking the stick at the random point V_1.
- The stick has now length $1 - V_1$.
- The second weight w_2 is obtained by breaking a proportion V_2 from the remaining stick.
- The process continues and generates the whole sequence of weights $w_1, w_2, \ldots$.

4. Let F be the discrete distribution that puts mass w_j at s_j, that is, $F = \sum_{j=1}^{\infty} w_j \delta_{s_j}$ where δ_{s_j} is a point mass at s_j.

After we observe the data $X = (X_1, \ldots, X_n)$, we are interested in the posterior distribution. The same idea applies here, instead of writing down a formula we describe an algorithm to sample for the posterior distribution. To sample from the posterior, we need the following theorem. Let F_n be the empirical distribution.

Theorem 8.3.1 ([12]). *Let $X_1, \ldots, X_n \sim F$. Let F have the prior $\pi = DP(\alpha, F_0)$. Then the posterior π for F given $X_1, \ldots, X_n$ is $DP(\alpha + n, \bar{F}_n)$ where*

$$\bar{F}_n = \frac{n}{n+\alpha} F_n + \frac{\alpha}{n+\alpha} F_0.$$

Since the posterior is again a Dirichlet process, we can sample from it the same way as we did for the prior. We only replace α with $\alpha + n$ and F_0 with $\bar{F}_n$. Thus the posterior $\bar{F}_n$ is a convex combination of the empirical distribution F_n and the prior guess F_0. To explore the posterior distribution, we could draw many random distribution functions form the posterior. We could then numerically construct two functions L_n and U_n such that

$$\pi(L_n(x) \leq F(x) \leq U_n(x) \text{ for all } x \mid X_1, \ldots, X_n) = 1 - \alpha$$

This is a Bayesian credible interval for F. When n is large then $\bar{F}_n \approx F_n$.

8.3.2 Example

Unfortunately, most people will be affected by lumber back pain (LBP) during the coarse of their lives. We classify the pain into acute and chronicle pain. The acute pain is most commonly caused by muscle strain or ligament sprain. The chronicle pain is most commonly cause by a disc tear, facet joint disorder, or sacroiliac joint disfunction.

In addition to the individual suffering of every patient, the society as a whole pays a big price for the treatment of LBP. For instance, the direct costs of LBP are estimated at 2.6 billion Euro, 6.1% of the total healthcare expenditure in Switzerland [37], which results in a total economic burden between 1.6 and 2.3% of Swiss GDP.

Despite the enormous burden on individual patients and the society, the geometric variability of deformations of the spine are still underexplored. For example, it has not been reported whether scoliosis patients (sidewise curvature of the spine) suffer more often from LBP than normal patients (Fig. 8.3).

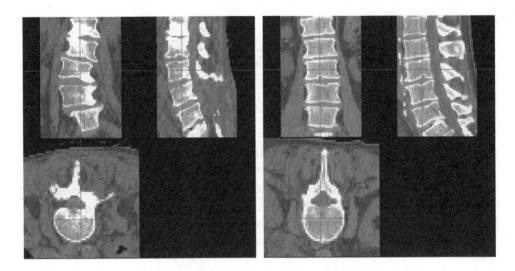

Figure 8.3 Left: Lumbar back pain patient with scoliosis. Right: Abdominal pain patient.

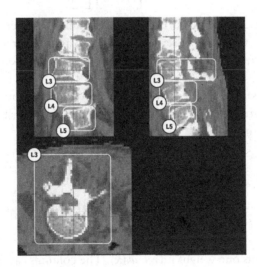

Figure 8.4 Parcellations in computational anatomy.

To explore this issue, we will investigate geometric differences between LBP and abdominal patients. We expect to see regional differences between the two patient groups. We propose to estimate regions using a BN clustering method that allows to incorporate geometric prior information. This clustering algorithm will find spatially contiguous voxel clusters (Fig. 8.4) without knowing in advance the number of clusters.

8.3.3 Model

Deformation fields can be found by registration methods as described in the first part of this book chapter. Additionally in this part we will assume that deformations are diffeomorphic, which means that they are differentiable and their inverses are differentiable. The input to our clustering algorithm are deformation fields encoded as Stationary Velocity Fields (SVF), which can be obtain through various registration algorithms [5, 15,34,20]. The SVF v is the unique solution to the Ordinary Differential Equation (ODE) $\partial\phi(x,t)/\partial t = v(\varphi(x,t))$ with the initial condition $\varphi(x,0) = \text{identity}$. The reason that ODEs are useful for image registration is that we can generate a diffeomorphic mapping of a patient image $\mathbf{I}_k$ to a template image $\mathbf{I}_0$ with $\mathbf{I}_0(x) = \mathbf{I}_k(\varphi_k(x))$, spatial position $x \in \mathbb{R}^3$, intensity image $\mathbf{I} : \mathbb{R}^3 \mapsto \mathbb{R}$, and diffeomorphic mapping $\varphi : \mathbb{R}^3 \mapsto \mathbb{R}^3$. This assumption makes sense for spines in the absence of fractures and collapse of tissue.

We model the observed velocity fields as a linear combination of linear transformations

$$v(x) = \sum_{i=1}^{p} w_i(x) \begin{bmatrix} L_i & q_i \\ 0 & 0 \end{bmatrix} \begin{bmatrix} x \\ 1 \end{bmatrix} + \varepsilon(x)$$

with an affine part L_i, translational part q_i, and additive independent and identically distributed voxelwise Gaussian noise $\varepsilon(x)$. Our goal is to infer both the number of parcels p and shape $w_i(x)$ with the assumption that $w_i(x)$ are non-overlapping binary weight images.

We formulate this in terms of a BN model by vectorizing both the image matrix and the linear transformations, and by introducing the binary matrix $\mathbf{W}$ that assigns n voxels to p parcels

$$\begin{bmatrix} v_1 \\ \vdots \\ v_n \end{bmatrix} = \mathbf{W} \begin{bmatrix} \text{Vectorize}(L_1) \\ q_1 \\ \vdots \\ \text{Vectorize}(L_p) \\ q_p \end{bmatrix} + \begin{bmatrix} \varepsilon_1 \\ \vdots \\ \varepsilon_n \end{bmatrix}.$$

Each column represents one weight image $w_i(x)$. The columns can grow in size as more observations become available (higher resolution images). The parameters of interest in this BN models are $\mathbf{W}$, L_i, and q_i. The BN part of this model is the matrix $\mathbf{W}$ because it is not fixed in column size.

Creating parcels in an image is similar to clustering voxels. We can formulate clustering as a density estimation problem by using a mixture model to approximate densities. Each component of the mixture model defines a cluster. Positions that are close to the mode of one component are assigned the same label. In BN we can perform density estimation using an extension of the Dirichlet process prior to the Dirichlet process mixture model. We now give a short introduction to density estimation in BN.

Consider that we observe $X_1, \ldots, X_n \sim F$ from a distribution F with density f. Without loss of generality we can assume that $X_i \in \mathbb{R}$. Our goal is to estimate the unknown density function f. The Dirichlet process is not an appropriate prior for this problem because it produces discrete distributions and densities are continuous. An intuitive way to construct the nonparametric estimation procedure is by starting with a parametric model and letting the number of parameters go to infinity. Consider the Gaussian mixture model

$$f(x) = \sum_{j=1}^{k} w_j f(x; \theta_j),$$

where $f(x; \theta_j)$ is normal and each component is parametrized with its mean and variance $\theta_j = (\mu_j, \sigma_j^2)$. In this model, we would have to estimate the number of components k, weights w_j, and parameters μ_j, σ_j^2. In the Bayesian approach, we have to put priors on k, w_j, and μ_j, σ_j. One option is to separate the estimation task into two parts by comparing the quality of a fixed set of models $k = 1, \ldots, K$. Recently, it became popular to use an infinite mixture model

$$f(x) = \sum_{j=1}^{\infty} w_j f(x; \theta_j),$$

which trades the finite model comparison problem into a more continuous problem with possibly infinitely many components k. Nevertheless, as we will see, we still need to pick a parameter that controls the number of components k indirectly. As a prior for the parameters we could take $\theta_1, \theta_2, \ldots$ to be drawn from some F_0 and we could take $w_1, w_2, \ldots$ to be drawn from the stick breaking prior. This is known as the Dirichlet process mixture model and is an extension of the Dirichlet process prior $F \sim \mathrm{DP}(\alpha, F_0)$ with the difference that we replace the point mass distribution δ_{θ_j} in the original form $F = \sum_{j=1}^{\infty} w_j \delta_{\theta_j}$ by smooth densities $f(x; \theta_j)$. Combining everything, the model is

$$F \sim \mathrm{DP}(\alpha, F_0)$$
$$\theta_1, \ldots, \theta_n \mid F \sim F$$
$$X_j \mid \theta_j \sim f(x; \theta_j), \quad j = 1, \ldots, n.$$

It is important to note that the discreteness of F automatically creates a clustering of the parameters θ_js. This can be considered as an implicit prior for the number of components k. We can control the number of k indirectly by choosing an appropriate concentration parameter α. However, there is no free lunch and choosing α usually involves additional priors [10,11].

To complete the model, we also define Gaussian priors on transformation parameters L_i and q_i. We decompose locally linear transformations

$$A_i x + b_i = \exp\left(\begin{bmatrix} L_i & q_i \\ 0 & 0 \end{bmatrix}\right)\begin{bmatrix} x \\ 1 \end{bmatrix}$$

using the Jordan/Schur decomposition

$$L_i = \frac{1}{2}(L_i - L_i^\mathsf{T}) + \frac{1}{2}(L_i + L_i^\mathsf{T})$$

$$L_i = \text{rotation} + \text{scaling} = \theta\begin{bmatrix} 0 & -r_3 & r_2 \\ r_3 & 0 & -r_1 \\ -r_2 & r_1 & 0 \end{bmatrix} + \text{diag}(s)$$

to obtain a rotation axis $[r_1 \quad r_2 \quad r_3]^\mathsf{T}$ and a rotation angle θ. These rotation axis and rotation angle are better interpretable than a transformation matrix and a translation vector and allow us to define subjective priors. For instance, we may want to define a prior on the angle to favor deformations centered at $0°$ with standard deviation $30°$.

8.3.4 Gibbs Sampler

If we are only interested in cluster assignments $\mathbf{W}$ and ignore the transformation parameter, we can integrate out L_i and q_i, and sample from the remaining integral using the distant dependent Chinese Restaurant Process [7]. This process is a Gibbs sampler. Gibbs samplers are convenient whenever we wish to sample from a posterior that can be decomposed into conditional distributions for which fast ways of sampling are available. For instance, to draw sample from this joint distribution

$$\theta_1, \ldots, \theta_T \sim \pi(\theta^1, \theta^2 \mid \gamma)$$

we can iterate between their respective conditional distributions,
- Step 1: $\theta_i^1 \sim \pi(\theta_1 \mid \gamma, \theta_{i-1}^2)$
- Step 2: $\theta_i^2 \sim \pi(\theta_2 \mid \gamma, \theta_i^1)$

and repeating it many times to obtain samples from the joint distribution

$$\left(\theta_1^1, \theta_1^2\right), \ldots, \left(\theta_T^1, \theta_T^2\right) \sim \pi(\theta^1, \theta^2 \mid \gamma).$$

We use the distant dependent Chinese Restaurant Process to draw samples form the marginal distribution for LBP versus abdominal pain dataset as illustrated in Fig. 8.5. Details on the technical implementation can be found in our recent conference article [26].

Sample 10:

Sample 20:

Sample 30:

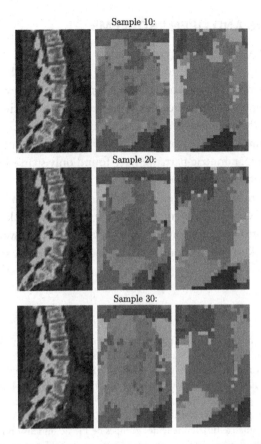

Figure 8.5 Colors are clusters. Left row: Template spine image. Middle row: Back pain patients. Right row: Abdominal pain patients. (For interpretation of the references to color in this figure, the reader is referred to the web version of this chapter.)

8.3.5 Software Implementations

An implementation of a variety of BN tools is available in the R package DPpackage. Our specific implementation for medical images can be found on the GitHub repository of the first author.[7]

8.3.6 Related Work

The usage of BN in computational anatomy is still in its infancy. Related work in the medical context are the detection of spatial activation patterns in fMRI [19] or tractography segmentation [35] using the Dirichlet processes.

---

[7] Code available: https://github.com/ChristofSeiler/BayesianNonparametrics.

8.4 CONCLUSIONS AND OPEN PROBLEMS

In this book chapter, we reviewed general concepts in Bayesian statistics and reported our experience with applying them to problems in computational anatomy. In the parametric part, our treatment focused on the *small deformation* framework. The translation of our work to *large deformation* framework, especially the translation of diagnostic tools for MCMC is currently open. A successful treatment of diagnostics for *large deformation* will most likely require an even stronger interplay between geometry and probability. As reported in the nonparametric part, we have found only sparse literature on applying BP ideas to computational anatomy problems.

Besides the theoretical developments, it will be paramount to provide the community with efficient software implementations in the form of R packages for reproducible research. The recent growing community around the STAN software [8] implementing HMC will hopefully facilitate a more routine usage of Bayesian statistics in computational anatomy.

REFERENCES

[1] S. Allassonnière, Y. Amit, A. Trouvé, Towards a coherent statistical framework for dense deformable template estimation, J. R. Stat. Soc., Ser. B, Stat. Methodol. 69 (1) (2007) 3–29.

[2] S. Allassonnière, E. Kuhn, A. Trouvé, Construction of Bayesian deformable models via a stochastic approximation algorithm: a convergence study, Bernoulli 16 (3) (2010) 641–678.

[3] J.L.R. Andersson, M. Jenkinson, S. Smith, Non-Linear Optimisation, Technical Report TR07JA1, FMRIB Analysis Group of the University of Oxford, 2007.

[4] V.I. Arnold, Sur la géométrie différentielle des groupes de Lie de dimension infinie et ses applications à l'hydrodynamique des fluides parfaits, Ann. Inst. Fourier (Grenoble) 16 (fasc. 1) (1966) 319–361.

[5] J. Ashburner, A fast diffeomorphic image registration algorithm, NeuroImage 38 (1) (2007) 95–113.

[6] J. Besag, Statistical analysis of non-lattice data, J. R. Stat. Soc., Ser. D, Stat. 24 (3) (1975) 179–195.

[7] D.M. Blei, P.I. Frazier, Distance dependent Chinese restaurant processes, J. Mach. Learn. Res. 12 (Aug) (2011) 2461–2488.

[8] B. Carpenter, A. Gelman, M. Hoffman, D. Lee, B. Goodrich, M. Betancourt, M.A. Brubaker, J. Guo, P. Li, A. Riddell, Stan: a probabilistic programming language, J. Stat. Softw. (2016) (in press).

[9] P. Diaconis, The Markov chain Monte Carlo revolution, Bull., New Ser., Am. Math. Soc. 46 (2) (2009) 179–205.

[10] M.D. Escobar, M. West, Bayesian density estimation and inference using mixtures, J. Am. Stat. Assoc. 90 (430) (1995) 577–588.

[11] M.D. Escobar, M. West, Computing nonparametric hierarchical models, in: Practical Nonparametric and Semiparametric Bayesian Statistics, Springer, 1998, pp. 1–22.

[12] T.S. Ferguson, A Bayesian analysis of some nonparametric problems, Ann. Stat. (1973) 209–230.

[13] U. Grenander, M.I. Miller, Computational anatomy: an emerging discipline, in: Current and Future Challenges in the Applications of Mathematics, Providence, RI, 1997, Q. Appl. Math. 56 (4) (1998) 617–694.

[14] M.P. Heinrich, I.J. Simpson, B.W. Papież, M. Brady, J.A. Schnabel, Deformable image registration by combining uncertainty estimates from supervoxel belief propagation, Med. Image Anal. 27 (2016) 57–71.

[15] M. Hernandez, M.N. Bossa, S. Olmos, Registration of anatomical images using geodesic paths of diffeomorphisms parameterized with stationary vector fields, in: ICCV 2007, IEEE, 2007, pp. 1–8.

[16] D.D. Holm, T. Schmah, C. Stoica, Geometric Mechanics and Symmetry: From Finite to Infinite Dimensions, Oxf. Texts Appl. Eng. Math., vol. 12, Oxford University Press, Oxford, 2009, With solutions to selected exercises by David C.P. Ellis.

[17] S. Holmes, S. Rubinstein-Salzedo, C. Seiler, Curvature and concentration of Hamiltonian Monte Carlo in high dimensions, preprint, arXiv:1407.1114, 2014.

[18] A. Joulin, Y. Ollivier, Curvature, concentration and error estimates for Markov chain Monte Carlo, Ann. Probab. 38 (6) (2010) 2418–2442.

[19] S. Kim, P. Smyth, H. Stern, A nonparametric Bayesian approach to detecting spatial activation patterns in fMRI data, in: MICCAI 2006, Part II, in: Lect. Notes Comput. Sci., Springer, Heidelberg, 2006, pp. 217–224.

[20] M. Lorenzi, N. Ayache, G.B. Frisoni, X. Pennec, LCC-demons: a robust and accurate diffeomorphic registration algorithm, NeuroImage 81 (2013) 470–483, http://dx.doi.org/10.1016/j.neuroimage.2013.04.114, http://www.sciencedirect.com/science/article/pii/S1053811913004825.

[21] R.M. Neal, MCMC using Hamiltonian dynamics, in: Handbook of Markov Chain Monte Carlo, in: Chapman & Hall/CRC Handb. Mod. Stat. Methods, CRC Press, Boca Raton, FL, 2011, pp. 113–162.

[22] P. Risholm, F. Janoos, I. Norton, A.J. Golby, W.M. Wells III, Bayesian characterization of uncertainty in intra-subject non-rigid registration, Med. Image Anal. 17 (5) (2013) 538–555.

[23] P. Risholm, S. Pieper, E. Samset, W.M. Wells III, Summarizing and visualizing uncertainty in non-rigid registration, in: International Conference on Medical Image Computing and Computer-Assisted Intervention, Springer, 2010, pp. 554–561.

[24] P. Risholm, E. Samset, W. Wells III, Bayesian estimation of deformation and elastic parameters in non-rigid registration, in: B. Fischer, B.M. Dawant, C. Lorenz (Eds.), Biomedical Image Registration, in: Lect. Notes Comput. Sci., vol. 6204, Springer, Berlin, Heidelberg, 2010, pp. 104–115.

[25] C. Robert, G. Casella, Introducing Monte Carlo Methods with R, Springer Science & Business Media, 2009.

[26] C. Seiler, X. Pennec, S. Holmes, Random spatial structure of geometric deformations and Bayesian nonparametrics, in: Geometric Science of Information, in: Lect. Notes Comput. Sci., vol. 8085, Springer, 2013, pp. 120–127.

[27] C. Seiler, S. Rubinstein-Salzedo, S. Holmes, Positive curvature and Hamiltonian Monte Carlo, in: Advances in Neural Information Processing Systems, NIPS, 2014, pp. 586–594.

[28] I. Simpson, M. Cardoso, M. Modat, D. Cash, M. Woolrich, J. Andersson, J. Schnabel, S. Ourselin, A.D.N. Initiative, et al., Probabilistic non-linear registration with spatially adaptive regularisation, Med. Image Anal. 26 (1) (2015) 203–216.

[29] I.J.A. Simpson, J.A. Schnabel, A.R. Groves, J.L.R. Andersson, M.W. Woolrich, Probabilistic inference of regularisation in non-rigid registration, NeuroImage 59 (3) (2012) 2438–2451.

[30] D.W. Thompson, On Growth and Form, Cambridge Univ. Press, 1942.

[31] A. Trouvé, L. Younes, Shape spaces, in: Handbook of Mathematical Methods in Imaging, Springer, 2011, pp. 1309–1362.

[32] M. Unser, Splines: a perfect fit for signal and image processing, IEEE Signal Process. Mag. 16 (6) (1999) 22–38.

[33] K. Van Leemput, Encoding probabilistic brain atlases using Bayesian inference, IEEE Trans. Med. Imaging 28 (6) (2009) 822–837.

[34] T. Vercauteren, X. Pennec, A. Perchant, N. Ayache, Diffeomorphic demons: efficient non-parametric image registration, NeuroImage 45 (1 Suppl.) (2009) S61–S72.

[35] X. Wang, E.E. Grimson, C.-F.F. Westin, Tractography segmentation using a hierarchical Dirichlet processes mixture model, NeuroImage 54 (1) (2011) 290–302.

[36] D. Wassermann, M. Toews, M. Niethammer, W. Wells III, Probabilistic diffeomorphic registration: representing uncertainty, in: Biomedical Image Registration, Springer, 2014, pp. 72–82.

[37] S. Wieser, B. Horisberger, S. Schmidhauser, C. Eisenring, U. Brügger, A. Ruckstuhl, J. Dietrich, A.F. Mannion, A. Elfering, O. Tamcan, U. Müller, Cost of low back pain in Switzerland in 2005, Eur. J. Health Econ. 12 (5) (2011) 455–467.

[38] X. Yang, M. Niethammer, Uncertainty quantification for LDDMM using a low-rank Hessian approximation, in: International Conference on Medical Image Computing and Computer-Assisted Intervention, Springer, 2015, pp. 289–296.

[39] L. Younes, Shapes and Diffeomorphisms, vol. 171, Springer Science & Business Media, 2010.

[40] M. Zhang, P.T. Fletcher, Finite-dimensional Lie algebras for fast diffeomorphic image registration, in: International Conference on Information Processing in Medical Imaging, Springer, 2015, pp. 249–260.

[41] M. Zhang, N. Singh, P.T. Fletcher, Bayesian estimation of regularization and atlas building in diffeomorphic image registration, in: Information Processing in Medical Imaging, IPMI, in: Lect. Notes Comput. Sci., Springer, 2013, pp. 37–48.

PART II

Open Source Implementation Examples

CHAPTER 9

Morpho and Rvcg – Shape Analysis in R
R-Packages for Geometric Morphometrics, Shape Analysis and Surface Manipulations

Stefan Schlager

Biological Anthropology, Albert-Ludwigs University Freiburg, Freiburg, Germany

Contents

9.1	Introduction	218
9.2	Preliminaries and Installation	218
9.3	Landmark Based Shape Analysis with Morpho	219
	9.3.1 Data Import/Export	220
	9.3.2 Imputation Methods for Landmark Data	220
	9.3.3 Interactive Outliers Detection	223
	9.3.4 Spatial Alignment and Procrustes Analysis	224
	9.3.4.1 Object Symmetry	*226*
	9.3.5 Principal Component Analysis and Relative Warp Analysis	227
	9.3.5.1 Determining Meaningful Principal Components	*229*
	9.3.6 Classification	230
	9.3.7 Covariation Between Shapes – Two-Block Partial Least-Squares (2B-PLS)	232
	9.3.8 Visualization	234
	9.3.9 Semilandmarks	237
	9.3.9.1 Semilandmarks on Symmetric Structures	*241*
	9.3.9.2 Transfer Semilandmarks from a Template to All Specimens in a Sample: A Simple Registration Approach	*242*
9.4	Manipulations on Triangular Meshes Using Rvcg (and Morpho)	244
	9.4.1 Rvcg	245
	9.4.2 Reading and Writing Triangular Surface Meshes (Table 9.9)	245
	9.4.3 Cleaning	245
	9.4.4 Retrieving Mesh Information	246
	9.4.5 Mesh Manipulations	246
	9.4.6 Spatial Queries: Closest Points and KD-Trees	248
	9.4.7 Vertex Clustering/Sampling	249
9.5	Beyond CRAN	250
	9.5.1 Surface Registration and Shape Models	250
9.6	Final Remarks	253
	References	255

Statistical Shape and Deformation Analysis
DOI: 10.1016/B978-0-12-810493-4.00011-0

9.1 INTRODUCTION

Over the last two decades, methods subsumed under the term Geometric Morphometrics (GM) have become indispensable for quantifying and analyzing shape variability in biological data [29,1]. These methods operate on 2D or 3D coordinates representing the geometric properties of a biological structure. The software presented in this chapter was created as a response to a lack of appropriate free (open source) tools that featured more advanced methods to capture and analyze 3D surface shapes, such as 3D sliding semilandmarks (see Section 9.3.9). Especially in the 3D domain, shape analysis is also closely linked to the ability to process and visualize digital 3D surfaces. While appropriate algorithms are often published in detail, implementations are often missing. To overcome this issue, the software presented in this chapter was written and published. The requirements to provide that sort of software were the openness of the underlying platform and ease of use, so the application can be shared with or extended by other researchers. The statistical/mathematical platform R [31] satisfies all these needs, provides a package manager, as well as a build-in documentation system and allows to analyze the data with a vast number of extensions (called *packages*), allowing sophisticated state-of-the-art statistical analyses. The fact that many scientists working with biological data are already familiar with R is also beneficial for sharing code and making it available to researchers with no or only little proficiency in software programming.

The functionality is split into two distinct packages: *Morpho*[1] (Section 9.3) for landmark based shape analysis and *Rvcg*[2] (Section 9.4) for computations involving 3D triangular meshes, i.e. surfaces represented by a concatenation of triangular cells. Both packages are multi-platform and available via the CRAN[3] repository (Section 9.2). This chapter will introduce to both packages and provide examples and code snippets allowing readers to readily reproduce the analyses. For documentation on using R, please refer to https://cran.r-project.org/ where a large amount of documentation for all proficiency levels is provided. I further assume the reader to have a basic understanding of geometric morphometrics and statistical shape analysis and to be familiar with triangular surface meshes [36]. While data processing and handling in R allows the use of a wide variety of statistical procedures, this chapter will concentrate on those that are shipped with the packages presented below.

9.2 PRELIMINARIES AND INSTALLATION

Installation is, thanks to the CRAN repository, quite easy. To install the packages presented in this chapter, issue the following command from your R-command line:

---

[1] https://cran.r-project.org/package=Morpho.

[2] https://cran.r-project.org/package=Rvcg.

[3] https://cran.r-project.org/.

```
> install.packages('Morpho')
```

This will also take care of the dependencies, one of which is the second package presented here: *Rvcg*.

For installing a development snapshot, the R-package *devtools* provides the functionality to install and build a package from git repositories such as github. As both packages contain code that needs compilation, appropriate compilers and toolchain need to be installed. While this will be already the case on most Linux systems, OS X users will need to install XCODE[4] and Windows users can use *Rtools*.[5] To install the latest development snapshot, using *devtools*, run

```
> devtools::install_github('zarquon42b/Morpho') ## or
> devtools::install_github('zarquon42b/Rvcg') ## respectively
```

As required by all packages distributed via CRAN, all public functions are documented and include self-contained working example code and data. The example landmark set provided by *Morpho* via the dataset *boneData* was placed in 3D Image data according to the definitions in [35, p. 33]. It consists of ten manually placed landmarks on the bone surface of 80 human noses. The digitization was performed on surfaces with a much higher resolution than the one also included in the example data (belonging to the first specimen) which is strongly decimated and some structures, such as sutures, are close to invisible. In all examples provided below, it is assumed that the required packages are already loaded (e.g. by issuing `require(Morpho)`). To get the full benefit of modern multi-threaded processors, it is recommended to install a BLAS implementation optimized for a specific machine, such as ATLAS[6] [39].

9.3 LANDMARK BASED SHAPE ANALYSIS WITH MORPHO

Morpho is targeting morphometricians dealing with landmark data and provides vast functionality covering data import/export, imputation methods and a variety of statistical analyses tailored for landmark data. Its main purpose is to perform landmark based statistical shape analysis, with emphasis on 3D configurations. It is intended to cover all bases starting from file input/output to imputation methods, statistical analyses, the handling of semilandmarks and basic registration routines that allow users to project configurations from a template to all specimens throughout the sample under observation. Due to the nature of R as a script based platform, as well as the wide range of tasks that can be managed, it is rather a collection of functions than a sleekly designed software aimed at one specific goal and a predefined workflow. Therefore, all provided examples

[4] https://developer.apple.com/xcode/.
[5] https://cran.r-project.org/bin/windows/Rtools/.
[6] http://math-atlas.sourceforge.net/.

are selected with respect to user requests and support emails that have reached me since the packages have been officially released on CRAN. Because analyses of large sample sizes and/or large amount of coordinates can be computational demanding, *Morpho* is profiled for speed and efficiency when processing both large amounts of landmarks as well as large sample sizes, making use of compiled C++ code, using the *Rcpp*[7] interface and parallel computing using *OpenMP*.[8]

9.3.1 Data Import/Export

The convention for storing and processing landmarks using *Morpho* is to store single landmark configurations of k landmarks in m dimensions as $k \times m$ matrices and a sample of size n as arrays of dimensionality $k \times m \times n$. That way, the landmark data can also be processed by other packages dealing with landmark configurations, like *shapes*[9] or *geomorph*.[10] To import/export landmarks into and from the R-workspace, *Morpho* provides a variety of import functions dealing with common file formats used to store landmark data (Table 9.1). As the IDAV-landmark editor[11] is still quite popular for digitizing landmarks, there are the functions read.pts to read single landmark configurations and read.lmdta for entire samples exported by landmark editor with the *dta* option enabled. As the type of landmark (standard landmark, semi-landmark on curves or surfaces) are encoded in the landmark name, the function cExtract allows for obtaining this information conveniently from data imported with the functions above.

For importing and exporting digital surface data, *Morpho* provides the functions ply2mesh, mesh2ply, obj2mesh, mesh2obj, that allow import and export of surface data stored in ASCII-based .ply/.obj file format. Additionally, file2mesh, a wrapper for the function vcgImport from the package *Rvcg* (see Section 9.4.2), allows importing *obj*, *ply* and *stl* files in both ASCII and binary format. The meshes are stored as objects of class *mesh3d*, as specified by the package *rgl*[12] which is also used for all 3D visualization purposes.

9.3.2 Imputation Methods for Landmark Data

Especially when dealing with paleological or prehistoric data, researchers are often confronted with incomplete structures and placing the same set of landmarks consistently throughout a sample is impossible. *Morpho* provides two approaches for imputing miss-

---

[7] https://cran.r-project.org/package=Rcpp.
[8] http://openmp.org/wp/.
[9] https://cran.r-project.org/package=shapes.
[10] https://cran.r-project.org/package=geomorph.
[11] http://graphics.idav.ucdavis.edu/research/projects/EvoMorph.
[12] https://cran.r-project.org/package=rgl.

Table 9.1 Functions dealing with landmark import and export

Function name	Purpose
write.fcsv	Reading/writing fiducials (a.k.a. landmarks) placed in 3DSlicer[a]
read.mpp	Read landmarks placed in meshlab[b]
readallTPS	Read landmark data acquired using James Rohlf's TPS series[c]
r2morphoj/r2morphologika	Save landmark data readable for MorphoJ[d] and Morphologika[e]
read.csv.folder	Batch import data from multiple csv files stored in the same folder
readLandmarks.csv	Import landmark data from csv files

[a] https://www.slicer.org/.
[b] http://meshlab.sourceforge.net/.
[c] http://life.bio.sunysb.edu/ee/rohlf/software.html.
[d] http://www.flywings.org.uk/morphoj_page.htm.
[e] https://sites.google.com/site/hymsfme/downloadmorphologica.

ing data: fixLMtps, operating on $k \times m \times n$ arrays and fixLMtps to estimate missing landmarks from their bilateral counterparts.

fixLMtps is based on a weighted nearest-neighbor interpolation combined with a thin-plate spline (TPS) deformation [6] (see Algorithm 9.1).

1 Determine missing landmarks for each specimen;
2 Calculate Procrustes Analysis (see Algorithm 9.2) on complete configurations;
3 **for** *each specimen* **do**
4   **if** *landmarks are missing* **then**
5   Align to mean computed from complete shapes;
6   Compute Procrustes distances to all complete shapes;
7   Use the *k*-closest complete shape and calculate weighted (by inverse distance) average;
8   Deform weighted average to defect specimen using a TPS interpolation calculated based on existing landmarks;
9   **end**
10 **end**

Algorithm 9.1: Outline of nearest-neighbor imputation of missing landmarks.

Example 9.1. This example shows the usage of fixLMtps to impute missing data using a weighted nearest neighbor approach.

We first load the example data and generate a copy.

```
> data(boneData)
> data <- boneLM
```

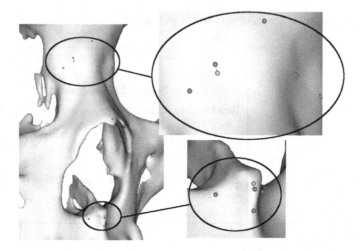

Figure 9.1 Imputation using fixLMtps, with estimated landmarks (red) compared to actual landmarks (cyan). (For interpretation of the references to color in this figure legend, the reader is referred to the web version of this chapter.)

Set the first and fifth landmark of first specimen to NA

```
> data[c(1,5),,1] <- NA
> repair <- fixLMtps(data,comp=4)
```

Both landmarks have been estimated quite reasonably (Fig. 9.1) with a displacement of 0.73 mm for the nasal spine and 1.14 mm for the *Nasomaxillofrontale*. In cases of missing bilateral data, there is also the function fixLMmirror (operating on single matrices and arrays) that estimates landmarks from their bilateral counterparts. Instead of using a weighted sample mean, the landmark configuration is mirrored, then left and right landmarks are relabeled and the missing data are imputed by deforming the mirrored version to the original one, again using a TPS deformation. In cases where a complete side is missing, the mirrored landmarks are aligned to the original ones by using unilateral landmarks only.

Example 9.2. Impute missing landmarks from their bilateral counterparts.

```
> data(boneData)
```

Determine corresponding landmarks on the left and right side. **Important**: keep order consistent

```
> left <- c(4,6,8)
> right <- c(3,5,7)
> pairedLM <- cbind(left, right)
> exampmat <- boneLM[,,1]
> exampmat[5,] <- NA #set 5th landmark to NA
> fixed <- fixLMmirror(exampmat, pairedLM=pairedLM)
```

Note: this function works, for obvious reasons, only on missing bilateral data where there is at least one side present. When confronted with a missing unilateral landmark, the output leaves it untouched and issues a warning. For fixing data with both unilaterally and bilaterally missing data, these functions can be combined easily, first imputing missing bilateral landmarks and then estimating the unilateral ones using fixLMtps.

9.3.3 Interactive Outliers Detection

Statistical analyses, especially with small sample sizes, can become strongly skewed by outliers. To identify those outliers in regard to the overall sample's distribution, *Morpho* provides an interactive tool (find.outliers) that allows conveniently browsing those specimens that are suspiciously different from the rest of the sample. If the difference is only caused by inconsistently ordered landmarks, it also allows for fixing this by manual reordering. find.outliers provides two metrics for detecting outliers: Procrustes distance and Mahalanobis distance. While the function can deal with singular covariance matrices using the general inverse, the option PCuse allows to perform an initial Principal Component Analysis (PCA) and then use the first *n* PCs to compute the Mahalanobis distances more robustly. After an initial alignment, distances (according to the selected metric) of all specimen to the sample consensus are calculated and sorted by decreasing values. Individual per-landmark differences to the mean are plotted in 3D or 2D, depending on the data dimensionality, to facilitate the identification of misplaced landmarks.

Example 9.3. Below is an example using the Mahalanobis distances computed from the first 10 PCs. At first, we create some outliers by swapping the left and right version of a bilateral landmark, a frequent real world mistake, of the first specimen.

```
> data(boneData)
> boneLM[7:8,,1] <- boneLM[8:7,,1]
> outliers <- find.outliers(boneLM, mahalanobis= TRUE, PCuse=10)
```

This leads to the following output, reporting the specimen's position in the array, its name (if available) and Mahalanobis distance, as well as the probability of *not* being an outlier, i.e. being a probable shape, based on the corresponding χ^2-distribution.

```
> outlier #1: 1 - skull_0144_ch_fe
Mahalanobis D^2 dist. to mean: 70.4909915979642
probability of specimen belonging to sample: 3.31793888087514e-11
add to outlierlist (y/N/s)? y=yes,n=no,s=switch landmarks: s
```

The plot (Fig. 9.2) correctly suggests the landmarks 7 and 8 to be mislabeled which we are going to fix by swapping their positions. After doing so by following the instructions, we get:

```
new distance to mean: 9.97066149724801
probability of specimen belonging to sample: 0.443071024719963
```

After swapping landmarks, the specimen is well within the range of the distribution.

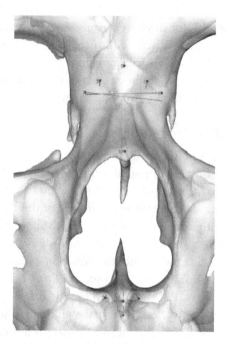

Figure 9.2 Graphical output of `find.outliers`. Blue spheres show the sample mean, with the red lines being the deviation of the observed specimen. For better comprehensibility, the surface mesh on which the landmarks where digitized is added in this figure. (For interpretation of the references to color in this figure legend, the reader is referred to the web version of this chapter.)

Note: `find.outliers` browses through *all* specimens, leaving it to the user to decide when distances to the mean become tolerable or whether to look at the entire sample.

9.3.4 Spatial Alignment and Procrustes Analysis

Removing differences owed to different positions in space and, optionally, scale and reflection, i.e. Procrustes superimposition, is an essential prerequisite of statistical shape analysis [18,14]. This means the original data recorded in an x–y–z coordinate system needs to be spatially transformed to make two or more shapes statistically comparable. *Morpho* provides functionality to compute a variety of transformations of which a subset is required for Procrustes superimposition.

The following spatial transformations can be computed in *Morpho*: rigid alignment, similarity transformation (rigid alignment + scaling), affine transformation (rigid alignment + scaling + shearing) as well as TPS interpolation. Additionally, there are functions allowing to rotate a 3D object around arbitrary axes in space.

Table 9.2 Functions dealing with spatial transformations

Function name	Purpose
computeTransform/ applyTransform	Wrapper for rotonto and tps3d to allow for computing and applying rigid, similarity, affine and TPS transformations
rotonto	Rotate/scale/translate one set of landmarks onto another
rotmesh.onto	Align a mesh based on two landmark configurations
tps3d	Deform a set of coordinates or a mesh using a TPS deformation based on two landmark configurations
rotaxis3d	Rotate an object around an arbitrary axis in 3D
rotaxisMat	Calculate a rotation matrix around an arbitrary axis through the origin in 3D
rotonmat	Rotate matrix of landmarks by using a rotation determined by two matrices
pcAlign	Align two pointclouds based on their principal axes – optionally with subsequent rigid icp alignment
ProcGPA	Perform General Procrustes Analysis
procSym	Perform General Procrustes Analysis, with support for bilateral landmarks, including projection into tangent space and subsequent PCA
icpmat	Match two pointclouds using an iterative closest point search

While the functionality is spread throughout different separate functions (Table 9.2), we concentrate on presenting a simplified interface for computing and applying transformations based on two sets of corresponding coordinates. At first we look at the transformations that are designed to handle the mapping between two configurations based on landmark information, or more general, between two sets of corresponding points. computeTransform operates on $k \times m$ matrices as well as on meshes of class *mesh3d* (assuming correspondence between their vertices). For affine transformations it returns a square transformation matrix, and for TPS transforms it returns an object of class *tps* that contains the coefficients used for TPS interpolation, as well as the original landmark coordinates. The resulting transformation can subsequently be applied using applyTransform to both matrices and meshes. The option invert allows to apply the inverse of the transformation – as there is no trivial inversion of TPS deformations, a TPS transformation will be computed by swapping target and reference coordinates.

For performing statistical analyses, a sample-wide alignment is needed. The most common method is the Generalized Procrustes Analysis (GPA) [18,14]. While there exists a closed form solution for the 2D case using imaginary numbers[13] [14], the 3D case uses an iterative procedure (Algorithm 9.2).

[13] As the results are not considerably different from the iterative procedure, *Morpho* only uses the latter.

1 Initialize variables *distance, threshold* with *distance > threshold*;
2 (Optional) scale all specimen to unit centroid size;
3 Initialize reference shape by selecting random specimen from the sample;

4 **while** *distance > threshold* **do**
5 | Align all specimens to current reference;
6 | (Optional) scale all specimens to minimize the general Procrustes distance (Mean Squared Procrustes distance to mean shape);
7 | Compute mean shape;
8 | Set reference to updated mean shape;
9 | Update: *distance* = Procrustes distance between old and new mean;
10 **end**

Algorithm 9.2: General Procrustes Analysis.

The functions `ProcGPA` and `procSym` both provide this functionality, also allowing for a variety of other options controlling the transformations, with `ProcGPA` being the workhorse and `procSym` also performing subsequent Euclideanization via orthogonal projection into the tangent space at the sample's mean, followed by a PCA. Additionally, a variety of useful statistics is reported. The most important parameters controlling the alignment are `scale` enabling scaling for minimizing Procrustes distances, `reflection` to allow/prohibit reflections and `CSinit` to perform an initial scaling to unit centroid size. Additionally, `ProcGPA` allows to assign weights (option `weights`) for each landmark and `centerweight` to allow the center of rotation to be the accordingly weighted centroid.

9.3.4.1 Object Symmetry

Biological structures often show intrinsic symmetry and there exist methods to deal with asymmetry in a GM context [26,21,20,19]. The general approach is to mirror all landmark configurations and relabel the bilateral landmarks accordingly. Using both the original and the mirrored and relabeled landmark configurations, a Procrustes Analysis is performed leading to a global alignment of all data. For each specimen the consensus of the mirrored and original landmarks is then calculated, leading to a perfectly symmetric shape, called the symmetric component of shape. The deviation of the actual shape from this symmetrized version can be viewed as the asymmetry inherent in the data. The function `procSym` implements this functionality, when the option `pairedLM` is provided with a two-column integer matrix where each row contains the indices of the corresponding bilateral landmarks.

Example 9.4. This example shows how to perform a Procrustes analysis for landmarks sets with object symmetry.

Table 9.3 Functions related to Principal Component Analysis

Function name	Purpose
prcompfast	Implementation of R's generic prcomp function, optimized for large datasets
procSym	Procrustes Analyses including a subsequent PCA
showPC	Retrieve shapes from all sorts of scores obtained from projections into a vector space
pcaplot3d	Visualize shape changes associated with a PCA calculated from 3D data

In the example data, rows 3, 5 and 7 contain the right hand landmarks with rows 4, 6 and 8 holding their left hand counterparts.

```
> left <- c(4,6,8)
```

Determine corresponding landmarks on the right side (**Important**: keep the order consistent)

```
> right <- c(3,5,7)
> pairedLM <- cbind(left,right)
> symproc <- procSym(boneLM, pairedLM=pairedLM)
```

The resulting object contains additional arrays named Asym and Sym, holding the per-specimen symmetric shape as well as the deviation from this symmetric shape. Additionally, PCAs are to be performed on each shape component separately. The resulting object is of class *symproc* and statistical significance for directional and fluctuating asymmetry can be assessed using a Procrustes ANOVA [21] by running the function procAOVsym on that object.

9.3.5 Principal Component Analysis and Relative Warp Analysis

Shape variables derived from superimposed landmark data are usually of high dimensionality, as each observation consists of $k * m$ variables. In statistical shape analysis, the most common procedure to reduce dimensionality is to perform a Principal Component Analysis (PCA) on the vectorized landmark data (Table 9.3). procSym takes care of this as well, returning not only the aligned coordinates but also the results of a PCA (eigenvectors and eigenvalues of the covariance matrix as well as the scores). A PCA describes the variability of a sample by parameterizing the variation around its mean looking for an orthonormal vector basis (principal components) optimally explaining this variation. Hereby, the first k principal components (PCs) are the best rank-k approximation to the covariance of shape displacements regarding to the mean. While standard PCA is a very useful tool for parameterizing the overall variability, it does not account for different types of transformations into which shape changes can be decomposed. If one is interested in such a decomposition, a Relative Warp Analysis (RWA)

[6,7] allows to separate affine and non-affine variation [32]. The decomposition is based on the TPS' property to address affine and non-affine deformations separately [6]. Both standard PCA and RWA are based on the decomposition of the data covariance matrix, but standard PCA does this with respect to Procrustes distance and an RWA with respect to bending energy, thus only addressing the non-affine variation. The affine (uniform) component of shape changes can be calculated using the methods proposed in [32]. That way, the shape space S can be decomposed into $S = U \oplus B$, with U being the subspace of uniform (affine) variation and B the subspace of pure bending (non-affine variation). Depending on the research question, the non-affine variation can be weighted either by the bending energy matrix or its inverse in order to emphasize large or small scale deformations (Fig. 9.7). Using *Morpho*, an RWA can be performed by the function `relWarps`, that returns both the uniform and non-uniform components of shape as well as the vector bases of the respective subspaces. However, due to the nature of the TPS and the matrix algebra involved, an RWA is not recommendable for shapes consisting of more than ∼5000 coordinates. If one is not explicitly interested in the basis of the vector space associated with non-affine variation, i.e. a basis of B (needed for restoring shape from relative warp scores), the option `computeBasis=FALSE` will suppress its computation, reducing calculation time and memory footprint.

Example 9.5. In this example, we are going to compare population differences in our example data using standard PCA, as well as two RWAs, one emphasizing small scale and one large scale deformations.

```
> data(boneData)
```

Compute a Procrustes analysis with subsequent PCA:

```
> procpca <- procSym(boneLM)
```

Perform a relative warp analysis emphasizing large scale variation

```
> relwarps1 <- relWarps(boneLM,alpha=1)
```

And finally, a relative warp analysis emphasizing small scale variation:

```
> relwarps2 <- relWarps(boneLM,alpha=-1)
```

Fig. 9.3 shows the differences between a standard PCA and RWA: While the PCA only shows that both groups are separated by the first PC, we can see from plotting the uniform component, that this is mostly owed to affine differences. Looking at the RW-scores, we then can see that the remaining differences can be better explained by small scale deformations. As landmark data are stored in $k \times m \times n$ arrays but most statistical analyses operate on matrices, *Morpho* provides the convenient function `vecx` to convert between arrays and matrices. In order to retrieve data from a projection into a vector space, such as PCA or RWA, `showPC` allows for easy retrieval of shapes from such data, provided the scores, the vector base and the corresponding mean shape by which the data was centered (see also Section 9.3.8).

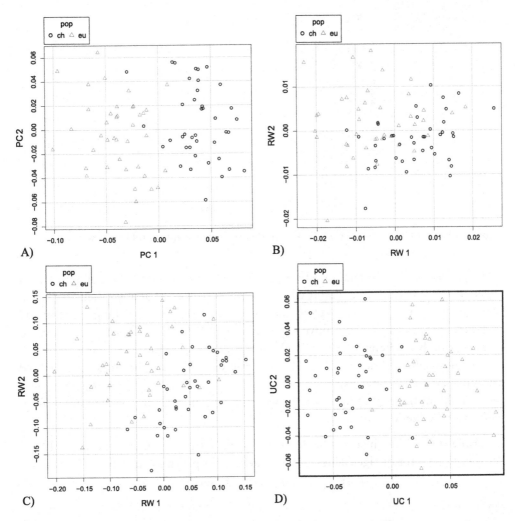

Figure 9.3 Comparison between standard PC-scores and relative warp/uniform component scores. A) Standard PC-scores; B) RW-scores emphasizing large scale bending; C) RW-scores emphasizing small scale bending; D) uniform component scores associated with affine differences.

9.3.5.1 Determining Meaningful Principal Components

In current research, one can often find interpretations of single PCs, treating the shape change along this PC as the effect of some predictor variable (e.g. age or sex). This, however, presumes that these PCs have a mathematical meaning as distinctive axes of an ellipsoid representing the sample's multivariate normal distribution. But if the analyzed axes belong to a glorified sphere rather than a "real" ellipsoid, these interpretations are meaningless because one could have used any other direction, leading to entirely dif-

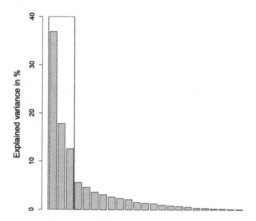

Figure 9.4 Variance explained by each PC. PCs inside the rectangle are considered meaningful.

ferent patterns and interpretations. Looking at these axis is only meaningful if one does not look at a spherical distribution where each axis is as good as any other. Bookstein [9] suggests an approach that exploits the likelihood ratio between the geometric and arithmetic mean of two consecutive eigenvalues (see [25,9] for details). This allows to estimate whether differences between those values are above an expected value, indicating a non-spherical distribution. The function `getPCtol` determines the threshold based on sample size and expected value, while `getMeaningfulPCs` operates on a vector holding the eigenvalues (sorted decreasingly) and the size of the sample from which these values originate.

Example 9.6. Using the example data from above, we determine the meaningful PCs. We specify a vector of eigenvalues and provide the sample size (the depth of the array).

```
> getMeaningfulPCs(procpca$eigenvalues,n=dim(boneLM)[3])
$tol
[1] 1.372847

$good
[1] 1 2 3
```

This tells us that, based on the specific sample size, a PC must be about 1.38 times the value of its successor to be considered meaningful and that this criterion is only met by the first three PCs. A barplot (Fig. 9.4) visualizes this comprehensively: while the first three eigenvalues stand for distinct directions, the ellipsoid becomes more spherical for the subsequent eigenvalues.

9.3.6 Classification

Classification, i.e. reliably assigning specimens based on their properties to distinct groups, is a common task in biological and medical research. A widely used statisti-

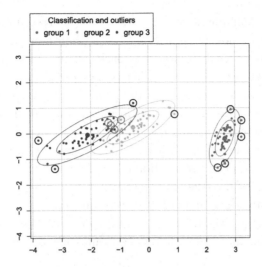

Figure 9.5 Classification using Mahalanobis distances to group centroids. Black circled dots: outliers not associated with any group; red circled: misclassified specimens. (For interpretation of the references to color in this figure legend, the reader is referred to the web version of this chapter.)

cal technique for classification is Canonical Variate Analysis (CVA). Hereby, the latent space is sought within which the between-group distances are maximized. The basis of this space (canonical axes) are the principal components of a PCA computed from the among-groups covariance matrix premultiplied by the inverse pooled within-group covariance matrix [3,4,11,22]. The canonical scores, based on which group assignment is determined, are the projection of the (grand mean) centered data into this space. The function CVA provides such functionality, optionally running permutation tests (both on Euclidean and Mahalanobis distances) for assessing statistical significance of pairwise between-group distances. If one is interested in the typicality probabilities or to classify unknown specimens, the functions typprobClass and typprob provide the required functionality. Hereby, it is assumed that the data within each group follow a multivariate normal distribution, either with common or distinct covariance structure. For each specimen to be classified, the probability of belonging to each group is then estimated based on its Mahalanobis distance to the group centroids [41]. The classification is performed by assigning each specimen to the group for which the highest probability is computed. If there is no group with a likely association (above a given threshold), the specimen is tagged as unclassifiable (see Fig. 9.5). It is also useful to calculate the typicality probabilities for CV-scores to find outliers. Common discriminant procedures, such as the above mentioned CVA, always assign a specimen to a specific group without determining whether this decision is likely. When dealing with unknown specimen, such as specimens of extinct taxa, a misclassification by assigning a specimen to an unlikely group can lead to erroneous conclusions.

Although CVA is appropriate for classifications, interpretations of shape changes along canonical axes are not straightforward: while the canonical axes are orthogonal, this is usually not the case for their projection back into the shape space, thus all such interpretations have to be considered carefully [28]. For assessing between-group structure and associated shape trajectories without distorting the shape space, a between-group PCA [10,28] provides a good compromise in many situations. Hereby, a PCA is performed on the groups' centroids and the entire sample, centered by the grand mean, is then projected onto the resulting PC-axes. Using *Morpho*, this can be accomplished by groupPCA, which also runs a permutation test on Euclidean between-group distances. As there is no scaling of the data space involved, the interpretation of shape changes along the resulting axes is straightforward, referring directly to the shape variation explaining between group variability. groupPCA and CVA both operate on landmark arrays as well as on data matrices with rows containing per-observation variables.

9.3.7 Covariation Between Shapes – Two-Block Partial Least-Squares (2B-PLS)

For exploring patterns of covariation between two sets of coordinates (or two sets of variables in general), a two-block partial least squares analysis (2B-PLS) [33] can be employed. Hereby, those linear combinations maximizing the covariation between the two sets of variables are sought, treating both sets symmetrically – i.e. no distinction is made between predictor and response. The general approach is to decompose (using a singular value decomposition) the submatrix of the joint covariance matrix that holds the information about the covariation between both sets. *Morpho* provides a series of functions allowing calculation, prediction and visualization of such analyses. The function pls2B performs the calculation of the basis vectors of the latent space for both sets of variables, accepting both data matrices and landmark arrays as input – subsequent predictions will also be returned in the same fashion. In order to compute PLS-scores for new data, the function getPLSscores can be used. predictPLSfromScores returns predictions for precomputed PLS-scores, whereas predictPLSfromData performs a projection into the latent space beforehand. In order to visualize the effect of a specific set of latent variables, plsCoVar predicts data along a given score in positive and negative direction showing how this variable effects the covariation for both sets of variables.

Example 9.7. Using the example data of the hard-tissue of the nose, we are going to evaluate the covariation between the nasal bones and the region around the nasal spine. The visualization can be seen in Fig. 9.6.

First load the data and perform a full GPA

```
> data(boneData)
> proc <- procSym(boneLM)
```

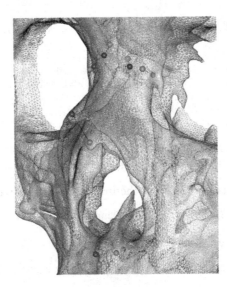

Figure 9.6 Effect of first latent variable computed via 2B-PLS. It can be interpreted that less prominent nasal bones (landmarks on the nasal bones) covary with a less profiled nasal spine.

The first four landmarks are placed around the nasal spine and landmarks 5–10 are placed on the nasal bones. Based on this information, we compute a 2B-PLS for both sets of variables. To assess statistical significance for each dimension, we run a permutation test on the singular values by comparing the actual value to those obtained by randomly combining both sets of landmarks. As both sets of variables are part of the same configuration, each permutation is accompanied by a full Procrustes alignment.

```
> plsBone <- pls2B(proc$rotated[1:4,,],proc$rotated[5:10,,],
  rounds = 1000,same.config = T)
> plsBone
 Covariance explained by the singular values

 singular value % total covar. Corr. coefficient p-value
5.572750e-04   81.119459655       0.7672251    0.001
2.319731e-04   14.055985483       0.8184755    0.001
7.979283e-05    1.663082043       0.4539590    0.001
7.567017e-05    1.495668646       0.4577120    0.001
5.564339e-05    0.808747689       0.4553959    0.001
[truncated]
```

The output reports that the first five latent variables are significantly related to explaining covariation between those two sets of landmarks, however, the first two account for most of the actual covariation, explaining ~81.1% and ~14.1% respectively. In order to visualize the shape changes associated with the first dimension, we predict the shapes for ±2 standard deviations of the PLS-scores for both sets.

```
covar <- plsCoVar(plsBone,i=1,sdx=2,sdy=2)
```

Now show the landmark displacements and display a warped mesh for better comprehensibility

```
> deformGrid3d(covar$x[,,1],covar$x[,,2],size=0.01,col1=4,col2=2)
> deformGrid3d(covar$y[,,1],covar$y[,,2],add = T,size=0.01,col1=4,
  col2=2)
> require(rgl) ## load rgl package
> wire3d(tps3d(skull_0144_ch_fe.mesh,boneLM[,,1],
  rbind(covar$x[,,1],covar$y[,,1])),col=4)
> wire3d(tps3d(skull_0144_ch_fe.mesh,boneLM[,,1],
  rbind(covar$x[,,2],covar$y[,,2])),col=2)
```

In order to predict a specific specimen (the first in our array), we can use

```
> predictPLSx <- predictPLSfromData(plsBone,x=proc$rotated[1:4,,1])
```

to predict values for the variables named x above, and

```
> predictPLSy <- predictPLSfromData(plsBone,x=proc$rotated[5:10,,1])
```

to predict values for the variables named y.

By optimizing the underlying routines for speed and memory efficiency, pls2B can
also be used on dense point clouds with a large amount of coordinates.

9.3.8 Visualization

R comes with a large variety of packages that allow sophisticated data visualizations.
Morpho calls plot from the built-in *graphics* package and uses the OpenGL based package *rgl* [2] for 3D visualizations. Table 9.4 shows the most important functions for
visualizing shape differences.

One of the main tasks of visualization is relating the result of statistical procedures
such as regressions that operate on PC–scores (or other latent spaces) to shape changes in
the configuration space. To visualize the effects, a projection back into the configuration
space is required. If x is a vector of scores resulting from a regression, for example, then
the shape can be retrieved by $\mu + \sum_i x_i v_i$, where v_i denote the vector space basis (i.e.
the PCs) and μ denotes a constant vector which has been subtracted from the data
beforehand (usually the sample mean). The function showPC allows for a convenient
retrieval of shapes from such data. Its arguments are the scores from which the shapes
should be predicted, a matrix containing the basis of the vector space (PCs) and the
constant (mean) shape in matrix notation (rows containing the landmark coordinates).

For visualizing shape differences between two sets of coordinates, the functions
deformGrid3d and deformGrid2d allow for comprehensible visualization of per-
landmark displacement. Additionally, the effect that TPS deformation, computed from
the landmark differences, has on a quadratic/cubic grid can be visualized. In the following example, we are going to retrieve the shapes associated with the first relative warp

Table 9.4 Functions related to visualization

Function name	Purpose
`warpmovie2d/warpmovie3d`	Create a sequence of images showing predefined steps of warping two meshes or landmark configurations (2D and 3D) into each other
`meshDist`	Calculate and visualize distances between surface meshes or 3D coordinates and a surface mesh
`deformGrid2d/deformGrid3d`	Visualize differences between two superimposed sets of 2D/3D landmarks by deforming a square/cubic grid based on a thin-plate spline interpolation

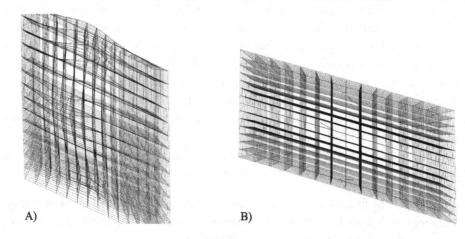

A)   B)

Figure 9.7 Visualization of affine and non-affine deformations. A) RW1 from Example 9.5; B) Uniform component 1 from the same example.

and the first uniform component from the example above. The shape change is visualized along the first RW and the first uniform component, using ±3 standard deviations of the respective scores.

Example 9.8. Visualize the uniform component and RW1 from Example 9.5.

```
> bescores <- as.matrix((c(-3,3))*sqrt(relwarps1$Var$eigenvalues[1]))
> predictRW1 <- showPC(bescores,relwarps1$bePCs[,1],
  mshape = relwarps1$mshape)
> deformGrid3d(predictRW1[,,1],predictRW1[,,2],ngrid = 15,add=F)
> ucscores <- as.matrix(c(-3,3))*sd(relwarps1$uniscores)
> predictUC1 <- showPC(ucscores,relwarps1$uniPCs[,1],
  mshape = relwarps1$mshape)
> deformGrid3d(predictUC1[,,1],predictUC1[,,2],ngrid = 15)
```

As to be expected, Fig. 9.7 shows the effects of the relative warp as an elastic deformation, while the uniform component results in an affine deformation of the cubic grid.

Table 9.5 Important parameters of `meshDist`

Parameter	Purpose
`from,to,tol`	Determine range to consider creation of the heat map, distances with absolute values below `tol` will be bright green
`steps`	Integer: determines break points for color ramp: n steps will produce $n-1$ colors
`distvec`	Vector defining custom per-vertex scalars

For visualizing distances between meshes, the function `meshDist` allows to create heat-maps, based on vertex to mesh distances and stores the result as per-vertex colors. This is done by calculating the closest distance from the reference vertices to the target surface. Hereby, the user can select between ordinary closest distances or along rays (stored as per-vertex normals in the reference mesh). The main options to control the creation of the heat map can be found in Table 9.5. A wide variety of parameters allow to modify the number of steps or range of colors in the heat map, or to specify a vector containing custom per-vertex scalars to compute the heat map from. The result can be stored and rendered with the same options, without recomputing distances, using the function `render`. Or it can be saved to disk as a colored mesh in *ply* format with the color scale saved as png file using the function `export`.

Example 9.9. The following example creates a deformed version of a mesh and renders the distances with a variety of options. Despite its name, `meshDist` does not only operate on meshes but on pointclouds, too (Fig. 9.8).

```
> data(nose)##load example data into workspace
```

Warp a mesh onto another landmark configuration and create an object of class *meshDist*

```
> longnose.mesh <- tps3d(shortnose.mesh, shortnose.lm, longnose.lm)
> mD <- meshDist(longnose.mesh, shortnose.mesh)
```

Now change the color ramp:

```
> render(mD,rampcolors = c("white","red"))
```

Use unsigned distances and render distances < 1 mm green

```
> render(mD,tol=1,sign = FALSE)
```

Finally, we create a vector of the same length as there are vertices, mimicking as some scalar (in this example the distance to the tip of the nose)

```
error <- sqrt(colSums((longnose.mesh$vb[1:3,]-longnose.lm[4,])^2))
meshDist(longnose.mesh,distvec=error)
```

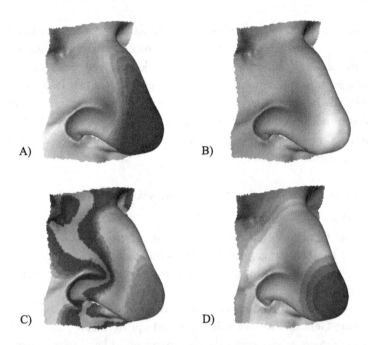

Figure 9.8 Per-vertex scalars (distances) rendered using meshDist: A) default parameters; B) using distances ranging from white to red; C) using unsigned distances and color all values < 1 as green; D) colored using a custom vector containing the distance to the tip of the nose. (For interpretation of the references to color in this figure legend, the reader is referred to the web version of this chapter.)

9.3.9 Semilandmarks

Since the introduction of semilandmarks on 2D curves [8] and their extension to 3D curves and surfaces [16], they have become an indispensable tool for quantifying shape on structures lacking traditional landmarks [15]. When digitizing semilandmarks, there is always a considerable amount of uncertainty involved about the exact position along a curve or on a surface. This uncertainty manifests itself as noise in the data suggesting variance that is simply owed to the lack of homology. For removing this noise, the semilandmarks are allowed to slide along the curves/surfaces minimizing a given metric. For obtaining a linear equation system despite the curved surfaces, the latter are approximated locally. For semilandmarks on surfaces this approximation is achieved by using tangent planes, orthogonal to the normal vector at that point. For semilandmarks on curves the direction vector of the neighboring points along the curve is used. As the sliding along this approximated surfaces may lead to the coordinates slipping off the actual surface, they are projected back onto the surfaces after each relaxation step.

The most common metric to be minimized by the sliding is bending energy (of a TPS deformation) but Procrustes distance can also be used. Both metrics have their advantages and disadvantages: While minimizing Procrustes distance is computationally cheap and can be applied to large amounts of semilandmarks, minimizing bending energy has its limits as the linear equation system based on the bending energy matrix grows quadratically to the amount of semilandmarks involved. For structures with large shape differences, however, minimizing Procrustes distance can lead to unwanted distortions, while the bending energy method ensures a smooth displacement in all cases. It has to be stressed, however, that the homology enforced via this procedure is only valid for the data included in the sliding process. An outline of the sliding procedure can be found in Algorithm 9.3.

---

1 Initialize *distance, threshold* with *distance > threshold*;

2 **while** *distance > threshold* **do**

3   Compute General Procrustes Analysis and compute mean;

4   Estimate tangents/tangent planes for each coordinate;

5   Relax all specimens against mean allowing them to slide along tangent (planes);

6   Project relaxed semilandmarks back onto surface;

7   Compute General Procrustes Analysis and update mean;

8   Update: *distance* = Procrustes distance between old and new mean;

9 **end**

Algorithm 9.3: Outline of the sliding process to enforce homology in semilandmarks.

To allow the minimization of bending energy to be as fast and memory efficient as possible, the matrix rearrangement suggested by Demetris Halazonetis[14] is used, which allows the processing of >3000 semilandmarks within a reasonable time.[15] The function slider3d deals with sample wide sliding of 3D semilandmarks and relaxLM allows to relax one specific set of semilandmarks against another. For 2D configurations, procSym allows semilandmarks to slide along curves. The parameter setup in relaxLM and slider3d is very similar (Table 9.6).

After setting up the parameters controlling the sliding, the surfaces have to be specified in order to estimate the tangent planes (via the mesh normals). There are basically three ways to point to the surface meshes in slider3d:

1. The meshes are stored on disk (less memory usage for large meshes/large samples), supported file types are *obj*, *stl* and *ply*.

   a. The dimnames of the array holding the coordinates contain the file names (without file extension): the surfaces are automatically loaded by specifying the

---

[14] http://www.dhal.com/downloads/CompactSlidingSemilandmarks.pdf.

[15] See my blog entry on that matter: http://zarquon42b.github.io/2014/10/31/slidingtweaks/.

Table 9.6 Functions allowing sliding of semilandmarks

Function name	Purpose
slider3d	Sample-wide relaxation of semilandmarks
relaxLM	Relax reference configuration against target
procSym	Relaxation of 2D semilandmarks along curves

Table 9.7 Important parameters of slider3d and relaxLM

Parameter	Purpose
SMvector	Vector containing indices of landmarks that are allowed to slide. Setting deselect=TRUE, the vector contains those landmarks that are ***not*** allowed to slide.
outlines	Vector or list of vectors containing the indices of the coordinates that constitute curves; these indices can refer both to semilandmarks as well as traditional landmarks – with the latter being useful to sensibly determine the start and end points of curves. For closed curves, the first and last entry must be identical.
surp	Vector of all landmarks that are allowed to slide on surfaces.
bending	If TRUE, bending energy is used as metric, Procrustes distance otherwise.
stepsize	For values <1, this parameter dampens the amount of sliding: If X is the matrix of original positions and D is the matrix of displacement vectors from sliding, then the displaced coordinates are $X + stepsize * D$.

path to the containing folder using the option sur.path and defining the file extension (e.g. *ply*) via the option sur.type.

    b. The array is unnamed: the parameter sur.name can be used to define a vector containing the surface file names.

2. Meshes are loaded into the workspace as objects of class *mesh3d* (faster but uses more memory):

- Store them in a list in the same order as the specimens in the array containing the landmark data.

The functionality also caters to cases where no underlying surface information is available:

1. No surfaces are available at all because the data was digitized directly on the physical object (e.g. using a Microscribe®device):

- In the this case, simply do not specify any meshes, the tangent planes will be approximated by estimating per-landmark normals based on the surrounding semilandmarks – the denser the semilandmarks are placed, the more accurate the normals will be. As there is, for obvious reasons, no surface to project the coordinates back onto, it is recommendable to restrict the incremental sliding along the tangent planes using the parameter stepsize (Table 9.7).

2. The surfaces are defect and some of the data are imputed using the methods above (see Section 9.3.2) but lack underlying surfaces. Surface semilandmarks with the underlying structure missing are relaxed freely in 3D.

    - The indices of the missing landmarks have to be specified per specimen, for an entire sample, the function createMissingList creates an appropriate list and for each specimen with imputed coordinates, a vector containing their indices has to be defined.
    - For relaxLM, where only one specimen is addressed, the option missing allows to specify a vector with indices of coordinates lacking underlying surface information.

Example 9.10. Below are examples for sliding operations on meshes that are already stored in the current workspace.

```
> data(nose)
```

Combine two sets of landmarks to one array

```
> data <- bindArr(shortnose.lm, longnose.lm, along=3)
```

Create an example mesh for the second landmark set by deforming shortnose.mesh (for demonstration purposes only)

```
> longnose.mesh <- tps3d(shortnose.mesh,shortnose.lm,longnose.lm)
```

Aggregate both meshes in a list

```
> meshlist <- list(shortnose.mesh,longnose.mesh)
```

Specify landmarks that are *not* allowed to slide (usually manually placed traditional landmarks)

```
> fix <- c(1:5,20:21)
```

Define two vectors containing indices of distinct outlines and store them in a list

```
> outline1 <- c(304:323)
> outline2 <- c(604:623)
> outlines <- list(outline1,outline2)
```

Select all coordinates allowed to slide along surfaces (simply select all indices and remove all previously selected)

```
> surp <- c(1:623)[-c(fix,outline1,outline2)]
```

Let it slide

```
> slideWithCurves <- slider3d(data, SMvector=fix, deselect=TRUE,
  surp=surp,meshlist=meshlist,iterations=3,
  outlines=outlines)
```

The resulting displacement is visualized in Fig. 9.9.

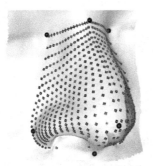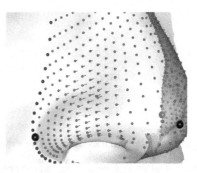

Figure 9.9 Left: Surface with traditional landmarks (black spheres); semilandmarks on curves (blue spheres); and semilandmarks on surfaces (red spheres). The black squares are the tangent planes at each coordinates. Right: the lines visualize the displacement after the sliding. (For interpretation of the references to color in this figure legend, the reader is referred to the web version of this chapter.)

9.3.9.1 Semilandmarks on Symmetric Structures

For assessing asymmetric shape variations using semilandmarks, an additional step is required in the sliding routine. In many cases, semilandmarks are defined on a template surface and then registered to all specimens in a sample (see Section 9.3.9.2). The asymmetry inherent in the template might then be transferred to all specimens and, by sliding to the sample consensus, subsequent analyses will be biased and falsely assume this to be a sample wide effect. Even if the template was perfectly symmetric, directional asymmetry might be overestimated because the relaxation will force a similar expression of asymmetry in all specimens toward the biased mean [35, p. 36]. Such artificially induced asymmetry can easily be removed by relaxing all specimens against a symmetrized sample mean. Therefore, the sample mean and its mirrored and relabeled version are calculated to obtain a perfectly symmetric shape. The sliding now minimizes the bending energy/Procrustes distance toward this symmetric shape, making the resulting coordinates as symmetric as the underlying surfaces allows [34,35]. Using slider3d, this can be accomplished by specifying the parameter pairedLM (according to Example 9.4). For relaxing a single specimen against a symmetric shape (e.g. its own symmetrized version), the function symmetrize computes a perfectly symmetric consensus between a set of landmarks and its mirrored version.

Example 9.11. Below we remove asymmetric noise from a bilateral semilandmark configuration.

```
> data(nose)
```

Define left and right landmarks.

```
> fix <- 6:23
> left <- c(2,fix[fix %% 2 == 1],324:623)
> right <- c(1,fix[fix %% 2 == 0],24:323)
> pairedLM <- cbind(left,right)
```

Symmetrize the landmark configuration

```
> shortnosesym.lm <- symmetrize(shortnose.lm,pairedLM = pairedLM)
```

Setup parameters for sliding:

```
> fixLM <- c(1:5,20:21)
> surp <- (1:623)[-fixLM]
```

Relax original semilandmarks against the symmetrized version

```
> symslide <- relaxLM(shortnose.lm,reference = shortnosesym.lm,
  mesh=shortnose.mesh,SMvector=fixLM,
  deselect=T,surp=surp)
```

9.3.9.2 Transfer Semilandmarks from a Template to All Specimens in a Sample: A Simple Registration Approach

Placing traditional landmarks provides useful information and incorporates biological information and concepts and therefore should preferably performed by a trained expert [15]. Placing semilandmarks, however, can be performed in an automated fashion to decrease processing time. For curves, one has to carefully decide: depending on the structure they run along, it is preferable to place them manually, however, the procedure outlined below, also allows for placing them semi-automatically.

Usually, a template specimen is selected (or computed as an average from the sample to avoid individual bias) on which surface landmarks are then placed. Gunz et al. [16] propose a procedure that simply warps the semilandmarks from the template to the target via a TPS deformation based on the traditional landmarks. This will, however, only work for very simple shapes and only in single layered meshes. For taking into account more complicated structures, *Morpho* provides basic functionality to perform a more sophisticated registration in a semi-automated and user friendly fashion (Table 9.8). The main assumptions are: all semilandmarks are placed on the *outside* of an oriented mesh, i.e. the faces of the mesh point are oriented consistently, and that the meshes are stored on disk in *ply*, *obj* or *stl* format. The procedure is outlined in Algorithm 9.4 (for more details see [35, pp. 37–41]). Based on a template, the semilandmarks are then transferred to all other specimens. An array will be returned containing complete sets of coordinates, consisting of the transferred semilandmarks and the manually placed landmarks, which can be processed by slider3d (see Example 9.10).

Example 9.12. This is a simple example for transferring semilandmarks from a template to all specimens in a sample.

Load data and create mesh for *longnose.lm*

```
> data(nose)
> longnose.mesh <- tps3d(shortnose.mesh,shortnose.lm,longnose.lm)
```

Table 9.8 Functions allowing sliding of semilandmarks

Function name	Purpose
createAtlas	Sets up a template to be used with placePatch, defining fixed landmarks, manually defined curves and semilandmarks to be transferred to all specimens in the sample
placePatch	Transfer semilandmarks from a template to all specimens in a sample
checkLM	Visually inspect the result of placePatch

1 **for** *all specimens in sample* **do**
2   **if** *Curves are placed throughout the sample* **then**
3   Relax all curves against template
4   **end**
5   Deform template to target via TPS based on those landmarks placed throughout the sample (including semilandmarks on curves if available);
6   Compute surface normals at semilandmark positions;
7   Inflate: Project semilandmarks outward along normals;
8   Deflate: Project them back to the closest intersection with the surface;
9   Compute deviation of original normals and surface normals at new positions;
10   **for** *all semilandmarks* **do**
11   **if** *deviation of normal direction* $> \frac{\pi}{2}$ **then**
12   tag coordinate as misplaced;
13   **end**
14   **end**
15   Relax semilandmarks against template and allow relaxation of tagged semilandmarks in 3D;
16   Project semilandmarks on target surface;
17 **end**

Algorithm 9.4: Outline of template registration using placePatch.

Create atlas

```
> fix <- c(1:5,20:21)
> atlas <- createAtlas(shortnose.mesh,
  landmarks = shortnose.lm[fix,],
  patch=shortnose.lm[-fix,])
```

View atlas

```
> plotAtlas(atlas)
```

Create named landmark array with only fixed landmarks

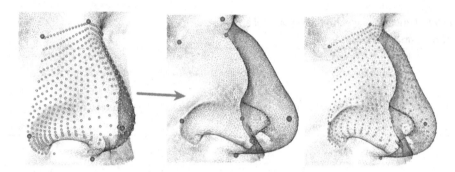

Figure 9.10 Transfer surface semilandmarks (green spheres) defined on template (left) to target surface (center) based on manually placed anatomical landmarks (red spheres). Right: result of placePatch from Example 9.12. (For interpretation of the references to color in this figure legend, the reader is referred to the web version of this chapter.)

```
> data <- bindArr(shortnose.lm[fix,], longnose.lm[fix,], along=3)
> dimnames(data)[[3]] <- c("shortnose", "longnose")
```

Write meshes to disk

```
> mesh2ply(shortnose.mesh, filename="shortnose")
> mesh2ply(longnose.mesh, filename="longnose")
```

Transfer semilandmarks from template to target (result is depicted in Fig. 9.10)

```
> patched <- placePatch(atlas, data, path="./", inflate=2,
  rhotol=pi/2)
```

A more sophisticated method also suitable for more complex shapes can be used by combining Example 9.16 and Example 9.17.

9.4 MANIPULATIONS ON TRIANGULAR MESHES USING RVCG (AND MORPHO)

As can be seen from the previous section, working with surface meshes is an essential task when dealing with 3D shape analysis. While the package *rgl* provides rendering capabilities for 3D structures, R was lacking proper tools for efficiently manipulating triangular meshes. The package *Rvcg* is designed to close this gap and to provide fast and efficient meshing operations from within R, allowing users to integrate these routines into their processing pipelines. Additionally, there are also some useful functions in *Morpho* complementing those from *Rvcg*. This section introduces a selection of functions from both packages. All functions calling *vcglib* routines are prefixed with *"vcg"* (e.g. vcgSmooth).

Table 9.9 Important functions for reading and writing triangular meshes

Function name	Purpose
vcgImport	Read obj, stl, ply files, both ASCII and binary
vcgPlyWrite	Save objects of class mesh3d in ply format
vcgStlWrite	Save objects of class mesh3d in stl format – ASCII and binary

9.4.1 Rvcg

Rvcg is built using the headers-only C++ library *vcglib*,[16] best known by being the code basis of **meshlab**.[17] To allow optimal compatibility with the 3D rendering package *rgl*, triangular meshes to be processed with *Rvcg* are stored in R as objects of class *mesh3d*, which is basically a list of matrices containing vertices, faces, normals and material properties. As all functionality of *Rvcg* and *Morpho* is targeting triangular meshes but the class *mesh3d* can also represent surfaces consisting of quadrilateral cell shapes, the helper function quad2trimesh (*Morpho*) allows a conversion to triangular ones. To allow the meshing operations contained in *vcglib* to operate on these meshes, *Rvcg* provides a wrapper, converting *mesh3d* objects into *vcglib's TriMesh* class and back. This allows to implement all functionality from *vcglib* which is then accessible from within R. To increase speed, some routines, such as closest point and/or KD-Tree searches are sped up using *OpenMP*.

9.4.2 Reading and Writing Triangular Surface Meshes (Table 9.9)

The most versatile function for importing and exporting 3D triangular meshes is vcgImport which can handle files stored in *stl*, *obj* and *ply* format, both the ASCII and binary versions. When loading a file, vertex normals are optionally recomputed (setting updateNormals=TRUE) and additional material such as vertex normals or texture can be processed when setting readcolor=TRUE. The option clean enables the removal of unreferenced vertices and duplicate faces – so it should be disabled when reading files containing point clouds.

9.4.3 Cleaning

In order to clean and smooth surface meshes, *Rvcg* provides three functions that implement a variety of algorithms. vcgClean allows for efficient cleaning such as removal of duplicated, unreferenced (i.e. not part of the triangles forming the surface) and non-manifold vertices as well as duplicated, non-manifold or degenerate faces. The different cleaning options are coded numerically and the parameter sel allows to specify a vector with multiple selections. Because one operation can cause the creation of another

[16] http://vcg.sf.net/.

[17] http://meshlab.sourceforge.net/.

unwanted property – for example merging close vertices can lead to degenerate faces – the option `iterate` will run the function with the selected cleaning operations, until no further cleaning is required. Often, meshes generated from clinical CT-data do not consist of manifolds but of a set of unconnected meshes. `vcgIsolated` can either extract the largest component(s) (those above a given diameter/number of faces) or split it into separate disjunct meshes (stored in a list). For smoothing a mesh, `vcgSmooth` provides several algorithms, including common ones like those proposed by Vollmer et al. [38] and Taubin [37].

9.4.4 Retrieving Mesh Information

To obtain information about a mesh's surface topology, the functions `vcgGetEdge`, `vcgNonBorderEdge`, `vcgCurve` and `vcgMeshres` report all edges (or those being border/non-border edges), information about per-face/per-vertex curvature or mesh resolution (i.e. average edge length). Additionally, `vcgMetro` is an implementation of the command line tool *Metro* [12] to evaluate and report differences (e.g. the Hausdorff distance) between two meshes.

9.4.5 Mesh Manipulations

There are three possible ways to change the resolution of a mesh: simplification (less triangles), complexification (more triangles) and a complete remeshing of the surface. All three can be accomplished using *Rvcg*. For high quality mesh simplification, `vcgQEdecim` implements a quadric edge collapse decimation, allowing the user to specify same parameters as the GUI of *meshlab* with the standard settings leading to good results in most cases. To increase the resolution of a mesh, i.e. to refine the surface, `vcgSubdivide` allows a triangle subdivision using a butterfly interpolation or loop subdivision scheme [42]. To create an entirely new mesh from an existing one, `vcgUniformRemesh` calculates a distance volume and creates a surface at the value zero using a marching cube algorithm.

Example 9.13. This example demonstrates how to simplify (to 20% the amount of its faces), subdivide (until an edge length of 1 is reached) and remesh an example surface (created from a surface scanner and included in *Rvcg's* example data) (Fig. 9.11).

```
> require(Rvcg) #load the package.
> data(humface)
```

Simplify

```
> decimface <- vcgQEdecim(humface, percent=0.2)
```

Subdivide

```
> subdivide <- vcgSubdivide(humface,type="Loop",
  looptype="regularity",threshold=1)
```

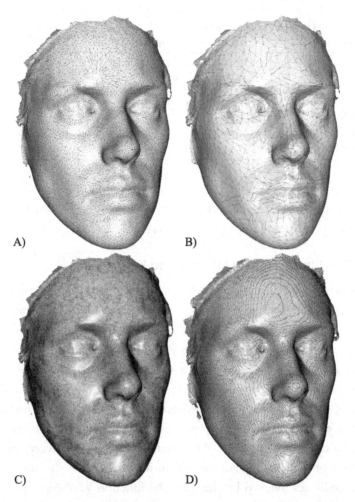

A)   B)

C)   D)

Figure 9.11 Meshing operations in *Rvcg*: A) original mesh; B) a mesh decimated to 20% of the amount its original faces; C) a mesh refined using a loop subdivision scheme optimizing regularity; D) a uniformly remeshed surface.

Remesh

```
> remesh <- vcgUniformRemesh(humface,voxelSize=1,multiSample = TRUE)
```

Additionally, *Morpho* contains a variety of functions for removing vertices (rmVertex), to reorient all faces of a mesh (invertFaces), to concatenate multiple meshes into one (mergeMeshes) or to align two meshes by their principal axes (pcAlign). For analyses dealing with asymmetry, the function mirror or mirror2plane can be of interest: the first one mirrors a mesh and tries to align it to the original one using a rigid iterative closest point procedure, while the second allows to

Table 9.10 Functions related to closest point queries

Function name	Purpose
`vcgClost/vcgClostKD`	Find closest points on surfaces
`vcgRaySearch`	Find intersections on mesh along rays
`vcgKDtree/vcgCreateKDtree/` `vcgSearchKDtree`	Create and search KD-Trees
`vcgCreateKDtreeFromBarycenters/` `vcgClostOnKDtreeFromBarycenters`	Closest points on meshes with reusable search structures

mirror a mesh at an arbitrary plane, defined by three coordinates on a plane or by one coordinate on a plane and the plane's normal vector.

9.4.6 Spatial Queries: Closest Points and KD-Trees

Finding closest points, either on a surface or in a point cloud, is an essential tasks in 3D shape analysis – be it for assessing distances (Example 9.9), registration tasks (Section 9.3.9.2) or projecting semilandmarks back onto the surface (Section 9.3.9). *Rvcg* implements a variety of approaches (Table 9.10): using a spatial grid (`vcgClost`), the search along rays (`vcgRaySearch`) and a variety of approaches based on KD-Trees. Here, emphasis will be put on KD-Trees, as *Rvcg* allows to create and reuse KD-Trees, which can significantly reduce processing time when closest points are sought multiple times on the same surface/point cloud (e.g. in iterative registration procedures). The function `vcgCreateKDtree` sets up a KD-Tree accepting a matrix of 3D coordinates or a mesh of class *mesh3d* (using the vertices) as input. For finding the closest coordinates indexed by this tree, the function `vcgSearchKDtree` can be used. If storing of the KD-Tree is not required, `vcgKDtree` can be used to combine both steps. A special case is the query of closest points on surface meshes: Here, the closest points cannot be determined from a finite set of indexed coordinates but by computing the distances to the mesh's faces (or a subset hereof). To avoid time consuming brute force searches running over an entire mesh, a set of faces has to be selected that are probable candidates for containing the closest points and to limit the search to this subset. *Rvcg* uses KD-Trees based on the barycenters of a mesh's faces to determine such likely candidates (see Algorithm 9.5). A reusable search structure can be set up using `vcgCreateKDtreeFromBarycenters`, which then can be reused multiple times by calling `vcgClostOnKDtreeFromBarycenters`. Analogous to `vcgKDtree`, if reusability is not an issue, the function `vcgClostKD` handles both steps. All KD-Tree queries are multi-threaded on systems supporting *OpenMP*.

1 Compute all barycenters on target mesh;

2 Create KD-Tree for barycenters;

3 Initialize $k > 0$;

4 **for** *all vertices of reference mesh* **do**

5 | Find k closest barycenters;

6 | Initialize *distance, closestPoint*;

7 | **for** *all faces associated with k closest barycenters* **do**

8 | | *newClosestPoint* = closest point on *face*;

9 | | *newDistance* = distance to closest point on face;

10 | | **if** *newDistance < distance* **then**

11 | | | *distance = newDistance*;

12 | | | *closestPoint = newClosestPoint*;

13 | | **end**

14 | **end**

15 **end**

Algorithm 9.5: Closest point search on triangular meshes using KD-Trees.

9.4.7 Vertex Clustering/Sampling

Subsampling surfaces can be a useful approach for sampling semilandmarks on a structure, e.g. for generating a template. `vcgSample` allows to subsample coordinates from a surface using three different approaches: Poisson Disk sampling, K-means clustering and Monte-Carlo sampling [17,13]. For selecting semilandmarks, the first approach leads to an equally distributed pattern, while the second one represents the vertex distribution patterns, as it creates spatial clusters of vertices and uses the cluster centers (projected onto the surface) as samples.

Example 9.14. We perform a Poisson Disk sampling on our example surface.

```
> data(humface)
> subsamplePD <- vcgSample(humface,SampleNum = 500, type="pd")
```

Please note, that due to the implemented algorithm the requested sample size of 500 will vary around this value

```
> nrow(subsamplePD)
[1] 744
```

In this case, we have an excess of 244 coordinates. If we want to insist on the exact number, the parameter `strict` enforces subsampling of the 744 points, using K-means clustering.

If one is interested in spatial clustering of point clouds, this can be efficiently obtained using `vcgKmeans` which is designed to fast and efficiently handle 2D or 3D coordinates – while the generic function for K-means clustering (kmeans) in R can be very slow when dealing with large numbers of coordinates.

9.5 BEYOND CRAN

The software packages presented so far, contain a lot of functionality for statistical shape analysis and surface mesh processing but lack features like more sophisticated registration and shape modeling algorithms. I will shortly address this issue, pointing to R-packages that incorporate some of this functionality, but are not published on CRAN. While the CRAN repository allows to conveniently distribute R-packages providing extensive formal checks of code quality, it also puts strong restrictions to how the software has to be designed and what can be expected being installed on the users' machines as a system requirement, in order to successfully use a package.

The package *mesheR*[18] contains more functionality when dealing with surface meshes but its most notable features are implementations of elastic surface registration algorithms (Table 9.11), optionally allowing to fit shape models generated and saved in the *statismo* [23] format. For dealing with those models, *RvtkStatismo* provides an R-interface to the excellent C++ library *statismo* [23], also adding some high-level functions for creating and modifying statistical shape models. Unfortunately, the compilation of *statismo* and the required build-dependencies make it currently impossible to distribute the package on CRAN, due to the above mentioned restrictions. As *mesheR* relies on some functionality of *RvtkStatismo*, this also excludes this package from being published that way.

The installation of *mesheR* is straightforward and can be directly installed from *github* using the command:

```
devtools::install_github('zarquon42b/mesheR')
```

RvtkStatismo, however, only runs on Linux or OSX, with the *statismo* library already installed. Detailed installation instructions can be found on the package's github page.[19]

9.5.1 Surface Registration and Shape Models

This subsection contains two examples on how to register two surfaces, with and without the aid of shape models. First, we look at the simplest case, registering two (single-layered) surfaces representing the same structure (a human nose). We can use gaussMatch for this task, a (tweaked) implementation of the algorithm proposed by Moshfeghi et al. [30]. Additionally, it allows an initial affine alignment based on landmarks, optionally followed by an affine iterative closest point (ICP) registration.

---

[18] https://github.com/zarquon42b/mesheR.
[19] https://github.com/zarquon42b/RvtkStatismo.

Table 9.11 Surface registration in `mesheR`

Function name	Purpose
`icp`	Align two meshes by an ICP, using rigid, similarity or affine transformations
`gaussMatch`	Fit a mesh/shape model to a target surface, using smooth displacement fields
`AmbergRegister`	Fit a mesh/shape model to a target surface, penalizing topological distortions using the method proposed by [5]
`modelFitting`	Fits a statistical shape model to an already aligned surface mesh, using an LBFSG-optimizer

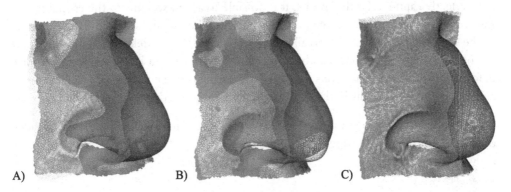

A)   B)   C)

Figure 9.12 Steps fitting a surface mesh to a target. White mesh: reference mesh; red mesh: target mesh. A) Initial state; B) after affine ICP; C) final result of elastic registration after 20 iterations. (For interpretation of the references to color in this figure legend, the reader is referred to the web version of this chapter.)

Example 9.15. Once again, we are using the example dataset from *Morpho*, this time to elastically register the short nose to the long nose.

```
> require(Morpho)
> data(nose)##load data
> longnose.mesh <- tps3d(shortnose.mesh,shortnose.lm,longnose.lm)
```

Set up the parameters for the affine ICP to prealign the meshes before the elastic registration

```
> affine <- list(iterations=20, subsample=100, rhotol=pi/2,
  uprange=0.9
```

We now register both surfaces by minimizing the symmetric distance (i.e. from and to the target mesh) using a displacement field, smoothed by a Gaussian kernel considering neighborhoods of 400 vertices. The parameter `visualize` allows to view the steps of each iteration (results can be seen in Fig. 9.12).

```
> matchNose <- gaussMatch(shortnose.mesh,longnose.mesh,
  lm1=shortnose.lm, lm2=longnose.lm, gamma=2,
  iterations=20,nh=400, angtol=pi/2,
  affine=affine,sigma=20,visualize=T)
```

The following example demonstrates an elastic surface registration procedure that incorporates a Gaussian Process model (GPM): First, a GPM is created from the reference mesh to model smooth deformations. The landmarks provided will be used to align the shape to the model. The model is then constrained based on the given landmark information [40,24] and, using the mean of the constrained model, additional ICPs using both similarity and affine transforms are computed. As the target shape usually is not perfectly captured by the model, it is possible to assign weights to the estimation by the model and to compute a weighted average from the model estimation and the result of the smoothed displacement, to allow shapes outside the model's variability. That way the benefits of using a model can be combined with those of a less restrictive free-form deformation. As large distortions mostly occur during the first iterations when reference and target are still very dissimilar, it is recommended to let the weights decrease slowly.

Example 9.16. We are going to register two femurs, using the example data provided on Marcel Lüthi's github page[20]

```
> require(RvtkStatismo)
> require(Rvcg)
> require(Morpho)
> require(mesheR)
> download.file(url="http://tinyurl.com/j5hfazv/VSD001_femur.vtk",
  "./VSD001_femur.vtk",method="w")
> download.file(url="http://tinyurl.com/j5hfazv/VSD002_femur.vtk",
  "./VSD002_femur.vtk",method="w")
> download.file(url="http://tinyurl.com/j5hfazv/VSD001-lm.csv",
  "./VSD001-lm.csv",method="w")
> download.file(url="http://tinyurl.com/j5hfazv/VSD002-lm.csv",
  "./VSD002-lm.csv",method = "w")
> ref <- read.vtk("VSD001_femur.vtk")
> tar <- read.vtk("VSD002_femur.vtk")
> ref.lm <- as.matrix(read.csv("VSD001-lm.csv",row.names=1,header=F))
> tar.lm <- as.matrix(read.csv("VSD002-lm.csv",row.names=1))
```

We now generate a Gaussian Process model based on a Gaussian kernel with a bandwidth of 50 scaled up by factor 50 and some isotropic scaling around the centroid of the model's mean.

```
> Gkernel <- GaussianKernel(50,50) ## Gaussian kernel
> Skernel <- IsoKernel(scale=0.1,x=ref) ##Isotropic kernel
```

---

[20] Links are shortened using tinyurl, the actual location is https://raw.githubusercontent.com/marcelluethi/statismo-shaperegistration/master/data.

```
> kernel <- SumKernels(Gkernel,Skernel) ## combine kernels
> mymod <- statismoModelFromRepresenter(ref,kernel=kernel,
  ncomp = 100)
```

Create an object of class *Bayes*, containing several parameters controlling the fitting of the shape model. The parameter shrinkfun allows to specify a function of the initial weight (wt) of the model's estimate at the *i*-th iteration. Let S_{disp} be the shape suggested by the smoothed displacement field and S_{model} be its projection into the model space. If wt is specified, the shape for each iteration is computed as $S_{disp} + wt[i] * S_{model}$. sdmax sets the probabilistic boundaries for shapes accepted in each iteration – in this case we allow shapes with probabilities associated with seven standard deviations extrapolated to multivariate data using Mahalanobis distances.

```
> Bayes <- createBayes(mymod,sdmax = rep(7,50),wt=1.5,
  shrinkfun = function(x,i){ x <- x*0.93^i })
```

Setup parameters for similarity and affine ICPs

```
> similarity = list(iterations=10,rhotol=pi/2)
> affine = list(iterations=10,rhotol=pi/2)
```

Run the fitting process with 30 iterations, putting constraints on the allowed normal deviations (set to $\frac{\pi}{2}$) and only considering closest points within a distance of 30 mm. The initial bandwidth of our smoothing kernel is 100 (sigma=100), and the neighborhood to be considered is again set to 400. The resulting steps can be seen in Fig. 9.13.

```
matchGP <- gaussMatch(Bayes,tar,lm1 = ref.lm,lm2=tar.lm,
 iterations = 35,sigma = 100,gamma=2,
 toldist = 30,angtol = pi/2,nh=400,
 visualize = T,similarity = similarity,
 affine = affine)
```

The registration can then optionally be used to transfer points defined on the reference to the target surface, exploiting the unified mesh topology. The function transferPoints from *mesheR*, does this based on barycentric coordinates [27]. The resulting coordinates can then be used as semilandmarks.

Example 9.17. In this final example, we first sample 500 coordinates from the reference and transport them to the target using barycentric coordinates and the fitted surface from the last example.

```
> coords <- vcgSample(ref,SampleNum=500)
> coord_trans <- mesheR::transferPoints(coords, ref, matchGP)
```

9.6 FINAL REMARKS

This chapter intended to give a brief introduction to methods of statistical shape analysis, operations on surface meshes as well as surface registration procedures that can

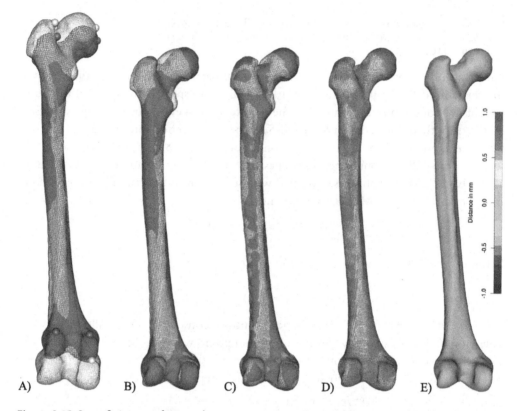

Figure 9.13 Steps fitting a surface mesh to a target using a Gaussian Process model. White mesh: reference mesh; red mesh: target mesh. A) Initial state; B) after affine ICP; C) result after 10 registration steps; D) final result of elastic registration after 20 iterations; E) heatmap visualizing differences between fitted and target surface. (For interpretation of the references to color in this figure legend, the reader is referred to the web version of this chapter.)

be performed using R. The main objective hereby was to provide researchers, dealing with shape analysis and shape modeling, with an overview of the current functionality. For those unfamiliar with or new to R, this may well serve as an entry point to doing shape analysis in R. One of the great advantages of R, as a computational platform, is its openness and extensibility, allowing to provide new algorithms to researches and developers alike. Thanks to the constant increase of functionality guaranteed by the ongoing development of contributed R-packages, and of course the platform itself, all data within the R-workspace can be subjected to a vast amount of state-of-the-art data processing methods and statistical procedures. Hopefully, those readers unfamiliar with R feel encouraged by this tutorial to start using it, which eventually will lead to an enrichment of the existing toolkit by future contributions.

REFERENCES

[1] D.C. Adams, F.J. Rohlf, D.E. Slice, A field comes of age: geometric morphometrics in the 21st century, Hystrix, Ital. J. Mammal. 24 (2013) 7–14, http://www.italian-journal-of-mammalogy.it/article/view/6283.

[2] D. Adler, D. Murdoch, et al., rgl: 3D visualization using OpenGL, https://CRAN.R-project.org/package=rgl, 2016, r package version 0.95.1441.

[3] G.H. Albrecht, Multivariate analysis and the study of form, with special reference to canonical variate analysis, Integr. Comp. Biol. 20 (1980) 679–693, http://icb.oxfordjournals.org/citmgr?gca=icbiol;20/4/679.

[4] G.H. Albrecht, Assessing the affinities of fossils using canonical variates and generalized distances, Hum. Evol. 7 (1992) 49–69, http://dx.doi.org/10.1007/BF02436412.

[5] B. Amberg, Editing Faces in Videos, Ph.D. thesis, University of Basel, 2011, http://edoc.unibas.ch/1415/.

[6] F.L. Bookstein, Principal warps: thin-plate splines and the decomposition of deformations, IEEE Trans. Pattern Anal. Mach. Intell. 11 (1989) 567–585.

[7] F.L. Bookstein, Morphometric Tools for Landmark Data: Geometry and Biology, Cambridge University Press, Cambridge, 1991.

[8] F.L. Bookstein, Landmark methods for forms without landmarks: morphometrics of group differences in outline shape, Med. Image Anal. 1 (1997) 225–243, http://dx.doi.org/10.1016/S1361-8415(97)85012-8.

[9] F.L. Bookstein, Measuring and Reasoning: Numerical Inference in the Sciences, Cambridge University Press, 2014.

[10] A.L. Boulesteix, A note on between-group PCA, Int. J. Pure Appl. Math. 19 (2005) 359–366.

[11] N.A. Campbell, W.R. Atchley, The geometry of canonical variate analysis, Syst. Zool. 30 (1981) 268–280, http://www.jstor.org/stable/2413249.

[12] P. Cignoni, C. Rocchini, R. Scopigno, Metro: measuring error on simplified surfaces, in: Computer Graphics Forum, Wiley Online Library, 1998, pp. 167–174.

[13] R.L. Cook, Stochastic sampling in computer graphics, ACM Trans. Graph. 5 (1986) 51–72.

[14] I.L. Dryden, K.V. Mardia, Statistical Shape Analysis, John Wiley and Sons, Chichester, 1998.

[15] P. Gunz, P. Mitteroecker, Semilandmarks: a method for quantifying curves and surfaces, Hystrix, Ital. J. Mammal. 24 (2013) 103–109, http://www.italian-journal-of-mammalogy.it/article/view/6292.

[16] P. Gunz, P. Mitteroecker, F. Bookstein, Semilandmarks in three dimensions, in: D. Slice (Ed.), Modern Morphometrics in Physical Anthropology, in: Developments in Primatology: Progress and Prospects, Kluwer Academic/Plenum Publishers, Chicago, IL, 2005, pp. 73–98.

[17] J.A. Hartigan, M.A. Wong, Algorithm as 136: a k-means clustering algorithm, J. R. Stat. Soc., Ser. C, Appl. Stat. 28 (1979) 100–108.

[18] D. Kendall, A survey of the statistical theory of shape, Stat. Sci. 4 (1989) 87–120.

[19] C.P. Klingenberg, Analyzing fluctuating asymmetry with geometric morphometrics: concepts, methods, and applications, Symmetry 7 (2015) 843–934, http://www.mdpi.com/2073-8994/7/2/843/htm.

[20] C.P. Klingenberg, M. Barluenga, A. Meyer, Shape analysis of symmetric structures: quantifying variation among individuals and asymmetry, Evolution 56 (10) (2002) 1909–1920.

[21] C.P. Klingenberg, G.S. McIntyre, Geometric morphometrics of developmental instability: analyzing patterns of fluctuating asymmetry with Procrustes methods, Evolution 52 (1998) 1363–1375, http://www.jstor.org/stable/2411306.

[22] C.P. Klingenberg, L.R. Monteiro, Distances and directions in multidimensional shape spaces: implications for morphometric applications, Syst. Biol. 54 (4) (2005) 678–688.

[23] M. Lüthi, R. Blanc, T. Albrecht, T. Gass, O. Goksel, P. Büchler, M. Kistler, H. Bousleiman, M. Reyes, P. Cattin, T. Vetter, Statismo – a framework for PCA based statistical models, http://hdl.handle.net/10380/3371, 2012.

[24] M. Lüthi, C. Jud, T. Gerig, T. Vetter, Gaussian process morphable models, arXiv:1603.07254, 2016.

[25] K.V. Mardia, J.T. Kent, J.M. Bibby, Multivariate Analysis, Academic Press, 1979.

[26] K.V. Mardia, A.N. Walder, Shape analysis of paired landmark data, Biometrika 81 (1994) 185–196.

[27] M. Meyer, A. Barr, H. Lee, M. Desbrun, Generalized barycentric coordinates on irregular polygons, J. Graph. Tools 7 (2002) 13–22.

[28] P. Mitteroecker, F. Bookstein, Linear discrimination, ordination, and the visualization of selection gradients in modern morphometrics, Evol. Biol. 38 (2011) 100–114, http://dx.doi.org/10.1007/s11692-011-9109-8.

[29] P. Mitteroecker, P. Gunz, Advances in geometric morphometrics, Evol. Biol. 36 (2009) 235–247, http://dx.doi.org/10.1007/s11692-009-9055-x.

[30] M. Moshfeghi, S. Ranganath, K. Nawyn, Three-dimensional elastic matching of volumes, IEEE Trans. Image Process. 3 (1994) 128–138, http://dx.doi.org/10.1109/83.277895, http://www.ncbi.nlm.nih.gov/pubmed/18291914.

[31] R Core Team, R: A Language and Environment for Statistical Computing, R Foundation for Statistical Computing, Vienna, Austria, 2015, https://www.R-project.org/.

[32] F.J. Rohlf, F.L. Bookstein, Computing the uniform component of shape variation, Syst. Biol. 52 (2003) 66–69, http://dx.doi.org/10.1080/10635150390132759, http://sysbio.oxfordjournals.org/content/52/1/66.abstract, http://sysbio.oxfordjournals.org/content/52/1/66.full.pdf+html.

[33] F.J. Rohlf, M. Corti, Use of two-block partial least-squares to study covariation in shape, Syst. Biol. 49 (2000) 740–753.

[34] S. Schlager, Sliding semi-landmarks on symmetric structures in three dimensions, Am. J. Phys. Anthropol. 147 (2012) 261, http://dx.doi.org/10.1002/ajpa.21502.

[35] S. Schlager, Soft-Tissue Reconstruction of the Human Nose: Population Differences and Sexual Dimorphism, Ph.D. thesis, Universitätsbibliothek Freiburg, 2013, http://www.freidok.uni-freiburg.de/volltexte/9181/.

[36] P.J. Schneider, D. Eberly, Geometric Tools for Computer Graphics, Elsevier Science Inc., New York, NY, 2002.

[37] G. Taubin, Curve and surface smoothing without shrinkage, in: Fifth International Conference on Computer Vision, Proceedings, 1995, pp. 852–857.

[38] J. Vollmer, R. Mencl, H. Müller, Improved Laplacian smoothing of noisy surface meshes, Comput. Graph. Forum 18 (1999) 131–138, http://dx.doi.org/10.1111/1467-8659.00334.

[39] R.C. Whaley, J.J. Dongarra, Automatically tuned linear algebra software, in: Proceedings of the 1998 ACM/IEEE Conference on Supercomputing, IEEE Computer Society, 1998, pp. 1–27.

[40] C.K. Williams, C.E. Rasmussen, Gaussian Processes for Machine Learning, MIT Press, Massachusetts Institute of Technology, 2006.

[41] S. Wilson, On comparing fossil specimens with population samples, J. Hum. Evol. 10 (1981) 207–214, http://dx.doi.org/10.1016/S0047-2484(81)80059-0, http://www.sciencedirect.com/science/article/pii/S0047248481800590.

[42] D. Zorin, P. Schröder, T. DeRose, L. Kobbelt, A. Levin, W. Sweldens, Subdivision for modeling and animation, ACM SIGGRAPH Course Notes 12 (2000).

CHAPTER 10

ShapeWorks
Particle-Based Shape Correspondence and Visualization Software

Joshua Cates[*,†,‡], **Shireen Elhabian**[*,†,¶], **Ross Whitaker**[*,§]

[*]Scientific Computing and Imaging Institute, University of Utah, Salt Lake City, USA
[†]Biomedical Image and Data Analysis Core, University of Utah, Salt Lake City, USA
[‡]Comprehensive Arrhythmia Research and Management Center, University of Utah, Salt Lake City, USA
[§]School of Computing, University of Utah, Salt Lake City, USA
[¶]Faculty of Computers and Information, Cairo University, Cairo, Egypt

Contents

10.1	Introducing ShapeWorks	258
	10.1.1 The Potential and the Challenge	258
	10.1.2 Rising to the Challenge	259
	10.1.2.1 Particle Systems for a Flexible Shape Representation	*259*
	10.1.2.2 Optimized Correspondences Address the Model Selection Problem	*260*
	10.1.3 ShapeWorks: An Open-Source Implementation of PBM	261
10.2	Particle-Based Modeling	262
	10.2.1 Overview	262
	10.2.2 Surface Representation	263
	10.2.3 Adaptive Distributions on Surface Features	265
	10.2.4 Surface Constraint	266
	10.2.5 The Kernel Width σ for PDF Estimation	267
	10.2.6 Numerical Considerations	267
	10.2.7 Correspondence: Entropy Minimization in Shape Space	268
	10.2.8 Setting Parameters	270
	10.2.9 Illustration of the Properties and Interpretability of the PBM Optimization	271
10.3	PBM Extensions	273
	10.3.1 Modeling Shape with Open Surfaces	274
	10.3.2 Modeling Shape Complexes	276
	10.3.3 Correspondence Based on Functions of Position	277
	10.3.4 Correspondence with Regression Against Explanatory Variables	279
	10.3.5 Dense PBM Correspondence Models	281
10.4	ShapeWorks Software Implementation and Workflow	284
	10.4.1 The ShapeWorks Shape Modeling Workflow	284
	10.4.1.1 Segmentation Preprocessing and Alignment	*285*
	10.4.1.2 Initialization and Optimization	*286*
	10.4.1.3 Analysis	*287*
	10.4.2 The PBM Code Library	288
	10.4.3 The ShapeWorks Software Tool Suite	289
10.5	ShapeWorks in Biomedical Applications	290

Statistical Shape and Deformation Analysis
DOI: 10.1016/B978-0-12-810493-4.00012-2

10.6 Conclusions and Future Work   292
References   292

10.1 INTRODUCING SHAPEWORKS

10.1.1 The Potential and the Challenge

A revolution in shape analysis is underway. While morphometrics have been important for the study of biology and medicine for 100 years, recent advances in multivariate shape representation and statistics [1–8], coupled with the increased availability of computed tomography (CT) and magnetic resonance imaging (MRI), have enabled a whole new generation of morphometric tools. Characterized by the use of modern computational techniques to automatically construct detailed three-dimensional shape representations, these new morphometrics are called Statistical Shape Modeling, and they have the potential to measure anatomy and its variability with an unprecedented level of precision and statistical power.

Statistical Shape Modeling (SSM) is beginning to impact a wide spectrum of basic scientific and clinical applications, including the study of mechanisms of disease in neurobiology [9–11], the design of optimal patient-specific implants and bone substitutes [12–16], reconstruction of anatomical structures from both two- and three-dimensional medical images [9,17–27], computer-aided surgeries through pre- and postoperative surgical planning [28–31,20,19], and reconstructive surgery [32–39]. In addition, research involving large cohorts of image data, an emerging "big data" problem, is also seeing benefit from population-level SSM to better understand disease etiology, monitor pathology progression, and study treatment [16,40–43].

While SSM is poised to revolutionize morphometry, its widespread adoption by the biological and medical research communities has been hindered by a lack of both effective software implementations and generic approaches that can be applied across a wide variety of anatomies. The relative complexity of SSM algorithms makes it challenging to engineer software implementations that are user friendly. The increased computational requirements of these approaches also make it difficult to deploy SSM algorithms on a standard desktop computer, which is the only hardware widely available in research labs. Thus, to maximize the scientific and clinical impact of SSM, we need approaches that are designed to be both simple to use, from an algorithmic perspective, and that are scalable to commodity computer hardware. Furthermore, approaches that are robust to a variety of anatomies will avoid costly development and validation of multiple custom solutions. General approaches also support a larger user base, which leads to more standardization and acceptance of SSM methodologies within research communities.

10.1.2 Rising to the Challenge

It is in response to the preceding challenges that we introduce ShapeWorks. Shape-Works is an open source software implementation of a surface correspondence approach to SSM called Particle Based Modeling (PBM). PBM is designed to be simple to use, robust, and applicable to general anatomy. Simplicity and generalizability are achieved through the use of a particle system representation of shape, instead of relying on specific shape parameterizations. Correspondence points are modeling as interacting sets of particles that redistribute themselves under an energy optimization. The optimization finds correspondence configurations that minimize the entropy of the model, which is a metric of information content. Thus, the optimization learns the shape parameters that are the most efficient descriptors of the geometry of the anatomy, which increases the robustness and statistical power of the model. This energy minimization is then balanced by a parameter-free regularization strategy that maximizes entropy of correspondence positions, in order to ensure good shape representations through efficient surface sampling.

10.1.2.1 Particle Systems for a Flexible Shape Representation

The choice of a specific shape representation defines, and may also limit, the class of shapes that can be modeled. To be applicable to the full range of shape analysis problems in biomedicine, a modeling methodology must be capable of representing different topological classes of shape. A correspondence-based model that relies on spherical parameterizations of shape, for example, can only represent manifold surfaces with spherical topologies [44,10]. By contrast, many important structures in the body, such as the heart, for example, consist of multiple interconnected chambers, with shared boundaries and open surfaces. The boundaries of other structures may be somewhat arbitrarily defined by specific landmarks or regions of interest. Many problems in orthopedics, for example, are concerned only with variability in specific areas of bone or the interaction of bone and cartilage surfaces at joints. In short, human anatomy can be very complex, and the simplifying assumptions of parametric models may even lead to erroneous or misleading results. Another important consideration is that medical or biological shapes are typically derived from the interfaces between organs or tissue types and usually defined implicitly in the form of segmented volumes, rather than explicit parameterizations, triangulations, or surface point samples. Such representations therefore require additional preprocessing steps that may limit the fidelity of the model and introduce error, especially as hypotheses regarding shape become more complex.

Instead of a parametric shape representation, PBM uses the idea of the particle system surface representation first proposed by Witkin and Heckbert, who introduced the idea of modeling a point set as a system of interacting particles that are constrained to lie on an implicit surface. Particles interact with one another with mutually repelling forces, such as electrostatic charge, so that they find distributions that optimally cover,

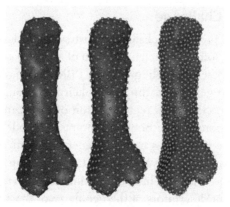

Figure 10.1 Femur shape representation using increasing numbers of particles.

and therefore describe, the surface geometry [45]. Meyer et al. proposed numerically robust extensions to this approach, including a new class of radial-basis energy functions and methods for curvature-adaptive surface sampling [46]. PBM adapts the numerical approaches of Meyer et al. for correspondence models by using a set of interacting particle systems, one for each shape in the sample, to produce optimal sets of surface correspondences. Adopting a point-based surface representation avoids many of the limitations and complexities inherent in parametric representations, such as the limitation to specific topologies and processing steps necessary to construct parameterizations. Another advantage is that, unlike representations that rely on surface meshes, particles do not have fixed neighbors and are free to move past one another to form different neighborhood configurations during the optimization process. This property means that the result is less constrained by the initialization and can potentially produce a less biased, more fully optimized model.

Fig. 10.1 illustrates the concept of a particle system representation of an implicit surface on a femur bone shape. The panels from left to right show an increasing number of particles placed on the surface and the resulting surface reconstruction from the particles. The surface reconstruction is done using the method for unorganized sets of points given by Hoppe et al. [47]. The number of particles doubles in each panel (256, 512, and 1024). As the particle count increases, so does the detail of the corresponding surface reconstruction.

10.1.2.2 Optimized Correspondences Address the Model Selection Problem

Correspondences offer a flexible shape representation and are intuitively the computational extension of traditional landmark models. However, the problem of how to automatically choose correspondence positions is difficult and ill-posed because of the fact that there are potentially an infinite number of possible configurations that can be chosen for the correspondences on the shapes in the sample set. Thus, an important advance in SSM technology is the idea of choosing correspondence positions based on the minimization of an energy function of their positions (or associated features). Such optimized correspondence models are motivated by the Occam's razor principle of parsimony: given a number of possible models for the data, choose the simplest model. This idea has historically been widely applied to model selection problems in statistics, in order to find models with minimal numbers of parameters and greater predictive power [48]. The idea of an optimized correspondence model of shape was first proposed by Kotcheff and Taylor, who developed an algorithm that minimizes the magnitude of the covariance of the correspondences. Davies, Cootes and Taylor later expanded on this idea, using an information-theoretic cost function of correspondence positions based on minimum description length (MDL) [49,50]. The PBM approach to optimized correspondence uses a minimization based on entropy, which is a related measure to minimum description length, in that it seeks to minimize information content of the model.

A major consideration in the optimization process is to avoid overfitting to the data. If correspondence placement is completely unconstrained, for example, the optimal solution with respect to the information content is to place all correspondences as close to one another as possible. For this reason, optimized SSM approaches usually incorporate a constraint that ensures that correspondences faithfully represent the geometries of the samples. Several regularization strategies in the basic MDL formulation have been proposed that entail additional free parameters and assumptions about the quality of the initial parameterizations. One strategy, for example, constrains the solution so that it remains close to an anchor shape [50]. Such approaches, however, artificially limit the minimization process and bias the solution toward the anchor shape. To avoid assumptions about the initial quality of sample surface representations and ad-hoc regularizations, the PBM algorithm instead explicitly constructs good shape representations during the optimization procedure by maximizing an entropy measure on their distributions.

10.1.3 ShapeWorks: An Open-Source Implementation of PBM

Developed at the University of Utah, the underlying methods and theory of PBM have been described in a series of papers over the last 10 years [51–55]. The scientific and clinical effectiveness of ShapeWorks has been demonstrated in a range of applications including neuroscience [53,56–59], biological phenotyping [60,61], orthopedics [61–63] and cardiology [64,65]. The ShapeWorks software itself consists both a flexible C++ code library for the PBM optimization and an evolving suite of software applications that implement workflows for applying PBM to image data. The latter includes command line executables suitable for batch processing large cohorts of image data, as well as desktop applications that support the complete shape analysis workflow, from preprocessing of image segmentations to analysis and visualization of PBM shape models.

The remainder of this chapter describes the theory and implementation of ShapeWorks, starting with the mathematical formulation and numerical implementation of the PBM algorithm and its extensions (Sections 10.2–10.3). Following the PBM development, we describe the ShapeWorks code distribution, which includes a C++ library of PBM code, and the workflow that is implemented in the ShapeWorks software applications (Section 10.4). Finally, we describe some scientific and clinical applications of ShapeWorks (Section 10.5) and conclude with a brief discussion of future work and ShapeWorks developments (Sections 10.6).

10.2 PARTICLE-BASED MODELING

10.2.1 Overview

The Particle Based Modeling (PBM) approach to SSM constructs a correspondence-point model of shape, which describes shape variation by choosing a discrete set of corresponding points on shape surfaces whose relative positions can be statistically analyzed. The correspondence model is analogous to a dense landmark model and is defined as follows. Consider a statistical sample of N surface representations drawn from a population of surfaces. The surface representations are embedded in a d–dimensional Cartesian space (typically, $d = 2$ or $d = 3$). A model for shape variation is constructed by choosing a set of M, d-dimensional points on each of the N surfaces. Each of the points is called a *correspondence* point. Collectively, the set of M points is known as the *configuration*, after Dryden and Mardia [5], and the space of all possible configurations is the *configuration space*. The *configuration matrix*, **C**, is the $M \times d$ matrix of Cartesian coordinates in a configuration. The ordering of the points in the N configurations and, equivalently, the rows in the configuration matrices explicitly define the correspondences among the surfaces. Row $k <= M$ in configuration matrix i, for example, corresponds to row k in configuration matrix j. The variation of the positional information encoded in the rows of the configuration matrices describe geometric variation in shape. Each configuration can be mapped to a single point X in a $d \times M$-dimensional *shape space* by concatenating the correspondence coordinate positions into a single vector. The mapping to the dual shape space is invertible. The sample set forms a distribution in shape space, whose statistical properties can be estimated.

The PBM algorithm models the correspondence positions as sets of dynamic particles that are constrained to lie on the surface of the sample set, as in the surface sampling methods described in Section 10.1.2.1. The optimization is based on the idea of treating correspondence position in configuration space as a random variable, while simultaneously treating correspondence configuration as a random variable. Correspondence positions are optimized by gradient descent on an energy function that balances the negative entropy of the distribution of particles in configuration space with the positive entropy of the distribution of the configurations in shape space. The method is to consider $\mathbf{z}_k \in \Re^{dM}$, $k = \{1, 2, \ldots, N\}$ both as observations on a $dM \times 1$ vector random variable $\mathbf{Z}$ and as N samples of M observations on N, $d \times 1$ vector random variables $\mathbf{X}_k$. The optimization to establish correspondence minimizes the energy function

$$Q = H(\mathbf{Z}) - \sum_{k=1}^{N} H(\mathbf{X}_k), \tag{10.1}$$

where H is an estimation of differential entropy. Minimization of the first term in Q produces a compact distribution of samples in shape space, while the second term seeks uniformly-distributed correspondence positions on the shape surfaces for accurate shape representation. Each term is given in commensurate units of entropy, avoiding the need for a separate regularization strategy. Because correspondence points in this formulation are not tied to a specific parameterization, the method operates directly on volumetric data and extends easily to arbitrary shapes, even nonmanifold surfaces.

10.2.2 Surface Representation

Consider a single configuration for a shape surface $\mathcal{S} \subset \Re^d$. The configuration consists of a discrete set of M points, which are the correspondence positions. The PBM formulation represents these positions with a set of particles, whose positions are considered a sample on a vector random variable $\mathbf{X} \in \Re^d$, with an associated probability density function describing their distribution. This probability density function $p(\mathbf{X} = \mathbf{x})$ gives the probability of an observation $\mathbf{x}$ on $\mathbf{X}$, denoted as $p(\mathbf{x})$. In the limit, the amount of information contained in the sample on $\mathbf{X}$ is the differential entropy of $p(\mathbf{X})$,

$$H(\mathbf{X}) = -\int_S p(\mathbf{X}) \log p(\mathbf{X}) \, dx = -E\{\log p(\mathbf{X})\}, \tag{10.2}$$

where $E\{\cdot\}$ is the expectation. When there are a sufficient number of points sampled from p, the expectation can be approximated by the sample mean [66], which gives

$$H(\mathbf{X}) \approx -\frac{1}{M} \sum_{i=1}^{M} \log p(\mathbf{x}_i). \tag{10.3}$$

The PBM algorithm manipulates particle positions using a gradient-descent optimization on a cost function C, that is an approximation of negative entropy,

$$C(\mathbf{x}_1, \ldots, \mathbf{x}_M) \approx -H(\mathbf{X}). \tag{10.4}$$

The optimization problem is given by

$$\mathbf{z} = \arg \min_{\mathbf{z}} C(\mathbf{x}_1, \ldots, \mathbf{x}_M) \text{ s.t. } \mathbf{x}_1, \ldots, \mathbf{x}_M \in \mathcal{S}, \tag{10.5}$$

and uses a Gauss–Seidel update with forward differences. Each particle therefore moves with a time parameter and positional update,

$$\mathbf{x}_i \leftarrow \mathbf{x}_i - \gamma \frac{\partial C}{\partial \mathbf{x}_i}, \tag{10.6}$$

where γ is a time step. The partial gradient of C for particle i is

$$\frac{\partial C}{\partial \mathbf{x}_i} = \frac{\partial}{\partial \mathbf{x}_i} \frac{1}{M} \sum_{j=1}^{M} \log p(\mathbf{x}_j) = \frac{1}{M} \sum_{j=1}^{M} \frac{\frac{\partial}{\partial \mathbf{x}_i} p(\mathbf{x}_j)}{p(\mathbf{x}_j)}. \tag{10.7}$$

The gradient requires estimates of the probability $p(\mathbf{X} = \mathbf{x}_j)$. For distributions of particles on surfaces, a probability density function may be quite complex, which suggests a nonparametric, kernel-based approach. The PBM algorithm uses a Parzen windowing density estimation [67] that is based on the particle configurations. The probability of the position of a particle in this formulation is given by the mixture of multivariate Gaussian kernels,

$$p(\mathbf{x}, \sigma) \approx \frac{1}{M} \sum_{j=1}^{M} G(\mathbf{x} - \mathbf{x}_j, \sigma), \tag{10.8}$$

where $G(\mathbf{x} - \mathbf{x}_j, \sigma)$ is a d-dimensional, isotropic Gaussian with standard deviation σ. When $j = i$ in (10.7), the partial derivative of p with respect to particle position is

$$\frac{\partial}{\partial \mathbf{x}_i} p(\mathbf{x}_i, \sigma_i) = \frac{1}{\sigma_i^2 M} \sum_{j=1}^{M} G(\mathbf{x}_i - \mathbf{x}_j, \sigma_i)(\mathbf{x}_i - \mathbf{x}_j). \tag{10.9}$$

When $i \neq j$, the derivative is

$$\begin{aligned}
\frac{\partial}{\partial \mathbf{x}_i} p(\mathbf{x}_j, \sigma_j) &= \frac{1}{M} \left[\frac{\partial}{\partial \mathbf{x}_i} G(\mathbf{x}_j - \mathbf{x}_1, \sigma_j) + \frac{\partial}{\partial \mathbf{x}_i} G(\mathbf{x}_j - \mathbf{x}_2, \sigma_j) + \ldots \right. \\
&\quad \left. + \frac{\partial}{\partial \mathbf{x}_i} G(\mathbf{x}_j - \mathbf{x}_i, \sigma_j) + \cdots + \frac{\partial}{\partial \mathbf{x}_i} G(\mathbf{x}_j - \mathbf{x}_M, \sigma_j) \right] \\
&= \frac{1}{M} \left[0 + 0 + \cdots - \sigma_j^{-2} G(\mathbf{x}_i - \mathbf{x}_j, \sigma_j)(\mathbf{x}_i - \mathbf{x}_j) + \cdots + 0 \right] \\
&= \frac{1}{\sigma_j^2 M} G(\mathbf{x}_j - \mathbf{x}_i, \sigma_j)(\mathbf{x}_i - \mathbf{x}_j).
\end{aligned} \tag{10.10}$$

Substituting (10.9) and (10.10) into (10.7) gives

$$\frac{\partial C}{\partial \mathbf{x}_i} = \frac{1}{M} \sum_{j=1}^{M} \frac{G(\mathbf{x}_i - \mathbf{x}_j, \sigma_i)(\mathbf{x}_i - \mathbf{x}_j)}{\sigma_i^2 p(x_i, \sigma_i)} + \frac{1}{M} \sum_{j=1}^{M} \frac{G(\mathbf{x}_j - \mathbf{x}_i, \sigma_j)(\mathbf{x}_i - \mathbf{x}_j)}{\sigma_j^2 p(x_j, \sigma_j)}. \tag{10.11}$$

The computational complexity for Eq. (10.11) is $\mathcal{O}(M^2)$, since the entire density function p must be recomputed for each particle update. To simplify the computation, the PBM formulation instead considers p to be fixed for a given particle update: for $j \neq i$ in (10.11), the estimation of the density function at j is allowed to lag behind the update of particle position i. Under this assumption, $\frac{\partial}{\partial \mathbf{x}_i} p(\mathbf{x}_j, \sigma_j) = 0$, and the second term in (10.11) drops out, simplifying the gradient computation to only $\mathcal{O}(M)$.

After dropping the second term, the final approximation to the gradient of particle positional entropy is given by

$$\begin{aligned}
\frac{\partial C}{\partial \mathbf{x}_i} &\approx \frac{1}{M} \sum_{j=1}^{M} \frac{G(\mathbf{x}_i - \mathbf{x}_j, \sigma_i)(\mathbf{x}_i - \mathbf{x}_j)}{\sigma_i^2 p(x_i, \sigma_i)} \\
&= \frac{1}{M} \sum_{j=1}^{M} \frac{G(\mathbf{x}_i - \mathbf{x}_j, \sigma_i)}{\sigma_i^2 \frac{1}{M} \sum_{k=1}^{M} G(\mathbf{x}_i - \mathbf{x}_k, \sigma_i)} (\mathbf{x}_i - \mathbf{x}_j) \\
&= \frac{1}{M} \sum_{j=1}^{M} w_{ij}(\mathbf{x}_i - \mathbf{x}_j),
\end{aligned} \tag{10.12}$$

where w_{ij} are Gaussian weights based on interparticle distance and $\sum_j w_{ij} = 1$. To minimize C, the particles must move away from each other. Thus, we have a set of particles moving under a repulsive force and constrained to lie on the surface, with $\gamma < \sigma^2$ in (10.6) for stability. The motion of each particle is away from all of the other particles, but interactions are effectively local for sufficiently small σ, where w_{ij} vanishes with increasing interparticle distance.

10.2.3 Adaptive Distributions on Surface Features

The preceding minimization produces a uniform sampling of a surface. For some applications, a strategy that samples adaptively in response to higher order shape information is more effective for several reasons. From a numerical point of view, the minimization strategy relies on a degree of regularity in the tangent planes between adjacent particles, which argues for sampling more densely in high curvature regions. An adaptive sampling strategy also produces a more efficient representation of geometric detail by reducing redundant samples in flatter regions. Adaptive sampling with PBM is implemented by modifying the Parzen windowing in Eq. (10.8) as follows:

$$\tilde{p}(x_i) \approx \frac{1}{M} \sum_{j=1, j \neq i}^{M} G\left(\frac{\mathbf{x}_i - \mathbf{x}_j}{k_j}, \sigma_i\right) \tag{10.13}$$

where k_j is a scaling term proportional to the curvature magnitude computed at each neighbor particle j. The effect of this scaling is to expand space in response to local curvature. A uniform sampling based on maximum entropy in the warped space translates into an adaptive sampling in unwarped space, where points pack more densely in higher curvature regions. The extension of Eq. (10.12) to incorporate the curvature-adaptive Parzen windowing is straightforward to compute. Since k_j is not a function of x_i, the modified gradient is

$$\frac{\partial C}{\partial \mathbf{x}_i} \approx \frac{1}{M} \sum_{j=1}^{M} \frac{G((\mathbf{x}_i - \mathbf{x}_j)/k_j, \sigma_i)(\mathbf{x}_i - \mathbf{x}_j)}{\sigma_i^2 k_j p(x_i, \sigma_i)}. \tag{10.14}$$

There are many possible choices for the scaling term k. Meyer et al. [68] describe an adaptive surface sampling that uses the scaling

$$k_i = \frac{1 + \rho \kappa_i(\frac{s}{2\pi})}{\frac{1}{2}s \cos(\pi/6)}, \tag{10.15}$$

where κ_i is the root sum-of-squares of the principal curvatures at surface location x_i. The user-defined variables s and ρ specify the ideal distance between particles on a planar surface and the ideal density of particles per unit angle on a curved surface, respectively. Note that the scaling term in this formulation could easily be modified to include surface properties other than curvature.

10.2.4 Surface Constraint

The surface constraint in both the uniform and adaptive optimizations is specified by the zero set of a scalar function $F(x)$. This constraint is maintained, as described in several papers [46], by projecting the gradient of the cost function onto the tangent plane of the surface, as prescribed by the method of Lagrange multipliers. The projection operator is given by

$$\mathbf{I} - \mathbf{n} \otimes \mathbf{n}, \tag{10.16}$$

where $\mathbf{I}$ is the identity matrix, $\mathbf{n}$ is the normal to the surface, and $\otimes$ denotes the outer, or tensor, product. The tangent-plane projection is followed by iterative reprojection of the particle onto the nearest root of F by the Newton–Raphson method. Principal curvatures are computed analytically from the implicit function, as described in [69].

10.2.5 The Kernel Width σ for PDF Estimation

Finally, the kernel width σ of the Parzen windowing estimation of particle density must be chosen at each particle. This is done automatically, before the positional update, using a maximum likelihood optimality criterion. The contribution to C of the ith particle is simply the probability of that particle position. Optimizing that quantity with respect to σ therefore gives a maximum likelihood estimate of σ for the current particle configuration. Using the Newton–Raphson method, the strategy is to find σ such that

$$\partial p(\mathbf{x}, \sigma)/\partial\sigma = 0, \tag{10.17}$$

which typically converges to machine precision in several iterations. For the adaptive sampling case, we find σ such that

$$\partial \tilde{p}(\mathbf{x}, \sigma)/\partial\sigma = 0, \tag{10.18}$$

so that the optimal σ is scaled locally based on the curvature. The iteration is given by

$$\sigma^{t+1} \leftarrow \sigma^t + \frac{\frac{\partial p}{\partial\sigma}}{\frac{\partial^2 p}{\partial\sigma^2}}, \tag{10.19}$$

and the first derivative of p with respect to σ, from (10.8), is

$$\frac{\partial}{\partial\sigma} \sum_{j=1}^{M} \frac{1}{M} G(\mathbf{x} - \mathbf{x}_j, \sigma) = \frac{\partial}{\partial\sigma} \sum_{j=1}^{M} \frac{1}{M(2\pi)^{d/2}\sigma^d} e^{\frac{-r_j}{2\sigma^2}}$$
$$= \frac{1}{M(2\pi)^{d/2}\sigma^{d+3}} \sum_{j=1}^{M} e^{\frac{-r_j}{2\sigma^2}} (r_j - d\sigma^2), \tag{10.20}$$

where $r_j = (\mathbf{x} - \mathbf{x}_j)^T(\mathbf{x} - \mathbf{x}_j)$ is the distance from x to x_j. The second derivative follows from (10.20), and is given by

$$\frac{\partial^2}{\partial\sigma^2} \sum_{j=1}^{M} \frac{1}{M} G(\mathbf{x} - \mathbf{x}_j, \sigma) = \frac{1}{M(2\pi)^{d/2}\sigma^{d+6}}$$
$$\times \left[\sum_{j=1}^{M} e^{\frac{-r_j}{2\sigma^2}} (r_j^2 - (3 + 2d)\sigma^2 r_j + d(1 + d)\sigma^4) \right]. \tag{10.21}$$

10.2.6 Numerical Considerations

There are a few important numerical considerations in computing the particle-based surface representation. First, the Gaussian kernels must be truncated. Typically, we truncate kernels so that $G(x, \sigma) = 0$ for $|x| > 3\sigma$. This means that each particle has a finite radius of influence, and a spatial binning structure to identify neighboring particles can be used to reduce the computational burden associated with particle interactions. A second consideration is the case where σ for a particle is too small to allow the particle to interact with its neighbors, and updates of σ or position cannot be computed. When σ is small, kernel size is updated using $\sigma \leftarrow 2 \times \sigma$, until σ is large enough for the particle to interact with its neighbors. A final numerical consideration is that the system must include bounds σ_{min} and σ_{max} to account for anomalies such as bad initial conditions or too few particles. These are not critical parameters, and as long as they are set to include the minimum and maximum resolutions, the system operates reliably.

One final aspect of the particle formulation to consider is that it computes the Euclidean distance between particles, rather than the geodesic distance on the surface. The PBM algorithm therefore assumes sufficiently dense samples so that nearby particles lie in the tangent planes of the zero sets of F. This is an important consideration; in cases where this assumption is not valid, such as highly convoluted surfaces, the distribution of particles may be affected by neighbors that are outside of the true manifold neighborhood. Limiting the influence of neighbors whose normals differ by some threshold value (e.g., 90 degrees) does limit these effects.

10.2.7 Correspondence: Entropy Minimization in Shape Space

A sample set, $\mathcal{E}$, is a collection of N surfaces, each with their own set of M particles mapped to a single, dM-dimensional vector in shape space, i.e. $\mathcal{E} = \mathbf{z}^1, \ldots, \mathbf{z}^N$. The sample set in vector form may be collected into a single matrix $\mathbf{P} = \mathbf{z}_j^k$, with particle positions along the rows and shape samples across the columns. Modeling $\mathbf{z}^k \in \Re^{dM}$ as an instance of random variable $\mathbf{Z}$, the PBM method for correspondence minimizes the combined sample and shape cost function

$$Q = H(\mathbf{Z}) - \sum_{k=1}^{N} H(\mathbf{X}^k), \tag{10.22}$$

which favors a compact representation of the sample, and is balanced against a uniform distribution of particles on each surface.

For this discussion we assume that the complexity of each shape is greater than the number of samples, and so normally $dM > N$. Given the low number of examples relative to the dimensionality of the space, the density estimation requires some assumptions. The PBM algorithm therefore assumes a normal distribution and models $p(\mathbf{Z} = \mathbf{z})$ parametrically using an anisotropic Gaussian with covariance $\mathbf{\Sigma}$. The entropy is then given by

$$H(\mathbf{Z}) \approx \frac{1}{2} \log |\mathbf{\Sigma}| = \frac{1}{2} \sum_{j=1}^{dM} \log \lambda_j, \tag{10.23}$$

where $\mathbf{e}_k, \lambda_k, j = 1, \ldots, dM$ are the eigenvalues of $\mathbf{\Sigma}$.

In practice, $\mathbf{\Sigma}$ will not have full rank, in which case the entropy is not finite. The problem must therefore be regularized with the addition of a diagonal matrix $\alpha\mathbf{I}$ to introduce a lower bound on the eigenvalues. The covariance is estimated from the data, and is given by

$$\mathbf{\Sigma} = (dMN-1)^{-1}\mathbf{YY}^{T}, \tag{10.24}$$

where

$$\mathbf{y}^{k} = \mathbf{z}^{k} - \boldsymbol{\mu}, \text{ and } \boldsymbol{\mu} = \frac{1}{N}\sum_{k=1}^{N}\mathbf{z}^{k}. \tag{10.25}$$

Thus, $\mathbf{Y}$ denotes the matrix of sample vectors $\mathbf{P}$ minus the sample mean $\boldsymbol{\mu}$, i.e. $\mathbf{Y} = \mathbf{P} - \boldsymbol{\mu}\mathbf{1}^{T}$, where $\mathbf{1}$ is a $dM \times 1$ vector of ones. Because $N < dM$, the eigenanalysis in (10.23) is done on the dual space of the $N \times N$ covariance matrix $\mathbf{\Sigma}^{T} = (dMN-1)^{-1}\mathbf{Y}^{T}\mathbf{Y}$. The nonzero eigenvalues of $\mathbf{\Sigma}$ can be obtained from $\mathbf{\Sigma}^{T}$ by noting the following relationships (see also [50]). For eigenvalues and eigenvectors $\{\mathbf{e}_{k}, \lambda_{k}\}$ of $\mathbf{\Sigma}$, $\mathbf{\Sigma}\mathbf{e}_{k} = \lambda_{k}\mathbf{e}_{k}$. Similarly, for eigenvalues and eigenvectors $\{\mathbf{e}'_{k}, \lambda'_{k}\}$ of $\mathbf{\Sigma}^{T}$, $\mathbf{\Sigma}^{T}\mathbf{e}'_{k} = \lambda'_{k}\mathbf{e}'_{k}$. Substituting for $\mathbf{\Sigma}^{T}$, we have

$$(dMN-1)^{-1}\mathbf{Y}^{T}\mathbf{Y}\mathbf{e}'_{k} = \lambda'_{k}\mathbf{e}'_{k}, \tag{10.26}$$

and premultiplying each side by $\mathbf{P}$ gives

$$(dMN-1)^{-1}\mathbf{YY}^{T}\mathbf{Y}\mathbf{e}'_{k} = \lambda'_{k}\mathbf{Y}\mathbf{e}'_{k}, \tag{10.27}$$

which is equivalent to

$$\mathbf{\Sigma}(\mathbf{Y}\mathbf{e}'_{k}) = \lambda'_{k}(\mathbf{Y}\mathbf{e}'_{k}). \tag{10.28}$$

Thus, $\mathbf{e}_{k} = \mathbf{Y}\mathbf{e}'_{k}$, and $\lambda_{k} = \lambda'_{k}$, for nonzero eigenvectors of $\mathbf{\Sigma}$. The covariances $|\mathbf{\Sigma}|$ and $|\mathbf{\Sigma}^{T}|$ are therefore equivalent (up to a constant factor of α), and the final cost function G associated with the sample entropy is given by

$$G(\mathbf{P}) = \frac{1}{2}\log|\mathbf{\Sigma}| = \frac{1}{2}\log\left|\frac{1}{dMN-1}\mathbf{Y}^{T}\mathbf{Y} + \alpha\mathbf{I}\right|. \tag{10.29}$$

To compute the gradient of G, we follow a logic similar to that used in the derivation of (10.11), and allow the estimation of the mean $\boldsymbol{\mu}$ of the distribution $\mathbf{Z}$ to lag behind the updates $\frac{\partial G}{\partial \mathbf{P}}$. This allows for the simplifying assumption $\frac{\partial G}{\partial \mathbf{P}} \approx \frac{\partial G}{\partial Y}$. This approximation becomes more accurate as the number of shape samples is increased, and changes in individual particle positions have increasingly less of an effect on the sample mean. The matrix of partial derivatives of G with respect to $\mathbf{Y}$ is derived as follows.

$$\frac{\partial G}{\partial \mathbf{Y}} = \frac{\partial}{\partial \mathbf{Y}} \left(\frac{1}{2} \log \frac{1}{(dMN-1)^n} + \frac{1}{2} \log |\mathbf{Y}^T \mathbf{Y}| \right)$$
$$= 0 + \frac{1}{2} |\mathbf{Y}^T \mathbf{Y}|^{-1} \frac{\partial}{\partial \mathbf{Y}} |\mathbf{Y}^T \mathbf{Y}| \qquad (10.30)$$
$$= |\mathbf{Y}^T \mathbf{Y}|^{-1} |\mathbf{Y}^T \mathbf{Y}| \mathbf{Y} (\mathbf{Y}^T \mathbf{Y})^{-1}$$
$$= \mathbf{Y} (\mathbf{Y}^T \mathbf{Y})^{-1}.$$

Adding the regularization to the covariance, we have the following equation for the updates

$$\frac{\partial G}{\partial \mathbf{P}} \approx \mathbf{Y} (\mathbf{Y}^T \mathbf{Y} + \alpha \mathbf{I})^{-1}. \qquad (10.31)$$

The regularization α on the inverse of $\mathbf{Y}^T \mathbf{Y}$ can now be seen to account for the possibility of a diminishing determinant. The negative gradient $-\partial G/\partial \mathbf{P}$ gives a vector of updates for the entire system, which is recomputed once per system update. This term is added to the shape-based updates described in the previous section to give the update of each particle:

$$\mathbf{z}_j^k \leftarrow \gamma \left[-\partial G / \partial \mathbf{z}_j^k + \partial E^k / \partial \mathbf{z}_j^k \right]. \qquad (10.32)$$

The stability of this update places an additional restriction on the time steps, requiring γ to be less than the reciprocal of the maximum eigenvalue of $(\mathbf{Y}^T \mathbf{Y} + \alpha I)^{-1}$, which is bounded by α. Thus, we have $\gamma < \alpha$ and note that α has the practical effect of preventing the system from slowing too much as it tries to reduce the thinnest dimensions of the sample distribution. This also suggests an annealing approach for computational efficiency in which α starts off somewhat large (e.g., the size of the shapes) and is incrementally reduced as the system iterates.

10.2.8 Setting Parameters

The SSM approach described thus far is a self-tuning system of particles that distribute themselves across the shape surface using repulsive forces to achieve optimal distributions. Particles may also optionally adjust their sampling frequency locally in response to surface curvature. Free parameters of the system are limited to the choice of the number of particles (M), and the parameters s and ρ from (10.15), if adaptive sampling is used. In practice, adaptivity parameters are typically determined empirically based on the data under analysis. The number of particles is also typically chosen empirically by adding particles until the representation is deemed to capture enough details for the given application.

Figure 10.2 PBM correspondences on example tori from a random distribution on *r* and *R*. Colors indicate correspondence. (For interpretation of the references to color in this figure, the reader is referred to the web version of this chapter.)

In order to explicitly manage the tradeoff between model compactness and the geometric regularization, an additional free parameter β may be introduced into Eq. (10.1) as follows:

$$Q = H(\mathbf{Z}) - \beta \sum_{k=1}^{N} H(\mathbf{X}_k). \tag{10.33}$$

Empirical results, however, suggest that the two terms in this function are already well balanced and $\beta = 1$ represents a good default setting.

10.2.9 Illustration of the Properties and Interpretability of the PBM Optimization

Finally, we present two examples that illustrate the properties of the optimization. The first is an experiment on a class of nonspherical shapes, for which the PBM optimization recovers the optimal ground truth shape parameters. The second example illustrates how PBM is able to discover the underlying mode of variation in a box ensemble with a moving bump in comparison to diffeomorphism-based shape modeling.

To illustrate PBM optimization results for ground truth synthetic data, we applied the algorithm to sample set of 40 randomly generated tori, which are nonspherical shapes that can be described by exactly two shape parameters, a small radius r and the large radius R (see also the example given in Fig. 10.3). Tori were randomly chosen from a distribution with mean $r = 1$, $R = 2$ and $\sigma_r = 0.30$, $\sigma_R = 0.15$. A rejection policy was used to exclude invalid tori (e.g., $r > R$). Correspondences were optimized using 1024 particles per shape, and a uniform sampling (no adaptivity). Fig. 10.2 shows the particle system distribution across several of the torus shapes in the sample set with 1024 correspondences. Correspondence positions are indicated by spherical glyphs and correspondence across shapes is indicated by the color of the glyph. Surface reconstructions for each sample were done using the correspondence positions and the algorithm given by Hoppe et al. for collections of unorganized points [47]. A principal component analysis (PCA) of the resulting correspondence positions indicates that the particle system method discovered two pure modes of variation. PCA mode 1 contains 69.7870% of

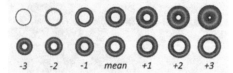

Figure 10.3 PCA modes 1 and 2 illustrated for the PBM torus shape model at -3 to $+3$ standard deviations from the mean shape. PCA 1 (top row) corresponds to r and PCA 2 (bottom row) correspondence to R.

total variation and PCA mode 2 contains 30.2076% of total variation. Less than 0.006% of total variation remains in the smaller, "error" modes.

Empirical observation of PCA modes 1 and 2 suggests that they correspond well to variation in r and R, respectively, from the parametric model that was used to generate the sample data. Fig. 10.3 shows the mean correspondence positions from the model moved along each of the top two PCA modes. Torus shapes along each mode are reconstructed from the learned PBM model parameters at -3 to $+3$ standard deviations from the mean. The top row illustrates variation in PCA 1, which corresponds to r, and the bottom row indicates variation in PCA 2, which corresponds to R. In this experiment, the PBM method appears to have estimated the true orthogonal modes of variation of the torus shape sample.

In the second experiment, we constructed an ensemble of 15 three-dimensional "box-bump" shapes with a bump at a varying location (see top row of Fig. 10.4). Each shape was constructed as a fast-marching distance transform of a union of a rounded-corner cuboid and an ellipsoid representing a bump added at a random location along the top side of the cuboid. This example is interesting because we would, in principle, expect a correspondence algorithm that is minimizing information content to discover this single mode of variability in the sample set. We used the PBM method to optimize 1024 particles per shape under uniform sampling on the "box-bump" shapes (see middle row of Fig. 10.4).

As opposed to optimized correspondences on shape surfaces, the shape geometry can be embedded in the image intensity values at pixels or voxels and then nonlinear registration can be used to map all sample images to a reference image, or atlas. The variation in shape is then considered to be captured by the nonlinear registration parameters. Of the image registration methods, the diffeomorphic methods are in most widespread use. Hence, it is important to show whether modeling shape variations using diffeomorphic warps are able to recover the known mode of variation (i.e., a moving bump) as compared to the analysis of the optimized correspondence model. In this regard, we generated an unbiased atlas of the "box-bump" ensemble as proposed in [70] and implemented in AtlasWerks [71]. We then constructed a parametric statistical model of shape variation using PCA on the deformation fields that map each box-bump sample to the estimated unbiased atlas (see bottom row of Fig. 10.4).

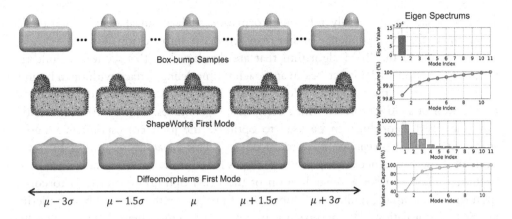

Figure 10.4 The box-bump experiment: PBM-based and diffeomorphic-based shape models.

Fig. 10.4 illustrates the mean and three standard deviations of the first mode of the PBM- and diffeomorphic-based models. Shapes from the particle method remain more faithful to those described by the original training set, even out to three standard deviations where the diffeomorphic description breaks down. In particular, one can observe that PCA identified a single dominant mode of variation for the PBM method (see middle row of Fig. 10.4). However, diffeomorphic warps recovered incorrect shape model in which the mean shape showed a box with two bumps rather than a single bump and the first mode of variation represented the relative height of the two "artificial" bumps. Further, diffeomorphic-based shape modeling showed five dominants shape modes in an ensemble of a single mode.

10.3 PBM EXTENSIONS

This section describes several mathematical extensions of the particle-based modeling (PBM) algorithm that are designed to make it more robust to realistically complex shape analysis problems. This work is motivated by the needs of the biomedical research community for tools to model more complicated anatomical shapes and statistical designs, such as joint variability of multiple structures and optimization with respect to explanatory variables. Cardiac anatomy, for example, consists of multiple interconnected chambers with shared openings, valve annuli, and septa. In orthopedics, researchers are often concerned with the mechanical interactions of multiple bone surfaces, in order to understand dysfunction in joints. In other cases, the geometric features of an anatomical object are not sufficient to properly establish correspondence. Some anatomy is highly variable across subjects and additional information, such as functional data, is helpful in determining how surface regions correspond. The cortical surface of the brain represents one such example. Cortical folding patterns are highly variable among individuals, and neuroanatomists typically rely on information such as sulcal depth and vascular connectivity for correspondence, rather than geometric information alone.

Extensions to the PBM algorithm that are developed in this section include the following. Section 10.3.1 describes an approach for modeling surface with open boundaries. The open surface method allows an arbitrary boundary to be defined as the intersection of a closed surface with a set of shape primitives. Section 10.3.2 describes how the PBM algorithm can be used to optimize the joint correspondence among shapes that consist of *multiple* anatomical objects. This capability is important in the study of shape covariance among anatomy that is functionally or structurally correlated. Section 10.3.3 describes a generalization of the PBM optimization criteria to correspondence in arbitrary, multivariate functions of position, rather than only considering positional information. This approach is useful for problems where there is data other than geometric information that indicates correspondence, such as multimodal imaging studies, and studies with functional imaging data. Finally, Section 10.3.4 describes a methodology for including a regression model on independent variables into the PBM correspondence optimization. Shape regression modeling can improve statistical power when controlling for correlations between shape and factors such as age or other clinical variables.

10.3.1 Modeling Shape with Open Surfaces

Conceptually, there are two ways to handle a surface boundary when optimizing correspondences. The first approach is to explicitly represent and model the boundary, which requires that correspondences must be allowed to lie on the boundary, and the optimization must track particle movement on and off of the boundary. This approach is appropriate for applications where the boundary shape is of specific interest to the problem, such as specifically modeling the shapes of valve annuli or the ostia of vascular openings in conjunction with chamber shape in the heart. In many cases, however, it is not important, or even desirable, to model the variation in the shape of the boundary. A segmentation, for example, may contain noise in the boundary shape due to ambiguities in its specification during the segmentation process. Such examples often arise in orthopedics, for example, where only the proximal or distal end of a bone may be of interest. In this situation, where the boundary is considered noisy, it can simply be treated as a constraint on the particle optimization, which is the approach currently implemented for ShapeWorks. Explicit modeling of open shape boundaries is left for future work.

The PBM algorithm for correspondence on open surfaces represents the surface boundary as the intersection of a closed surface (e.g., S in Section 10.2.2) with a set of geometric primitives, such as cutting planes and spheres. The boundary representation is then used to influence the entropy maximization of the PBM algorithm particle position (Section 10.2.2), so that it indirectly constrains the positions of particles to lie within

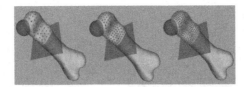

Figure 10.5 Increasing numbers of PBM particles on an open surface, where the boundary is defined by the intersection of an implicit bone surface with a sphere and a cutting plane.

the surface boundary. The goal of the open surface modeling algorithm is to formulate particle interactions with the boundaries so that the positions of the constraints have as little influence as possible on the statistical shape model. This approach is consistent with the idea that the boundary shape may contain noise, and we wish to minimize the influence of this noise on the model.

The algorithm proceeds as follows. For each geometric primitive in the surface boundary representation, the algorithm constructs a *virtual* particle distribution that consists of all of the closest points on its surface to the particles with positions $\mathbf{x}_j$ on $\mathcal{S}$. During the gradient descent optimization, particles $\mathbf{x}_j$ interact with the virtual particles, and are therefore effectively repelled from the geometric primitives, and thus from the open surface boundary. The virtual distributions are updated after each iteration, as the particles on $\mathcal{S}$ redistribute under the optimization. Because the virtual particles are allowed to factor into the Parzen windowing kernel size estimation (Eq. (10.8)), particles $\mathbf{x}_i$ maintain a distance from the boundary proportional to their density on the surface $\mathcal{S}$. In this way, features near the boundary may be sampled, but particles are never allowed to lie on the boundary itself, limiting the effect of errors in the boundary specification on the configuration. Note that the virtual particle distributions are also not used in the correspondence optimization term (the sample entropy from Eq. (10.23)) and therefore do not directly affect the distribution of samples in shape space.

Fig. 10.5 illustrates a particle configuration using the method outlined above for open surfaces, and shows the effect of increasing the number of particles. In the figure, the open surface boundary is defined by the intersection of an implicit bone surface, a cutting plane, and a sphere. As the number of particles is increased, the distribution samples regions of the bone become closer and closer to the surface boundary. Note, however, that the particle distribution never touches or crosses the boundaries.

Further examples are given in Fig. 10.6. Fig. 10.6A illustrates placement of particles on a neonatal head surface (from MRI), with a cutting plane defined by anatomical landmarks at the tip of the nose and the center of the ears, and spheres placed at those same landmarks to exclude both nose and ears from the analysis. This approach was used in work characterizing shape change during neonatal head development [60]. Another example is that of the left atrium, which is shown in Fig. 10.6B. The left atrium is a complex structure with openings to the mitral valve and the pulmonary veins. For

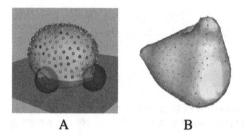

A   B

Figure 10.6 Examples of PBM optimization on open surfaces, including pediatric head shape (A) and the left atrium (B).

studies involving left atrial shape in atrial fibrillation [64], we masked out the mitral valve region using a collection of exclusion spheres.

10.3.2 Modeling Shape Complexes

In general, biological function and phenotype is not explained by any single anatomical structure, but are instead the result of complex systems of functionally- or structurally-related anatomy. This section presents an extension of the PBM algorithm for shape modeling of multiple, disconnected anatomical surfaces, or what we will call shape complexes. A *multiobject complex* is defined as a set of solid shapes, each representing a single, connected biological structure. The complex of structures are assembled into a scene within a common coordinate frame. Each structure in a multiobject complex contains shape, pose, scale, and positional information. Some examples include the segmentations of multiple brain structures from a single MRI of a patient and sets of bones segmented from a CT scan. The proposed correspondence method for establishing correspondence on multiobject complexes is novel in that it optimizes correspondence positions in the full, joint shape space of the object complex. Researchers have previously only considered the correspondence problem separately for each structure, thus ignoring the interstructural shape correlations in the optimization process. By explicitly modeling the correlations among variabilities, however, optimization in the joint space may produce more compact distributions for correspondences, resulting in fewer model parameters and greater statistical power.

The particle-based correspondence method described in Section 10.2 can be directly applied to multiobject complexes by treating all of the objects in the complex as defining a single surface. However, if the objects themselves have distinct identities (i.e., object-level correspondence is known a priori), we can assign each particle to a specific object, decouple the spatial interactions between particles on different shapes, and constrain each particle to its associated object. In this way, each correspondence is guaranteed to stay on a particular anatomical structure, and the surface sampling is not influenced by regions where structures in the complex happen to be near to one another. The shape-space statistics using this method, however, remain coupled, and the covariance Σ (Eq. (10.23)) includes all particle positions across the entire complex, so that optimization takes place on the joint, multiobject model.

As with the single-object framework, any set of implicitly defined surfaces is appropriate as input to the multiobject framework, with similar preprocessing considerations as those discussed in Section 10.4.1. In the case of binary segmentations, the input is now a set of N segmentations of K-object complexes, which contains $N \times K$ distinct, volumetric label masks.

10.3.3 Correspondence Based on Functions of Position

Geometric features of an anatomical object are often not sufficient to properly establish correspondence. The basic PBM algorithm described in Section 10.2 only considers particle position information in the optimization, which only represents the geometric, or structural, information of the shape surface. Here we describe an extension to the PBM algorithm to establish correspondence by minimization of the entropy of arbitrary, vector-valued *functions* of position. This more general method is useful in cases where the notion of correspondence is not well defined by the surface geometry, but can be described by other metrics.

The extension to the PBM algorithm to incorporate functional data, which we refer to as the *generalized* PBM algorithm, is straightforward. It consists of substituting the entropy estimation of the matrix of particle positions with an entropy estimation on an arbitrary, vector-valued function of the particle position. From Section 10.2, the energy term for the basic PBM optimization is given by

$$Q = H(\mathbf{Z}) - \beta \sum_{k=1}^{N} H(\mathbf{X}_k), \qquad (10.34)$$

where H is an estimation of entropy, X_k is a vector random variable with the distribution of particle configuration k, and Z is the vector random variable with the distribution of the shape samples in the dM-dimensional shape space. The extension to the generalized PBM algorithm only modifies the correspondence term $H(\mathbf{Z})$. The entropy associated with individual correspondence configurations, $H(\mathbf{X}_k)$, is not modified, and still operates on positional information. In other words, particles are still constrained to lie on the surface of the shape and distribute themselves across shape surfaces using the maximization of positional entropy, but their correspondence is established using a function of positional information. Note that a function of position could be designed to also include particle position, so that both structural and functional data influence the correspondences.

Recall from Section 10.2.7, that the entropy estimation of the sample distribution in shape space is given by

$$H(\mathbf{Z}) \approx \frac{1}{2} \log |\Sigma|, \text{ and } \Sigma = (dMN - 1)^{-1}\mathbf{YY}^T, \tag{10.35}$$

where Σ is the covariance matrix, and $\mathbf{Y}$ is the $dM \times N$ data matrix P of sample vectors $\mathbf{z}^k$, $k = \{1, \ldots, N\}$, minus the sample mean $\boldsymbol{\mu}$, and each vector $\mathbf{z}^k$ consists of the positional information from M particles on the shape surface k. In the case of computing entropy of vector-valued functions of the correspondence positions, the extension to functional data considers the more general case where columns of the data matrix are instead given by

$$\tilde{\mathbf{p}}^k = \begin{bmatrix} f(\mathbf{x}_0^k) \\ f(\mathbf{x}_1^k) \\ \vdots \\ f(\mathbf{x}_j^k) \\ \vdots \\ f(\mathbf{x}_{M-1}^k) \end{bmatrix}, \tag{10.36}$$

where $\mathbf{x}_j^k$ is the positional information of particle j for shape k, and $f : \Re^d \to \Re^q$.

The matrix $\mathbf{Y}$ now becomes a matrix $\tilde{\mathbf{Y}}$ of the function values at the particle points, minus the means of those functions at the points. Columns of $\tilde{\mathbf{Y}}$ are given by

$$\tilde{\mathbf{y}}^k = \begin{bmatrix} f(\mathbf{x}_0^k) - \frac{1}{N}\sum_{i=1}^{N} f(\mathbf{x}_0^i) \\ f(\mathbf{x}_1^k) - \frac{1}{N}\sum_{i=1}^{N} f(\mathbf{x}_1^i) \\ \vdots \\ f(\mathbf{x}_{M-1}^k) - \frac{1}{N}\sum_{i=1}^{N} f(\mathbf{x}_{M-1}^i) \end{bmatrix}. \tag{10.37}$$

The new cost function $\tilde{G}$ is the estimation of entropy of the samples $\tilde{\mathbf{y}}^k$. With the same assumption of a Gaussian distribution in shape space, by the same logic as for the derivation of the cost function G in (10.29), we have

$$\tilde{G}(\tilde{\mathbf{z}}) = \log \left| c\tilde{\mathbf{Y}}^T\tilde{\mathbf{Y}}, \right|, \tag{10.38}$$

with c a constant.

Let $Q = (\tilde{\mathbf{Y}}^T\tilde{\mathbf{Y}} + \alpha\mathbf{I})^{-1}$. By the chain rule, the partial derivative of $\tilde{G}$ with respect to the data $\mathbf{y}^k$ becomes

$$-\frac{\partial \tilde{G}}{\partial \tilde{\mathbf{P}}^k} = \mathbf{J}_k^T \mathbf{Q}^k, \tag{10.39}$$

where $\mathbf{J}_k$ is the Jacobian of the functional data for shape k. The matrix $\mathbf{J}_k$ has the structure of a block diagonal matrix with $M \times M$ blocks, with diagonal blocks the $q \times d$ submatrices of the function gradients at particle j. Specifically, for each shape k, we have function data

$$\mathbf{y}^k = \left[f_1^0, f_2^0, \dots, f_q^0, f_1^1, f_2^1, \dots, f_q^1, \dots, f_1^{M-1}, f_2^{M-1}, \dots, f_q^{M-1} \right]^T, \tag{10.40}$$

and a diagonal submatrix block of the Jacobian $\mathbf{J}_k = \nabla_{\mathbf{z}^k} \mathbf{y}^k$ has the structure

$$\begin{bmatrix} \partial f_1^j / \partial x_{dj+1} & \partial f_1^j / \partial x_{dj+2} & \dots & \partial f_1^j / \partial x_{dj+d} \\ \partial f_2^j / \partial x_{dj+1} & \partial f_2^j / \partial x_{dj+2} & \dots & \partial f_2^j / \partial x_{dj+d} \\ & & \vdots & \\ \partial f_q^j / \partial x_{dj+1} & \partial f_q^j / \partial x_{dj+2} & \dots & \partial f_q^j / \partial x_{dj+d} \end{bmatrix}, \tag{10.41}$$

where $j = \{0, 1, 2, \dots, M - 1\}$ is the block number, which corresponds to a single particle, and $\{x_1, x_2, \dots, x_{dM}\}$ are the directional components of the full set of M particles. The correspondence optimization proceeds by gradient descent, as described in Section 10.2, with the substitution of the gradient of the new cost function $\tilde{G}$ for the original cost function G in Eq. (10.32).

In summary, the generalized PBM algorithm replaces the entropy of positional information with entropy of an arbitrary function of positional information. This modification offers a much more generalized framework for optimizing the statistical properties of an ensemble of shapes. Note that the standard PBM algorithm from Section 10.2 is now just a special case of the generalized PBM algorithm, where $f(\mathbf{z}) = \mathbf{z}$.

10.3.4 Correspondence with Regression Against Explanatory Variables

In general, the design of a scientific study in biology or medicine cannot control for all confounding variables. The variability in shape due to such factors as age, differential growth rates, or clinical variables, for example, must be accounted for during the analysis phase. In other cases, this variability is the specific focus of the study, and researchers want to examine the correlation of an explanatory variable with shape. A typical experiment, for example, might examine the correlation of disease progression with the shape of anatomical structures or the change in the shape of anatomy with age. If such correlations can be established, they may lead to new diagnostic protocols or interventional planning.

This section extends the PBM algorithm to the problem of establishing correspondence in the presence of confounding variables and examining the correlation of shape with explanatory variables. Like in the previous section, this method allows for a more general notion of correspondence that takes into account additional information about the data under study. The algorithm works by expanding the point-based correspondence model from Section 10.2 to include a regression against the independent variables. The optimization of correspondence position is then done on the *residual* to the regression model. Of course, the alternative is simply to use statistical methods in the analysis of post-optimization PBM shape parameters. However, the motivation to instead optimize the parameters of the residual model itself is the same principle of parsimony behind the basic PBM formulation: to further minimize model parameters and further maximize statistical power.

Under the assumption of a Gaussian distribution for the random variable $\mathbf{Z}$ from Eq. (10.33), which is the distribution of shape samples in shape space, we can write the generative statistical model

$$\mathbf{z} = \mu + \epsilon, \ \epsilon \sim \mathcal{N}(\mathbf{0}, \Sigma) \tag{10.42}$$

for particle correspondence positions, where ϵ is normally-distributed error. Replacing μ in this model with a function of an explanatory variable t gives the more general, regression model

$$\mathbf{z} = f(t) + \hat{\epsilon}, \ \hat{\epsilon} \sim \mathcal{N}(\mathbf{0}, \hat{\Sigma}). \tag{10.43}$$

The optimization described for the basic PBM algorithm minimizes the entropy associated with ϵ, which is the difference from the mean. In this section, the goal is to optimize correspondences under the regression model in Eq. (10.43) by instead minimizing entropy associated with $\hat{\epsilon}$, the residual from the regression model. For the simple case where particle correspondence is a linear function of t, given as $f(t) = \mathbf{a} + \mathbf{b}t$, parameters $\mathbf{a}$ and $\mathbf{b}$ must be estimated to compute $\hat{\epsilon}$. These parameters are estimated with a least-squares fit to the correspondence data,

$$\underset{\mathbf{a},\mathbf{b}}{\arg\min} E(\mathbf{a}, \mathbf{b}) = \frac{1}{2} \sum_k \left[(\mathbf{a} + \mathbf{b}t_k) - \mathbf{z}_k\right]^T \Sigma^{-1} \left[(\mathbf{a} + \mathbf{b}t_k) - \mathbf{z}_k\right]. \tag{10.44}$$

Setting $\frac{\delta E}{\delta \mathbf{a}} = \frac{\delta E}{\delta \mathbf{b}} = 0$ and solving for $\mathbf{a}$ and $\mathbf{b}$, we have

$$\mathbf{a} = \frac{1}{n}\left(\sum_k \mathbf{z}_k - \sum_k \mathbf{b}t_k\right), \tag{10.45}$$

and

$$\mathbf{b} = \left(\sum_k t_k \mathbf{z}_k - \sum_k \mathbf{z}_k \sum_k t_k\right) \Big/ \left(\sum_k t_k^2 - (\sum t_k)^2\right). \tag{10.46}$$

The proposed regression model optimization algorithm proceeds as follows. Correspondences are first optimized under the nonregression model (Eq. (10.42)) to minimize

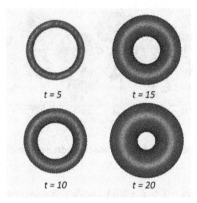

Figure 10.7 Shape regression on an explanatory variable *t* correlated with the small radius of a set of tori with $r \sim \mathcal{U}(5, 20)$.

the entropy associated with the total error ϵ. This process also establishes an initial estimate for **a** and **b**. The next step is to optimize under the regression model, which proceeds by gradient descent on $H(\mathbf{Z}) \approx \frac{1}{2} \log |\hat{\Sigma}| + H(P^k)$. In other words, the method follows the same optimization procedure as the basic PBM framework (Section 10.2), but replaces the covariance of the model with the covariance of the underlying residual, relative to the generative model. The two estimation problems are interwoven: the parameters **a** and **b** are re-estimated after each iteration of the gradient descent on the particle positions.

As an example of a correspondence optimization, consider the regression method applied to a set of $N = 40$ tori. To generate each torus, the large radius R was randomly drawn from a Gaussian distribution $R \sim \mathcal{N}(35, 3)$, and small radius r randomly drawn from a uniform distribution $r \sim \mathcal{U}(5, 20)$. An explanatory variable $t_i = r_i + \epsilon$, with $\epsilon \sim \mathcal{N}(0, .3)$, was assigned to each shape sample $i \in \{1, \ldots, 40\}$ to establish a good correlation with variation in the small torus radius. Correspondences were optimized using 1024 particles per shape and the PBM regression algorithm outlined above. In the resulting correspondences, variation in the residuals to the regression line exhibits one major mode that empirically corresponds to r. Empirical observation of the regression line, which is shown in Fig. 10.7 suggests good correlation with R.

10.3.5 Dense PBM Correspondence Models

PBM yields relatively sparse correspondence models that may be inadequate to reconstruct thin structures and high curvature regions of the underlying anatomical surfaces. However, for many applications, we require a denser correspondence model, for example, to construct better surface meshes, make more detailed measurements, or conduct biomechanical or other simulations on mesh surfaces. One option for denser modeling

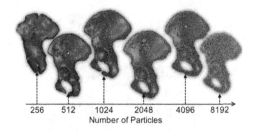

Figure 10.8 ShapeWorksView: pelvis surface reconstruction as a function of the number of particles.

is to increase the complexity of the PBM model via increasing its number of particles per shape sample. However, this approach necessarily increases the computational overhead, especially when modeling large clinical cohorts. Fig. 10.8 shows a pelvis surface reconstruction with an increased number of particles in which poor reconstructions are observed with smaller number of particles, especially along the pubic arch and iliac crest. Hence, an extension is needed to recover anatomically plausible and accurate 3D shapes at both the population and sample levels from a sparse set of particles.

In this extension, we adopt a template-deformation approach to establish an inter-sample dense surface correspondence, given a sparse set of optimized particles. To avoid introducing bias due to the template choice, we propose an unbiased framework for template mesh construction that includes three steps. First, generalized Procrustes alignment [72] is used to define the mean particle system from the PBM model, while estimating the rigid transformation that maps the sample-level particle system to the population-level counterpart. Second, the distance transform (DT) of each shape is deformed based on a nonlinear warping function that is built using the sample's particle system and the mean particle system as control points. Third, the warped DTs are then averaged to compute an average DT whose zero level set represents the geometry and topology of the mean shape in the population space. The dense *template mesh* is then constructed by triangulating the isosurface of this mean DT. This unbiased strategy will preserve the topology of the desired anatomy by taking into account the shape population of interest. In order to recover a sample-specific surface mesh, a warping function is constructed using the sample-level particle system and the mean/template particle system as control points. This warping function is then used to deform the template dense mesh to the sample space. A core ingredient of this PBM-based surface reconstruction is the form of the warping function being guided by the sparse particle system. One option is to use a thin-plate spline (TPS) [73] that defines a spatial mapping with global support, i.e., any perturbation in a single correspondence affects the whole warping function. Fig. 10.9 shows a sample PBM-based pelvis surface reconstruction using TPS with 512 particles. Compared to the current ShapeWorksView, our proposed reconstruction is able to recover the pubic arc and iliac crest with few particles. Nonetheless,

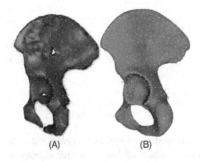

                    (A)   (B)

Figure 10.9 Surface reconstruction of a pelvis mean shape with 512 particles using (A) Shape-WorksView and (B) the proposed warping-based reconstruction (with thin plate splines).

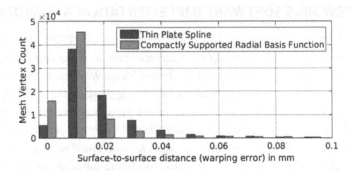

Figure 10.10 TPS vs. RBF: histogram of vertex-wise average distance (in mm) between original patient-specific femur meshes and the warped template femur mesh.

TPS poses timing and memory challenges to process PBM models with more particles. To speed up the reconstruction process while minimizing the memory footprint, we propose the use of compactly supported radial basis functions (RBF) [74], which results in a sparse matrix that can be solved using sparse solvers. For preliminary results, we used a cohort of 70 femur shapes with 0.7 mm resolution. The computation time for template mesh construction was reduced from 4 hours using TPS to 9 minutes using RBF. Sequentially deforming the template mesh to the space of each subject was reduced from 4.5 hours using TPS to 5 minutes using RBF. To ensure that we are not sacrificing accuracy for this speed-up, Fig. 10.10 shows histograms of the warping error (in mm) between the groundtruth and reconstructed patients meshes in which both TPS and RBF attain similar trends and the majority of error is less than a seventh of voxel size. With a femur cohort at a resolution of 0.24 mm, Fig. 10.11 shows the RBF-based warping error being quantified at different numbers of particles where *less than half voxel size* warping error is achieved with 1K particles or more. This PBM-based warping can be further used to deform clinical measurements from patient to population space and vice versa [63].

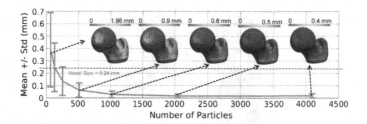

Figure 10.11 Mean and standard deviation of RBF-based warping error in mm as a function of number of particles. Vertex-wise mean error along all the 70 femur shapes was computed and visualized as colormaps on the template dense mesh.

10.4 SHAPEWORKS SOFTWARE IMPLEMENTATION AND WORKFLOW

ShapeWorks is an open-source software distribution of the PBM approach to SSM. The software includes a careful numerical implementation of the approaches developed in Sections 10.2–10.3 in a modular C++ library. ShapeWorks also includes a suite of software applications for applying PBM to image segmentations. The software distribution consists of a set of command line tools for preprocessing binary segmentations (ShapeWorksGroom) and computing landmark-based shape models (ShapeWorksRun). It also includes a simple user interface to analyze and visualize the optimized shape models (ShapeWorksView). The current shape modeling pipeline for establishing shape correspondences from a set of binary segmentation image volumes is outlined in Fig. 10.12, with reference to the software tools that implement each step. In addition to the command line tools and visualization software, ShapeWorks also includes a full graphical user interface called ShapeWorksStudio, which can be used to interactively run all of the steps in the shape modeling pipeline. All ShapeWorks source code, binaries, and user documentation are freely available from the ShapeWorks website (www.sci.utah.edu/software/shapeworks.html) under the MIT license that is General Public License (GPL) compatible. According to Google Scholar, ShapeWork technology has received over 260 citations and over 2500 downloads since its initial release in 2009.

The remainder of this section describes each of the elements of the ShapeWorks software implementation in more detail, including the PBM code framework, an overview of the steps in the ShapeWorks SSM workflow, and an introduction to the ShapeWorks software applications.

10.4.1 The ShapeWorks Shape Modeling Workflow

A typical workflow for establishing shape correspondence from binary image volume inputs is outlined in Fig. 10.12. The preprocessing steps in the pipeline establish an initial alignment of the segmentations and generate suitable distance transforms. The

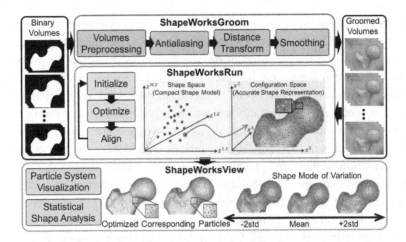

Figure 10.12 ShapeWorks shape modeling pipeline.

optimization phase consists of initializing the particle system, running the PBM optimization, and optionally refining the alignment using the Procrustes algorithm to iteratively remove residual nonshape data. Iterations of the correspondence optimization are interleaved with alignment steps until convergence. The remainder of this section discusses each of these steps in more detail.

10.4.1.1 Segmentation Preprocessing and Alignment

Any set of implicitly defined surfaces, such as a set of binary segmentations, is appropriate as input to the PBM algorithm. The algorithm, however, can be applied directly to binary segmentation volumes, which are often the output of a manual or automated segmentation process. Binary volumes contain an implicit shape surface at the interface of the labeled pixels and the background. Any suitably accurate distance transform from that interface may be used to form the implicit surface necessary for the particle optimization.

Segmentation data typically requires some processing to remove aliasing artifacts in the binary mask. Aliasing artifacts can adversely affect numerical approximations of surface features and the computations required to maintain the surface constraint in the PBM algorithm. One effective method for antialiasing binary volumes is given by Whitaker in [75], who describes a method for fitting an antialiased, level-set surface to a binary volume through an iterative relaxation process. The process uses curvature flow of the surface, with constraints on the flow dictated by the binary voxel locations of the segmentation. Another effective antialiasing method is the r-tightening algorithm given by Williams et al. [76]. The surface tightening method follows a similar approach to that of Whitaker, but constrains the level-set relaxation process using binary volumes that result from morphological opening and closing of the targeted binary surface. This method has proven to be particularly effective at removing aliasing artifacts without compromising the precision of the segmentation. As a final preprocessing step, the distance transform is typically followed by a slight Gaussian blurring to remove the high-frequency artifacts that can occur as a result of numerical approximations.

A collection of shape segmentations must often be aligned in a common coordinate frame for modeling and analysis. Where no information exists to specify a correct alignment, one approach is to first align segmentations with respect to their centers of mass and the orientation of their first principal eigenvectors. Then, during the optimization, the PBM method may optionally further align shapes with respect to rotation, translation, and scale using generalized Procrustes analysis (GPA) [77]. The GPA alignment is applied at regular intervals after particle updates in order to remove any residual, non-shape information from the model. GPA alignment during the optimization process is only enabled once the full set of M particles have been initialized on all surfaces. Where the true shape alignments are known, however, the GPA iterations may be omitted. A subset of the GPA alignment parameters may also be applied, such as only the rotational and translational components, leaving the scale unaffected.

10.4.1.2 Initialization and Optimization

There are number of possibilities for initializing the particle systems on the sample shapes, including manual specification of points and regular surface sampling. For spherical topologies, Paniagua et al. have proposed initialization of PBM with parametric SPHARM-PDM models [10]. One effective approach for general categories of shape, however, is to use an iterative, *particle splitting* strategy, in which the full set of particles is initialized in a multiscale fashion as follows. First, the PBM system is initialized with a single particle on each shape that finds the nearest zero of the implicit surface. This single particle is then split to produce a new, nearby particle. The two-particle (per shape) system is then optimized for correspondence until a steady state is reached. The splitting process, followed by optimization, is then repeated until a specific number of particles have been produced. Thus, the initialization proceeds simultaneously with the optimization in a multiscale fashion, generating progressively more detailed correspondence models with each split.

Typically, we set the numerical parameters for the PBM optimization automatically as follows. The numerical parameter σ_{min} is set to machine precision and σ_{max} is set to the size of the domain. The annealing parameter α starts with a value roughly equal to the diameter of an average shape and is reduced to machine precision over several hundred iterations. Particles are initialized on each shape using the splitting procedure described above. These default settings have been found to produce reliably good results that are very robust to the initialization, although some degree of parameter tuning is typically warranted in practice.

Processing time for the PBM algorithm on a modern desktop computer averages around 1/16,000 second/particle per iteration. This translates to full optimization times that scale linearly with the number of particles in the system and are on the order of minutes for small systems of a few thousand particles to several hours for larger systems of tens-of-thousands of particles. Optimizations of very large systems of hundreds-of-thousands to millions of particles may take processing times of several dozen hours.

10.4.1.3 Analysis

For analysis, sets of configurations are usually aligned within a common d-dimensional coordinate frame by a rotation, translation, and scaling to remove the geometric information unrelated to shape variation. Goodall's model of shape [77,78] describes each of these nonshape components and the residual variation around the mean correspondence configuration. For configuration matrix C_i, the model is given by

$$\mathbf{C}_i = a_i \boldsymbol{\mu} + \mathbf{E}_i \mathbf{R}_i + \mathbf{1} \mathbf{t}_i, \tag{10.47}$$

where a_i is a scalar representing the relative size of specimen i relative to the mean size, $\mathbf{E}_i$ are the residuals from the mean configuration μ, $\mathbf{R}_i$ is a rotation matrix describing the orientation of sample i, $\mathbf{1}$ is a d-dimensional vector of 1s, and $\mathbf{t}_i$ is a translation vector describing the locational information for sample i.

The most common method for estimating μ, and the nonshape components $\mathbf{R}_i$, $\mathbf{t}_i$, and a_i is generalized Procrustes analysis (GPA) [7,72,79,5]. When transformed using GPA, correspondences are said to be in *Procrustes space*. Statistical analysis is commonly done in Procrustes space because, for reasonably similar sets of shapes, distance measures between Procrustes coordinates have been shown to be good linear approximations to the geodesic distances in Kendall's shape space [5,6].

The remaining geometric variability of a correspondence model after GPA is often summarized as a set of shape parameters that are the orthogonal directions of a principal components analysis (PCA) of the correspondence point positions. A complete mathematical description of this process can be found in, for example, [77,80], with application to PBM in [81]. PCA-based shape parameters allow us to compress the very large amount of geometric information into a much smaller representation of shape that is suitable for traditional statistics, while still retaining most of the geometric information of the shapes. Typically, we choose a finite number of shape parameters m for analysis either empirically, or by picking a set that accounts for most of the variability in the model. A more objective approach for PCA model selection is to use a method called parallel analysis to automatically determine a finite set of PCA modes that are distinguishable from Gaussian noise in the model [82].

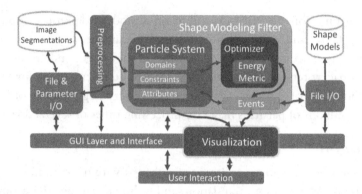

Figure 10.13 Abstract implementation of PBM shape modeling.

10.4.2 The PBM Code Library

A numerical implementation of the PBM algorithms from Sections 10.2–10.3 is available as an open source library of C++ code (github.com/joshcates/ITKParticleShape Modeling). This library is part of the ShapeWorks distribution and is used in all Shape-Works applications. The PBM code library is built using the Insight Toolkit (ITK, www.itk.org) and can be easily compiled with ITK as an external module. The PBM codebase also conforms to ITK coding standards and includes Doxygen-based documentation (www.doxygen.org) and unit regression tests for all C++ classes. The PBM code library is templated on dimensionality, so that the methods operate equally for two- and three-dimensional image segmentation data.

Fig. 10.13 depicts an abstract view of the main code modules in the PBM library and how they are combined to implement a software application for shape analysis, including the data flow among these components. With reference to Fig. 10.13, the PBM library includes ITK filters for preprocessing image segmentations and converting them to suitable distance transform inputs (see Section 10.4.1.1). Distance transforms are input into a PBM Shape Modeling Filter object, which manages the construction of the multiple Particle Systems and executes the optimization process using a suitable Optimizer and Energy Metric. The main output of the Shape Modeling Filter is the set of correspondence point positions for all N input shapes. The Particle System data container is central to the processing. It is implemented as a facade class [83] that stores and manipulates all of the point-based representations of the input implicit shape surfaces, their local coordinate domains, and the mappings between those domains. The Particle System class, along with most other objects that maintain state in the PBM framework, can make use of ITK's command/observer framework to allow state changes to trigger attached processes at the application level. For example, visualization code can be written to listen for particle position changes that are broadcast from the Particle System object and then trigger graphical updates in response.

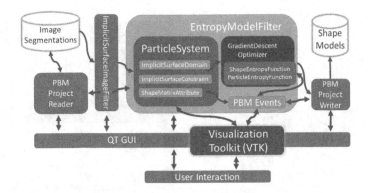

Figure 10.14 Concrete implementation of PBM shape modeling.

Fig. 10.14 shows the specific instantiation of the abstract framework from Fig. 10.13 that implements a basic PBM correspondence optimization. The generic class types have been replaced with specific instantiations in each box. Alternative PBM modeling approaches can be implemented by substituting alternative implementations of the various modular components in this figure. For example, the PBM regression optimization described in Section 10.3.4 can be implemented by simply substituting the "RegressionShapeMatrixAttribute" for the more general "ShapeMatrixAttribute" class in Fig. 10.14. For more details on the PBM code library and its use, the reader is referred to the documentation included with the ShapeWorks distribution.

10.4.3 The ShapeWorks Software Tool Suite

With reference to Fig. 10.12, the ShapeWorks software suite includes the following applications, which collectively implement the complete workflow described in Section 10.4.1. **ShapeWorksGroom** is a command line tool for batch processing binary segmentations, as described in Section 10.4.1.1. It can be used to perform simple alignment of segmentations, basic quality control, and to generate appropriate distance transform inputs for the PBM optimization. **ShapeWorksRun** implements the PBM correspondence optimization algorithm described in Section 10.2, including multiscale initialization via particle splitting and iterative GPA (Section 10.4.1.2). ShapeWorksRun also includes the extensions to PBM described in Sections 10.3.1–10.3.4. **Shape-WorksView** is an application for visualizing the results of the PBM optimization. It is built using the open-source Visualization Toolkit (VTK, www.vtk.org) and includes visualization of correspondence positions on shape surfaces, reconstructions of the mean shape of the correspondence model, and reconstructions of shapes along PCA modes (Section 10.4.1.3) and regression lines (Section 10.3.4).

ShapeWorks Studio is a new desktop application that encapsulates the entire processing pipeline from Fig. 10.12 under a single user interface. The different steps of the

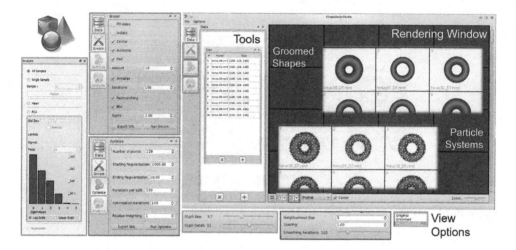

Figure 10.15 The ShapeWorks Studio desktop application.

pipeline are organized under separate tabs, as shown in Fig. 10.15. ShapeWorks Studio also includes postprocessing to produce dense correspondence models, as described in Section 10.3.5. ShapeWorks Studio is built using the PBM C++ library, ITK, Qt (www.qt.io) for graphical user interface elements, and VTK for visualization components. The design of the software and its dataflow are shown in Fig. 10.14.

10.5 SHAPEWORKS IN BIOMEDICAL APPLICATIONS

ShapeWorks has been applied to a wide variety of problems in medicine and biology, including neurobiology, biological phenotyping, orthopedics, and cardiology. In this section, we review some of the applications in these areas.

The PBM algorithm and many of its extensions were originally developed in the context of neurobiology. The method of modeling ensembles of shape complexes (Section 10.3.2), for example, was developed with application to a study of the shape of subcortical brain structures in pediatric autism. In that study, we identified statistically significant differences between the joint mean shape of 10 subcortical structures in autism patients and those of normal controls [60]. The PBM regression modeling approach outlined in Section 10.3.4 was used to describe a longitudinal model for neonatal head shape development [58]. Oguz et al. have applied the generalized method of PBM correspondence (Section 10.3.3) for human cortical surface correspondence [56,57], and have shown improved correspondence with respect to cortical thickness and sulcal depth over more commonly used approaches such as FreeSurfer (http://www.freesurfer.net).

Gene targeting is one of the most important tools for genetic study, and is widely used to examine the role that specific genes play in human development and disease.

Because gene targeting studies often rely on metrics of shape to quantify phenotypic expression, more comprehensive and detailed representations of shape from SSM allow for observations of genetic expression that have not been possible with traditional morphometrics. We applied ShapeWorks to quantify phenotype of the forepaw in mice deficient in the *Hoxd11* gene and compare to normal control mice. *Hoxd11* is known to play a role in the normal patterning of the appendicular skeleton. Our results showed both significant gross shape changes associated with bone length and thickness, but also more subtle local shape abnormalities at the distal end of the bones [60].

In orthopedics, ShapeWorks has been applied to the quantification of the spectrum of hip deformities associated with cam-type femoroacetabular impingement (FAI). We have analyzed variation in the thickness of the femur cortex between asymptomatic controls and cam-FAI patients, for whom we had quantitative evidence that repetitive impingement would induce bone hypertrophy [62,84,63]. We have further used ShapeWorks to study the resection of the lesion in cam-FAI patients to reduce the likelihood of damage to chondrolabral tissue and have developed guidelines to assist surgeons with resection of cam lesions as an effective intraoperative guide [85]. We established the limitations of radiographic measurements used in the clinical diagnosis of cam-FAI for which shape modeling greatly promise in deriving measurements to best describe cam-FAI deformities [86]. ShapeWorks has further helped in developing cost-effective patient-specific finite element (FE) models (which require hundreds of man-hours otherwise) of the cartilage and labrum to advance our understanding of contact mechanics and the pathogenesis of osteoarthritis [87].

In cardiology, ShapeWorks has been applied to study the maladaptive remodeling of the left atrium (LA) and clinical outcomes in atrial fibrillation (AF). PBM is uniquely suited for modeling the LA (and other heart structures) because its particle-system formulation can accommodate holes in a surface, such as the mitral valve openings and pulmonary vein openings. LA shape and size changes have previously been associated with AF progression and decreased response to catheter ablation, a first-line therapy for symptomatic drug refractory AF [88–91]. Using ShapeWorks models of the LA, we described the statistically significant differences in LA shape in populations with increasing severity of the disease [64]. In that same study, we also identified characteristic shape changes of the left-atrial appendage that are associated with an increased likelihood of thrombus, a major risk factor for cardioembolic stroke. These results suggest the possibility of a shape-based clinical measure for stroke risk, as well as other AF outcomes. In another study, we applied the PBM correspondence models in the LA to develop a population atlas of the distribution of fibrosis in AF sufferers [65]. Fibrosis is a well-known degradation of the tissue structure in AF that can be measured using late-gadolinium enhancement MRI [92].

10.6 CONCLUSIONS AND FUTURE WORK

Statistical shape modeling (SSM) represents a revolutionary new approach to morphometry, but the widespread adoption of these new tools will require careful engineering to make them accessible to the average user and applicable to a wide variety of biological shapes. ShapeWorks is an implementation of Particle Based Modeling (PBM), which is designed to be a general tool for SSM. By modeling surfaces nonparametrically as collections of dynamic particle systems, ShapeWorks is not limited to any particular topology and can even represent shapes with multiple disconnected surfaces and surfaces with arbitrary open boundaries. The particle system approach also avoids the algorithmic complexity and parameter tuning associated with constructing parameterizations.

The ShapeWorks software suite implements a general SSM workflow that supports interactive processing on smaller datasets and offline processing of very large cohorts. ShapeWorks also includes preprocessing and analysis tools and, for many applications, can be a complete end-to-end SSM solution. The PBM C++ code library represents a careful numerical implementation of the particle system framework, and is extensible and reusable in any application supporting ITK. The ShapeWorks approach has been used successfully for investigation in many areas of biomedical investigation, including neurobiology, genetic phenotyping, orthopedics, and cardiology.

Despite proven success for many biomedical applications, the rapid advances and growing use of medical imaging technologies, and the associated need to model more complex morphological variations, require significant functionality and usability improvements of all modern SSM approaches, including ShapeWorks. In order to meet today's real-world shape modeling problems from a diverse biomedical community, we are developing new algorithmic extensions and new software versions. For example, motivated by studies of the highly variable anatomy of the atria of the heart, we are expanding our PBM approach to allow mixtures of Gaussians, in order to find natural clusterings of shapes in the populations and better discrimination when seeking shape-based indicators of treatment outcome. To support ever increasing cohort sizes and reduce ShapeWorks' run-times and memory requirements, we are developing parallel versions of the PBM optimization and modifying our algorithms to operate directly on mesh-based representations of surfaces. Finally, more comprehensive user interfaces for our software and domain-specific customizations of the PBM workflows are in progress, in order to make the PBM approach more accessible to the larger morphometrics community.

REFERENCES

[1] D.G. Kendall, The diffusion of shape, Adv. Appl. Probab. 9 (1977) 428–430.
[2] D.G. Kendall, Shape-manifolds, procrustean metrics and complex projective spaces, Bull. Lond. Math. Soc. 16 (1984) 81–121.

[3] J.T. Kent, The complex Bingham distribution and shape analysis, J. R. Stat. Soc. B 56 (1994) 285–299.

[4] F.L. Bookstein, Biometrics, biomathematics, and the morphometric synthesis, Bull. Math. Biol. 58 (1996) 313–365.

[5] Ian Dryden, Kanti Mardia, Statistical Shape Analysis, John Wiley and Sons, 1998.

[6] D.G. Kendall, Shape and Shape Theory, Wiley, 1999.

[7] C.P. Klingenberg, Morphometrics and the role of the phenotype in studies of the evolution of developmental mechanisms, Gene 287 (2002) 3–10.

[8] D. Adams, F.J. Rolf, D. Slice, Geometric morphometrics: ten years of progress following the "revolution", Ital. J. Zool. 71 (2004) 5–16.

[9] Tobias Heimann, Hans-Peter Meinzer, Statistical shape models for 3D medical image segmentation: a review, Med. Image Anal. 13 (4) (2009) 543–563.

[10] Martin Styner, Ipek Oguz, Shun Xu, Christian Brechbühler, Dimitrios Pantazis, James J. Levitt, Martha E. Shenton, Guido Gerig, Framework for the statistical shape analysis of brain structures using SPHARM-PDM, Insight J. 1071 (2006) 242.

[11] Martin Styner, Jeffrey A. Lieberman, Dimitrios Pantazis, Guido Gerig, Boundary and medial shape analysis of the hippocampus in schizophrenia, Med. Image Anal. 8 (3) (2004) 197–203.

[12] Amir A. Zadpoor, Harrie Weinans, Patient-specific bone modeling and analysis: the role of integration and automation in clinical adoption, J. Biomech. 48 (5) (2015) 750–760.

[13] Daniel P. Nicolella, Todd L. Bredbenner, Development of a parametric finite element model of the proximal femur using statistical shape and density modelling, Comput. Methods Biomech. Biomed. Eng. 15 (2) (2012) 101–110.

[14] Rebecca Bryan, P. Surya Mohan, Andrew Hopkins, Francis Galloway, Mark Taylor, Prasanth B. Nair, Statistical modelling of the whole human femur incorporating geometric and material properties, Med. Eng. Phys. 32 (1) (2010) 57–65.

[15] Laura Belenguer Querol, Philippe Büchler, Daniel Rueckert, Lutz P. Nolte, Miguel Á. González Ballester, Statistical finite element model for bone shape and biomechanical properties, in: Medical Image Computing and Computer-Assisted Intervention – MICCAI 2006, Springer, 2006, pp. 405–411.

[16] Nina Kozic, Stefan Weber, Philippe Büchler, Christian Lutz, Nils Reimers, Miguel Á. González Ballester, Mauricio Reyes, Optimisation of orthopaedic implant design using statistical shape space analysis based on level sets, Med. Image Anal. 14 (3) (2010) 265–275.

[17] Heiko Seim, Dagmar Kainmueller, Hans Lamecker, Matthias Bindernagel, Jana Malinowski, Stefan Zachow, Model-based auto-segmentation of knee bones and cartilage in MRI data, in: Medical Image Analysis for the Clinic: A Grand Challenge, Beijing, 2010.

[18] Hans Lamecker, Thomas H. Wenckebach, Hans-Christian Hege, Atlas-based 3D-shape reconstruction from X-ray images, in: 18th International Conference on Pattern Recognition, ICPR 2006, vol. 1, IEEE, 2006, pp. 371–374.

[19] Primoz Markelj, Dejan Tomaževič, Bostjan Likar, Franjo Pernuš, A review of 3D/2D registration methods for image-guided interventions, Med. Image Anal. 16 (3) (2012) 642–661.

[20] Guoyan Zheng, Sebastian Gollmer, Steffen Schumann, Xiao Dong, Thomas Feilkas, Miguel A. González Ballester, A 2D/3D correspondence building method for reconstruction of a patient-specific 3D bone surface model using point distribution models and calibrated X-ray images, Med. Image Anal. 13 (6) (2009) 883–899.

[21] Nora Baka, B.L. Kaptein, Marleen de Bruijne, Theo van Walsum, J.E. Giphart, Wiro J. Niessen, Boudewijn P.F. Lelieveldt, 2D–3D shape reconstruction of the distal femur from stereo X-ray imaging using statistical shape models, Med. Image Anal. 15 (6) (2011) 840–850.

[22] Guoyan Zheng, Xiao Dong, Kumar T. Rajamani, Xuan Zhang, Martin Styner, Ramesh U. Thoranaghatte, Lutz-Peter Nolte, Miguel A. González Ballester, Accurate and robust reconstruction of a surface model of the proximal femur from sparse-point data and a dense-point distribution model for surgical navigation, IEEE Trans. Biomed. Eng. 54 (12) (2007) 2109–2122.

[23] Jalda Dworzak, Hans Lamecker, Jens von Berg, Tobias Klinder, Cristian Lorenz, Dagmar Kainmüller, Heiko Seim, Hans-Christian Hege, Stefan Zachow, 3D reconstruction of the human rib cage from 2D projection images using a statistical shape model, Int. J. Comput. Assisted Radiol. Surg. 5 (2) (2010) 111–124.

[24] Guoyan Zheng, Steffen Schumann, Steven Balestra, Benedikt Thelen, Lutz-P. Nolte, 2D–3D reconstruction-based planning of total hip arthroplasty, in: Computational Radiology for Orthopaedic Interventions, Springer, 2016, pp. 197–215.

[25] Moritz Ehlke, Heiko Ramm, Hans Lamecker, Hans-Christian Hege, Stefan Zachow, Fast generation of virtual X-ray images for reconstruction of 3D anatomy, IEEE Trans. Vis. Comput. Graph. 19 (12) (2013) 2673–2682.

[26] Aurélie Carlier, Liesbet Geris, Johan Lammens, Hans Van Oosterwyck, Bringing computational models of bone regeneration to the clinic, Wiley Interdiscip. Rev., Syst. Biol. Med. 7 (4) (2015) 183–194.

[27] P. Sami Väänänen, Lorenzo Grassi, Gunnar Flivik, Jukka S. Jurvelin, Hanna Isaksson, Generation of 3D shape, density, cortical thickness and finite element mesh of proximal femur from a DXA image, Med. Image Anal. 24 (1) (2015) 125–134.

[28] B. Reggiani, L. Cristofolini, E. Varini, M. Viceconti, Predicting the subject-specific primary stability of cementless implants during pre-operative planning: preliminary validation of subject-specific finite-element models, J. Biomech. 40 (11) (2007) 2552–2558.

[29] C. Dean Barratt, Carolyn S.K. Chan, Philip J. Edwards, Graeme P. Penney, Mike Slomczykowski, Timothy J. Carter, David J. Hawkes, Instantiation and registration of statistical shape models of the femur and pelvis using 3D ultrasound imaging, Med. Image Anal. 12 (3) (2008) 358–374.

[30] Lawrence M. Specht, Kenneth J. Koval, Robotics and computer-assisted orthopaedic surgery, Bull. Hosp. Joint Dis. Orthop. Inst. 60 (3–4) (2001) 168–172.

[31] T. Kumar Rajamani, Martin A. Styner, Haydar Talib, Guoyan Zheng, Lutz P. Nolte, Miguel A. González Ballester, Statistical deformable bone models for robust 3D surface extrapolation from sparse data, Med. Image Anal. 11 (2) (2007) 99–109.

[32] Stefan Zachow, Hans Lamecker, Barbara Elsholtz, Michael Stiller, Reconstruction of mandibular dysplasia using a statistical 3D shape model, in: International Congress Series, vol. 1281, Elsevier, 2005, pp. 1238–1243.

[33] Stefan Zachow, Hans-Christian Hege, Peter Deuflhard, Computer assisted planning in cranio-maxillofacial surgery, CIT, J. Comput. Inf. Technol. 14 (1) (2006) 53–64.

[34] Li Wang, Yi Ren, Yaozong Gao, Zhen Tang, Ken-Chung Chen, Jianfu Li, Steve G.F. Shen, Jin Yan, Philip K.M. Lee, Ben Chow, et al., Estimating patient-specific and anatomically correct reference model for craniomaxillofacial deformity via sparse representation, Med. Phys. 42 (10) (2015) 5809–5816.

[35] Kun Zhang, Wee Kheng Leow, Yuan Cheng, Performance analysis of active shape reconstruction of fractured, incomplete skulls, in: Computer Analysis of Images and Patterns, Springer, 2015, pp. 312–324.

[36] H. Lamecker, S. Zachow, H. Hege, H. Haberl, Surgical treatment of craniosynostosis based on a statistical 3D-shape model: first clinical application, Int. J. Comput. Assisted Radiol. Surg. 1 (10) (2006) 253.

[37] Carlos S. Mendoza, Nabile Safdar, Kazunori Okada, Emmarie Myers, Gary F. Rogers, Marius George Linguraru, Personalized assessment of craniosynostosis via statistical shape modeling, Med. Image Anal. 18 (4) (2014) 635–646.

[38] Mascha Hochfeld, Hans Lamecker, Ulrich-W. Thomale, Matthias Schulz, Stefan Zachow, Hannes Haberl, Frame-based cranial reconstruction: technical note, J. Neurosurg. Pediatrics 13 (3) (2014) 319–323.

[39] Stefan Zachow, Computational planning in facial surgery, Facial Plast. Surg. 31 (5) (2015) 446.

[40] Francis Galloway, Max Kahnt, Heiko Ramm, Peter Worsley, Stefan Zachow, Prasanth Nair, Mark Taylor, A large scale finite element study of a cementless osseointegrated tibial tray, J. Biomech. 46 (11) (2013) 1900–1906.

[41] Jeffrey E. Bischoff, Yifei Dai, Casey Goodlett, Brad Davis, Marc Bandi, Incorporating population-level variability in orthopedic biomechanical analysis: a review, J. Biomech. Eng. 136 (2) (2014) 021004.

[42] Rebecca Bryan, Prasanth B. Nair, Mark Taylor, Use of a statistical model of the whole femur in a large scale, multi-model study of femoral neck fracture risk, J. Biomech. 42 (13) (2009) 2171–2176.

[43] C. Merle, W. Waldstein, J.S. Gregory, S.R. Goodyear, R.M. Aspden, P.R. Aldinger, D.W. Murray, H.S. Gill, How many different types of femora are there in primary hip osteoarthritis? An active shape modeling study, J. Orthop. Res. 32 (3) (2014) 413–422.

[44] Ch. Brechbühler, Guido Gerig, Olaf Kübler, Parametrization of closed surfaces for 3-D shape description, Comput. Vis. Image Underst. 61 (2) (1995) 154–170.

[45] Andrew P. Witkin, Paul S. Heckbert, Using particles to sample and control implicit surfaces, in: Proceedings of the 21st Annual Conference on Computer Graphics and Interactive Techniques, ACM Press, 1994, pp. 269–277.

[46] Miriah D. Meyer, Pierre Georgel, Ross T. Whitaker, Robust particle systems for curvature dependent sampling of implicit surfaces, in: Proceedings of the International Conference on Shape Modeling and Applications, June 2005, pp. 124–133.

[47] H. Hoppe, T. DeRose, T. Duchamp, J. McDonald, W. Stuetzle, Surface reconstruction from unorganized points, in: ACM SIGGRAPH 1992 Conference Proceedings, 1992, pp. 71–78.

[48] Mark Hansen, Bin Yu, Model selection and the principle of minimum description length, J. Am. Stat. Assoc. 96 (454) (June 2001) 746–774.

[49] Rhodri H. Davies, Carole J. Twining, Timothy F. Cootes, John C. Waterton, Christopher J. Taylor, A minimum description length approach to statistical shape modeling, IEEE Trans. Med. Imaging 21 (5) (2002) 525–537.

[50] Rhodri H. Davies, Carole J. Twining, Timothy F. Cootes, John C. Waterton, Christopher J. Taylor, 3D statistical shape models using direct optimisation of description length, in: ECCV (3), 2002, pp. 3–20.

[51] Joshua Cates, Miriah Meyer, Thomas Fletcher, Ross Whitaker, et al., Entropy-based particle systems for shape correspondence, in: 1st MICCAI Workshop on Mathematical Foundations of Computational Anatomy: Geometrical, Statistical and Registration Methods for Modeling Biological Shape Variability, 2006, pp. 90–99.

[52] Joshua Cates, P. Thomas Fletcher, Martin Styner, Martha Shenton, Ross Whitaker, Shape modeling and analysis with entropy-based particle systems, in: Information Processing in Medical Imaging, Springer, Berlin, Heidelberg, 2007, pp. 333–345.

[53] Joshua Cates, P. Thomas Fletcher, Martin Styner, Heather Cody Hazlett, Ross Whitaker, Particle-based shape analysis of multi-object complexes, in: Medical Image Computing and Computer-Assisted Intervention – MICCAI 2008, Springer, Berlin, Heidelberg, 2008, pp. 477–485.

[54] Manasi Datar, Yaniv Gur, Beatriz Paniagua, Martin Styner, Ross Whitaker, Geometric correspondence for ensembles of nonregular shapes, in: Medical Image Computing and Computer-Assisted Intervention – MICCAI 2011, Springer, Berlin, Heidelberg, 2011, pp. 368–375.

[55] M. Datar, P. Muralidharan, A. Kumar, S. Gouttard, J. Piven, G. Gerig, R.T. Whitaker, P.T. Fletcher, Mixed-effects shape models for estimating longitudinal changes in anatomy, in: Stanley Durrleman, P. Thomas Fletcher, Guido Gerig, Marc Niethammer (Eds.), Spatio-Temporal Image Analysis for Longitudinal and Time-Series Image Data, in: Lect. Notes Comput. Sci., vol. 7570, Springer, Berlin/Heidelberg, 2012, pp. 76–87.

[56] Ipek Oguz, Joshua Cates, Thomas Fletcher, Ross Whitaker, Derek Cool, Stephen Aylward, Martin Styner, Cortical correspondence using entropy-based particle systems and local features, in: 5th IEEE International Symposium on Biomedical Imaging: From Nano to Macro, ISBI 2008, IEEE, 2008, pp. 1637–1640.

[57] Ipek Oguz, Marc Niethammer, Josh Cates, Ross Whitaker, Thomas Fletcher, Clement Vachet, Martin Styner, Cortical correspondence with probabilistic fiber connectivity, in: Information Processing in Medical Imaging, Springer, Berlin, Heidelberg, 2009, pp. 651–663.

[58] Manasi Datar, Joshua Cates, P. Thomas Fletcher, Sylvain Gouttard, Guido Gerig, Ross Whitaker, Particle based shape regression of open surfaces with applications to developmental neuroimaging, in: Medical Image Computing and Computer-Assisted Intervention – MICCAI 2009, Springer, Berlin, Heidelberg, 2009, pp. 167–174.

[59] Manasi Datar, Ilwoo Lyu, SunHyung Kim, Joshua Cates, Martin A. Styner, Ross Whitaker, Geodesic distances to landmarks for dense correspondence on ensembles of complex shapes, in: Medical Image Computing and Computer-Assisted Intervention – MICCAI 2013, Springer, 2013, pp. 19–26.

[60] Joshua Cates, P. Thomas Fletcher, Zachary Warnock, Ross Whitaker, A shape analysis framework for small animal phenotyping with application to mice with a targeted disruption of hoxd11, in: 5th IEEE International Symposium on Biomedical Imaging: From Nano to Macro, ISBI 2008, IEEE, 2008, pp. 512–515.

[61] Kevin B. Jones, Manasi Datar, Sandhya Ravichandran, Huifeng Jin, Elizabeth Jurrus, Ross Whitaker, Mario R. Capecchi, Toward an understanding of the short bone phenotype associated with multiple osteochondromas, J. Orthop. Res. 31 (4) (2013) 651–657.

[62] Michael D. Harris, Manasi Datar, Ross T. Whitaker, Elizabeth R. Jurrus, Christopher L. Peters, Andrew E. Anderson, Statistical shape modeling of cam femoroacetabular impingement, J. Orthop. Res. (2013).

[63] Penny B. Atkins, Shireen Elhabian, Praful Agrawal, Mike Harris, Jeffery Weiss, Chris Peters, Ross Whitaker, Andrew Anderson, Quantitative comparison of cortical bone thickness using correspondence-based shape modeling in patients with cam femoroacetabular impingement, J. Orthop. Res. (2016).

[64] Joshua Cates, Erik Bieging, Alan Morris, Gregory Gardner, Nazem Akoum, Eugene Kholmovski, Nassir Marrouche, Christopher McGann, Rob S. MacLeod, Computational shape models characterize shape change of the left atrium in atrial fibrillation, Clin. Med. Insights, Cardiol. 8 (Suppl. 1) (2014) 99.

[65] Gregory Gardner, Alan Morris, Koji Higuchi, Robert MacLeod, Joshua Cates, A point-correspondence approach to describing the distribution of image features on anatomical surfaces, with application to atrial fibrillation, in: 2013 IEEE 10th International Symposium on Biomedical Imaging (ISBI), IEEE, 2013, pp. 226–229.

[66] T. Cover, J. Thomas, Elements of Information Theory, Wiley and Sons, 1991.

[67] Emanuel Parzen, On estimation of a probability density function and mode, Ann. Math. Stat. 33 (3) (1962) 1065–1076.

[68] M. Meyer, B. Nelson, R.M. Kirby, R. Whitaker, Particle systems for efficient and accurate high-order finite element visualization, IEEE Trans. Vis. Comput. Graph. 13 (5) (2007) 1015–1026.

[69] G. Kindlmann, R. Whitaker, T. Tasdizen, T. Moller, Curvature-based transfer functions for direct volume rendering, in: Proceedings of IEEE Visualization, 2003, pp. 512–520.

[70] Sarang Joshi, Brad Davis, Matthieu Jomier, Guido Gerig, Unbiased diffeomorphic atlas construction for computational anatomy, NeuroImage 23 (2004) S151–S160.

[71] SCI Institute, AtlasWerks: an open-source (BSD license) software package for medical image atlas generation, Scientific Computing and Imaging Institute (SCI), 2016, available from http://www.sci.utah.edu/software/atlaswerks.html.

[72] John C. Gower, Generalized Procrustes analysis, Psychometrika 40 (1) (1975) 33–51.

[73] Fred L. Bookstein, Principal warps: thin-plate splines and the decomposition of deformations, IEEE Trans. Pattern Anal. Mach. Intell. 11 (6) (1989) 567–585.

[74] Shengxin Zhu, Compactly supported radial basis functions: how and why?, SIAM Rev. (2012).

[75] Ross Whitaker, Reducing aliasing artifacts in iso-surfaces of binary volumes, in: IEEE Volume Visualization and Graphics Symposium, 2000, pp. 22–23.

[76] J. Williams, J. Rossignac, Tightening: curvature-limiting morphological simplification, in: Proc. Ninth ACM Symposium on Solid and Physical Modeling, 2005, pp. 107–112.

[77] C. Goodall, Procrustes methods in the statistical analysis of shape, J. R. Stat. Soc. B 53 (1991) 285–339.

[78] F.J. Rohlf, Bias and error in estimates of mean shape in geometric morphometrics, J. Hum. Evol. 44 (2003) 665–683.

[79] J.M.F. Ten Berge, Orthogonal Procrustes rotation for two or more matrices, Psychometrika 42 (1977) 267–276.

[80] T.F. Cootes, C.J. Taylor, D.H. Cooper, J. Graham, Active shape models – their training and application, Comput. Vis. Image Underst. 61 (1) (January 1995) 38–59.

[81] Joshua Cates, Thomas Fletcher, Ross Whitaker, et al., A hypothesis testing framework for high-dimensional shape models, in: 2nd MICCAI Workshop on Mathematical Foundations of Computational Anatomy, 2008, pp. 170–181.

[82] L.W. Glorfeld, An improvement on Horn's parallel analysis methodology for selecting the correct number of factors to retain, Educ. Psychol. Meas. 55 (1995) 377–393.

[83] Erich Gamma, Design Patterns: Elements of Reusable Object-Oriented Software, Pearson Education India, 1995.

[84] Penny Atkins, Prateep Mukherjee, Shireen Elhabian, Sumedha Singla, Michael Harris, Jeffery Weiss, Ross Whitaker, Andrew Anderson, Proximal femoral cortical bone thickness in patients with femoroacetabular impingement and normal hips analyzed using statistical shape modeling, in: Summer Biomechanics, Bioengineering and Biotransport Conference, 2015, oral presentation.

[85] Penny Atkins, Shireen Elhabian, Praful Agrawal, Ross Whitaker, Jeffery Weiss, Stephen Aoki, Chris Peters, Andrew Anderson, Can the sclerotic subchondral bone of the proximal femur cam lesion be used as a surgical resection guide? An objective analysis using 3D computed tomography and statistical shape modeling, in: International Society of Hip Arthroscopy Annual Scientific Meeting, 2016, submitted abstract.

[86] Penny Atkins, Shireen Elhabian, Praful Agrawal, Ross Whitaker, Jeffery Weiss, Chris Peters, Stephen Aoki, Andrew Anderson, Which radiographic measurements best identify anatomical variation in femoral head anatomy? Analysis using 3D computed tomography and statistical shape modeling, in: International Society of Hip Arthroscopy Annual Scientific Meeting, 2016, submitted abstract.

[87] Penny Atkins, Prateep Mukherjee, Shireen Elhabian, Sumedha Singla, Ross Whitaker, Jeffery Weiss, Andrew Anderson, Warping of template meshes for efficient subject-specific FE mesh generation, in: International Symposium of Computer Methods in Biomechanics and Biomedical Engineering, 2015, oral presentation.

[88] Toshiya Kurotobi, Katsuomi Iwakura, Koichi Inoue, Ryusuke Kimura, Yuko Toyoshima, Norihisa Ito, Hiroya Mizuno, Yoshihisa Shimada, Kenshi Fujii, Shinsuke Nanto, et al., The significance of the shape of the left atrial roof as a novel index for determining the electrophysiological and structural characteristics in patients with atrial fibrillation, Europace 13 (6) (2011) 803–808.

[89] Luigi Di Biase, Pasquale Santangeli, Matteo Anselmino, Prasant Mohanty, Ilaria Salvetti, Sebastiano Gili, Rodney Horton, Javier E. Sanchez, Rong Bai, Sanghamitra Mohanty, et al., Does the left atrial appendage morphology correlate with the risk of stroke in patients with atrial fibrillation? Results from a multicenter study, J. Am. Coll. Cardiol. 60 (6) (2012) 531–538.

[90] Felipe Bisbal, Esther Guiu, Naiara Calvo, David Marin, Antonio Berruezo, Elena Arbelo, José Ortiz-Pérez, Teresa María Caralt, José María Tolosana, Roger Borràs, et al., Left atrial sphericity: a new method to assess atrial remodeling. Impact on the outcome of atrial fibrillation ablation, J. Cardiovasc. Electrophysiol. 24 (7) (2013) 752–759.

[91] Teresa S.M. Tsang, Marion E. Barnes, Kent R. Bailey, Cynthia L. Leibson, Samantha C. Montgomery, Yasuhiko Takemoto, Pauline M. Diamond, Marisa A. Marra, Bernard J. Gersh, David O. Wiebers, et

al., Left atrial volume: important risk marker of incident atrial fibrillation in 1655 older men and women, in: Mayo Clin. Proc., vol. 76, Elsevier, 2001, pp. 467–475.

[92] Christopher J. McGann, Eugene G. Kholmovski, Robert S. Oakes, Joshua J.E. Blauer, Marcos Daccarett, Nathan Segerson, Kelly J. Airey, Nazem Akoum, Eric Fish, Troy J. Badger, et al., New magnetic resonance imaging-based method for defining the extent of left atrial wall injury after the ablation of atrial fibrillation, J. Am. Coll. Cardiol. 52 (15) (2008) 1263–1271.

PART III

Applications

CHAPTER 11

Applications of Statistical Deformation Model

Yipeng Hu*, Xiahai Zhuang[†]
*Centre for Medical Image Computing, Department of Medical Physics and Biomedical Engineering, University
College London, London, United Kingdom
[†]School of Data Science, Fudan University, Shanghai, China

Contents

11.1	Image-Guided Prostate Intervention	301
	11.1.1 Segmentation of Multiple Organs in MR	302
	11.1.2 Simplified Finite Element Models	303
	11.1.2.1 Meshing	*306*
	11.1.2.2 Material Properties	*308*
	11.1.2.3 Boundary Conditions	*309*
	11.1.3 Statistical Motion Model	311
	11.1.4 The Use of the Statistical Motion Models	312
	11.1.4.1 Manual Point Features	*314*
	11.1.4.2 Point Distance	*315*
	11.1.4.3 SMM as a Transformation Model	*316*
	11.1.5 A Population-Based Model Generation Method	317
11.2	Whole Heart Segmentation	319
	11.2.1 Definition, Applications, Challenges and Methodologies	320
	11.2.2 Deformable Registration for Atlas-Based Segmentation	322
	11.2.3 Improving Segmentation Using Statistical Modeling of the Atlas	323
References		327

11.1 IMAGE-GUIDED PROSTATE INTERVENTION

In this section, we briefly describe a method to build a statistical motion model (SMM)
for the purpose of registering a preoperative MR image to the intraoperative 3D transrectal ultrasound image (TRUS), in order to guide TRUS-guided procedures widely
adopted for prostate cancer biopsy and therapies. The relevant work and clinical background have been presented in previous publication [3,4,6,5,7], we focus on one
particular implementation in this chapter. Fig. 11.1 provides an illustration of two of
most widely adopted variations of these procedures.[1]

---

[1] This is a description of a prototype version of the algorithm implemented in a commercial guidance
system (smarttarget.co.uk).

Statistical Shape and Deformation Analysis
DOI: 10.1016/B978-0-12-810493-4.00014-6

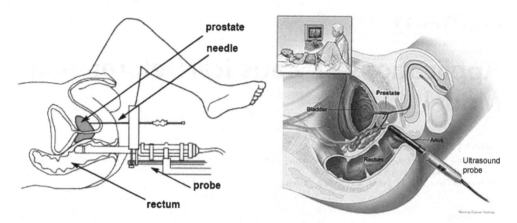

Figure 11.1 Left: An illustration of a TRUS-guided freehand biopsy. Right: An illustration of a TRUS-guided template biopsy, where the biopsy needle goes through the template that is parallel to the patient's perinea. The template, illustrated as a block attached to the TRUS probe, provides relative location information.

The aim of the statistical motion model (SMM) is to predict and therefore compensate the prostate gland motion, primarily caused by movement of the ultrasound probe. The estimated deformation then can locate the tumor, identified from MR, in the TRUS coordinates, which can be targeted for biopsy or localized treatment during procedures. An overview of the method is provided in Fig. 11.2.

11.1.1 Segmentation of Multiple Organs in MR

Diagnostic multi-parametric MR images were segmented by manually defining contours on transverse slices using an image analysis GUI[2] (see Fig. 11.3). This is to obtain the patient-specific geometric information of prostate gland and surrounding organs. Other regions of interest, such as the location of the tumor, can also be segmented to form a part of the procedural plan. The segmentation process was time-consuming (typically taking 45 minutes per patient), but was still considered as ground truth available for segmenting pelvic anatomy. The outer surface of the prostate gland capsule was segmented, and the gland itself divided into the central and peripheral zones, which are usually clearly visible in T2-weighted MR images. The pelvic bone, the rectum, and the bladder at the base of the prostate were also segmented (see Fig. 11.3, Fig. 11.4 and Fig. 11.5).

---

[2] There are a number of open-source packages providing capacities to load and view DICOM images with simple denotation and manipulation functionalities, notably, 3D Slicer (www.slicer.org) and OsiriX (www.osirix-viewer.com) among others.

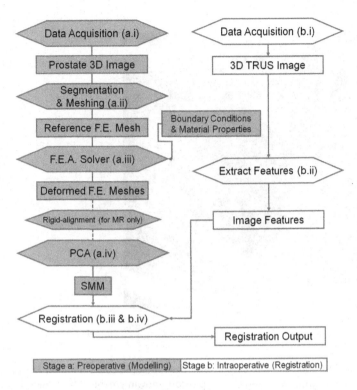

Figure 11.2 A flowchart of the modeling method (shaded) and the registration method (clear).

11.1.2 Simplified Finite Element Models

The aim of the multiple FE simulations is to provide training data for later statistical analysis. In particular, building MR-derived SMMs requires patient-specific segmentation of multiple organs. Although a number of semi- and fully-automatic segmentation algorithms have been proposed, it is still a significant challenge to simultaneously obtain accuracy segmentations for all these organs. This section describes an alternative to the automatic segmentation for reducing the burden of manual delineation and therefore making patient-specific SMM more clinically practical. Specifically, a geometrically simplified FE mesh, in which, some anatomical structures are replaced by equivalent structures with a simplified geometry, or omitted completely, is described. This strategy is inspired by the observation that since a PCA-based SMM trained using a set of deformed FE meshes captures the statistical variation in mesh node displacements, adopting a geometrically simplified mesh may not affect the characteristic parameters (i.e. the mean and variance of a Gaussian distribution) of this distribution significantly. Therefore, the accuracy of the final MR-derived SMM may not be compromised significantly by adopting a simplified mesh when generating the training data.

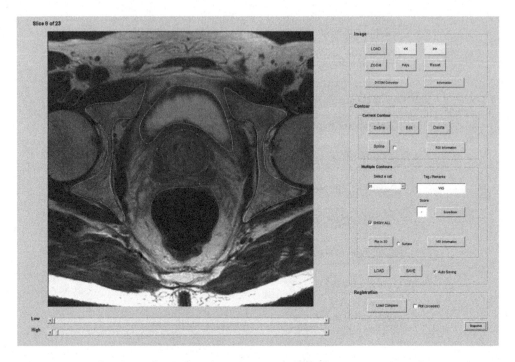

Figure 11.3 An illustration the GUI used to segment multiple contours in an MR slice, a transverse slice through a T2-weighted MR image of the prostate showing manually delineated contours used to segment the prostate gland, the rectum, the pelvis and the bladder.

The change of material properties and/or boundary conditions is expected to affect significantly the individual simulation. However, the impact of the simplified FE models is investigated by comparing the accuracy of MR–derived prostate SMMs built using different simplified FE mesh geometries with a reference model built from training data simulated using an FE mesh in which the geometry of organs are all accurately defined. The details of these comparisons and the results to assess the effect of the SMMs based on simplified geometries are presented in [3].

For each patient, a *fully-specified* FE mesh was used to build a reference (control) SMM (see Fig. 11.5, Fig. 11.6 and Fig. 11.7). This fully-specified FE model was based on FE simulations that consider the prostate gland, pelvic bone, rectum, and bladder segmented fully from an MR scan as distinct, homogeneous structures, as shown in Fig. 11.6. The corresponding SMM was generated by randomly assigning boundary conditions and elastic properties for each tissue type, and computing the subsequent deformations.

SMMs based on FE simulation data were also built by reducing the number of soft-tissue compartments in the model (equivalent to assigning identical material properties

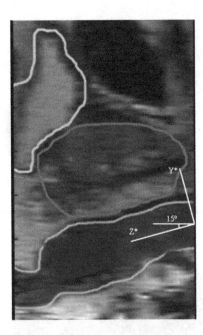

Figure 11.4 An illustration of the local TRUS probe coordinate system shown on a sagittal prostate MR image. The prostate gland, rectal wall, probe and bladder are shown in red, green, blue and yellow, respectively. The local reference coordinate system was defined with the z-axis orientated at 15 degree relative to the cranial–caudal axis. (For interpretation of the references to color in this figure legend, the reader is referred to the web version of this chapter.)

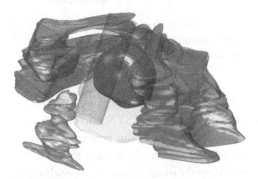

Figure 11.5 An illustration of surface meshes obtained by segmenting an MR image. The TRUS probe (with sheath), approximated by a cylinder is shown in blue. The prostate gland, the pelvis, and the bladder are shown in red, gray and yellow, respectively. (For interpretation of the references to color in this figure legend, the reader is referred to the web version of this chapter.)

to adjacent compartments in the fully-specified model) and/or simplifying the geometry of the pelvic bone. Either of these simplifications directly reduces the amount of prerequisite segmentation required to build an SMM from FE simulations.

Figure 11.6 An illustration of surface meshes of gland (red), bladder (yellow), rectum (green) and pelvis (gray). (For interpretation of the references to color in this figure legend, the reader is referred to the web version of this chapter.)

Figure 11.7 An illustration of TRUS probe (blue cylindrical structure) position relative to the prostate gland and the pelvis. (For interpretation of the references to color in this figure legend, the reader is referred to the web version of this chapter.)

11.1.2.1 Meshing

Using the refinement tool available in ANSYS, the region around the rectum was re-meshed to obtain high element density in this region. This enabled the TRUS probe – or more precisely, the fluid-filled sheath placed over the TRUS probe, approximated by a cylinder – to be modeled directly in each simulation without the need for re-meshing.

As shown in Fig. 11.8, the prostate gland surface was represented by an SH surface fitted to the transverse contours. The contours are first converted into a binary volume. Surface points then are sampled from the binary image and fit an SH surface using the method that is described by Tutar et al. [8]. To obtain a smooth and uniformly sampled surface, suitable for generating a high quality mesh for FEA, the following adaptive sampling scheme, similar to the one described in Zhou et al. [10], was implemented. The SH surface was first filtered in the frequency domain using a trapeziform low-pass function before being meshed into triangles by projecting a uniformly triangulated

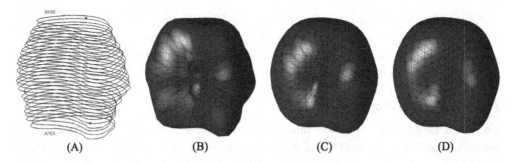

Figure 11.8 Reconstruction of a smooth, triangulated SH surface from manually drawn prostate contours: (A) original contours with apex and base points; (B) initial fitted SH surface; (C) filtered SH surface; and (D) surface in (C) following mesh refinement.

sphere template. Refinement of the triangulated mesh was performed by maximizing the sum of a triangle quality measure over the surface using a quasi-Newton numerical optimization algorithm, implemented in MATLAB. The smoothness of the final surface mesh is controlled by the degree of the SH, the coefficients of the filter, and the density of the mesh. Values for these parameters were set experimentally such that the surface appeared smooth visually with the constraint that the maximum absolute distance (MAD) between the reconstructed surface and the original contour points was less than 1.0 mm. (Note that, as reported in Tutar et al. [8], the maximum inter-observer error was found to be significantly greater than this value.) Definition of the apex and base points was found to be important for producing a geometrically accurate surface. These were defined immediately adjacent to the available contours to maintain a topologically correct gland surface. An example of a smoothed SH representation of a prostate gland, reconstructed from TRUS contours, is illustrated in Fig. 11.8D.

The above surface meshing process can be replaced by the iso-surface algorithm based methods. And again, the quality of the mesh needs to be assessed and the fitting error of the meshing process needs to be validated by computing the distance error between the segmentation and the final mesh.

As described in [3], a zonal structure of prostate gland that is visible in TRUS images is different from that in MR images. The segmentation obtained from MR image (see Fig. 11.3) was used to separate central and peripheral zones for MR-derived SMM.

The surrounding tissue uses the MR image sections which adequately cover the same region. The surface meshes and the block structure were imported into a commercial FEA software package (ANSYS). A linear four-node tetrahedral FE mesh was then constructed automatically using trimmed parametric surfaces and Delaunay tessellation techniques provided by the software. Only the solid meshing tool of ANSYS was used in this work; although ANSYS is a general purpose FEA package, an alternative fast FE solver was employed to compute mesh node displacements. For each patient case, the mesh comprised approximately 35,000–100,000 elements. Volumetric regions corresponding to the inner prostate gland, the outer prostate gland, the rectal wall, and surrounding tissue were individually labeled and attributed with different material properties, as described in the next section.

11.1.2.2 Material Properties

The element groups, which are modeled using different material properties and/or are assigned different types of nodal displacements, are referred as compartments. These compartments include the central and peripheral zones of the prostate gland, the rectum, the surrounding tissue, the pelvic bones and the TRUS probe. These different compartments are segmented as described in Section 11.1.1. In practice, elements within the same compartment of the FE model were labeled according to the corresponding tissue type. Different material property can then easily be assigned to the corresponding compartment. All the organs are assumed to be geometrically connected to the surrounding tissue. The pelvis provides a geometrically realistic, rigid constraint, which balances the driving force exerted by the movement of the TRUS probe.

All the materials were assumed to be linear in initial work, and later were changed to be nonlinear. In both cases, a nonlinear solver, such as the fully nonlinear total Lagrangian explicit finite element formulation,[3] is essential, as a larger deformation breaks the linear assumption of geometry despite the linear material model. Because there is a lack of studies in the literature pointing to any particular model suitable for modeling the prostate gland motion, a simple linear elastic model and a hyperelastic neo-Hookean model with two parameters were used. The neo-Hookean model provides a relatively simple formulation to predict the nonlinear strain–stress behavior of hyperelastic material undergoing large deformation. Although exact behavior of soft tissue is expected to be complex and nonlinear, it may be argued that the exact formulation of the material model is not important in this application where only the variance of the motion is of interest.

To partially account for this, tissue material properties were included as variable parameters in the generation of training data for the prostate SMM. Given the variability and uncertainty associated with published material properties, sample values are sampled from a relatively wide range, consistent with the range normally applied for soft-tissue modeling [1]. It is important to note, however, that the prediction of displacements in the FEA is only dependent on the ratio of elasticity moduli assigned to different compartments and not on their absolute values. The uniform ranges used in this thesis are given in Table 11.1.

The usual condition of incompressibility ("equivalent" to Poisson's ratio = 0.5) was not assumed because it can be argued that this is not appropriate for organs such as the

[3] sourceforge.net/projects/niftysim/.

Table 11.1 Material properties and boundary conditions used for FE simulations. The ranges of the material properties and the boundary conditions have been assigned based on plausible values as well as on the observations

Description	Parameter(s)	Range	Reference value(s)	DOF
Balloon radius	R	$[0.9R_0, 1.5R_0]$[a]	R_0[a]	1
Balloon translation	Tb_x, Tb_y, Tb_z	$[-5, 5]$ mm	$Tb_x = Tb_y = Tb_z = 0$ mm	3
Balloon rotation	$\theta b_x, \theta b_y, \theta b_z$	$[-10, 10]°$	$\theta b_x = \theta b_y = \theta b_z = 0°$	3
Pelvis scaling[d]	S	$[0.8, 1.2]$	$S = 1$	1
Pelvis translation[d]	Tp_x, Tp_y, Tp_z	$[-10, 10]$ mm	$Tp_x = Tp_y = Tp_z = 0$ mm	3
Pelvis rotation[d]	$\theta p_x, \theta p_y, \theta p_z$	$[-15, 15]°$	$\theta p_x = \theta p_y = \theta p_z = 0°$	3
Shear modulus	G_1, G_2, G_4[b]	$[3.36, 76.9]$ kPa	—	4
	G_3[c]	$[3.36, 67.1]$ kPa		
Bulk modulus	K_1, K_2, K_4	$[8.33, 3.33 \times 10^3]$ kPa	—	4
	K_3[c]	$[0.17, 3.33]$ GPa		

[a] R_0 denotes the radius of the balloon measured from the source image.
[b] The subscripts 1–4 correspond to the prostate central zone/inner gland (1), the peripheral zone/outer gland (2), the rectal wall (3), and the surrounding tissue (4), respectively.
[c] The rectal wall in contact with the balloon is assumed to be nearly incompressible.
[d] For the generic pelvic model used in TRUS-derived SMMs only.

prostate, rectum and bladder, which are compressible due to gain and loss of blood and other fluids, as well as the presence of cavities.

Material properties were assigned in two ways, depending on number of organs that need to be segmented as follows:

MP1: In the first case, material properties are assigned independently to each of the segmented soft-tissue five regions – i.e., the rectal wall, the bladder, the central and peripheral zones of the prostate gland, and the surrounding tissue (assumed to be homogeneous).

MP2: In the second, simpler case, (i) the prostate gland and (ii) surrounding organs and tissue are treated as two single homogeneous materials. This is one example of an SMM with simplified geometry.

11.1.2.3 Boundary Conditions

Two sources of prostate deformation were considered: the expansion of the TRUS balloon and a change in the pose of the TRUS probe/balloon. Further boundary constraints were imposed on the pelvic bone surface so that the magnitude of these nodes had a displacement of zero. Slippages between organ surfaces and between the TRUS probe and rectal wall were not modeled.

Therefore, two types of boundary conditions were applied: the rigid constraint imposed by the pelvic bone and the position and orientation of the TRUS probe. In one

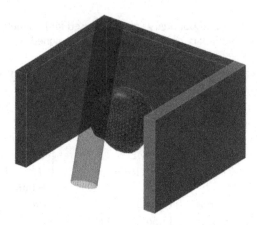

Figure 11.9 An illustration of the three-plates-structure (gray plates) relative to the positions of prostate gland (red mesh) and TRUS probe (blue cylinder). (For interpretation of the references to color in this figure legend, the reader is referred to the web version of this chapter.)

configuration of the simplified FE model investigated, the pelvic bone was approximated by three boundary planes, as shown in Fig. 11.9 and Fig. 11.10. This choice of representation was motivated by the need for a clinically practical method for approximating the bony constraints within the pelvis.

The positions of these planes for an individual patient were determined by measuring two distances, d_x and d_y, in the approximately mid-gland transverse plane of the MR image, as shown in Fig. 11.10. Assuming that the prostate capsule has been segmented, d_x is the average of the two distances measured along the left–right axis from center of mass of the prostate gland to the nearest intersections with the axis on the left and right sides of the pelvis. Distance d_y is the distance from the prostate center of mass to the nearest point on the posterior side of the pubis along the anterior–posterior axis. In this thesis, these distances were computed automatically using the segmentation of the pelvic bone, but, importantly, both can be easily measured without needing to segment the pelvis. The displacement at each mesh node of the pelvic bone, or alternatively the surrogate planes, was fixed to zero for all simulations.

In the experiments described below, three different pelvic boundary conditions – referred to as BC1, BC2, and BC3 – were used. These are defined as follows:

BC1: An anatomically realistic, patient-specific pelvic bone. This requires complete segmentation of the bone on MR.

BC2: Three planes placed according the patient-specific measurements, as described above. This requires only simple measurements from an MR image.

BC3: Three planes placed at fixed positions determined by the average measurement calculated for the remaining 6 patients in the test dataset. Setting this boundary condi-

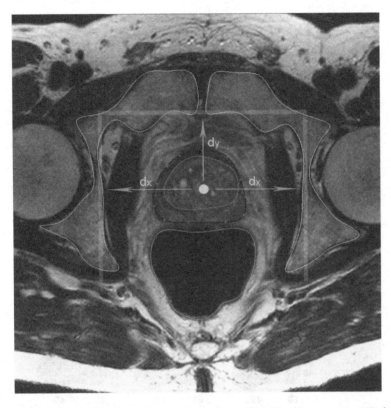

Figure 11.10 Illustration of the three-plates-structure (green lines) in the transverse MR slice, and the distance measures d_x and d_y are also demonstrated. (For interpretation of the references to color in this figure legend, the reader is referred to the web version of this chapter.)

tion only requires segmentation of the prostate capsule (in order to compute the center of mass of the gland).

As the driving force for the prostate motion, the size and 3D motion of the TRUS probe were specified in terms of the diameter of the water-filled sheath surrounding the probe, and the motion with respect to a local 3D coordinate system, defined with respect to an initial reference position (see details in Section 11.1.1).

11.1.3 Statistical Motion Model

As there is no common reference position between MR and TRUS coordinates, before extracting the principal components, a rigid alignment was applied using Procrustes analysis, introduced in [9], to eliminate the variance due to change of pose in the model.

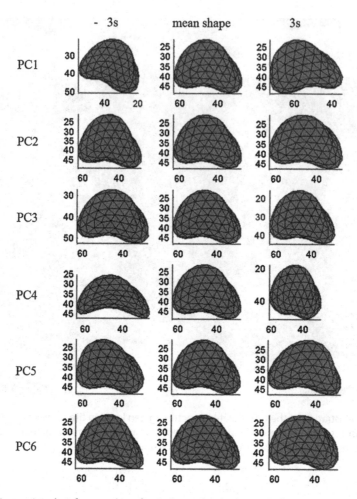

Figure 11.11 Instantiated surface meshes after independently changing the weights corresponding to the first six principal components, PC1–PC6. The left and right columns show the shapes after changing each weight to ±3s, respectively, where s is the standard deviation of the weight over the training data. The middle column shows the mean shape for comparison.

11.1.4 The Use of the Statistical Motion Models

The use of conventional statistical shape models has been summarized in [2]. The variance of the motion, in this case, is computed with respect to the mean shape which is computed by averaging the rigid-transformation-excluded node positions, please refer to Fig. 11.11 for illustration. Therefore, in the instantiation of the model, and possibly in later optimization as well, an extra rigid component should be added to compensate for this.

However, though the surfaces are identified as corresponding features, more detailed point-to-point correspondences remain unknown. The model-to-image registration problem therefore is converted into a feature registration problem, where the features are the surface points and/or surface normal vectors sampled from the gland. There are methods to formulate the feature registration problem as a probabilistic maximum likelihood (ML) problem. An overview of this class of problems is presented in [3], where only the point features are considered. The previous work has extended this method 1) to incorporate the SMM as a constrained transformation model; 2) to use the additional orientation vectors representing the surface normal.

The detailed correspondence can be optimized with respect to the maximization of a specifically designed similarity measure. For instance, the RMS of weighted distances between individual model feature and all the image features may be adopted, whereas the weighting of each image feature can be calculated according to the distance between single image feature and single model feature. The most simplified solution is to use only the closest image feature, i.e. the weightings of others are set zeros.

A general-purpose optimizer may be used to optimize this RMS distance, which is referred as to the *direct optimization approach*. On the other hand, if the similarity measure is defined as the likelihood function of a normalized PDF, the optimization problem becomes an ML problem. The latter is referred to as the *probabilistic approach* and may be solved via classical statistical methods, such as the expectation maximization (EM) algorithm.

The distance between features, either the spatial locations and/or the surface normal orientations, is first extracted from the model (the surface of SMM), which could be trivial as the model is usually of certain mathematically convenient representation, and extracted from the image, which requires fast and minimal human interaction in order to enable a rapid and efficient intraoperative procedure.

Table 11.2 summarizes the framework of registration between a model and an image developed. In theory, all the combinations, between different model features, image features and different optimization procedures can be applied using SMMs or any other transformation model. The work in [3] describes two instances as the items shaded in black and gray in Table 11.2. The first implementation (black background) employs an MR-derived SMM with manually picked surface points from an intraoperative TRUS image and a direct (general purpose) optimization algorithm to assess the statistical model; the second (gray background) automatically extracts the image vector features from a TRUS image to align with the model in order to solve the MR-to-TRUS image registration problem. The details of the second approach was not discussed here and can be found in [4]. A variation of the first approach is described.

The use of the biomechanically-constrained SMM enables registration to be achieved rapidly, which is particularly important for time-critical applications such as image-guided prostate cancer interventions. Furthermore, although only the capsule

Table 11.2 Summary of the model-to-image registration framework and its implementations

DIRECT OPTIMIZATION APPROACH		
PROBABILISTIC APPROACH		

Feature Extraction	Spatial Location Only			Spatial Location and Orientation	
Model Feature	Sampled surface point locations			Surface points & surface normal vectors	
Image Feature	Manually defined points	Voxels having high gradient magnitude	Voxels having high sheetness response	Voxels with gradient vectors	Voxels with normal vectors of sheetness

surface is aligned during the registration, the displacement of every voxel within the gland can be calculated as the model captures motion based on the displacement of every node in the FE mesh. This is particularly advantageous because the location of clinically important features, such as tumors, which are usually only visible in MR images, can be predicted within the TRUS volume. It also overcomes the problem of lack of corresponding intra-prostatic features visible in both MR and TRUS images, as discussed above.

11.1.4.1 Manual Point Features

In many circumstances, manually-defined points remain the most accurate and reliable features to identify the prostate capsule in the TRUS data. In this case, typically 5–20 points on each 2D slice are identified at the position where the operator believes the surface of the prostate is located. Fig. 11.12 illustrates the GUI for defining such points. A cubic spline can be fitted to these points to form a closed contour in each slice.

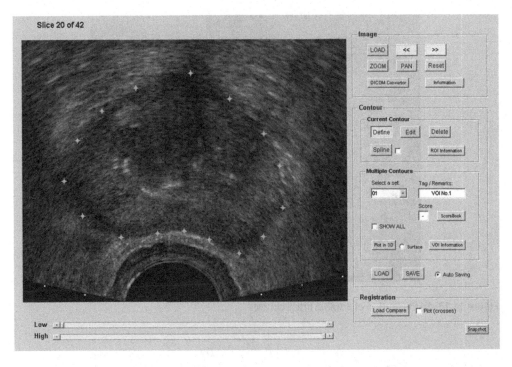

Figure 11.12 A snapshot of the software for defining points on a 2D image slice.

Evenly spaced points can then be sampled from the spline as illustrated in Fig. 11.13 and Fig. 11.14. The re-sampled contour points form a point set that describes the surface.

11.1.4.2 Point Distance

The spatial distance from the model surface to the target point is simplified by the closest point between densely sampled points (from the surface) to the closest target point. The target points can be the manually defined points, voxels having a large magnitude (e.g. of the gradient defined) or voxels having a high sheetness response, depending on the method used to extract the target features. Therefore, the overall similarity measure between two source (model) and target (image data) point sets, $\mathbf{x}_j$ $(j = 1\ldots J)$ and $\mathbf{y}_i$ $(i = 1\ldots I)$, respectively, is given by the RMS of the Euclidean distances:

$$
D_{spa} = \begin{cases}
\sqrt{\frac{1}{J} \sum_{j=1}^{J} \|\mathbf{y}_j^* - \mathbf{x}_j\|^2}, & \text{if } J \ll I \\[2ex]
\sqrt{\frac{1}{I} \sum_{i=1}^{I} \|\mathbf{x}_i^* - \mathbf{y}_i\|^2}, & \text{if } J \gg I \\[2ex]
\sqrt{\frac{1}{I+J} \left(\sum_{j=1}^{J} \|\mathbf{y}_j^* - \mathbf{x}_j\|^2 + \sum_{i=1}^{I} \|\mathbf{x}_i^* - \mathbf{y}_i\|^2 \right)}, & \text{if } J \approx I
\end{cases} \tag{11.1}
$$

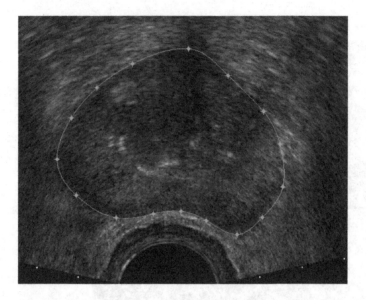

Figure 11.13 An illustration of a spline fitted contour.

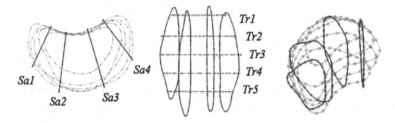

Figure 11.14 Three views of defined contours in four sagittal views (solid contours), five transverse views (dotted contours) and sampled 25 target points (circle-knots) on each contour.

where $\mathbf{y}_j^*$ is the closest point to point $\mathbf{x}_j$ from point set $\mathbf{y}_i$, and $\mathbf{x}_i^*$ is the closest point to $\mathbf{y}_i$ from $\mathbf{x}_j$. This RMS distance reflects the overall similarity measure between two point sets.

11.1.4.3 SMM as a Transformation Model

First, the rigid transformation may be applied solely to compensate motion when the remaining nonrigid motion is negligible. Then, together with the shape model, SMM can be considered as a nonrigid transformation model where rigid transformation and shape parameters need to be estimated in order to achieve the MR-to-TRUS registration. This may be solved by a direct optimization approach with a general purpose nonlinear numerical optimization optimizers, such as Levenberg–Marquardt algorithm and other gradient based algorithms. We provide a simple algorithm similar to the iterative closest point algorithm (ICP):

1. Initialize the prostate surface model by aligning the apex and base (note that it cannot be solved by SVD-based approach as it only forms an underdetermined linear system, therefore a simple rotation based approach is more appropriate to recover the other rotations without changing the rotation about the apex–base axis);
2. Check how many point pairs should be used depending on which point sets (between model surface and manual points) has more points and pre-defined outlier level;
3. Find the closest point pairs between model surface and manual points;
4. Compute the rigid transformation;
5. Compute the shape transformation using a constrained least-squares, with respect to a bound constraint that ensures the estimated shape parameters do not go beyond three times of the corresponding standard deviation on principal component;
6. Repeat from 3 until converge.

A more principled interpretation to the algorithm described is to use the probabilistic model-to-image framework introduced in [ref], only using truncated posterior probabilities, i.e. the closest features. The detailed derivation of this interpretation however is omitted here.

11.1.5 A Population-Based Model Generation Method

In this relatively standalone section [7], an alternative organ motion modeling method is proposed that is particularly suited to applications such as modeling prostate deformation where a surrogate motion signal (such as a respiratory or cardiac signal) does not exist to establish temporal correspondence between different subjects; the proposed method enables a subject-specific SSM that describes shape variation due to motion to be built without knowing the motion correspondence between subject subspaces. It also requires only limited subject-specific geometric data – for example, a reference shape based on the segmentation of a single (static) MR image – to predict the organ motion for a new (i.e. unseen) subject. The method is also potentially very useful when subject-specific shape training data is too expensive or practically difficult to obtain on each new subject. In this case, the proposed population-based model provides a means of predicting subject-specific motion with the only requirement being a single reference shape that specifies one instance of the shape of the subject's organ. We demonstrate the application of this method for nonrigid registration of MR and TRUS images of the prostate.

The underlying concept of the proposed method is that variations in organ shape due to motion can be expressed with respect to a 'mixed-subject' – i.e. population-based – SSM that is built using training data from multiple subjects and multiple shapes for each subject. The resulting SSM captures shape variation both *between* and *within* individuals.

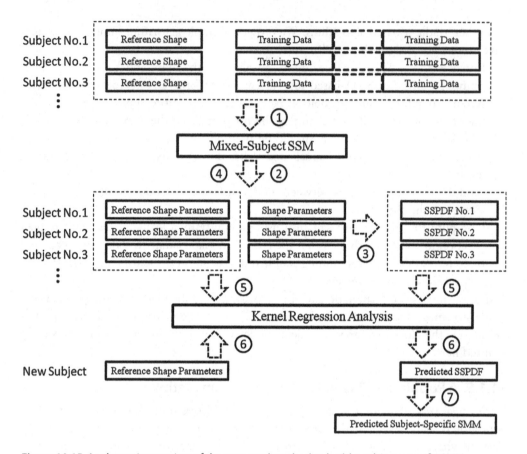

Figure 11.15 A schematic overview of the proposed method to build a subject-specific SMM.

Kernel regression analysis provides a powerful method for expressing the multivariate subject-specific probability density function (SSPDF), which represents the distribution of *shape parameters* (also known as component scores or weights) related to intra-subject organ motion, as a function of the parameters of a pre-chosen *reference shape*. Once this relationship has been established, the SSPDF that describes the expected organ motion for a new (i.e. unseen) subject can be estimated from new reference shape data for that particular subject. The resulting SSPDF can then be used to construct a subject-specific SMM for the new subject.

A schematic overview of the method used to build a subject-specific SMM is shown in Fig. 11.15. The steps involved are as follows:

1. Build a mixed-subject SSM using all available training data.
2. Obtain the shape parameters for each training dataset with respect to the mixed-subject SSM (e.g. by projection for the case of a linear model).

3. Estimate the SSPDF for each set of shape parameters corresponding to the different training shapes for each subject. The SSPDF may itself be expressed in parametric form and represented by a number of *distribution parameters* (e.g. the mean and variance of a Gaussian distribution).

4. Identify a reference shape for each subject. For example, the reference shape may describe an organ in its 'resting', or un-deformed state, or in general at a time corresponding to a particular physiological event. The reference shape is then represented by its shape parameters.

5. Perform kernel regression analysis between the parameters that characterize each SSPDF and the shape parameters that specify the reference shape.

6. Given the reference shape for a new (unseen) subject, calculate the SSPDF for the new subject using regression analysis.

7. Finally, construct a subject-specific SMM for the new subject by using the predicted SSPDF.

The resulting subject-specific SMM is an alternative to a subject-specific SMM built directly from training data available for this subject (including image-based and simulated training data). Therefore, the subject-specific SMM estimated using this method can be compared directly with one generated using the conventional method. In this section, an illustration of implementing these steps is provided using the example of building a subject-specific SMM of the prostate that captures deformation caused by the placement of a TRUS probe in the rectum.

Fig. 11.16 shows examples of random shape instances generated using the biomechanically-based SMM (used here as the ground-truth), the model-predicted subject-specific SMM of a prostate for the same subject, and the mixed-subject SSM (which captures the general shape variation over the training population of 36 patient prostates). By comparing the general form of the shapes generated using the three methods (see Fig. 11.16), it is visually evident that the subject-specific SMM generates shapes which look more physically realistic than those generated by the mixed-subject SSM, and are closer in appearance to those obtained from the ground-truth biomechanically-based SMM. (It should be noted that because the shape instances shown in Fig. 11.16 are based on random sampling, they are purely illustrative of the form of shapes generated by each SMM, and therefore should be compared group-wise, between rows, and not down each column.)

11.2 WHOLE HEART SEGMENTATION

In this section we introduce the whole heart segmentation (WHS) and the application of statistical modeling of atlases to improve the performance of the atlas-based segmentation. The relevant work and clinical background have been presented in previous publications [16,13,15,12,17,14]. We first introduce the background of the WHS

Biomechanically-based Subject-Specific SMM

Model-Predicted Subject-Specific SMM

Mixed-Subject SSM

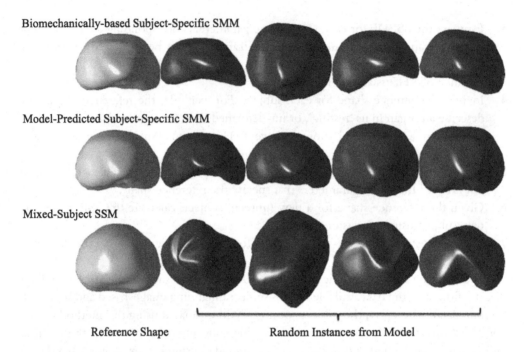

Reference Shape   Random Instances from Model

Figure 11.16 Top row: the randomly sampled prostate glands from the ground-truth biomechanically-based SMM of a test subject (as in the leave-one-out validation). Middle row: samples from the model-predicted subject-specific SMM, which are constructed from data excluding the test subject. Bottom row: samples from the mixed subject SSM which includes both intra- and inter-subject shape variations in the training data. The first column shows the reference shape from each model.

in Section 11.2.1, then describe the registration method for the atlas-based WHS in Section 11.2.2, and finally show the roadmap of developing the single atlas-based segmentation, to multi–atlas segmentation and multiple-path propagation and segmentation, and eventually to the application of the deformation regression model to the subject–specific atlas-based segmentation.

11.2.1 Definition, Applications, Challenges and Methodologies

WHS aims to extract the volume and shapes of all the substructures of the heart, commonly including the blood cavities of the four chambers, left ventricle myocardium, and sometimes the great vessels as well if they are of interest [17]. The substructures are extracted into separate individuals in WHS. Fig. 11.17 provides an example.

WHS is a useful technology and the results can be used in a number of clinical studies and applications. First, one can extract the functional indices of the heart such as the chamber volumes, ejection fraction (EF), and myocardial mass or myocardial thickness from the segmentation results. The EF of a chamber determines how well the heart is

Figure 11.17 An example of cardiac MRI (left) and the label image providing the corresponding whole heart segmentation result (right). The numbers in the label image indicate the names of local regions: left ventricle (1), right ventricle (2), left atrium (3), right atrium (4), myocardium (5), pulmonary artery (6), and aorta (7).

pumping out blood and is important in diagnosing and tracking heart failure. It is also anticipated that the functional analysis of the whole heart may detect subtle functional abnormalities or changes of the heart, and this is important to patients who otherwise have normal systole in ventricles but are suspected to have abnormal functions in other regions. Also, the segmentation enables 3D surface rendering of the whole heart. This not only can be used in the morphological studies of the congenital heart diseases by investigating the malformations, but also, the latent definition of the landmarks and the geometrical information of the whole heart can provide important information and assistance in the interventional procedures. One example is the radio frequency ablation surgery, where the surgeons can obtain a better relative position of the interventional catheter tip to the heart by setting certain transparency to the 3D surface model. Also, the landmarks from the segmentation result enable fast navigation of the catheter tip during surgeries, such as accessing the pulmonary veins in the radio frequency ablation surgery.

Albeit desired, manually delineating all the substructures of the whole heart is labor-intensive and tedious. For example, WHS from one volume can take up to 10 hours for a well-trained observer [16,12]. Moreover, manual delineation may produce inconsistent results due to the intra- and inter-observer variations. Therefore, automating WHS has become increasingly popular in recent years, particularly when a large number of images are presented. However, fully automatic WHS is difficult. Firstly, the heart organ, with multiple chambers and great vessels, is complex in geometry. The heart shape can vary significantly in different subjects or from the same subject at different cardiac conditions. This shape variation is especially evident when pathological cases are involved. The segmentation using a prior model constructed from a training set of certain pathologies may perform poorly on the test data from other pathologies. *The shape variation represents the major challenge for the automated WHS. It is practically difficult to capture all possible heart shapes from different pathologies using a prior model trained by a limited training dataset.*

Other challenges include the indistinct boundaries between anatomical substructures and the artefacts and low quality of images due to the complex motions and blood flow within the heart. All these contribute to the unsmooth and suboptimal delineation of the boundaries. Therefore, smoothness constraints and shape regularization are generally adopted for WHS.

Automatic segmentation of complex organs, such as the WHS, is generally formulated as a fitting process from a prior model to the target images [11]. The prior models contain the segmentation information and are used to guide the segmentation procedure, and the fitting procedure is generally implemented in a hierarchy manner. The prior models, either atlases or deformable mesh models, contain information of the heart shapes and important texture features of the images. In the atlas-based segmentation, one can use multiple atlases, of which each has a different heart shape and texture pattern, and design a multiple classification strategy to improve the segmentation. In the deformable model-based segmentation, the statistical shape model is commonly used and when the statistical information of image appearance is used, the model is also known as an active appearance model. Either in the atlas-based or the deformable model-based segmentation, the fitting process from the prior to the target image is the key to propagate the segmentation. In WHS, this fitting process, also referred to as registration, is particularly difficult due to the challenges we described above. Therefore, a hierarchical scheme is generally adopted, where the degree-of-freedom in each level is gradually increased. In the deformable fitting, shape regularization is required to maintain a realistic and smooth heart shape in the resultant segmentation.

11.2.2 Deformable Registration for Atlas-Based Segmentation

In this section we describe the application of deformable registration for automatic segmentation, by fitting pre-constructed atlases to the target images for segmentation propagation. This method is referred to as the atlas-based segmentation.

The main difficulty of the atlas-based segmentation frameworks is to estimate the spatial transformation from the target image to the atlas. Usually, the standard registration scheme uses a global affine registration to localize the heart, and then apply a fully deformable registration, such as the FFD registration, to refine the local details. However, one common issue is the relative sensitivity of the nonrigid registration to initialization of the substructures of the whole heart, making the algorithms less robust to large shape variability, which is however commonly seen in pathological data.

The locally affine registration method (LARM) was proposed and applied to obtain a robust initialization of the different substructures of the heart such as the four chambers and the major great vessels [15,14]. The resultant transformation globally deforms the atlas but locally maintains the shape of the predefined substructures. Particularly, the optimization of LARM, using local driving forces while maintaining the global intensity relation of images, improves the robustness of registration by neglecting the disturbing

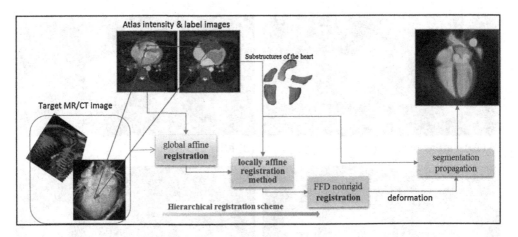

Figure 11.18 The workflow of the atlas-based segmentation.

driving forces from the background artefacts and organs such as the liver and abdomen, where the shape and intensity across subjects vary more significantly. After the initialization, an FFD registration is applied to refine the local details. Fig. 11.18 demonstrates the atlas-based WHS framework.

In WHS, it was suggested to adopt a hierarchical LARM procedure between the first-step global affine localization registration and the final-step FFD refinement registration. This hierarchical LARM generally includes three levels:

LARM-2: two locally affine registrations are associated with two regions, which are the lower ventricular region and the rest upper atrial and great vessel region.

LARM-4: after LARM-2, the ventricular region is further divided into the left ventricle and right ventricle, and the upper region is divided into right atrium and the rest region.

LARM-7: after LARM-4, seven affine registrations are used, of which each is assigned to either one of the four chambers, or the three great vessel trunks (the ascending aorta, the descending aorta, the pulmonary artery).

This hierarchical LARM scheme improves the robustness of the atlas-to-target registration. For example, Fig. 11.19 shows the segmentation result by using the FFD registration directly after the affine registration (E), which is poor. By contrast, the segmentation result by the scheme using the hierarchical LARM plus FFD registration is much better (F). The intermediate results of hierarchical LARM are provided in (B)–(D).

11.2.3 Improving Segmentation Using Statistical Modeling of the Atlas

In atlas-based segmentation, it is important to construct atlases with high quality and to use the atlases which have similar heart shape to the target image. Conventionally,

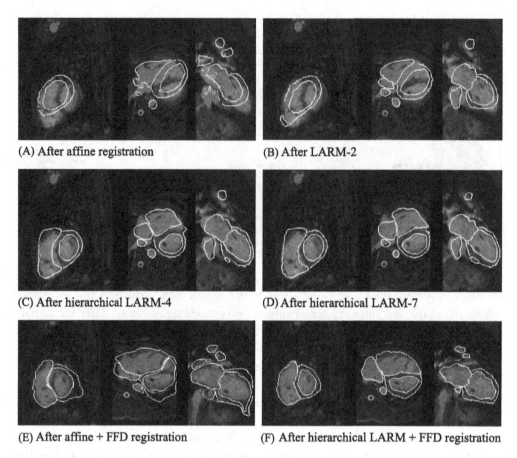

(A) After affine registration   (B) After LARM-2

(C) After hierarchical LARM-4   (D) After hierarchical LARM-7

(E) After affine + FFD registration   (F) After hierarchical LARM + FFD registration

Figure 11.19 The yellow contour lines of atlas superimposed on the target image. (For interpretation of the references to color in this figure legend, the reader is referred to the web version of this chapter.)

an atlas does not contain statistical information of the heart shape for enhancing the atlas–based segmentation of the whole heart images. A straightforward solution is to use the multiple atlas strategy, i.e. multi–atlas segmentation (MAS). MAS selects a set of *good* atlases, e.g. high image quality and high similarity to the target image in terms of heart shape, and uses each of the selected atlases to segment the target image. The final segmentation requires a process to combine all these atlas segmentation results using an algorithm referred to as label fusion. Fig. 11.20 (left) illustrates the framework of MAS. In WHS of cardiac MRI volumes, the conventional MAS did not demonstrate convincing improvement compared to the single atlas segmentation using a high-quality atlas [15]. An alternative to MAS is to use the multiple path propagation and segmentation (MUPPS) scheme to achieve the multiple classifications, where multiple atlases are related to a common atlas and a set of deformations [13]. The common atlas is a

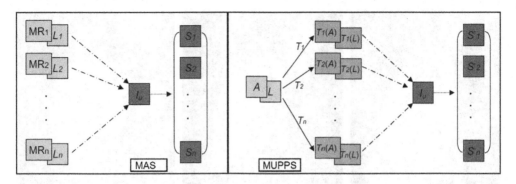

Figure 11.20 The two segmentation schemes: $\{MR_i\}$ are the MR images of each subject and $\{L_i\}$ are the corresponding label images, I_u indicates the unseen target image, and $\{S_i\}$ are the resultant segmentations; A and L are the intensity image and label image of a high-quality atlas, $\{P_i\}$ and $\{T_i\}$ are the propagation paths and corresponding transformations.

pre-constructed high-quality image and each of the deformations is used to apply to the image to generate a new atlas with high image quality. Fig. 11.20 (right) illustrates the framework of MUPPS.

Let $T_i \in \mathbb{T}$ be the random variable of deformations, $\mathbb{T}$ is the set of deformations. A value of the deformation variable is referred to as a path in MUPPS, which is used to generate an instance of the atlas variable $A_i \in \mathbb{A}$. Let A_r be the common atlas, thus $A_i = T_i(A_r)$. MUPPS resembles the MAS except that the atlases are generated using the common atlas and deformations, instead of the original cardiac MRI images, which may contain artefacts and low image quality.

The inspiration from MUPPS is that one can generate subject-specific atlases, instead of requesting more samples of atlases. Fig. 11.21 shows an example of a generated atlas for the segmentation of a patient. In Zhuang et al. [17], in vivo cardiac MRI data were used to demonstrate the advantage of the subject-specific atlas, which is the expectation of a common atlas related to the pathological condition of the subject. The data set consists of 5 healthy volunteer subjects and 19 pathological subjects, including myocardium infarction, atrial fibrillation, tricuspid regurgitation, aortic valve stenosis, aortic coarctation, Alagille syndrome, Williams syndrome, hypertrophic cardiomyopathy, and Tetralogy of Fallot. A reference atlas image was pre-constructed using the mean of 10 healthy volunteers [15]. For each target image I_u, a subject-specific atlas can be generated from a set of training atlases $\{A_i\}$ using the shape regression method,

$$A_u = T_u(A_r)\,|_{T_u=E(T|c_u)}\,. \tag{11.2}$$

The deformation can be expressed in terms of a constant velocity field, $T(\mathbf{x}) = \mathbf{x} + \exp v(\mathbf{x})$ and thus the Nadaraya–Watson kernel regression of the deformation is

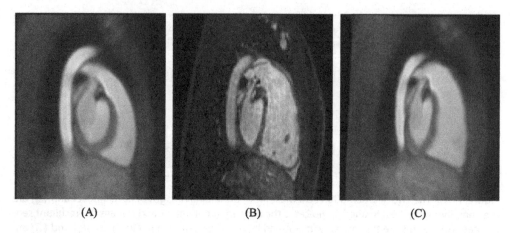

(A)   (B)   (C)

Figure 11.21 Example of generated atlas using MUPPS techniques. (A) The common atlas. (B) The pathological image showing a hypertrophic right ventricle. (C) The generated atlas from the common atlas with right ventricle hypertrophy.

computed from the locally weighted velocity fields,

$$v_u(\mathbf{x}) = \sum_{i=1}^{N} \frac{K_h(\mathrm{dist}(c_u, c_i))}{\sum_{j=1}^{N} K_h(\mathrm{dist}(c_u, c_j))} v_i(\mathbf{x}) \qquad (11.3)$$

where the velocity v_i is related to the deformation field, T_i, for deforming a training atlas A_i to the common space, $v_i(\mathbf{x}) = \log(T_i - \mathrm{Id})(\mathbf{x})$; the kernel function $K_h(\mathrm{dist}(c, c_i))$ is defined on the pathological distance between the target image and the atlas, and it can be estimated using the image similarity,

$$K_h\big(\mathrm{dist}(c_u, c_i)\big) = \begin{cases} a(\mathrm{NMI}(A_i, I_u) - b), & \text{if } \mathrm{NMI}(A_i, I_u) - b > \epsilon \\ 0, & \text{otherwise} \end{cases} \qquad (11.4)$$

where a will disappear when plugging this formula into the regression equation, b can be estimated using the minimal NMI value between the target image and all the training atlases.

The experiments in [17] showed that the generated subject-specific atlas could provide much better segmentation accuracy compared to the conventional atlas-based methods: The mean WHS surface distance on the 21 *in vivo* cases are respectively 1.86 ± 0.28 mm (MUPPS using multiple generated subject-specific atlases), 1.96 ± 0.38 mm (using one subject-specific atlas), 2.07 ± 0.62 mm (conventional MUPPS using existing atlases), 2.17 ± 0.70 mm (using the one best atlas selected from existing atlases), 2.47 ± 1.31 mm (the conventional single atlas-based segmentation). The latter three segmentation schemes do not use subject-specific atlas and achieved less accurate WHS results.

REFERENCES

[1] T.J. Carter, M. Sermesant, D.M. Cash, D.C. Barratt, C. Tanner, D.J. Hawkes, Application of soft tissue modelling to image-guided surgery, Med. Eng. Phys. 27 (2005) 893–909.

[2] T. Heimann, H.P. Meinzer, Med. Image Anal. 13 (2009) 543–563.

[3] Y. Hu, Registration of Magnetic Resonance and Ultrasound Images for Guiding Prostate Cancer Interventions, University College London, 2013.

[4] Y. Hu, H.U. Ahmed, Z. Taylor, C. Allen, M. Emberton, D. Hawkes, D. Barratt, Med. Image Anal. 16 (2012) 687–703.

[5] Y. Hu, R. Van Den Boom, T. Carter, Z. Taylor, D. Hawkes, H.U. Ahmed, M. Emberton, C. Allen, D. Barratt, Prog. Biophys. Mol. Biol. 103 (2010) 262–272.

[6] Y. Hu, T. Carter, H. Ahmed, Med. Imaging 30 (2011) 1887–1900.

[7] Y. Hu, E. Gibson, H.U. Ahmed, C.M. Moore, M. Emberton, D.C. Barratt, Med. Image Anal. 26 (2015) 332–344.

[8] I.B. Tutar, S.D. Pathak, L. Gong, P.S. Cho, K. Wallner, Y. Kim, IEEE Trans. Med. Imaging 25 (2006) 1645–1654.

[9] S. Umeyama, Least-squares estimation of transformation parameters between two point patterns, Pattern Anal. Mach. Intell. 13 (1991) 376–380.

[10] K. Zhou, H. Bao, J. Shi, Comput. Des. 36 (2004) 363–375.

[11] X. Zhuang, Challenges and methodologies of fully automatic whole heart segmentation: a review, J. Healthc. Eng. 4 (2013) 371–408.

[12] X. Zhuang, W. Bai, J. Song, S. Zhan, X. Qian, W. Shi, Y. Lian, D. Rueckert, Multiatlas whole heart segmentation of CT data using conditional entropy for atlas ranking and selection, Med. Phys. 42 (2015) 3822–3833.

[13] X. Zhuang, K. Leung, K. Rhode, R. Razavi, D. Hawkes, S. Ourselin, Whole heart segmentation of cardiac MRI using multiple path propagation strategy, in: Proc. MICCAI, 2010, pp. 435–443.

[14] X. Zhuang, K. Rhode, S. Arridge, R. Razavi, D. Hill, D. Hawkes, S. Ourselin, An atlas-based segmentation propagation framework using locally affine registration – application to automatic whole heart segmentation, in: Proc. MICCAI, 2008, pp. 425–433.

[15] X. Zhuang, K.S. Rhode, R.S. Razavi, D.J. Hawkes, S. Ourselin, A registration-based propagation framework for automatic whole heart segmentation of cardiac MRI, IEEE Trans. Med. Imaging 29 (2010) 1612–1625.

[16] X. Zhuang, J. Shen, Multi-scale patch and multi-modality atlases for whole heart segmentation of MRI, Med. Image Anal. 31 (2016) 77–87.

[17] X. Zhuang, W. Shi, H. Wang, D. Rueckert, S. Ourselin, Computation on shape manifold for atlas generation: application to whole heart segmentation of cardiac MRI, in: Proceedings of SPIE, 2013, 866941.

REFERENCES

CHAPTER 12

Statistical Shape and Deformation Models Based 2D–3D Reconstruction

Guoyan Zheng, Weimin Yu

Institute for Surgical Technology and Biomechanics, University of Bern, Bern, Switzerland

Contents

12.1 Introduction	329
12.2 Statistical Shape Model Based 2D–3D Reconstruction and Its Application in THA	331
12.2.1 Construction of Statistical Shape Models	331
12.2.2 Hierarchical 2D–3D Reconstruction	332
12.2.3 2D–3D Reconstruction Based Implant Planning	335
12.2.4 Experimental Results	338
12.3 Statistical Deformation Model Based 2D–3D Reconstruction	338
12.3.1 Training Process	339
12.3.2 Reconstruction Process	341
12.3.3 Experiments and Results	343
12.4 Final Remarks	346
References	347

12.1 INTRODUCTION

The common approach to derive three-dimensional (3D) model is to use imaging techniques such as computed tomography (CT) or magnetic resonance imaging (MRI). These have the disadvantages that they are expensive and/or induce high-radiation doses to the patient. An alternative is to reconstruct surface models from two-dimensional (2D) X-ray or C-arm images. Although single X-ray or C-arm image based solutions have been presented before for certain specific applications [25,41], it is generally agreed that in order to achieve an accurate surface model reconstruction, two or more images are needed. For this purpose, one has to solve three related problems, i.e., patient tracking/immobilization, image calibration, and 2D–3D reconstruction. Depending on the applications, different solutions have been presented before, which will be reviewed below.

Patient tracking/immobilization means to establish a coordinate system on the underlying anatomy and to co-register the acquired multiple images with respect to this common coordinate system. In literature, both external positional tracker based solutions and calibration phantom based solutions have been introduced [37,12,20,14,5,33].

Statistical Shape and Deformation Analysis
DOI: 10.1016/B978-0-12-810493-4.00015-8

The methods in the former categories usually require a rigid fixation of the so-called dynamic reference base (DRB) onto the underlying anatomy, whose position can be tracked in real-time by using an external positional tracker [37,12,20]. In contrast, the methods in the latter categories eliminate the requirement of using an external positional tracker [14,5,33]. In such a method, the calibration phantom itself acts as a positional tracker, which requires the maintenance of a rigid relationship between the calibration phantom and the underlying anatomy during image acquisition. Although not mentioned in the context of 2D–3D reconstruction, immobilization solutions [4,30] have been developed before to maintain such a rigid fixation.

The second related problem is image calibration, which means to determine the intrinsic and extrinsic parameters of an acquired image. The image is usually calibrated with respect to the common coordinate system established on the underlying anatomy. When an external positional tracker is used, this means the coordinate system established on the DRB [37,12,20]. When a calibration phantom acts as a positional tracker, this usually means a coordinate system established on the phantom itself [14,33]. Another issue is how to model the X-ray projection, which determines the way how the imaging parameters are calculated. No matter what kind of model is used, a prerequisite condition before the imaging parameters can be calculated is to establish correspondences between the 3D fiducials on the calibration phantom and their associated 2D projections which then facilitate the computation of the intrinsic and extrinsic parameters of an acquired image.

The third problem is related with the methods used to compute 3D models from 2D calibrated X-ray images. The available techniques can be divided into two categories [22]: those based on one generic model [24,18,19,23,13,39] and those based on statistical shape and/or deformation models [10,28,38,42,17,43,29,44,1,2,11,36,9,31,32,45]. The methods in the former category derive a patient-specific 3D model by deforming a generic model while the SSM based methods use an SSM to produce only the statistically likely types of models and to reduce the number of parameters to optimize. Hybrid methods, which combine the SSM based methods with the generic model based methods, have also been introduced. For example, Zheng et al. [44] presented a method that combines the SSM based instantiation with thin-plate spline based deformation.

This chapter introduces statistical shape and deformation model based 2D–3D reconstruction. The chapter is organized as follows. Section 12.1 will describe statistical shape model (SSM) based 2D–3D reconstruction and its application in planning of Total Hip Arthroplasty (THA). Section 12.2 will present statistical deformation model (SDM) based 2D–3D reconstruction and its application in estimating a patient-specific intensity volume of a proximal femur, followed by final remarks in Section 12.3.

12.2 STATISTICAL SHAPE MODEL BASED 2D–3D RECONSTRUCTION AND ITS APPLICATION IN THA

Statistical shape analysis is an important tool for understanding anatomical structures from medical images. Statistical models give efficient parametrization of the shape variations found in a collection of sample models of a given population. Model-based approaches are popular [8,6,7] due to their ability to robustly represent objects. In the last twenty years, constructing a patient-specific shape model from a limited number of calibrated X-ray images and an SSM has drawn more and more attention [10,28,42,43, 29,44,2,31,32]. One of the application areas is the THA planning based on the models reconstructed from 2D X-rays. In such an application, we are aiming to derive patient-specific 3D models of both pelvis and femur and then use the derived 3D models to plan implant placement. Below details about the construction of the associated statistical shape models and the SSM based 2D–3D reconstruction are presented.

12.2.1 Construction of Statistical Shape Models

The Point Distribution Model (PDM) [7] was chosen as the representation of the SSMs of both the pelvis and femur. The pelvic PDM used in this study was constructed from a training database of 114 segmented binary volumes with an equally distributed gender (57 male and 57 female) where the sacrum was removed from each dataset. After one of the binary volumes was chosen as the reference, diffeomorphic Demon's algorithm [35] was used to estimate the dense deformation fields between the reference binary volume and the other 113 binary volumes. Each estimated deformation field was then used to displace the positions of the vertices on the reference surface model, which was constructed from the reference binary volume, to the associated target volume, resulting in 114 surface models with established correspondences.

Following the alignment, the pelvic PDM was constructed as follows. Let $\mathbf{X}_i$, $i = 0, 1, ..., m - 1$ be m members in the aligned training population. Each member is described by a vector $\mathbf{X}_i$ containing N vertices:

$$\mathbf{X}_i = \{x_0, y_0, z_0, ..., x_{N-1}, y_{N-1}, z_{N-1}\} \tag{12.1}$$

The pelvic PDM is constructed by applying Principal Component Analysis (PCA) [7] on these aligned vectors:

$$\mathbf{D} = \frac{1}{(m-1)} \sum_{i=0}^{m-1} (\mathbf{x}_i - \bar{\mathbf{x}}) \cdot (\mathbf{x}_i - \bar{\mathbf{x}})^T$$
$$\mathbf{P} = (\mathbf{p}_0, \mathbf{p}_1, ..., \mathbf{p}_{m-2}); \quad \mathbf{D} \cdot \mathbf{p}_i = \sigma_i^2 \cdot \mathbf{p}_i \tag{12.2}$$

where $\bar{\mathbf{x}}$ and $\mathbf{D}$ represent the mean vector and the covariance matrix, respectively; $\{\sigma_i^2\}$ are non-zero eigenvalues of the covariance matrix $\mathbf{D}$, and $\{\mathbf{p}_i\}$ are the corresponding

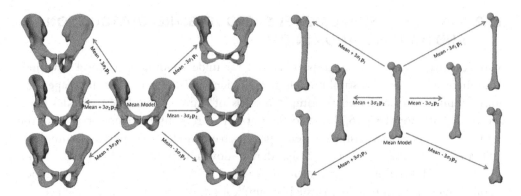

Figure 12.1 Statistical shape models of the pelvis (left) and the femur (right). In each image, the mean model (in the middle) as well as the plusminus three times variations along the first three eigenmodes (from top to down) is shown.

eigenvectors. The descendingly sorted eigenvalues $\{\sigma_i^2\}$ and the corresponding eigenvector $\mathbf{p}_i$ of the covariance matrix are the principal directions spanning a shape space with $\bar{\mathbf{x}}$ representing its origin.

A similar procedure was used to construct the femoral SSM from the 119 segmented binary volumes. Fig. 12.1 shows the mean models as well as the plusminus three times variations along the first three eigenmodes of the pelvic SSM (left) and the femoral SSM (right), respectively. Since usually an AP pelvic X-ray image only contains proximal femur part, we have accordingly derived an SSM of the proximal femur from the SSM of the complete femur by selecting only vertexes belonging to the proximal part.

12.2.2 Hierarchical 2D–3D Reconstruction

The existing feature-based 2D–3D reconstruction algorithms [17,23,15] have the difficulty in reconstructing concaving structures as they depend on the correspondences between the contours detected from the X-ray images and the silhouettes extracted from the PDMs. However, for THA, surgeons are interested not only in an accurate reconstruction of overall shape of the anatomical structures but also in an accurate reconstruction of the specific acetabular joint which consists of two surfaces: the acetabular surface and the proximal femur surface. The accuracy in reconstructing the acetabular joint will determine the accuracy of the pre-operative planning. Although the 2D–3D reconstruction scheme that we developed before can be used to reconstruct a patient-specific model of the proximal femur surface [44], its direct application to reconstruction of the acetabulum surface may lead to less accurate results.

To explain why we need to develop a new 2D–3D reconstruction scheme, the current 2D–3D correspondence establishment and thus the generation of the silhouette needs to be analyzed. Fig. 12.2 shows a calculated silhouette of the pelvis. From this

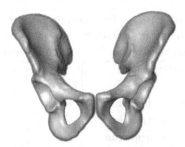

Figure 12.2 Limitations of the existing 2D–3D reconstruction scheme. This image shows an AP view of the 3D pelvic surface model and the calculated silhouettes. Certain parts, i.e., the anterior acetabulum rims do not contribute to the silhouette generation. Thus, they will not participate in the 2D–3D reconstruction process.

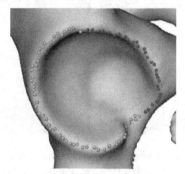

Figure 12.3 Anterior (gray points) and posterior (yellow points) acetabular rims defined on the statistical shape model as features. (For interpretation of the references to color in this figure legend, the reader is referred to the web version of this chapter.)

image it seems clear, that everything which does not generate a silhouette, will not be contributing to the finally found solution because it does not build any correspondence. To obtain more correspondences, more contours need to be drawn on the X-ray and accordingly identified on the SSM. This is realized with the so-called features [31] and patches [3] as explained below.

The features which build correspondences and should contribute to the reconstruction were selected in the SSM with an in-house developed SSM-Construction application. The acetabular rim was split into an anterior and posterior part. The boundary where the anterior part starts and ends was defined. Fig. 12.3 shows the selected feature points for the anterior and posterior part of the acetabular rim respectively. The rim points were chosen as feature because they are almost located at the same position for all view angles, due to a more or less sharp edge.

The left and right hemi-pelvis as well as the acetabular fossa were introduced as patches, whereas a patch is a subregion of a surface. A patch is handled differently in the

Table 12.1 List of features and patches used in our hierarchical 2D–3D reconstruction

Anatomical name	Feature	Patch
Left hemi-pelvis		X
Right hemi-pelvis		X
Left anterior rim	X	
Left posterior rim	X	
Right anterior rim	X	
Right posterior rim	X	
Left acetabular fossa		X
Right acetabular fossa		

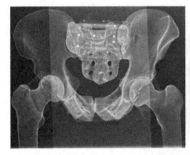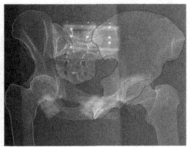

Figure 12.4 Contours assigned on an AP (left) and an oblique (right) X-ray images for the hierarchical 2D–3D reconstruction. Red: left hemi-pelvis contour, orange: right hemi-pelvis contour, light brown: leftright acetabular fossa, blue: leftright anterior acetabular rim, dark purple: leftright posterior acetabular rims, cyan: femur contours. (For interpretation of the references to color in this figure legend, the reader is referred to the web version of this chapter.)

developed application compared to a feature. As the patch describes a sub-region of the surface model, a silhouette can be calculated, depending on the viewing direction.

All additional features and patches are listed in Table 12.1. There is no differentiation on the X-ray images between contours building correspondences with a patch, a feature or just the model. However, the contours for the acetabular anterior and posterior rims are not drawn on all the X-ray images. Although the rims are on an edge, the contour cannot be well identified on the X-ray images as it is not necessarily the most outer contour. Therefore the decision was made to draw only the inner (more medial) rim-contours, except for the AP image. On the AP image all contours are assigned, because the most outer contour can be clearly assigned to the rim. An example of the AP- and the outlet-image with the assigned contours is shown in Fig. 12.4.

As the feature points do not generally coincide with the silhouette points, they are projected individually onto the image plane. To ensure that no wrong correspondences are found, the 2D contours are assigned specifically to a feature or a patch. Instead of

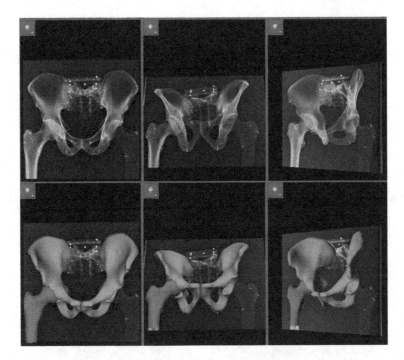

Figure 12.5 The contour definition (top) and the models (bottom) obtained with our hierarchical 2D–3D reconstruction algorithm.

trying to find a matching point pair in the whole contour dataset, it only considers the contour assigned with the feature by their name. As now all the related features and patches contribute to the 2D–3D reconstruction process, it is expected that more accurate reconstruction of the hip joint models will be obtained [31,3]. Fig. 12.5 shows a reconstruction example.

12.2.3 2D–3D Reconstruction Based Implant Planning

Once the surface models of the pelvis and the proximal femur are obtained, we can use morphological parameters extracted from these models to automatically plan THA. More specifically, we need not only to estimate the best-fit implants, their sizes, and positions but also to reconstruct leg length and the position of the center of rotation. In order to automatically plan the cup implant, we predefine the acetabular rim using the mean model of the pelvic SSM (see Fig. 12.6 for details). Then, based on known vertex correspondences between the mean model and the reconstructed pelvic model, the acetabular rim from the reconstructed pelvic model can be automatically extracted. After that, we fit a 3D circle to the extracted rim points. The normal of the plane where the fitted 3D circle is located, the center and the diameter of the fitted circle and the

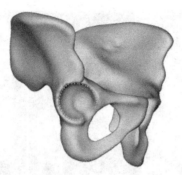

Figure 12.6 Illustration of predefining acetabular rim on the mean model of the pelvic SSM.

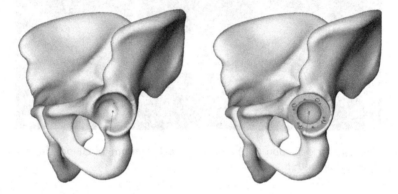

Figure 12.7 Left: estimate the normal (red line) to the plane where the acetabular rim is located, the acetabular center and the fossa apex (red dots). Right: Using the extracted morphological parameters of the reconstructed acetabulum to plan the cup implant. (For interpretation of the references to color in this figure legend, the reader is referred to the web version of this chapter.)

fossa apex of the reconstructed acetabulum will then be used to plan the cup implant (see Fig. 12.7, left).

The cup implant is planned as follows. First the cup size can be determined by the diameter of the fitted 3D circle. Second, a Computer-Aided Design (CAD) model of the selected cup can be automatically positioned using the fitted 3D circle and the fossa apex (see Fig. 12.7, right). More specifically, orientation of the cup can be adjusted by aligning the axis of the cup CAD model with the normal to the plane where the acetabular rim is located.

The planning of the stem implant is mostly done semi-automatically. Again, we first predefine the center of the femoral head, and the femoral neck axis as well as the femoral shaft axis from the mean model of the femoral SSM. Then, based on known vertex correspondences between the mean model and the reconstructed femoral model, the center of the femoral head as well as the two axes can be automatically computed

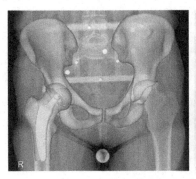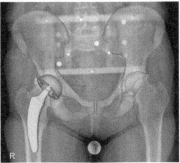

Figure 12.8 A combined 2D–3D view was used to fine-tune the type, size and positioning of the stem implant.

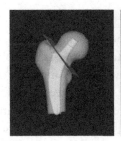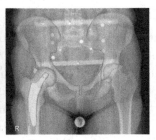

Figure 12.9 Left two images: automatically planning of femoral osteotomy plane; right image: fine-tuning of the femoral osteotomy plane with the help of a 2D–3D combined view.

from the reconstructed femoral model. An initial position of a selected stem implant can then be achieved by aligning the CAD model of the implant to above mentioned morphological features. After that, the best fit stem implant can only be achieved by a manual fine-tuning (see Fig. 12.8 for an example).

The next step for the THA planning is to determine the femoral osteotomy plane. Based on the morphological features extracted from the reconstructed femoral model, the system automatically suggests an osteotomy plane (Fig. 12.9, the left two images). Its optimal location and orientation can then be fine-tuned interactively using a combined 2D–3D view (Fig. 12.9, the right image). After that, virtual femoral osteotomy will then be conducted. Final step for planning of THA is to reconstruct leg length and the position of the center of rotation. The leg length reconstruction is achieved by interactively changing the position of the femoral bone after osteotomy along the femoral shaft axis until the leg length difference between two legs is eliminated. See Fig. 12.10 for an example.

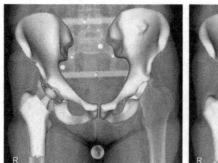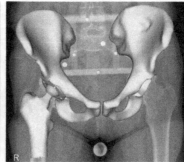

Figure 12.10 The leg length reconstruction is achieved by interactively changing the position of the femoral bone after osteotomy along the femoral shaft axis.

12.2.4 Experimental Results

To evaluate the accuracy of the reconstructions we conducted preliminary validation experiments based on calibrated X-ray radiographs. Three bones, i.e., two cadaveric hips (we named them as model #1 and #2, respectively) with each one cadaveric femur and one plastic hip containing two femurs with metallic coating (we named these two as model #3 and #4), are used in our experiment. Three calibrated X-ray images (AP, Oblique, Outlet) were acquired for each of the four hip joints and used as the input for the reconstruction algorithms. For model #1 we reconstructed the right hip joint and for model #2 the left hip joint. For the plastic bone we did a reconstruction of both left and right hip joints.

The present hierarchical 2D–3D reconstruction algorithm was compared with the 2D–3D reconstruction algorithm introduced in [44]. Surface models segmented from the CT scan of each bone were regarded as the ground truth. In order to evaluate the reconstruction accuracy, the surface models reconstructed from the X-ray images were transformed to the coordinate system of the associated ground truth models with a surface-based rigid registration before a surface-to-surface error can be computed.

When the 2D–3D reconstruction algorithm introduced in [44] was used, a mean surface reconstruction error of 1.1 ± 0.0 mm and 2.1 ± 0.3 mm was found for the femur and the pelvis, respectively. Using the hierarchical 2D–3D reconstruction algorithm led to a mean surface distance error of 0.8 ± 0.1 mm and 1.9 ± 0.2 mm for the femur and the pelvis, respectively.

12.3 STATISTICAL DEFORMATION MODEL BASED 2D–3D RECONSTRUCTION

Recently, intensity-based non-rigid 2D–3D registration has drawn more and more attention [29,1,11,36,21,39]. The reported techniques can be split into two main categories: those based on statistical shape and appearance models [29,1,36] and those based on one template image that is either derived from CT scan(s) [39] or from visual hull computation [21]. Methods in the former category, in comparison with methods in the latter category, are usually more efficient due to the lesser number of parameters to optimize. They are also more robust due to the statistical constraints applied by the shape and appearance models. In methods of both categories, the registration is conducted by iteratively comparing the reference 2D X-ray images with the floating simulation images called digitally reconstructed radiographs (DRR), which are obtained by ray casting a 3D volume data.

Here we introduce a non-rigid free-from 2D–3D registration approach for personalized reconstruction of the proximal femur from a limited number (e.g., 2) of 2D X-ray images. Unlike existing approaches, where statistical shape and appearance models [29,1, 11,36] are used, our approach uses b-spline-based statistical deformation model (SDM) introduced in [26]. The SDM is learned from a set of known deformations of proximal femur images to a given common template space. This SDM accounts for the mean and variability of the known deformations and thus encodes *a priori* information about the underlying anatomy. It has further advantages of constraining the 2D–3D registration procedure to produce only statistically likely types of warps and of reducing the number of parameters to optimize. The iterative registration of the 3D b-spline-based SDM to the 2D X-ray images requires a computationally expensive inversion of the instantiated deformation in each iteration. In this paper, we propose to solve this challenge with a fast B-spline pseudo-inversion algorithm that is implemented on graphics processing unit (GPU).

Our non-rigid free-from 2D–3D registration approach consists of two processes: the training process, where the SDM will be constructed from a set of training images, and the reconstruction process, where given X-ray images of an unseen subject, we will derive a patient-specific volume by non-rigidly matching the SDM to the input images. Note that the training process needs to be performed only once in order to be able to statistically register X-ray images of any unseen subject. Details about each process will be given in what follows.

12.3.1 Training Process

Following the idea introduced in [26], we construct the SDM from CT data of 40 left cadaveric proximal femurs based on a two-stage procedure. More specifically, in the first stage, we randomly chose one of the proximal femur from this given training population as the reference volume $\mathbf{V}_0^{1st}$. All other volumes $\{\mathbf{V}_i^{1st}, i = 1, ..., 39\}$ were aligned to this reference volume with similarity registrations. We then applied the b-spline-based free-from deformation (FFD) algorithm [27] as implemented in the registration toolbox 'elastix' [16] to establish correspondences between the reference volume and each one of the 39 floating volumes. Each time, the output from the b-spline-based FFD algorithm is a local displacement expressed as the 3D tensor product of the 1D cubic B-splines [27]:

$$T_i^{FFD}(\mathbf{x}) = \sum_{r=0}^{3}\sum_{s=0}^{3}\sum_{t=0}^{3} B_r(u)B_s(v)B_t(w)\mathbf{c}_{l+r,m+s,n+t} \tag{12.3}$$

where $\mathbf{c}$ denotes the B-spline coefficients for a number of control points that form a regular lattice size of $(L+3) \times (M+3) \times (N+3)$; l, m, n are the indexes of the control points satisfying $-1 \leqslant l \leqslant (L+1), -1 \leqslant m \leqslant (M+1), -1 \leqslant n \leqslant (N+1)$, and $0 \leqslant u, v, w < 1$ corresponds to the relative positions of $\mathbf{x}$ in lattice coordinates.

By concatenating $(L+3) \times (M+3) \times (N+3)$ 3D control points for each local displacement, we have 39 FFDs described as control point vectors $\mathbf{C}_1^{1st}, ..., \mathbf{C}_{39}^{1st}$, where $\mathbf{C}_i^{1st} = vec(\mathbf{c}_1, ..., \mathbf{c}_{(L+3) \times (M+3) \times (N+3)})$. Here we use the operator, vec, to represent control point vectorization, and the operator, vec^{-1}, as the inverse of vectorization. Similarly, we can create 39 non-rigidly deformed floating volumes $\{\mathbf{I}_i^{1st}, i=1, ..., 39\}$, where $\mathbf{I}_i^{1st}$ is a concatenation of gray values in the ith warped floating volume. From these data, we computed the average control point vector $\bar{\mathbf{C}}^{1st} = (39^{-1}) \cdot \sum_{i=1}^{39} \mathbf{C}_i^{1st}$ and the average intensity distribution $\bar{\mathbf{I}} = (40^{-1}) \cdot \sum_{i=0}^{39} \mathbf{I}_i^{1st}$, with $\mathbf{I}_0^{1st}$ standing for gray values of the reference data.

The purpose of the second stage is to remove the possible bias introduced by the reference volume selection. To achieve this goal, we applied the FFD generated from the average control point vector $\bar{\mathbf{C}}^{1st}$ to the reference volume $\mathbf{V}_0^{1st}$ to create a new volume $\mathbf{s}_0$ and assigned the average intensity distribution $\bar{\mathbf{I}}$ to this newly created volume. The new volume $\mathbf{s}_0$ was named as the atlas. It was used as the new reference volume in the second stage and all other 40 proximal femur volumes were regarded as the floating volumes. The b-spline-based FFD algorithm was used again to establish the correspondences between the atlas and the other 40 floating volumes. We thus obtained a set of 40 new control point vectors $\{\mathbf{C}_i; i=1, ..., 40\}$. We could then construct the SDM as:

$$\mathbf{S}_C = ((m-1)^{-1}) \cdot \sum_{i=1}^{m}(\mathbf{C}_i - \bar{\mathbf{C}})(\mathbf{C}_i - \bar{\mathbf{C}})^T$$

$$\bar{\mathbf{C}} = (m^{-1}) \cdot \sum_{i=1}^{m}\mathbf{C}_i \tag{12.4}$$

$$\mathbf{P}_C = (\mathbf{p}_C^1, \mathbf{p}_C^2, ...); \quad \mathbf{S}_C \cdot \mathbf{p}_C^i = (\sigma_C^i)^2 \cdot \mathbf{p}_C^i$$

$$\mathbf{C} = \bar{\mathbf{C}} + \sum_{k=1}^{M_C}\alpha_C^k \sigma_C^k \mathbf{p}_C^k$$

where $m = 40$ is the number of training samples; $\bar{\mathbf{C}}$ and $\mathbf{S}_C$ are the average and the covariance matrix of the control point vectors, respectively; $\{(\sigma_C^i)^2\}$ and $\{\mathbf{p}_C^i\}$ are the

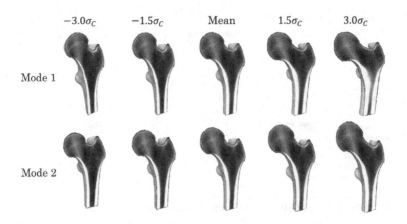

Figure 12.11 The mean and the first two modes of variations of the SDM.

descendingly ordered eigenvalues and associated eigenvectors, respectively; $\{\alpha_C^i\}$ are the model parameters; M_C is the cut-off points.

Fig. 12.11 shows the mean and the first two modes of variations of the SDM. Each instance of the SDM was generated by evaluating $\mathbf{I}_s = \bar{\mathbf{I}}(T^{FFD}(\mathbf{C}) \circ \mathbf{s}_0)$, where $T^{FFD}(\mathbf{C}) = \sum_{r=0}^{3} \sum_{s=0}^{3} \sum_{t=0}^{3} B_r(u) B_s(v) B_t(w) vec^{-1}(\mathbf{C})$ is the FFD generated from the instantiated control point vector $\mathbf{C} = \bar{\mathbf{C}} + \alpha_C \sigma_C^i \mathbf{p}_C^i$, with $\alpha_C \in \{-3.0, -1.5, 0, 1.5, 3.0\}$ and $i \in \{1, 2\}$ corresponding to the first two modes.

12.3.2 Reconstruction Process

For an unseen proximal femur, we assume that we have a set of $Q \geq 2$ X-ray images and that all images are calibrated and co-registered to a common coordinate system called $\mathbf{f}$. Given an initial estimation of the registration parameters, our algorithm iteratively generates a 3D image and updates the parameter estimation by minimizing the dissimilarity between the input 2D X-ray images and the associated DRRs that are created from the instantiated 3D image.

3D Image Instantiation and Alignment. The 3D image instantiation and alignment process is parametrized by two sets of parameters, i.e., the set of shape parameters $\mathbf{b} = (\alpha_C^1, \alpha_C^2, ..., \alpha_C^{M_C})^T$ determining a forward FFD from the atlas $\mathbf{s}_0$ to an instantiated 3D image $\mathbf{s}$ and the set of parameters $\mathbf{a} = (\Lambda, \beta, \gamma, \theta, t_x, t_y, t_z)^T$ determining a similarity transformation from the space of the instantiated image $\mathbf{s}$ to the common coordinate system $\mathbf{f}$, where Λ is the scaling parameter; β, γ, θ are rotational parameters and t_x, t_y, t_z are translational parameters. An instantiated 3D image that is aligned to the common coordinate system $\mathbf{f}$ is defined by the following equation:

$$\bar{\mathbf{I}}(x_f(\mathbf{a}, \mathbf{b})) = \bar{\mathbf{I}}(\mathbf{A}(\mathbf{a}) \circ T^{FFD}(\mathbf{b}) \circ x_0) \tag{12.5}$$

where $\mathbf{A}(\mathbf{a})$ is the similarity transformation and $T^{FFD}(\mathbf{b})$ is the forward FFD.

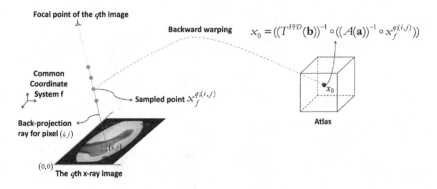

Figure 12.12 A schematic illustration of the backward warping.

Eq. (12.5) describes a forward warping. It is known that implementing this forward warping may result in holes in the aligned 3D image and a backward warping (see Fig. 12.12 for an illustration) should be used instead, as follows:

$$\bar{\mathbf{I}}(x_f^{q;(i,j)}) = \bar{\mathbf{I}}(x_0) = \bar{\mathbf{I}}\big((T^{FFD}(\mathbf{b}))^{-1} \circ \big((A(\mathbf{a}))^{-1} \circ x_f^{q;(i,j)}\big)\big) \tag{12.6}$$

where $x_f^{q;(i,j)}$ is a sampled discrete point (see Fig. 12.12) along the back–projection ray of a pixel (i,j) in the qth input X-ray images.

It is straightforward to compute the inverse of the similarity transformation $\mathbf{A}(\mathbf{a})$. However, it is computationally expensive to compute the inverse of the forward FFD $T^{FFD}(\mathbf{b})$. In this chapter, instead of computing the inverse of the forward FFD, we propose to compute a pseudo-inverse of the instantiated B-spline transformation using the B-spline pseudo-inverse algorithm introduced in [34]. The computed pseudo-inverse B-spline coefficients then allow us to compute a backward FFD in order to warp the atlas to the instantiated volume **s**. For details about this algorithm, we refer to [34]. To speed up the 3D B-spline pseudo-inverse computation, we have implemented the algorithm on GPU with the Compute Unified Device Architecture (CUDA) programming environment.

Registration Criterion. We chose to use the robust dissimilarity measure introduced in [40] to compare the floating DRRs to the associated reference X-ray images. This robust dissimilarity measure is defined as:

$$
\begin{aligned}
E(\mathbf{a}, \mathbf{b}) = \sum_{q=1}^{Q} \Bigg[& \lambda \sum_{i,j}^{I,J} D_{q;(i,j)}^2(\mathbf{a}, \mathbf{b}) \\
& + (1-\lambda) \sum_{i,j}^{I,J} \frac{1}{card(N_{i,j}^r)} \sum_{(i',j') \in N_{i,j}^r} (D_{q;(i,j)}(\mathbf{a}, \mathbf{b}) - D_{q;(i',j')}(\mathbf{a}, \mathbf{b}))^2 \Bigg]
\end{aligned}
\tag{12.7}
$$

where $I \times J$ is the size of each X-ray image; $D_q = \{D_{q;(i,j)}\}$ is the qth observed difference image; $N = \{N_{i,j}^r\}$ are the rth order neighborhood systems and $card(N_{i,j}^r)$ is the number of pixels in $N_{i,j}^r$. We refer interested readers to [40] for the details about how the difference images are computed and about the details of the above equation.

Optimization Strategy. Considering the least-squares form of Eq. (12.7), we decided to use Levenberg–Marquardt optimizer to minimize $E(\mathbf{a}, \mathbf{b})$. More specifically, the following two stages are executed until convergence.

- Similarity registration stage: The shape parameters are fixed to the current estimation $\mathbf{b}_t$ and the Levenberg–Marquardt optimizer is used to iteratively minimize the image dissimilarity energy $E(\mathbf{a}, \mathbf{b}_t)$ in order to obtain a new estimation of the similarity transformation parameters $\mathbf{a}_{t+1}$.

- Non-rigid registration stage: The similarity transformation parameters are fixed to $\mathbf{a}_{t+1}$ and the Levenberg–Marquardt optimizer is used again to iteratively estimate the new shape deformation parameters $\mathbf{b}_{t+1}$. At each iteration, the following two steps are performed.

  - Step 1: Following the Levenberg–Marquardt optimizer, compute the gradient and the regularized Hessian of Eq. (12.7) with respect to the shape parameters, and then calculate an additive update $\Delta\mathbf{b}_t$ of the shape parameters to get a new estimation $\mathbf{b}_{t+1} = \mathbf{b}_t + \Delta\mathbf{b}_t$.

  - Step 2: Based on $\mathbf{b}_{t+1}$, we first instantiate the control point vector $\mathbf{C} = \bar{\mathbf{C}} + \sum_{k=1}^{M_C} \alpha_C^k \sigma_C^k \mathbf{p}_C^k$. We then compute its pseudo-inverse. Based on the computed pseudo-inverse B-spline coefficients, we can compute a backward FFD, which will be used to warp the atlas to the instantiated image $\mathbf{s}$ to generate DRRs for the next iteration.

12.3.3 Experiments and Results

The atlas has a resolution of $192 \times 128 \times 192$ voxels with a voxel size of $0.664 \times 0.664 \times 1.0$ mm^3. The control point lattice has a size of $25 \times 18 \times 35$ points with a grid spacing of $6.0 \times 6.0 \times 6.0$ mm^3. For all the registration experiments, the cut-off point M_C was empirically chosen to be 9. Shape parameters $\mathbf{b}$ were initialized to zeros and pose parameters $\mathbf{a}$ were initialized with anatomical landmark based registration. Implemented on a laptop with 2.5 GHz Intel Core i5 processor and Nvidia GeForce GT 750M graphics card, it took about 7 minutes to register the SDM to 2 X-ray images with a resolution of 768×576 pixels.

Pseudo-Inverse Validation Experiment. Here we compared the backward FFD calculated from the results of our pseudo-inverse algorithm with the one computed by 'elastix' [16] via non-rigid registration. Please note that in 'elastix' a smaller grid spacing than the forward FFD is chosen for the inverse transform which prevents a direct comparison of the inverted B-spline coefficients generated by these two different methods.

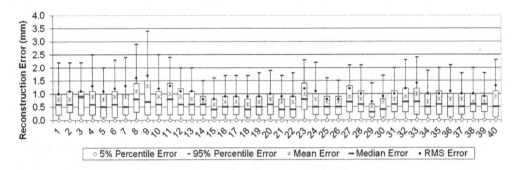

Figure 12.13 Boxplot of the reconstruction results of the leave-one-out experiment.

This experiment was conducted on the registration outputs of the first 10 proximal femurs in the second stage of the SDM construction. Correlation along each axis of the two backward FFDs, the mean and median magnitudes (*Mean_M* and *Median_M*) of the difference vectors at all voxels, and the computing time are presented in Table 12.2, which demonstrates that our b-spline pseudo-inverse algorithm is accurate and fast.

Leave-One-Out (LOO) Experiment. In this experiment, each time, we took one training dataset out and used the rest 39 datasets to construct an SDM. We then generated two DRRs of the left-out volume and used them as the input X-ray images to run our intensity-based non-rigid 2D–3D registration algorithm. Surface model extracted from the reconstructed volume was then compared to that extracted from the left-out volume, which was regarded as the ground truth. Results of the LOO experiment are shown in Fig. 12.13. The mean reconstruction errors range from 0.5 mm to 1.3 mm and an overall average error of 0.8 mm was found.

Experiment on C-Arm Images of 10 Cadaveric Femurs. In this experiment, 10 cadaveric femurs (none of them was used in the SDM construction) were used. For each femur, we acquired two calibrated C-arm images around the proximal region. The reconstruction accuracies were evaluated by randomly digitizing dozens points from the surface of each femur and then computing the distances from those digitized points to the associated surface model which was segmented from the reconstructed volume. Mean and median reconstruction errors for each femur are presented in Table 12.3. The mean reconstruction errors range from 1.0 mm to 1.6 mm and an average accuracy of 1.3 mm was found.

Experiment on X-Ray Images of 6 Cadaveric Femurs. In this experiment, X-ray images of another 6 cadaveric femurs (again, none of them was used in the SDM construction) were used, where 3 of them were part of complete hips. For each femur, the ground truth surface models were either obtained with a CT-scan reconstruction method (for 3 complete hips) or with a hand-held laser-scan reconstruction method (for others). The surface models segmented from the reconstructed volumes were then

Table 12.2 Pseudo-inverse validation results where R_x, R_y, and R_z are correlation coefficients of the two backward FFDs along x, y, and z axis, respectively

Quantity		#1	#2	#3	#4	#5	#6	#7	#8	#9	#10	Average
R_x		0.993	0.997	0.998	0.994	0.991	0.997	0.995	0.997	0.998	0.997	0.996
R_y		0.996	0.997	0.997	0.996	0.996	0.997	0.997	0.997	0.997	0.996	0.997
R_z		0.998	0.995	0.993	0.998	0.998	0.995	0.993	0.995	0.993	0.994	0.995
Mean_M (mm)		0.049	0.086	0.081	0.095	0.104	0.097	0.097	0.094	0.084	0.096	0.088
Median_M (mm)		0.014	0.024	0.023	0.023	0.025	0.027	0.027	0.023	0.021	0.027	0.023
Computing time (s)	elastix	202	200	199	201	204	205	205	204	202	203	202.5
	ours	1.33	1.22	1.02	1.06	1.01	1.09	1.05	1.03	1.09	0.98	1.09

Table 12.3 Results of experiment on C-arm images of 10 cadaveric femurs

Quantity	#1	#2	#3	#4	#5	#6	#7	#8	#9	#10	Average
Mean (mm)	1.5	1.0	1.1	1.0	1.4	1.6	1.4	1.3	1.1	1.1	1.3
Median (mm)	1.2	0.7	0.8	0.8	1.1	1.4	1.2	1.1	0.9	0.9	1.0

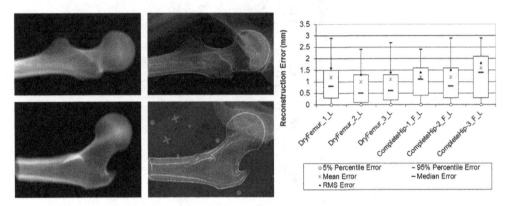

Figure 12.14 Results of applying the present approach to X-ray images of 6 cadaveric femurs. Left column shows DRRs of a reconstructed volume, middle column shows X-ray images (for visualization purpose, only regions around the proximal femur are shown) superimposed with edges extracted from the DRRs, and right image shows the errors of reconstructing all 6 cadaveric femurs.

compared to the associated ground truth models to evaluate the reconstruction accuracies. A reconstruction example as well as the errors of reconstructing volumes of all 6 femurs is shown in Fig. 12.14. An average mean reconstruction accuracy of 1.2 mm was found.

12.4 FINAL REMARKS

This chapter intends to present statistical shape and deformation models based 2D–3D reconstruction techniques as well as the applications of these techniques to derive patient-specific 3D surface models or 3D intensity volumes from 2D X-ray images. The selection of exact techniques will be application dependent. For example, for planning acetabular component in THA, the models derived from a statistical shape model based 2D–3D reconstruction technique will be enough. However, if one would like to plan stem component in THA, such models are not enough due to the missing information about proximal femur morphology and the intramedullary anatomy. Thus, a statistical deformation model based 2D–3D reconstruction should be used instead such that a patient-specific intensity volume can be derived from 2D X-ray images. Combining these two techniques, we are able to provide true 3D solutions for computer assisted planning of THA using only 2D X-ray radiographs, which is not only innovative but also cost-effective.

REFERENCES

[1] O. Ahmad, K. Ramamurthi, K. Wilson, K. Engelke, R. Prince, R. Taylor, Volumetric DXA (VXA): a new method to extract 3D information from multiple in vivo DXA images, J. Bone Miner. Res. 25 (2010) 2744–2751.

[2] N. Baka, B. Kaptein, M. de Bruijne, T. van Walsum, J. Giphart, W. Niessen, B. Lelieveldt, 2D–3D shape reconstruction of the distal femur from stereo X-ray imaging using statistical shape models, Med. Image Anal. 15 (2011) 840–850.

[3] S. Balestra, Statistical Shape Model-Based Articulated 2D–3D Reconstruction, Master's thesis, University of Bern, Switzerland, 2013.

[4] C. Carter, G. Hicken, Device for immobilizing a patient and compressing a patient's skeleton, joints and spine during diagnostic procedures using an MRI unit, CT scan unit or X-ray unit, US Patent 6860272, 2005.

[5] F. Cheriet, C. Laporte, S. Kadoury, H. Labelle, J. Dansereau, A novel system for the 3-D reconstruction of the human spine and rib cage from biplanar X-ray images, IEEE Trans. Biomed. Eng. 54 (2007).

[6] T. Cootes, A. Hill, C. Taylor, J. Haslam, Use of active shape models for locating structures in medical images, Image Vis. Comput. 12 (1994) 355–365.

[7] T. Cootes, C. Taylor, D. Cooper, J. Graham, Active shape models – their training and application, Comput. Vis. Image Underst. 61 (1995) 38–59.

[8] I. Dryden, K. Mardia, Statistical Shape Analysis, Wiley Series in Probability and Statistics, John Wiley & Sons, Inc., New York, 1998.

[9] M. Ehlke, H. Ramm, H. Lamecker, H. Hege, S. Zachow, Fast generation of virtual X-ray images for reconstruction of 3D anatomy, IEEE Trans. Vis. Comput. Graph. 19 (2013) 2673–2682.

[10] M. Fleute, S. Lavallée, Nonrigid 3D/2D registration of images using statistical models, in: Medical Image Computing and Computer-Assisted Intervention, in: Lect. Notes Comput. Sci., vol. 1496, 1998, pp. 138–147.

[11] G. Zheng, Personalized X-ray reconstruction of the proximal femur via intensity-based non-rigid 2D–3D registration, in: Medical Image Computing and Computer-Assisted Intervention, 2011, pp. 598–606.

[12] R. Hofstetter, M. Slomczykowski, M. Sati, L. Nolte, Fluoroscopy as an imaging means for computer-assisted surgical navigation, Comput. Aided Surg. 4 (1999) 65–76.

[13] L. Humbert, J. De Guise, B. Aubert, B. Godbout, W. Skalli, 3D reconstruction of the spine from biplanar X-rays using parametric models based on transversal and longitudinal inferences, Med. Eng. Phys. 31 (2009) 681–687.

[14] A. Jain, G. Fichtinger, C-arm tracking and reconstruction without an external tracker, in: Medical Image Computing and Computer-Assisted Intervention, MICCAI 2006, Springer, 2006, pp. 494–502.

[15] S. Kadoury, F. Cheriet, J. Dansereau, H. Labelle, Three-dimensional reconstruction of the scoliotic spine and pelvis from uncalibrated biplanar X-ray images, J. Spinal Disord. Tech. 20 (2007) 160–167.

[16] S. Klein, M. Staring, K. Murphy, M. Viergever, J. Pluim, Elastix: a toolbox for intensity-based medical image registration, IEEE Trans. Med. Imaging 29 (2010) 196–205.

[17] H. Lamecker, T. Wenckebach, H.C. Hege, Atlas-based 3D-shape reconstruction from X-ray images, in: Proceedings of the 2006 International Conference on Pattern Recognition, ICPR 2006, 2006, pp. 371–374.

[18] S. Laporte, W. Skalli, J. de Guise, F. Lavaste, D. Mitton, A biplanar reconstruction method based on 2D and 3D contours: application to the distal femur, Comput. Methods Biomech. Biomed. Eng. 6 (2003) 1–6.

[19] A. Le Bras, S. Laporte, V. Bousson, D. Mitton, J. De Guise, J. Laredo, W. Skalli, 3D reconstruction of the proximal femur with low-dose digital stereoradiography, Comput. Aided Surg. 9 (2004) 51–57.

[20] H. Livyatan, Z. Yaniv, L. Joskowicz, Robust automatic C-arm calibration for fluoroscopy-based navigation: a practical approach, in: Medical Image Computing and Computer-Assisted Intervention, Springer, 2002, pp. 60–68.

[21] B. Lucas, Y. Otake, M. Armand, R. Taylor, An active contour method for bone cement reconstruction from C-arm X-ray images, IEEE Trans. Med. Imaging 31 (2012) 860–869.

[22] P. Markelj, D. Tomazevic, B. Likar, F. Pernus, A review of 3D/2D registration methods for image-guided interventions, Med. Image Anal. 16 (2012) 642–661.

[23] D. Mitton, S. Deschênes, S. Laporte, B. Godbout, S. Bertrand, J. de Guise, W. Skalli, 3D reconstruction of the pelvis from bi-planar radiography, Comput. Methods Biomech. Biomed. Eng. 9 (2006) 1–5.

[24] D. Mitton, C. Landry, S. Véron, W. Skalli, F. Lavaste, J. De Guise, 3D reconstruction method from biplanar radiography using non-stereocorresponding points and elastic deformable meshes, Med. Biol. Eng. Comput. 38 (2000) 133–139.

[25] J. Novosad, F. Cheriet, Y. Petit, H. Labelle, Three-dimensional 3-D reconstruction of the spine from a single X-ray image and prior vertebra models, IEEE Trans. Biomed. Eng. 51 (2004) 1628–1639.

[26] D. Rueckert, A. Frangi, J. Schnabel, Automatic construction of 3D statistical deformation models using non-rigid registration, in: MICCAI 2001, in: Lect. Notes Comput. Sci., vol. 2208, 2001, pp. 77–84.

[27] D. Rueckert, L. Sonoda, C. Hayes, D. Hill, M. Leach, D. Hawkes, Nonrigid registration using free-form deformations: application to breast MR images, IEEE Trans. Med. Imaging 18 (1999) 721.

[28] S. Benameur, M. Mignotte, S. Parent, H. Labelle, W. Skalli, J. de Guise, 3D/2D registration and segmentation of scoliotic vertebra using statistical models, Comput. Med. Imaging Graph. 27 (2003) 321–337.

[29] O. Sadowsky, G. Chintalapani, R. Taylor, Deformable 2D–3D registration of the pelvis with a limited field of view, using shape statistics, in: Medical Image Computing and Computer-Assisted Intervention, in: Lect. Notes Comput. Sci., vol. 4792, 2007, pp. 519–526.

[30] B. Schmit, M. Keeton, B. Babusis, Restraining apparatus and method for use in imaging procedures, US Patent 6882878.

[31] S. Schumann, L. Liu, M. Tannast, M. Bergmann, L.P. Nolte, G. Zheng, An integrated system for 3D hip joint reconstruction from 2D X-rays: a preliminary validation study, Ann. Biomed. Eng. 41 (2013) 2077–2087.

[32] S. Schumann, Y. Sato, Y. Nakanishi, F. Yokota, M. Takao, N. Sugano, G. Zheng, Cup implant planning based on 2-D/3-D radiographic pelvis reconstruction-first clinical results, IEEE Trans. Biomed. Eng. 62 (2015) 2665–2673.

[33] S. Schumann, B. Thelen, S. Ballestra, L. Nolte, P. Buechler, G. Zheng, X-ray image calibration and its application to clinical orthopedics, Med. Eng. Phys. 36 (2014) 968–974.

[34] A. Tristan, I. Arribas, A fast b-spline pseudo-inversion algorithm for consistent image registration, in: CAIP 2007, in: Lect. Notes Comput. Sci., vol. 4473, 2007, pp. 768–775.

[35] T. Vercauteren, X. Pennec, A. Perchant, N. Ayache, Diffeomorphic demons: efficient non-parametric image registration, NeuroImage 45 (2009) S61–S72.

[36] T. Whitmarsh, L. Humbert, M. De Craene, L. Del Rio Barquero, A. Frangi, Reconstructing the 3D shape and bone mineral density distribution of the proximal femur from dual-energy X-ray absorptiometry, IEEE Trans. Med. Imaging 30 (2011) 2101–2114.

[37] Z. Yaniv, L. Joskowicz, A. Simkin, M. Garza-Jinich, C. Milgrom, Fluoroscopic image processing for computer-aided orthopaedic surgery, in: Proceedings of MICCAI 1998, in: Lect. Notes Comput. Sci., vol. 1496, Springer, 1998, pp. 325–334.

[38] J. Yao, R. Taylor, Assessing accuracy factors in deformable 2D/3D medical image registration using a statistical pelvis model, in: Proc. ICCV 2003, 2003, pp. 1329–1334.

[39] W. Yu, G. Zheng, 2D–3D regularized deformable b-spline registration: application to the proximal femur, in: ISBI 2015, 2015, pp. 829–832.

[40] G. Zheng, Effective incorporating spatial information in a mutual information based 3D–2D registration of a CT volume to X-ray images, Comput. Med. Imaging Graph. 34 (2010) 553–562.

[41] G. Zheng, Statistically deformable 2D/3D registration for estimating post-operative cup orientation from a single standard AP X-ray radiograph, Ann. Biomed. Eng. 38 (2010) 2910–2927.

[42] G. Zheng, M. Ballester, M. Styner, L. Nolte, Reconstruction of patient-specific 3D bone surface from 2D calibrated fluoroscopic images and point distribution model, in: Medical Image Computing and Computer-Assisted Intervention, 2006, pp. 25–32.

[43] G. Zheng, X. Dong, M. Gonzalez Ballester, Unsupervised reconstruction of a patient-specific model of a proximal femur from calibrated fluoroscopic images, in: Medical Image Computing and Computer-Assisted Intervention, 2007, pp. 834–841.

[44] G. Zheng, S. Gollmer, S. Schumann, X. Dong, T. Feilkas, M. González Ballester, A 2D/3D correspondence building method for reconstruction of a patient-specific 3D bone surface model using point distribution models and calibrated X-ray images, Med. Image Anal. 13 (2009) 883–899.

[45] G. Zheng, W. Yu, Non-rigid free-form 2D–3D registration using statistical deformation model, in: Proc. MLMI 2015, in: Lect. Notes Comput. Sci., vol. 9352, 2015, pp. 102–109.

CHAPTER 13

Statistical Shape Analysis for Brain Structures

Li Shen*,†, Shan Cong*,‡, Mark Inlow*

*Department of Radiology and Imaging Sciences, Center for Neuroimaging, Indiana University School of Medicine, Indianapolis, IN, USA
†Center for Computational Biology and Bioinformatics, Indiana University School of Medicine, Indianapolis, IN, USA
‡Department of Electrical and Computer Engineering, Indiana University–Purdue University Indianapolis, Indianapolis, IN, USA

Contents

13.1	Introduction	351
13.2	Surface Modeling and Registration	353
	13.2.1 SPHARM Surface Modeling	353
	13.2.1.1 Spherical Parameterization	*353*
	13.2.1.2 SPHARM Expansion	*355*
	13.2.2 SPHARM Surface Normalization	357
	13.2.2.1 SPHARM Normalization	*357*
	13.2.2.2 Subfield-Guided Registration	*359*
13.3	Statistical Inference on the Surface	362
	13.3.1 Surface Atlas and Signal Processing	363
	13.3.2 General Linear Model and Random Field Theory	364
	13.3.3 Statistical Parametric Mapping Distribution Analysis	364
13.4	An Example Application	369
13.5	Conclusions	372
	Acknowledgments	373
	References	373

13.1 INTRODUCTION

Recent advances in non-invasive scanning techniques have resulted in a prominent growth of research into the analysis of high quality 3D brain images. One fundamental problem in brain image analysis is identifying the morphological abnormalities of the neuroanatomy that are associated with a particular disorder to aid diagnosis and treatment. One approach is volumetric analysis (e.g., [1,2]), which is based on measuring the volume of a brain structure. Its major advantage is the simplicity; however, many structural differences may be overlooked. A newer approach, shape analysis (e.g., [3–26]), has the potential to provide valuable information beyond simple volume measurements. For example, it may help identify where the volume change is located, or characterize abnormalities in the absence of volume differences.

Statistical Shape and Deformation Analysis
DOI: 10.1016/B978-0-12-810493-4.00016-X

351

Over the past decades, statistical shape analysis [27,28] has emerged as a promising new field with applications to medicine, biology and other scientific domains. The pioneers are Kendall [29–31] and Bookstein [32–34], and their methods mostly focus on landmark data. Computational anatomy (CA) is a prominent shape analysis model proposed in the area of brain imaging [35–39]. The approach is based on creating and analyzing diffeomorphisms, which are smooth invertible mappings between geometric objects. The model consists of three steps: (1) computing deformation maps of individuals from a template, (2) computing probability laws of anatomical variations using deformations as shape descriptors, and (3) performing inferences for diseases and anomalies. Similar to the CA model, statistical shape analysis in general can be divided into two categories: (1) to establish a statistical shape model for one group of geometric objects by characterizing the mean and variability of the population; and (2) to identify shape changes between two groups of objects.

Here we concentrate on the second type of statistical shape analysis and its application to the morphometric analysis of brain structures extracted from the magnetic resonance imaging (MRI) scans. Among many MRI morphometric techniques in brain imaging, voxel-based morphometry (VBM) [40–42] and surface-based morphometry (SBM) [43–48] are two widely used methods. VBM aims to compare regional differences in relative tissue concentration or deformation and has been applied to many neuroanatomical studies. SBM can be used to quantify the amount of gray matter by estimating the cortical thickness [44,46], where it requires a preprocessing step of segmenting the inner and outer cortical surfaces and the distance between the two surfaces is defined as the cortical thickness. SBM has also been used in studying brain structures other than cortical surfaces, such as hippocampus [10,11,48], ventricles [49], thalamus, globus pallidus, and putamen [50]. Besides VBM and SBM, there are several other shape models used in biomedical imaging studies. Examples include landmarks [34,51], deformation fields mapping a template image to individual images [7,8,35,52,36–39], distance transforms [53,13], and medial axes [54,55,23].

In this chapter, we focus on the topic of surface-based morphometry (SBM) in brain imaging. We present typical shape analysis methods for modeling and analyzing 3D surface data. We use hippocampal morphometry in Alzheimer's disease (AD) as a test bed to demonstrate these methods. Our goal is to identify hippocampal shape abnormalities associated with AD or mild cognitive impairment (MCI, a prodromal stage of AD) in order to aid early diagnosis. We first present classic spherical harmonics (SPHARM) methods for modeling and registering 3D hippocampal surfaces [56], and then discuss advanced methods that take into consideration hippocampal subfield information [5]. After that, we describe techniques for shape analysis of registered surface models, including traditional general linear models (GLMs) [44,57] as well as a newly developed statistical parametric mapping distribution analysis (SPM-DA) method for performing surface-based morphometry [6]. Finally, we use an example neuroimaging application [6] to demonstrate the effectiveness of these techniques.

13.2 SURFACE MODELING AND REGISTRATION

13.2.1 SPHARM Surface Modeling

The *spherical harmonics* (SPHARM) technique [56] creates a parametric surface description using spherical harmonics [58] as basis functions. Spherical harmonics were first used as surface representation for radial surfaces ($r(\theta, \phi)$) [53], and later extended to more general shapes by describing a surface using three spherical functions [56]. At present SPHARM is a powerful surface modeling approach for arbitrarily shaped but simply-connected objects.

This section provides a short summary of the SPHARM surface modeling technique. An input object surface is assumed to be a *voxel surface*, which is a square surface mesh converted from a voxel representation; see Fig. 13.1A for an example. Two steps are involved in obtaining a SPHARM shape description for such a voxel surface: (1) spherical parameterization, and (2) SPHARM expansion. Below we discuss these two steps.

13.2.1.1 Spherical Parameterization

Spherical parameterization aims to create a continuous and uniform mapping from the object surface to the surface of a unit sphere so that each vertex on the object surface is assigned a pair of spherical coordinates (θ, ϕ). As a result of spherical parameterization, the surface of the unit sphere becomes our parameter space. To match the definition of spherical harmonics [58], the following convention for spherical coordinates (θ, ϕ) is employed (see also Fig. 13.2): θ is taken as the polar (colatitudinal) coordinate with $\theta \in [0, \pi]$, and ϕ as the azimuthal (longitudinal) coordinate with $\phi \in [0, 2\pi)$. Thus, the north pole has $\theta = 0$ and the south pole has $\theta = \pi$.

In order to create shape descriptors that can be compared across different 3D surfaces, we require an appropriate parameterization that has the following properties.
1. Bijective mapping: each vertex on the object surface must map to exactly one point on the sphere, and the inverse map must also be one-to-one.
2. Area preservation: each unit area on the object surface should be assigned to the same relative amount of area in parameter space.
3. Topology preservation: each square face on the object surface should map to a spherical quadrilateral in parameter space.
4. Minimal angular distortion: the spherical mapping of each square face should be as close to a "spherical surface square" as possible.

To achieve the above goals, we employ the spherical parameterization approach proposed by Brechbühler et al. [56]. This approach can be applied to a voxel surface, since

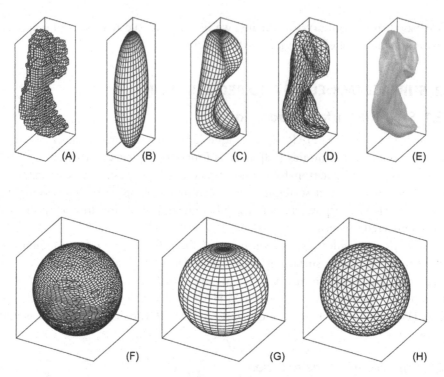

Figure 13.1 An example SPHARM reconstruction. (A) The voxel surface of a hippocampus. (B–E) SPHARM reconstructions of the hippocampal surface using coefficients up to degrees 1, 5, 10 and 15, respectively. (F) Spherical parameterization of the hippocampal surface shown in (A). (G) A regular spherical mesh grid used for SPHARM reconstructions shown in (B) and (C). (H) An icosahedral subdivision used for SPHARM reconstructions shown in (D) and (E).

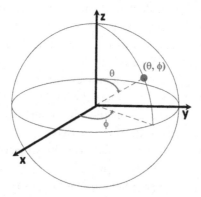

Figure 13.2 Rotational convention of spherical coordinates used in spherical harmonics. For the point (θ, ϕ) on the unit sphere, θ is taken as the polar (colatitudinal) coordinate, and ϕ as the azimuthal (longitudinal) coordinate.

it exploits the uniform quadrilateral structure of a square surface mesh. The approach consists of two steps: (1) initialization, and (2) optimization.

Step 1. Initialization. An initial parameterization is formed by constructing a harmonic map from the object surface to the parameter surface. For colatitude θ, two poles are selected by finding two surface vertices with the maximum and minimum z coordinates in the object space, respectively. Then, a Laplace equation (Eq. (13.1)) with Dirichlet conditions (Eq. (13.2) and Eq. (13.3)) is solved for colatitude θ:

$$\nabla^2 \theta = 0 \quad \text{(except at the poles)} \tag{13.1}$$

$$\theta_{north} = 0 \tag{13.2}$$

$$\theta_{south} = \pi \tag{13.3}$$

Given our case being discrete, we approximate Eq. (13.1) by assuming that each vertex's colatitude (except at the poles) equals the average of its neighbors' colatitudes. Thus, after assigning $\theta_{north} = 0$ to the north pole and $\theta_{south} = \pi$ to the south pole, we form a system of linear equations (one for each vertex) and obtain the solution by solving this linear system. For longitude ϕ, the same approach can be employed except that longitude is a cyclic parameter. To solve this problem, a "date line" is introduced. When crossing the date line, longitude is added or subtracted by 2π, depending on the crossing direction. After adjusting the linear system accordingly, the solution for longitude ϕ can be obtained.

Step 2. Optimization. The initial parameterization is refined to obtain an area preserving mapping. Brechbühler et al. [56] formulate this refinement process as a constrained optimization problem. They establish a few constraints for topology and area preservation and formulate an objective function for minimizing angular distortion. To solve this constrained optimization problem, an iterative procedure is developed to perform the following two steps alternately: (1) satisfying the constraints using the Newton–Raphson method [59], and (2) optimizing the objective function using a conjugate gradient method [59]. For more details about these steps, refer to [60,56].

13.2.1.2 SPHARM Expansion

Spherical harmonics are a natural and convenient choice of basis functions for representing any twice-differentiable spherical function [61,53,58]. They form the Fourier basis on the sphere, including an infinite set of spherical functions that are continuous, orthogonal, single-valued, and complete.

Using the notational convention for spherical coordinates described in Fig. 13.2, spherical harmonics $Y_l^m(\theta, \phi)$ of degree l and order m are defined as follows:

$$Y_l^m(\theta, \phi) = \sqrt{\frac{2l+1}{4\pi} \frac{(l-m)!}{(l+m)!}} \, P_l^m(\cos\theta) \, e^{im\phi}, \tag{13.4}$$

where $P_l^m(\cos\theta)$ are associated Legendre polynomials (with argument $\cos\theta$), and l and m are integers with $-l \le m \le l$. The associated Legendre polynomial P_l^m is defined by the differential equation

$$P_l^m(x) = \frac{(-1)^m}{2^l l!} (1-x^2)^{\frac{m}{2}} \frac{d^{l+m}}{dx^{l+m}} (x^2-1)^l. \tag{13.5}$$

Any twice-differentiable spherical function $r(\theta,\phi)$ can be represented by a linear combination of spherical harmonics $Y_l^m(\theta,\phi)$ as follows:

$$r(\theta,\phi) = \sum_{l=0}^{\infty} \sum_{m=-l}^{l} a_l^m Y_l^m(\theta,\phi), \tag{13.6}$$

where the coefficients a_l^m are uniquely determined by (see [62])

$$a_l^m = \int_0^\pi \int_0^{2\pi} Y_l^m(\theta,\phi)^* r(\theta,\phi) \sin\theta \, d\phi \, d\theta. \tag{13.7}$$

Here $Y_l^m(\theta,\phi)^*$ is the complex conjugate of $Y_l^m(\theta,\phi)$.

Spherical harmonics were first used for surface representation for radial or stellar surfaces $r(\theta,\phi)$ (e.g., [53,62]), where the radial function, $r(\theta,\phi)$, encodes the distance of surface points from a chosen origin. Brechbühler et al. [60,56] extended this spherical harmonics expansion technique to more general shapes by representing a surface using three spherical functions. This surface expansion technique has been referred to as the *SPHARM expansion* in previous studies (e.g., [10,49]). The SPHARM expansion technique can be applied to arbitrarily shaped but simply-connected objects. It is suitable for surface comparison and can deal with protrusions and intrusions. In this chapter, the SPHARM expansion is employed to describe 3D closed surfaces.

The SPHARM expansion requires a spherical parameterization performed in advance, as described in Section 13.2.1.1. The spherical parameterization has the following form:

$$\mathbf{v}(\theta,\phi) = \begin{pmatrix} x(\theta,\phi) \\ y(\theta,\phi) \\ z(\theta,\phi) \end{pmatrix}, \tag{13.8}$$

where $x(\theta,\phi)$, $y(\theta,\phi)$, and $z(\theta,\phi)$ are three functions defined on the sphere. Thus, the object surface can be described via expanding these three spherical functions using spherical harmonics.

The expansion takes the form:

$$\mathbf{v}(\theta,\phi) = \begin{pmatrix} x(\theta,\phi) \\ y(\theta,\phi) \\ z(\theta,\phi) \end{pmatrix} = \sum_{l=0}^{\infty} \sum_{m=-l}^{l} \begin{pmatrix} \mathbf{c}_{xl}^{m} \\ \mathbf{c}_{yl}^{m} \\ \mathbf{c}_{zl}^{m} \end{pmatrix} Y_{l}^{m}(\theta,\phi) = \sum_{l=0}^{\infty} \sum_{m=-l}^{l} \mathbf{c}_{l}^{m} Y_{l}^{m}(\theta,\phi), \qquad (13.9)$$

where

$$\mathbf{c}_{l}^{m} = \begin{pmatrix} c_{xl}^{m} \\ c_{yl}^{m} \\ c_{zl}^{m} \end{pmatrix}. \qquad (13.10)$$

The coefficients $\mathbf{c}_{l}^{m}$ up to a user-specified degree can be estimated by solving three sets of linear equations in a least squares fashion, given a pre-defined spherical parameterization. The object surface can be reconstructed using these coefficients, and using more coefficients leads to a more detailed reconstruction. Fig. 13.1 provides an example of an object surface (Fig. 13.1A) and its SPHARM reconstructions (Fig. 13.1B–E). Thus, a set of coefficients actually form an object surface description. Note that the degree one reconstruction is always an ellipsoid (Fig. 13.1B). Using more coefficients in a SPHARM expansion yields a more accurate reconstruction (Fig. 13.1C–E).

In our brain imaging applications, a hippocampus is originally represented by a voxel surface that may contain errors due to the voxel quantization and the limited voxel resolution. Using the first few degrees of SPHARM coefficients to describe the surface can help to smooth the object and reduce these errors, given that a hippocampal surface is assumed to be relatively smooth. In Fig. 13.1, we use coefficients up to degree 15, which can derive smooth but detailed reconstructions. Thus, for each expansion ($x(\theta,\phi)$, $y(\theta,\phi)$, or $z(\theta,\phi)$), there are $\sum_{l=0}^{15} \sum_{m=-l}^{l} 1 = 256$ coefficients. In total, $256 \times 3 = 768$ coefficients are extracted to describe a hippocampal surface.

13.2.2 SPHARM Surface Normalization

13.2.2.1 SPHARM Normalization

The previous section shows how the SPHARM expansion derives a set of SPHARM coefficients that describe a 3D closed surface. In order to compare different surfaces, these coefficients need to be normalized into a common reference system. We call this step *SPHARM normalization.*

In this chapter, we are mainly concerned with *shape* information, i.e.,

all the geometrical information that remains when location, scale, and rotational effects are filtered out from an object [27].

Thus, the goal of the SPHARM normalization is to create a shape descriptor for any given surface by removing the effects of translation, rotation, and scaling. The shape descriptor is formed by a set of normalized SPHARM coefficients that are designed to be comparable across different object surfaces.

The *translation effect* of a SPHARM model can be easily removed by ignoring the degree 0 coefficient. This revision centers the SPHARM reconstruction at the origin.

The *scaling effect* can be removed by dividing all the coefficients by a scaling factor f. For example, we can choose f so that the object volume is constant. Note that while the object volume information must be removed for pure shape analysis, in general morphometric analysis, the object volume can be treated as an additional feature and can be combined together with shape features. Another strategy is to keep the size information in the SPHARM model and perform morphometric analysis on the geometric configuration of *size and shape*.

The *rotation effect* can be removed by first creating a surface correspondence and then aligning the corresponding parts together in object space. This is achieved by aligning the degree one reconstruction, which is always an ellipsoid and often called first order ellipsoid (FOE). To *establish the surface correspondence*, the parameter net on the FOE (see Fig. 13.3A) is rotated to a canonical position such that the north pole is at one end of the longest main axis, and the crossing point of the zero meridian and the equator is at one end of the shortest main axis (see Fig. 13.3B). Surface correspondence is then defined by taking two points with the same parameter pair (θ, ϕ) on two different surfaces to be a corresponding pair. To align the corresponding parts together in object space, each surface is rotated in object space such that the main axes of its degree one ellipsoid coincide with the coordinate axes, putting the shortest axis along x and longest along z (see Fig. 13.3C).

The normalization technique described above works only if the FOE is a "real" ellipsoid (i.e., one with three different main axes), such as our hippocampal data, but not an ellipsoid of revolution or a sphere. In the latter case, higher degree coefficients might need to be involved for normalization. In addition, even for a "real" ellipsoid, it has rotational symmetry of order 2 with respect to each of its main axes. Therefore, in practice, manual inspection is involved to avoid any excessive 180 degree rotation that breaks the alignment.

After the above steps, a set of canonical coordinates (i.e., normalized coefficients) can be used as a shape descriptor for each surface, and these shape descriptors are comparable across objects. Assuming the scaling effect is properly handled, the procedure for removing translation and rotation effects, including *establishing surface correspondence*, is also called *registration*. Fig. 13.3 shows an example registration procedure: (A) original objects, (B) objects with aligned parameterization, and (C) objects registered in both parameter and object spaces.

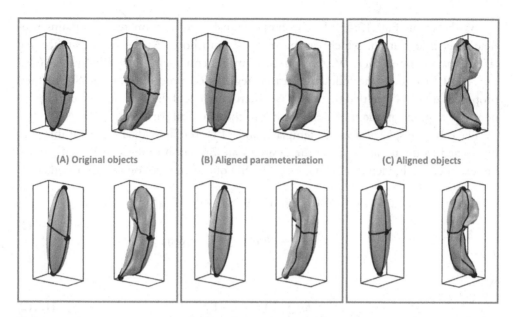

Figure 13.3 SPHARM registration using first order ellipsoids (FOEs). Each row shows one sample hippocampus. Each of (A)–(C) shows the FOE on left and degree 15 reconstruction on right. Parameterization is indicated by the lines on the surface, including equator $\theta = \pi/2$ and four longitudinal lines $\phi = -\pi/2, 0, \pi/2, \pi$. The north and south poles and the point $(\pi/2, 0)$ are shown as dots.

13.2.2.2 Subfield-Guided Registration

The hippocampus is composed of multiple subfields [63], and many hippocampal studies have indicated that subfields play an important role in brain functions [64,63,65]. There is an increased interest in recent literature in examining the subfields of the hippocampal formation using MRI [66–69,4,3,70,71]. However, the above SPHARM modeling framework does not take into account this critical subfield information. In particular, it is important to use the hippocampal subfield information to guide hippocampal surface registration.

One possible approach is to define landmarks on subfield boundaries, and use these landmarks to help surface registration via a landmark-guided approach as proposed in [72]. However, it becomes a challenging problem to identify anatomically meaningful landmarks that are comparable across surfaces. To overcome this limitation, we have proposed in [5] a landmark-free method. Specifically, we form a spherical label image by assigning subfield labels to the surface parameterization on the sphere. The subfield-aware surface registration problem can then be solved using a spherical image registration method to align subfield label information across surfaces. The spherical image registration method used in this chapter is the spherical demons (SD) framework proposed in [73].

As shown in Fig. 13.4A, B, the hippocampal segmentation result contains sub-field label information, which can be mapped onto both (A) the object surface and (B) the parameter surface. We use spherical images containing these label values (e.g., Fig. 13.4B) to guide the following surface registration procedure by the SD method proposed in [73]. Note that the major goal here is to establish better surface correspondence by aligning all the subfields together in the spherical parameter space.

Let F be the spherical image template, and M be the individual spherical image to be aligned to the template, Γ be the desired transformation to register M to F, and γ be intermediate hidden transformation. The SD objective function is formulated as:

$$(\gamma^*, \Gamma^*) = \underset{\gamma, \Gamma}{\mathrm{argmin}} \, \|\Sigma^{-1}(F - M \circ \Gamma)\|^2 \\ + \frac{1}{\sigma_x^2} dist(\gamma, \Gamma) + \frac{1}{\sigma_T^2} Reg(\gamma) \tag{13.11}$$

while:

$$dist(\gamma, \Gamma) = \|\gamma - \Gamma\|^2 \tag{13.12}$$

$$Reg(\gamma) = \|\nabla(\gamma - Id)\|^2 \tag{13.13}$$

Here Eq. (13.12) defines the geodesic distance between the hidden transformation γ and the optimized transformation Γ. Eq. (13.13) characterizes the regularization penalization on gradient magnitude of the displacement field $\gamma - Id$ of the hidden transformation γ. Our goal is to minimize not only the loss function defined by the first term of Eq. (13.11) but also the above geodesic distance and the regularization term. Parameter Σ is a diagonal matrix that specifies the feature variability of each vertex on the surface. Parameters σ_x and σ_T describe the trade-off between the similarity measure and the regularization of the objective function.

Following the spherical vector spline interpolation theory [74], the SD algorithm proposed in [73] implements the optimization procedure in two steps: (1) The first step is to formulate a nonlinear least-square problem and solve the first two terms in Eq. (13.11) by Gauss–Newton optimization. (2) The second step is to solve the last two terms in Eq. (13.11) using a single convolution of the displacement field Γ with a smoothing kernel [75]. The details of the SD algorithm are available in [73].

We employ a multi-resolution strategy at different levels in our adaptation of the SD algorithm. We pre-define four levels of icosahedral subdivisions meshes, including 2562 vertices at level 1, 10,242 vertices at level 2, 40,962 vertices at level 3, and 163,842 vertices at level 4. We start from the level with lowest resolution and apply the SD algorithm to register the individual to the template. At each level, we linearly interpolate both the template and the registered individual to the next higher level, and apply the SD algorithm again. We repeat this procedure until the highest level is reached. The complete

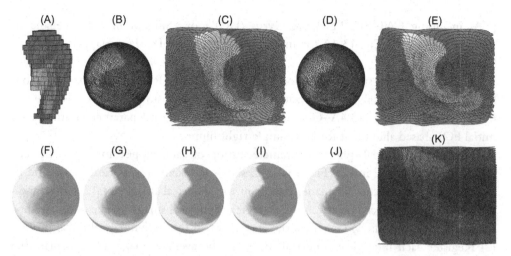

Figure 13.4 (A–E) Example result of spherical parameterization (A–C) and spherical demons (SD) registration (D–E). (A) Original object in Euclidean space. (B–C) Original spherical mapping and its unfolded version on the 2D plane. (D–E) Spherical mapping registered to the template subfield parameterization (shown in (J)–(K)) and its unfolded version on the 2D plane. (F–K) SD registration procedure for creating the template subfield parameterization. (F) Mean subfield image after spherical parameterization and FOE alignment. (G–J) Mean subfield image after 1st–4th iterations during SD registration respectively. (K) Spherical image in (J) unfolded to 2D space. The mean spherical image shown in (J)–(K) is the converging result of SD method and is chosen to be our resulting template subfield parameterization. Green, light green and yellow colors correspond to three subfield categories respectively: cornu ammonis (CA, including CA1, CA2 and CA3), dentate gyrus (DG), and subiculum and miscellaneous (SUB+MISC). (For interpretation of the references to color in this figure legend, the reader is referred to the web version of this chapter.)

algorithmic details of applying the SD method to subfield-guided hippocampal surface registration are available in [5].

We demonstrate the performance of the above subfield-guided hippocampal surface registration using a sample of 12 healthy control (HC) participants recruited at the Indiana Alzheimer's Disease Center (IADC). MRI scans were acquired on a Siemens MAGNETOM Prisma 3T MRI scanner. The scanning protocols include a T1-weighted (MPRAGE) whole-brain scan and a T2-weighted (TSE) partial-brain scan and an oblique coronal slice orientation (positioned orthogonally to the main axis of the hippocampus). See [70,76] for similar imaging protocols to collect high resolution hippocampal subfield data.

Hippocampal subfields were segmented by the Automatic Segmentation of Hippocampal Subfields (ASHS) software [70]. Topology fix was performed on segmentation results to ensure a spherical topology for each hippocampus. Subfields were assigned to the corresponding locations on the surface of hippocampal results. In this work, we include the following three subfield groups on the surface: "Cornu Ammonis

(CA, including CA1–3)", "Dentate Gyrus (DG)", or "Subiculum + Miscellaneous (SUB+MISC)". See Fig. 13.4A for an example hippocampal surface mapped with sub-field labels.

Each hippocampal surface was modeled using the SPHARM method described be-fore, including spherical parameterization, SPHARM expansion, and SPHARM align-ment using FOE. Fig. 13.4A–C shows the result of spherical parameterization with initial FOE-based alignment for an example right hippocampus.

Note that the initial spherical parameterization of each hippocampus can be rep-resented as a spherical image. Fig. 13.4B–C shows an example spherical image and its unfolded version on the 2D plane. Based on the spherical images of 12 subjects, we use the following procedure to create an average spherical image as our template sub-field parameterization: (1) Calculate the average of all the individual spherical images. (2) Register each individual spherical image to the average image. (3) Calculate the average of registered individual spherical images. (4) Repeat Steps 2 and 3 until the average image converges. The resulting average image is used as our template subfield parameterization.

Fig. 13.4F–K shows the SD registration procedure for creating the template sub-field parameterization for the right hippocampus. Fig. 13.4F shows the mean subfield image after spherical parameterization and FOE alignment. Here, we can see that the boundaries between three subfields are blurred, indicating that the initial alignment is not optimized for subfield registration.

Fig. 13.4G–J shows the mean subfield image after 1st–4th iterations during the SD registration respectively. Fig. 13.4K shows the spherical image in (J) unfolded to 2D space. The mean spherical image shown in Fig. 13.4J–K is the converging result of SD method and is chosen to be our resulting template subfield parameterization. Compared with the initial alignment shown in Fig. 13.4F, this resulting template demonstrates an improved alignment of three subfield regions.

Fig. 13.5 shows the root mean square distance (RMSD) at each iteration for each subject. The mean RMSDs of 12 subjects are 0.49 and 0.52 respectively for left and right hippocampi at the beginning. They reduce to 0.32 and 0.34 after first iteration, and then keep reducing until reaching 0.18 and 0.20 at the convergence.

13.3 STATISTICAL INFERENCE ON THE SURFACE

In this section, we first show how to extract surface signals for statistical shape analysis. After that, we describe two methods for performing statistical inference on the surface: (1) One which uses the classical general linear model (GLM) coupled with random field theory (RFT) for multiple comparison correction [44,57]. (2) The other which uses a newly proposed statistical parametric mapping distribution analysis (SPM-DA) [6].

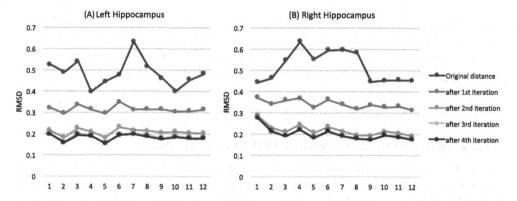

Figure 13.5 SD registration performance. Root mean square distances between each individual (labeled from 1 to 12) and the template (i.e., the mean subfield image) at each iteration are shown for both left and right hippocampal.

13.3.1 Surface Atlas and Signal Processing

To facilitate statistical shape analysis, we need to put all the shape objects into a common reference system and make them comparable. One common practice is to create a shape atlas and register all the individual shapes to the atlas, where the atlas can be defined by the average of all the aligned shape objects. In the SPHARM case, we can use the FOE approach to align the first order ellipsoid (FOE) of each individual SPHARM model to a canonical configuration in both parameter and object spaces, as shown in Fig. 13.3. After that, the shape atlas is simply the average of all the aligned individual SPHARM models. In the hippocampal case, if subfield information is available, we need to further adjust the spherical parameterization of all the individual shapes and the atlas. Following the procedure described in Section 13.2.2.2, the template parameterization becomes the spherical parameterization of the atlas, and the parameterization of each individual shape needs to be registered to this template parameterization.

After atlas generation and shape registration, all the SPHARM models are aligned to the same reference system (i.e., the atlas). This facilitates the subsequent analysis on the surface. In order to perform statistical shape analysis, we need to define signals or variables on the surface. Let x_t be the atlas. An individual shape x can be described by its deformation field $\delta(x) = x - x_t$ relative to the atlas x_t. However, for each vertex, three components (corresponding to x, y, z coordinates) in $\delta(x)$ are used to characterize the local shape change. For simplicity, similar to several previous studies [6,4,77,78], in this chapter, we focus only on the deformation component along the surface normal direction to reduce the number of variables considered for each surface location. This deformation component is defined as the surface signal for each shape.

To perform statistical analysis in a Euclidean domain, Gaussian kernel smoothing is often applied as a typical step for increasing the signal-to-noise ratio (SNR). Since the surface geometry of a brain structure is non-Euclidean, we cannot directly apply Gaussian kernel smoothing. Instead, we employ heat kernel smoothing, which generalizes Gaussian kernel smoothing to arbitrary Riemannian manifolds [44]. Heat kernel smoothing is implemented by constructing the kernel of a heat equation on manifolds that is isotropic in the local conformal coordinates. By smoothing the data on the surface, the SNR increases and consequently it becomes easier to localize shape changes.

13.3.2 General Linear Model and Random Field Theory

Let us start our discussion from an example neuroimaging study. In this study, we want to examine whether there is hippocampal shape change in late mild cognitive impairment (LMCI), a prodromal stage of Alzheimer's disease. In the analysis, we also want to remove the effect of certain covariates (e.g., age, gender). To achieve this goal, we consider the following general linear model (GLM):

$$\text{signal} \;=\; \beta_0 + \beta \cdot \text{group} + \beta_1 \cdot \text{age} + \beta_2 \cdot \text{gender} + \epsilon.$$

To make it general, let $N_{i,j}$ denote the surface signal at vertex/location j ($j = 1, 2, \ldots, m$) for subject i ($i = 1, 2, \ldots, n$), and we can describe the GLM as follows:

$$N_{i,j} = \beta_{0,j} + \beta_j x_i + \beta_{1,j} z_{1,i} + \beta_{2,j} z_{2,i} + \cdots + \beta_{p,j} z_{p,i} + \epsilon_{i,j},$$
$$i = 1, 2, \ldots, n, \; j = 1, 2, \ldots, m, \tag{13.14}$$

in which x is the predictor of interest (i.e., diagnostic group in our case) and $z_1, z_2, \ldots, z_p$ are nuisance covariates (e.g., age and gender in our case). At each vertex j, we fit Eq. (13.14) and compute the t-statistic for testing H_0: $\beta_j = 0$ versus H_1: $\beta_j \neq 0$. The resulting m t-statistics comprise the statistical parametric map (SPM) for analyzing the relationship between x and the hippocampal surface signals.

Random field theory (RFT) [44,57] is a widely used method for multiple comparison correction in general SPM analyses, and can be adapted to our surface-based morphometry study. Specifically, RFT first estimates the smoothness of the surface data, uses the smoothness values to determine statistical thresholds that control the family wise error rate (FWER), and then provides corrected p-values for the result. The details about how to apply RFT to surface-based morphometry are available in [44].

SurfStat [79] is a Matlab toolbox that implements both GLM and RFT for statistical inference on surfaces. Fig. 13.6 shows an example SurfStat result, where statistical shape differences are identified on the hippocampal surface between healthy controls and LMCI patients after removing the age and gender effects.

13.3.3 Statistical Parametric Mapping Distribution Analysis

As shown in Fig. 13.6, hippocampal shape change in LMCI can be detected by RFT. However, using the same strategy, we identified no significant difference between HC

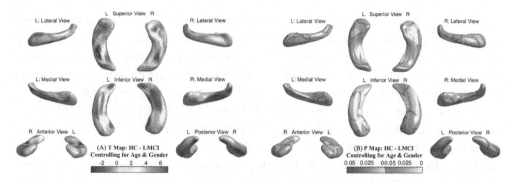

Figure 13.6 Effects of healthy control (HC) versus late mild cognitive impairment (LMCI). The *t*-map (A) and *p*-map (B) of the diagnostic effect (HC-LMCI) on surface signals after removing the effects of age and gender.

and early mild cognitive impairment (EMCI) subjects [4]. To bridge this gap, we recently developed a new and powerful SPM image analysis framework, *Statistical Parametric Mapping (SPM) Distribution Analysis* or *SPM-DA* [6]. Unlike RFT and permutation methods which focus on peak amplitude, clusters of supra-threshold statistics, or combinations of the two, SPM-DA detects relationships by analyzing the information provided by the entire distribution of the statistics comprising the SPM (*t*-statistics in our case), hence the phrase *Distribution Analysis*. By making greater use of the SPM information, SPM-DA can potentially achieve greater power than RFT methods. There are various ways to capture the SPM distribution information. Here we employ a simplified version of *Lindsey's Method* in which the distribution is estimated by first constructing a frequency histogram and then analyzing the frequencies (bin counts) using Poisson generalized linear models [80]. A key advantage of this method is that it converts density estimation into a regression problem [81]. In more detail, the resulting SPM-DA method can be described and implemented by the following four steps.

Step 1. Statistical parametric map (SPM). As mentioned earlier, our surface signal $N_{i,j}$ at vertex i of subject j is analyzed using the following regression model

$$N_{i,j} = \beta_{0,j} + \beta_j x_i + \beta_{1,j} z_{1,i} + \beta_{2,j} z_{2,i} + \cdots + \beta_{p,j} z_{p,i} + \epsilon_{i,j},$$
$$i = 1, 2, \ldots, n, \ j = 1, 2, \ldots, m,$$

(13.15)

in which x is the predictor of interest (e.g., group indicator) and $z_1, z_2, \ldots, z_p$ are nuisance covariates. In this step, at each vertex j, we fit the above model and compute the *t*-statistic for testing H_0: $\beta_j = 0$ versus H_1: $\beta_j \neq 0$. The resulting m *t*-statistics comprise the SPM for analyzing the relationship between x and the surface signals.

Step 2. Frequency histogram of the SPM *t*-statistics. Note that the RFT approach described above analyzes the SPM using either peak amplitude or cluster size statistics as implemented by SurfStat [79,57]. In contrast to this, our SPM-DA method captures approximately all the information provided by the SPM t-statistics by estimating their distribution with a frequency histogram. To facilitate analysis the histogram is constructed using n_h bins of unequal width: the bin boundaries are chosen so that each bin is equally likely under the null hypothesis. More specifically, we compute frequency F_k, $k = 1, \ldots, n_h$, where F_k is the number of SPM t-statistics in the bin interval $[Q_k, Q_{k+1})$ and the bin boundaries Q_l, $l = 1, \ldots, n_h + 1$, partition the real line into n_h intervals of equal probability with respect to the $t_{n-(p+2)}$ distribution.

Step 3. Regression analysis of the frequency histogram. The histogram bin frequencies are analyzed to detect departures from count uniformity using two regression models:

$$F_k = \beta_u w_{u,k} + \epsilon_k, \ k = 1, \ldots, n_h, \qquad (13.16)$$

$$F_k = \beta_l w_{l,k} + \epsilon_k, \ k = 1, \ldots, n_h, \qquad (13.17)$$

in which F_k denotes the frequency of the k-th of n_h bins. In our analyses we used $n_h = 12$. This choice ensured that the bin frequencies for our data were large enough so that the F_k, and thus the ϵ_k, were approximately normally distributed by the Central Limit Theorem. For the first model in Eq. (13.16), we let

$$w'_u = (0, 0, 0, 0, 0, 0, 0, 1, 2, 3, 4, 5)$$

be our predictor. Thus, the coefficient β_u will be positive when positive SPM statistics (right-tail values) are enriched in the result. This indicates a positive relationship between surface signal values and the predictor of interest. Similarly, for the second model in Eq. (13.17), we let

$$w'_l = (5, 4, 3, 2, 1, 0, 0, 0, 0, 0, 0, 0)$$

be our predictor. Thus, the coefficient β_l will be positive when negative (left-tail) values are enriched in the result. This indicates a negative relationship. Fig. 13.7 shows an example of the Eq. (13.16) predictor data (i.e., the solid line) and the bin counts computed from the actual and permuted hippocampal surface data (i.e., the "o" and "+" values respectively).

Step 4. SPM-DA permutation test statistic. To detect whether the surface shape is related to the predictor of interest, we test the following composite hypotheses:

$$H_0\text{: } \beta_u \leq 0 \text{ and } \beta_l \leq 0 \text{ versus } H_1\text{: } \beta_u > 0 \text{ or } \beta_l > 0.$$

Using Bonferroni, a p-value for testing these hypotheses is given by $p = 2\min(p_l, p_u)$ where p_u is the p-value for testing H'_0: $\beta_u \leq 0$ versus H'_1: $\beta_u > 0$ and p_l is defined similarly. For various reasons, in particular correlation between the bin counts F_k, p_l and p_u cannot be computed via the usual regression t-tests. Therefore we compute p_l and

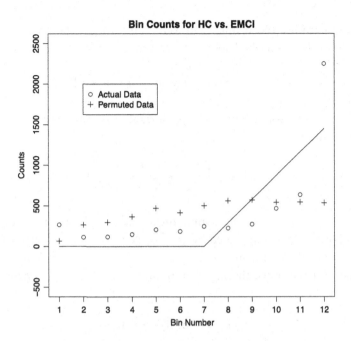

Figure 13.7 Bin counts for HC vs. EMCI. Our linear model in Eq. (13.16) aims to use the values on the solid line to predict the "+" values (for permuted data) or the "○" values (for actual data). Note that the significance of the group difference is driven mainly by the "○" value on the 12th bin.

p_u using permutation tests. As a result, we avoid the strict and often unmet assumptions of RFT methods and obtain a distribution-free method: the only requirement for valid permutation inference is that the data are exchangeable [82]. In order to achieve exchangeability in the presence of covariates we use the Smith ("orthogonalization") method in which the predictor of interest (here x) is orthogonalized with respect to all covariates prior to permutation [83]. Thus, in more detail, we compute p_l and p_u as follows:

a. Execute Steps 1–3 described above on the unpermuted data to obtain the least squares estimates $\hat{\beta}_u$ and $\hat{\beta}_l$.

b. Implement the Smith method by computing $\tilde{x}$ where $\tilde{x}_i = x_i - (\hat{B}_0 + \hat{B}_1 z_{1,i} + \hat{B}_2 z_{2,i} + \cdots + \hat{B}_p z_{p,i})$ and the $\hat{B}_i$ are the least squares coefficient estimates for the model

$$x_i = B_0 + B_1 z_{1,i} + B_2 z_{2,i} + \cdots + B_p z_{p,i} + \epsilon_i, \quad i = 1, 2, \ldots, n.$$

Note that $\tilde{x}$ is the zero mean orthogonalization of x with respect to the covariates $z_1, z_2, \ldots, z_p$, i.e., the n-dimensional vector $\tilde{x}$ is perpendicular to the vectors $z_1, z_2, \ldots, z_p$.

c. Randomly permute the elements of $\tilde{x}$ N times to generate $\tilde{x}_k^*$, $k = 1, 2, \ldots, N$. Execute Steps 1–3 described above on each $\tilde{x}_k^*$ to generate permutation coefficient estimates $\hat{\beta}_{u,1}^*, \ldots, \hat{\beta}_{u,N}^*$ and $\hat{\beta}_{l,1}^*, \ldots, \hat{\beta}_{l,N}^*$.

d. Compute the p-value p_u for testing H_0': $\beta_u \leq 0$ versus H_1': $\beta_u > 0$ (a positive relationship between the surface signals $N_{i,j}$ and x) as follows:

- If the distribution of the $\hat{\beta}_u^*$'s is normal, compute the p-value using the t-distribution,

$$p_u = P\left(t_{N-1} \geq \frac{\hat{\beta}_u - \bar{X}}{S\sqrt{1 + 1/N}} \right),$$

in which $\bar{X}$ and S are the sample mean and sample standard deviation of the $\hat{\beta}_u^*$'s. Note that S is increased by $\sqrt{1 + 1/N}$ to account for using $\tilde{x}$ instead of a constant in the numerator.

- If the distribution of the $\hat{\beta}_u^*$'s is nonnormal compute the p-value in the usual manner,

$$p_u = (\# \text{ of } \hat{\beta}_u^*\text{'s} \geq \hat{\beta}_u)/N.$$

e. Similarly compute p_l for testing H_0'': $\beta_l \leq 0$ versus H_1'': $\beta_l > 0$ (a negative relationship between the surface normals $N_{i,j}$ and x).

Simulation studies. We perform simulation studies to compare the performance of the proposed SPM-DA method with that of traditional RFT peak and cluster methods. We first create a random data set of 72 subjects on a template surface with 652 vertices, using the following model:

$$
\begin{aligned}
S_{i,j} &= \beta x_i + \epsilon_{i,j}, \quad i = 1, \ldots, 72, \; j = 1, \ldots, 126, \\
&= \epsilon_{i,j}, \quad i = 1, \ldots, 72, \; j = 127, \ldots, 652,
\end{aligned}
$$

where $S_{i,j}$ represents the surface value at location j for subject i. We conduct two studies involving simulated two-sample data with x_i equal to -1 for $i = 1, \ldots, 36$ and 1 for $i = 37, \ldots, 72$. Corresponding to these x values and the model the signal extends across 126 contiguous locations and is constant with a magnitude determined by β. In both studies, β takes on the values 0, 1/12, 1/6, and 1/3. In the first study, the random errors $\epsilon_{i,j}$ are independent normal ($\mu = 0, \sigma^2 = 1$) pseudorandom numbers. In the second study, the $\epsilon_{i,j}$ are also independent normal ($\mu = 0, \sigma^2 = 1$) but are smoothed prior to the signal being added, where the smoothing is applied using the heat kernel smoothing method [44]. The resulting data sets are analyzed using our SPM-DA method (implemented in R [84]) and RFT peak and cluster statistics (implemented by SurfStat [79,57]). For each combination of β and choice of unsmoothed or smoothed random errors, 100 simulated surface data sets are generated and analyzed.

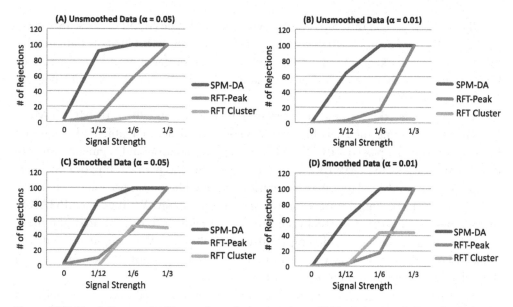

Figure 13.8 Simulation results. The number of rejections (out of 100 runs) at $\alpha = 0.05$ and 0.01 for the SPM-DA, RFT Peak, and RFT Cluster methods.

The results of our simulation studies are shown in Fig. 13.8, which plots the number of rejections (out of 100 runs) of the SPM-DA, RFT Peak, and RFT Cluster methods for two significance levels: $\alpha = 0.05$ and $\alpha = 0.01$. For the null (signal strength $= 0$) scenarios, all three methods have type I error rates at or below α. For all non-null scenarios the SPM-DA method outperforms the RFT Cluster method, demonstrating substantially greater power at all signal strengths. SPM-DA also outperforms the RFT Peak method for all but the strongest signal case where both methods reject H_0 on every run. In particular, its power is at least eight times greater than RFT Peak for the weakest signals.

13.4 AN EXAMPLE APPLICATION

In this section, we use a real world neuroimaging application to demonstrate the shape modeling and statistical analysis methods described in this chapter. The hippocampus is an important brain structure responsible for learning and memory and is widely studied in Alzheimer's Disease (AD). MRI-based hippocampal measures have been shown as effective biomarkers for detecting the status of AD or LMCI [85–87,70]. In this application [6], we demonstrate that our SPM-DA method, coupled with the SPHARM modeling, can identify hippocampal shape differences in EMCI that are undetected by standard RFT methods.

The real data used in this study were downloaded from the ADNI database [88]. One goal of ADNI has been to test whether a serial MRI, positron emission tomography, other biological markers, and clinical and neuropsychological assessment can be combined to measure the progression of MCI and early AD. For up-to-date information, see www.adni-info.org.

We downloaded the relevant data consisting of 172 HC and 267 early MCI (EMCI) participants, including their baseline 3T MRI scans and demographic and diagnostic information. We employed FIRST [89], a surface registration and segmentation tool developed as part of the FMRIB Software Library (FSL), to perform hippocampal segmentation. We conducted topology fix on the segmented hippocampal volumes to make sure that each hippocampal surface has the spherical topology. We used the SPHARM method described earlier to model each surface and registered it to the atlas (the mean of all the HC individuals) by aligning its first order ellipsoid. We computed the deformations along the surface normal direction of the atlas at each surface location for each individual shape and used these as our surface signals.

We evaluated the age and gender effects on the surface signals, using the following two GLMs respectively:

$$\text{signal} = \beta_0 + \beta \cdot \text{age} + \epsilon$$
$$\text{signal} = \beta_0 + \beta \cdot \text{gender} + \epsilon$$

At each surface vertex/location, we fit either of the above models and computed the t-statistic for testing H_0: $\beta = 0$ versus H_1: $\beta \neq 0$. The maps of the resulting t-statistics are shown in Fig. 13.9A, C. The maps of the corresponding p-values corrected by RFT are shown in Fig. 13.9B, D.

We evaluated the diagnostic effect (HC-EMCI) on surface signals, without and with removing the effects of age and gender, using the following two GLMs respectively:

$$\text{signal} = \beta_0 + \beta \cdot \text{group} + \epsilon$$
$$\text{signal} = \beta_0 + \beta \cdot \text{group} + \beta_1 \cdot \text{age} + \beta_2 \cdot \text{gender} + \epsilon$$

At each surface vertex/location, we fit either of the above models and computed the t-statistic for testing H_0: $\beta = 0$ versus H_1: $\beta \neq 0$. The maps of the resulting t-statistics are shown in Fig. 13.10A, C. The map of the corresponding p-values corrected by RFT for the first model is shown in Fig. 13.10B. The p-value map for the second model is identical to that of the first model, and thus is not shown in Fig. 13.10.

We applied the SPM-DA method to the comparison of HC vs. EMCI after removing the effects of age and gender. Fig. 13.10D shows the SPM-DA bin value map, where 12 bins were used in the analysis. The p-value generated by SPM-DA is 0.009, indicating that the hippocampal shape difference between HC and MCI is statistically different

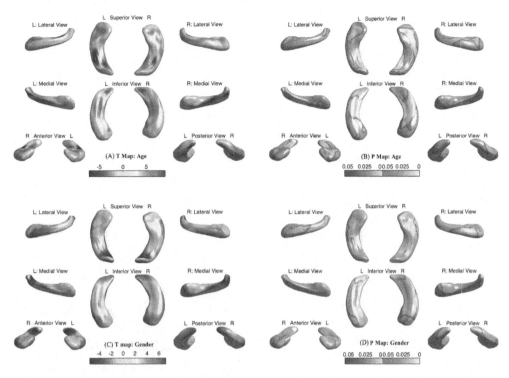

Figure 13.9 Effects of age and gender. (A–B) The *t*-map and *p*-map of the age effect on surface signals. (C–D) The *t*-map and *p*-map of the gender effect on surface signals.

at level $\alpha = 0.01$. The *p*-values generated by RFT peak and RFT cluster methods are both > 0.05, and thus neither method detected significant hippocampal shape changes in EMCI. It is encouraging that the SPM-DA method identified hippocampal shape changes in EMCI which were not detected by standard RFT methods. This demonstrates the promise of SPM-DA for early diagnosis of the prodromal stage of Alzheimer's disease.

The histogram patterns (i.e., bin counts) captured by SPM-DA are shown as o's for the actual data and +'s for an example of permuted data in Fig. 13.7. For the actual data, it is obvious that there are trends toward an enrichment of SPM values in the upper tail of the distribution, which matches the Eq. (13.16) predictor data (shown as the solid line) better than the Eq. (13.17) predictor data (not shown). This pattern suggests there is hippocampal atrophy in EMCI compared with HC.

In sum, we have used a real world neuroimaging study to demonstrate almost all the shape modeling and statistical analysis methods described in this chapter, except the subfield-guided registration method. Since subfield-guided registration is a newly

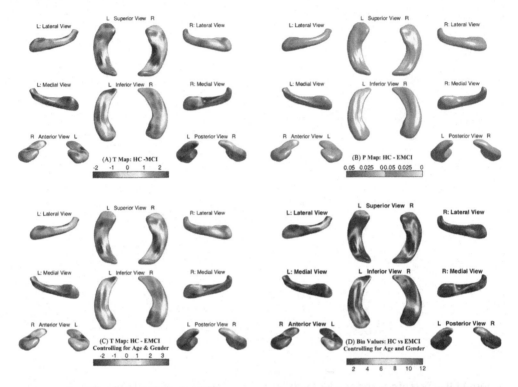

Figure 13.10 Effects of HC-EMCI. (A–B) The *t*-map and *p*-map of the diagnostic effect (HC-EMCI) on surface signals. (C) The *t*-map of the diagnostic effect (HC-EMCI) on surface signals after removing the effects of age and gender. The corresponding *p*-map is identical to (B) and thus not shown here. (D) The SPM-DA bin value map for the comparison of HC vs. EMCI after removing the effects of age and gender. In *t*-maps (A, C), red/blue colors respectively indicate expansion/contraction in HC compared with EMCI. (For interpretation of the references to color in this figure legend, the reader is referred to the web version of this chapter.)

proposed method [5], we are currently working on applying it to the analyses of the hippocampal data available in ADNI (used here) and other independent cohorts.

13.5 CONCLUSIONS

Statistical shape analysis is a fundamental topic in biomedical image computing, and plays important roles in numerous applications in brain imaging. This chapter describes a surface–based morphometry framework for modeling and analyzing 3D surface data in brain imaging studies. These studies examine 3D brain structures, and aim to identify morphometric abnormalities associated with a particular condition in order to aid diagnosis and treatment. We have presented a classic shape description method based on spherical harmonics (SPHARM) for modeling arbitrarily shaped but simply connected

3D brain structures. We have discussed a traditional SPHARM-based shape registration method via aligning first order ellipsoids, as well as a newly developed registration method that takes into account subfields on the surface. After that, we have described two techniques for shape analysis of registered surface models. The first is the traditional general linear model (GLM), coupled with random field theory (RFT) for multiple comparison correction, to perform vertex-wise surface-based morphometry. The second is a newly proposed and powerful analysis approach, Statistical Parametric Mapping (SPM) Distribution Analysis, or SPM-DA, which provides a distribution-free alternative to RFT methods. We have demonstrated these shape modeling and analysis methods using a real world brain imaging application of identifying hippocampal shape changes in early mild cognitive impairment. These methods successfully identified promising shape-based imaging biomarkers for early detection of Alzheimer's disease.

ACKNOWLEDGMENTS

This work was supported in part by NIH R01 LM011360, R01 EB022574, U01 AG024904, RC2 AG036535, R01 AG19771, P30 AG10133, R01 AG040770, UL1 TR001108, R01 AG042437, and R01 AG046171; DOD W81XWH-14-2-0151, W81XWH-13-1-0259, and W81XWH-12-2-0012; NCAA 14132004; and IUPUI RSFG.

REFERENCES

[1] A. Goldszal, C. Davatzikos, D.L. Pham, M.X.H. Yan, R.N. Bryan, S.M. Resnick, An image processing system for qualitative and quantitative volumetric analysis of brain images, J. Comput. Assist. Tomogr. 22 (5) (1998) 827–837.

[2] Michael D. Nelson, Andrew J. Saykin, Laura A. Flashman, Henry J. Riordan, Hippocampal volume reduction in schizophrenia as assessed by magnetic resonance imaging, Arch. Gen. Psychiatry 55 (1998) 433–440.

[3] Shan Cong, Maher Rizkalla, et al., Building a surface atlas of hippocampal subfields from MRI scans using FreeSurfer, FIRST and SPHARM, in: 2014 IEEE 57th International Midwest Symposium on Circuits and Systems (MWSCAS), IEEE, 2014, pp. 813–816.

[4] Shan Cong, Maher Rizkalla, Paul Salama, John West, Shannon Risacher, Andrew Saykin, Li Shen, Surface-based morphometric analysis of hippocampal subfields in mild cognitive impairment and Alzheimer's disease, in: 2015 IEEE 58th International Midwest Symposium on Circuits and Systems (MWSCAS), IEEE, 2015, pp. 1–4.

[5] Shan Cong, Maher Rizkalla, Paul Salama, Shannon L. Risacher, John D. West, Yu-chien Wu, Liana Apostolova, Eileen Tallman, Andrew J. Saykin, Li Shen, ADNI, Building a surface atlas of hippocampal subfields from high resolution T2-weighted MRI scans using landmark-free surface registration, in: MWSCAS'16: The IEEE 58th International Midwest Symposium on Circuits and Systems, Abu Dhabi, United Arab Emirates, 2016.

[6] Mark Inlow, Shan Cong, Shannon L. Risacher, John West, Maher Rizkalla, Paul Salama, Andrew J. Saykin, Li Shen, A New Statistical Image Analysis Approach and Its Application to Hippocampal Morphometry, Springer International Publishing, 2016, pp. 302–310.

[7] J.G. Csernansky, S. Joshi, L. Wang, J.W. Halleri, M. Gado, J.P. Miller, U. Grenander, M.I. Miller, Hippocampal morphometry in schizophrenia by high dimensional brain mapping, Proc. Natl. Acad. Sci. USA 95 (1998) 11406–11411.

[8] J.G. Csernansky, L. Wang, D. Jones, D. Rastogi-Cruz, J.A. Posener, G. Heydebrand, J.P. Miller, M.I. Miller, Hippocampal deformities in schizophrenia characterized by high dimensional brain mapping, Am. J. Psychiatr. 159 (2002) 2000–2006.

[9] Rhodri H. Davies, Carole J. Twining, P. Daniel Allen, Timothy F. Cootes, Christopher J. Taylor, Shape discrimination in the hippocampus using an MDL model, in: Christopher J. Taylor, J. Alison Noble (Eds.), IPMI 2003: 18th International Conference on Information Processing in Medical Imaging, Ambleside, UK, in: Lect. Notes Comput. Sci., vol. 2732, 2003, pp. 38–50.

[10] G. Gerig, M. Styner, Shape versus size: improved understanding of the morphology of brain structures, in: Proc. MICCAI'2001: 4th International Conference on Medical Image Computing and Computer-Assisted Intervention, Utrecht, The Netherlands, in: Lect. Notes Comput. Sci., vol. 2208, 2001, pp. 24–32.

[11] G. Gerig, M. Styner, M. Chakos, J.A. Lieberman, Hippocampal shape alterations in schizophrenia: results of a new methodology, in: 11th Biennial Winter Workshop on Schizophrenia, 2002.

[12] P. Golland, B. Fischl, M. Spiridon, N. Kanwisher, R.L. Buckner, M.E. Shenton, R. Kikinis, A. Dale, W.E.L. Grimson, Discriminative analysis for image-based studies, in: Proc. of MICCAI'2002: 5th International Conference on Medical Image Computing and Computer Assisted Intervention, Tokyo, Japan, in: Lect. Notes Comput. Sci., vol. 2488, 2002, pp. 508–515.

[13] P. Golland, W.E.L. Grimson, M.E. Shenton, R. Kikinis, Small sample size learning for shape analysis of anatomical structures, in: Proc. MICCAI'2000: 3rd International Conference on Medical Image Computing and Computer-Assisted Intervention, Pittsburgh, Pennsylvania, USA, in: Lect. Notes Comput. Sci., vol. 1935, 2000, pp. 72–82.

[14] P. Golland, W.E.L. Grimson, M.E. Shenton, R. Kikinis, Deformation analysis for shaped based classification, in: Proc. IPMI'2001: 17th International Conference on Information Processing and Medical Imaging, in: Lect. Notes Comput. Sci., vol. 2082, 2001, pp. 517–530.

[15] A.J. Saykin, L.A. Flashman, T. McHugh, C. Pietras, T.W. McAllister, A.C. Mamourian, R. Vidaver, L. Shen, J.C. Ford, L. Wang, F. Makedon, Principal components analysis of hippocampal shape in schizophrenia, in: International Congress on Schizophrenia Research, Colorado Springs, Colorado, USA, 2003.

[16] Li Shen, James Ford, Fillia Makedon, Laura Flashman, Andrew Saykin, Organization for human brain mapping, surface-based morphometric analysis for hippocampal shape in schizophrenia, NeuroImage 19 (2) (2003) S1004.

[17] Li Shen, James Ford, Fillia Makedon, Andrew Saykin, Effective classification of 3D closed surfaces: application to modeling neuroanatomical structures, in: International Conference on Computer Vision, Pattern Recognition and Image Processing (CVPRIP) in conjunction with Seventh Joint Conference on Information Sciences (JCIS), Association for Intelligent Machinery, Cary, North Carolina, USA, 2003, pp. 708–711.

[18] Li Shen, James Ford, Fillia Makedon, Andrew Saykin, Hippocampal shape analysis: surface-based representation and classification, in: M. Sonka, J.M. Fitzpatrick (Eds.), Medical Imaging 2003: Image Processing, San Diego, CA, USA, in: Proc. SPIE, vol. 5032, 2003, pp. 253–264.

[19] Li Shen, James Ford, Fillia Makedon, Andrew Saykin, A surface-based approach for classification of 3D neuroanatomic structures, Intell. Data Anal. 8 (5) (2004).

[20] Li Shen, James Ford, Fillia Makedon, Yuhang Wang, Tilmann Steinberg, Song Ye, A. Saykin, Morphometric analysis of brain structures for improved discrimination, in: MICCAI: Medical Image Computing and Computer Assisted Intervention, Montreal, Quebec, Canada, in: Lect. Notes Comput. Sci., vol. 2879, 2003, pp. 513–520.

[21] Li Shen, Fillia Makedon, Andrew Saykin, Shape-based discriminative analysis of combined bilateral hippocampi using multiple object alignment, in: J.M. Fitzpatrick, M. Sonka (Eds.), Medical Imaging 2004: Image Processing, San Diego, CA, USA, in: Proc. SPIE, vol. 5370, 2004.

[22] M.E. Shenton, G. Gerig, R.W. McCarley, G. Szekely, R. Kikinis, Amygdala–hippocampal shape differences in schizophrenia: the application of 3D shape models to volumetric MR data, Psychiatry Res. Neuroimaging 115 (2002) 15–35.

[23] M. Styner, G. Gerig, J. Lieberman, D. Jones, D. Weinberger, Statistical shape analysis of neuroanatomical structures based on medial models, Med. Image Anal. 7 (3) (2003) 207–220.

[24] M. Styner, G. Gerig, S. Pizer, S. Joshi, Automatic and robust computation of 3D medial models incorporating object variability, Int. J. Comput. Vis. 55 (2/3) (2003) 107–122.

[25] M. Styner, J. Lieberman, G. Gerig, Boundary and medial shape analysis of the hippocampus in schizophrenia, in: MICCAI 2003, Medical Image Computing and Computer Assisted Intervention, Montreal, Quebec, Canada, 2003.

[26] S.J. Timoner, P. Golland, R. Kikinis, M.E. Shenton, W.E.L. Grimson, W.M. Wells III, Performance issues in shape classification, in: Proc. MICCAI'2002: 5th International Conference on Medical Image Computing and Computer-Assisted Intervention, Tokyo, Japan, in: Lect. Notes Comput. Sci., vol. 2488, 2002, pp. 355–362.

[27] I.L. Dryden, K.V. Mardia, Statistical Shape Analysis, John Wiley and Sons, New York, 1998.

[28] C.G. Small, The Statistical Theory of Shape, Springer, New York, 1996.

[29] D.G. Kendall, The diffusion of shape, Adv. Appl. Probab. 9 (1977) 428–430.

[30] D.G. Kendall, Shape manifolds, Procrustean metrics and complex projective spaces, Bull. Lond. Math. Soc. 16 (1984) 81–121.

[31] David G. Kendall, A survey of the statistical theory of shape, Stat. Sci. 4 (2) (1989) 87–99.

[32] Fred L. Bookstein, The Measurement of Biological Shape and Shape Change, Lect. Notes Biomath., vol. 24, Springer-Verlag, Berlin, New York, ISBN 0387089128, 1978, viii+191 pp.

[33] Fred L. Bookstein, Morphometric Tools for Landmark Data: Geometry and Biology, Cambridge University Press, Cambridge, New York, 1991, xvii+435 pp.

[34] F.L. Bookstein, Shape and the information in medical images: a decade of the morphometric synthesis, Comput. Vis. Image Underst. 66 (2) (1997) 97–118.

[35] Ulf Grenander, Michael I. Miller, Computational anatomy: an emerging discipline, Q. Appl. Math. LVI (4) (1998) 617–694.

[36] Michael I. Miller, Computational anatomy: shape, growth, and atrophy comparison via diffeomorphisms, NeuroImage 23 (Suppl. 1) (2004) S19–S33, http://www.sciencedirect.com/science/article/B6WNP-4DD95D7-2/2/1464f473b6e671929e545888cb6ea8ba.

[37] Michael I. Miller, M. Faisal Beg, Can Ceritoglu, Craig Stark, Increasing the power of functional maps of the medial temporal lobe by using large deformation diffeomorphic metric mapping, Proc. Natl. Acad. Sci. USA 102 (27) (2005) 9685–9690, http://dx.doi.org/10.1073/pnas.0503892102, http://www.pnas.org/cgi/content/abstract/102/27/9685.

[38] Anqi Qiu, Benjamin J. Rosenau, Adam S. Greenberg, Monica K. Hurdal, Patrick Barta, Steven Yantis, Michael I. Miller, Estimating linear cortical magnification in human primary visual cortex via dynamic programming, NeuroImage 31 (1) (2006) 125–138, http://www.sciencedirect.com/science/article/B6WNP-4J7300B-1/2/8b52c730f16dcfa477de2cf91d4e4c68.

[39] Lei Wang, J. Philip Miller, Mokhtar H. Gado, Daniel W. McKeel, Marcus Rothermich, Michael I. Miller, John C. Morris, John G. Csernansky, Abnormalities of hippocampal surface structure in very mild dementia of the Alzheimer type, NeuroImage 30 (1) (2006) 52–60, http://www.sciencedirect.com/science/article/B6WNP-4HCMSHW-3/2/0ac679d0a9d5b6c5f6f94c26ff0f1b14.

[40] J. Ashburner, K.F. Friston, Voxel-based morphometry – the methods, NeuroImage 11 (2000) 805–821.

[41] J. Ashburner, K. Friston, Why voxel-based morphometry should be used, NeuroImage 14 (2001) 1238–1243.

[42] C. Davatzikos, A. Genc, D. Xu, S.M. Resnick, Voxel-based morphometry using the ravens maps: methods and validation using simulated longitudinal atrophy, NeuroImage 14 (2001) 1361–1369.

[43] M.K. Chung, K.M. Dalton, L. Shen, et al., Weighted Fourier series representation and its application to quantifying the amount of gray matter, IEEE Trans. Med. Imaging 26 (4) (2007) 566–581.

[44] M.K. Chung, S. Robbins, K.M. Dalton, R.J. Davidson, A.L. Alexander, A.C. Evans, Cortical thickness analysis in autism via heat kernel smoothing, NeuroImage 25 (2005) 1256–1265.

[45] M.K. Chung, K.J. Worsley, S. Robbins, T. Paus, J. Taylor, J.N. Giedd, J.L. Rapoport, A.C. Evans, Deformation-based surface morphometry applied to gray matter deformation, NeuroImage 18 (2003) 198–213.

[46] B. Fischl, A.M. Dale, Measuring the thickness of the human cerebral cortex from magnetic resonance images, Proc. Natl. Acad. Sci. USA 97 (2000) 11050–11055.

[47] M.I. Miller, A.B. Massie, J.T. Ratnanather, K.N. Botteron, J.G. Csernansky, Bayesian construction of geometrically based cortical thickness metrics, NeuroImage 12 (2000) 676–687.

[48] Li Shen, James Ford, et al., A surface-based approach for classification of 3D neuroanatomic structures, Intell. Data Anal. 8 (6) (2004) 519–542.

[49] G. Gerig, M. Styner, D. Jones, D. Weinberger, J. Lieberman, Shape analysis of brain ventricles using SPHARM, in: Proc. IEEE Workshop on Mathematical Methods in Biomedical Image Analysis, 2001, pp. 171–178.

[50] A. Kelemen, G. Szekely, G. Gerig, Elastic model-based segmentation of 3-D neuroradiological data sets, IEEE Trans. Med. Imaging 18 (10) (1999) 828–839.

[51] T.F. Cootes, C.J. Taylor, D.H. Cooper, J. Graham, Active shape models – their training and application, Comput. Vis. Image Underst. 61 (1995) 38–59.

[52] S.C. Joshi, M.I. Miller, U. Grenander, On the geometry and shape of brain sub-manifolds, Int. J. Pattern Recognit. Artif. Intell. 11 (8) (1997) 1317–1343 (Special Issue on Magnetic Resonance Imaging).

[53] D.H. Ballard, C.M. Brown, Computer Vision, Prentice-Hall, NJ, 1982.

[54] S.M. Pizer, D.S. Fritsch, P. Yushkevich, V. Johnson, E. Chaney, Segmentation, registration, and measurement of shape variation via image object shape, IEEE Trans. Med. Imaging 18 (10) (1999) 851–865.

[55] M. Styner, G. Gerig, S. Pizer, S. Joshi, Automatic and robust computation of 3D medical models incorporating object variability, Int. J. Comput. Vis. 55 (2–3) (2003) 107–122.

[56] Ch. Brechbühler, G. Gerig, O. Kubler, Parametrization of closed surfaces for 3D shape description, Comput. Vis. Image Underst. 61 (2) (1995) 154–170.

[57] K.J. Worsley, S. Marrett, P. Neelin, A.C. Vandal, K.J. Friston, A.C. Evans, A unified statistical approach for determining significant signals in images of cerebral activation, Hum. Brain Mapp. 4 (1996) 58–73.

[58] Eric W. Weisstein, Spherical harmonic, in: MathWorld – A Wolfram Web Resource, http://mathworld.wolfram.com/SphericalHarmonic.html.

[59] William H. Press, Numerical Recipes in C: The Art of Scientific Computing, second ed., Cambridge University Press, Cambridge, New York, 1992.

[60] Ch. Brechbühler, Description and Analysis of 3-D Shapes by Parametrization of Closed Surfaces, Ph.D. thesis, IKT/BIWI, ETH, Zurich, 1995.

[61] George B. Arfken, Mathematical Methods for Physicists, third ed., Academic Press, Orlando, 1985.

[62] David Ritchie, Graham Kemp, Fast computation, rotation, and comparison of low resolution spherical harmonic molecular surfaces, J. Comput. Chem. 20 (1999) 383–395.

[63] Susanne G. Mueller, L.L. Chao, et al., Evidence for functional specialization of hippocampal subfields detected by MR subfield volumetry on high resolution images at 4T, NeuroImage 56 (3) (2011) 851–857.

[64] Thorsten Bartsch, Juliane Döhring, et al., CA1 neurons in the human hippocampus are critical for autobiographical memory, mental time travel, and autonoetic consciousness, Proc. Natl. Acad. Sci. USA 108 (42) (2011) 17562–17567.

[65] Marcus Rössler, Rosemarie Zarski, Jürgen Bohl, Thomas G. Ohm, Stage-dependent and sector-specific neuronal loss in hippocampus during Alzheimer's disease, Acta Neuropathol. 103 (4) (2002) 363–369.

[66] Julie Winterburn, Jens C. Pruessner, et al., High-resolution in vivo manual segmentation protocol for human hippocampal subfields using 3T magnetic resonance imaging, J. Vis. Exp. 105 (2015) e51861.

[67] Paul A. Yushkevich, Robert S.C. Amaral, et al., Quantitative comparison of 21 protocols for labeling hippocampal subfields and parahippocampal subregions in in vivo MRI: towards a harmonized segmentation protocol, NeuroImage 111 (2015) 526–541.

[68] Bernd Merkel, Christopher Steward, et al., Semi-automated hippocampal segmentation in people with cognitive impairment using an age appropriate template for registration, J. Magn. Reson. Imaging 42 (6) (2015) 1631–1638.

[69] Michael R. Hunsaker, David G. Amaral, A semi-automated pipeline for the segmentation of rhesus macaque hippocampus: validation across a wide age range, PLoS ONE 9 (2) (2014) e89456.

[70] Paul A. Yushkevich, John B. Pluta, et al., Automated volumetry and regional thickness analysis of hippocampal subfields and medial temporal cortical structures in mild cognitive impairment, Hum. Brain Mapp. 36 (1) (2015) 258–287.

[71] Juan Eugenio Iglesias, Jean C. Augustinack, et al., A computational atlas of the hippocampal formation using ex vivo, ultra-high resolution MRI: application to adaptive segmentation of in vivo MRI, NeuroImage 115 (2015) 117–137.

[72] Li Shen, Sungeun Kim, Andrew J. Saykin, Fourier method for large-scale surface modeling and registration, Comput. Graph. 33 (3) (2009) 299–311.

[73] B.T. Thomas Yeo, Mert R. Sabuncu, et al., Spherical demons: fast diffeomorphic landmark-free surface registration, IEEE Trans. Med. Imaging 29 (3) (2010) 650–668.

[74] Joan Glaunès, Marc Vaillant, et al., Landmark matching via large deformation diffeomorphisms on the sphere, J. Math. Imaging Vis. 20 (1–2) (2004) 179–200.

[75] Pascal Cachier, Eric Bardinet, et al., Iconic feature based nonrigid registration: the PASHA algorithm, Comput. Vis. Image Underst. 89 (2) (2003) 272–298.

[76] Susanne G. Mueller, Norbert Schuff, et al., Hippocampal atrophy patterns in mild cognitive impairment and Alzheimer's disease, Hum. Brain Mapp. 31 (9) (2010) 1339–1347.

[77] J. Wan, L. Shen, K.E. Sheehan, et al., Shape analysis of thalamic atrophy in multiple sclerosis, in: MIAMS 2009: MICCAI Workshop on Medical Image Analysis on Multiple Sclerosis, 2009, pp. 93–104.

[78] Li Shen, Andrew J. Saykin, et al., Morphometric analysis of hippocampal shape in mild cognitive impairment: an imaging genetics study, in: IEEE 7th Int. Symp. on Bioinformatics and Bioengineering, Boston, MA, 2007, pp. 211–217.

[79] K.J. Worsley, SurfStat, http://www.math.mcgill.ca/keith/surfstat.

[80] J.K. Lindsey, Construction and comparison of statistical models, J. R. Stat. Soc. B 36 (1974) 418–425.

[81] Bradley Efron, Large-Scale Inference: Empirical Bayes Methods for Estimation, Testing, and Prediction, Institute of Mathematical Statistics Monographs, Cambridge University Press, Cambridge, ISBN 9780511761362, 2010, xii+263 pp.

[82] Thomas E. Nichols, Andrew P. Holmes, A unified statistical approach for determining significant signals in images of cerebral activation, Hum. Brain Mapp. 15 (2001) 1–25.

[83] A.M. Winkler, G.R. Ridgway, M.A. Webster, S.M. Smith, T.E. Nichols, Permutation inference for the general linear model, NeuroImage 92 (2014) 381–397.

[84] R Core Team, R: A Language and Environment for Statistical Computing, R Foundation for Statistical Computing, Vienna, Austria, 2014, http://www.R-project.org/.

[85] John Pluta, Paul Yushkevich, et al., In vivo analysis of hippocampal subfield atrophy in mild cognitive impairment via semi-automatic segmentation of T2-weighted MRI, J. Alzheimer's Dis. 31 (1) (2012) 85–99.

[86] Li Shen, Andrew J. Saykin, Sungeun Kim, Hiram A. Firpi, John D. West, Shannon L. Risacher, Brenna C. McDonald, Tara L. McHugh, Heather A. Wishart, Laura A. Flashman, Comparison of manual and automated determination of hippocampal volumes in MCI and early AD, Brain Imaging Behav. 4 (1) (2010) 86–95.

[87] C. Testa, M.P. Laakso, et al., A comparison between the accuracy of voxel-based morphometry and hippocampal volumetry in Alzheimer's disease, J. Magn. Reson. Imaging 19 (3) (2004) 274–282, http://dx.doi.org/10.1002/jmri.20001, http://www.ncbi.nlm.nih.gov/pubmed/14994294.

[88] M.W. Weiner, D.P. Veitch, et al., The Alzheimer's Disease Neuroimaging Initiative: a review of papers published since its inception, Alzheimer's Dement. 9 (5) (2013) e111–e194, http://dx.doi.org/10.1016/j.jalz.2013.05.1769, http://www.ncbi.nlm.nih.gov/pubmed/23932184.

[89] Brian Patenaude, Stephen M. Smith, David N. Kennedy, Mark Jenkinson, A Bayesian model of shape and appearance for subcortical brain segmentation, NeuroImage 56 (3) (2011) 907–922.

CHAPTER 14

Statistical Respiratory Models for Motion Estimation

Christoph Jud*, Philippe C. Cattin*, Frank Preiswerk*,†
*Department of Biomedical Engineering, University of Basel, Allschwil, Switzerland
†Department of Radiology, Brigham and Women's Hospital, Harvard Medical School, Boston, MA, USA

Contents

14.1	Background	379
14.2	4-Dimensional MR Imaging	381
	14.2.1 Acquisition Protocol	382
	14.2.2 Retrospective Stacking	383
14.3	Motion Model Building	384
	14.3.1 Statistical Shape Model	384
	14.3.2 Statistical Motion Model	385
	14.3.3 Inference on Statistical Models	386
	14.3.3.1 Sparse Reconstruction	*386*
	14.3.3.2 Statistical Model Regression	*387*
14.4	Establishment of Correspondence	388
	14.4.1 Inter-Subject Correspondence	388
	14.4.2 Inter-Subject Correspondence by Registration	390
	14.4.3 Intra-Subject Correspondence	391
14.5	Statistical Motion Modeling	391
14.6	Bayesian Reconstruction from Sparse Data	392
14.7	Applications of Population-Based Statistical Motion Models to Motion Reconstruction	394
14.8	Reconstruction by Regression	401
	14.8.1 Average Breathing Cycle	401
	14.8.2 Motion Model Prediction	402
14.9	Conclusion	404
	Acknowledgments	404
	References	404

14.1 BACKGROUND

The quasi–perpetual motion of organs caused by the respiratory cycle is a complicating factor in many medical contexts. In particular during imaging, for example using Magnetic Resonance Imaging (MRI) or Computed Tomography (CT), respiratory motion is problematic as it is responsible for motion blur in the resulting images. But respiratory motion is not only problematic during the diagnostic stage, it also complicates surgical treatment and tumour ablation in the abdomen. There exist three approaches often used

Statistical Shape and Deformation Analysis
DOI: 10.1016/B978-0-12-810493-4.00017-1

in these applications: (1) breath holding or high-frequency low tidal ventilation, (2) respiratory gating where the patient is only imaged/treated around exhalation or (3) target tracking with implantable markers such as gold beads or active markers. Common to these approaches is that they either prolong treatment time or they are of invasive nature, thereby complicating the procedure.

As cancer is a leading cause of death worldwide, better motion mitigation approaches are required. Cancer accounted for 13% of all deaths in 2014 [36]. The main types of lethal cancer are lung, stomach, liver, colon, and breast tumors. All these sites are subject to respiratory motion, which is a complicating factor for treatment. A well known and established method for the treatment of non-resectable tumors is radiation therapy, which uses ionizing radiation to destroy tumor cells. The art of radiation therapy is to deliver a lethal dose to all cancerous tissue while sparing as much healthy tissue as possible, in particular organs that are radiation sensitive or for which the consequences of side effects can be severe. Consequently, accurate target localization and delivery is one of the main challenges in radiation therapy.

There are a number of different factors contributing to potential inaccuracies in irradiation. Of these, respiratory motion is one of the main problems in the thorax and in the abdomen and has been shown to have a large dosimetric impact on conventional radiotherapy [41,13,21]. Besides quasi-periodic respiratory motion, which includes variation in breathing depth and speed, the organs undergo also other modes of deformation, called secondary modes [40]. The secondary modes are caused, for example, by the cardiac motion, digestive activity, gravity, muscle relaxation, or filling of the bladder. The quasi-periodic and secondary modes constitute the total organ motion seen during treatment.

Recent advances in three-dimensional planning and advanced treatment technologies, such as intensity modulated radiotherapy (IMRT), intensity modulated proton therapy (IMPT) [19] and high-frequency focused ultrasound (HIFU) [7–9], have brought about new possibilities in delivering highly conformal dose distributions. The advantage of highly localized treatment makes these techniques sensitive to organ motion, which represents a limiting factor for exploiting their full potential [40,27].

Countless works on handling organ motion, more specifically respiratory motion, have been published in recent years. A comprehensive overview including practical guidelines can be found in a report of the American Association of Physicists in Medicine [1,15].

A simple method to avoid respiratory motion is to completely interrupt breathing while the therapy beam is on [25,23,11,4,16]. In gated treatment the beam is only turned on during a certain period of the breathing cycle, for example during exhalation [13,24,18,10,12]. Although the aforementioned approaches compensate respiratory motion to some extent, they require reproducibility of the organ position for the selected breathing phase [40] and prolong the treatment time. More importantly, they only compensate for the perpetual respiratory motion and as such are completely oblivious to all other modes of organ motion. Thus they are only accurate in a short window of a couple of minutes after patient set-up, as was show in [40]. Hence, organ motion during radiotherapy continues to be a problem [9] and much research effort is being put into understanding and addressing this issue.

It would be desirable to keep the target and the treatment beam aligned throughout the entire breathing cycle. This technique is commonly referred to as tracking. Tracking is very demanding as it requires some prediction of the target motion [35,34, 39]. Although tracking is in principle designed to follow any target motion, it profits from a possibly regular breathing pattern because this simplifies the required short-time prediction of the motion trajectory. The high sampling rate and low lag of ultrasound (US) makes it a very attractive non-ionizing imaging modality to use for tracking of the target. However, only few tumors are visible under US and it completely fails for structures inside the lung due the air. Thus, conventional ultrasound imaging alone is rather limited in its capability as a tumor tracking modality. On the other hand, theoretical investigation on the liver, reported in [39], showed that the knowledge about the position of one or more surrogate 3D points (not the tumor directly) and a statistical motion model of the organ of interest allows predicting the position of the entire organ with good spatial precision.

Recently, a theoretical analysis of a yet novel motion mitigation approach has been proposed in [22]. In this work a so-called self-scanning approach is proposed that does not try to follow the breathing induced tumor motion but rather uses a fixed treatment beam and lets the respiratory motion move the tumor through the treatment beam. It was shown that the approach works surprisingly well for high-frequency focused ultrasound (HIFU). As no active following of the tumor is required, the HIFU transducers could be simplified too, however, at the cost of computationally more complex treatment planning that requires a statistical motion model of the organ.

14.2 4-DIMENSIONAL MR IMAGING

The basis for building statistical respiratory motion models is ground truth motion data. This motion data ideally has to show the moving organs in free-breathing not only with good 3-dimensional spatial resolution but also with a sufficiently high temporal frame rate of a couple of volumes per second. In other words, to build respiratory motion models, 3-dimensional plus time (3D+t) data is needed, or in short 4-dimensional data.

Today's imaging methods such as US, CT or MRI have limited ability to fulfill these criteria, for different reasons. While the required chest volume to capture is too big for US, there have been approaches proposed to capture 4D CT data [14]. While 4D CT has adequate spatial as well as temporal resolution, only a limited number of breathing cycles can be observed due to the involved X-ray exposure to the patient. This lim-

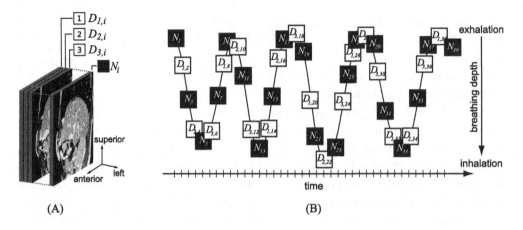

Figure 14.1 (A) Spatial arrangement of the slices and (B) their acquisition sequence over time.

its the quality and range of suitable data for motion models. MRI on the other hand, does not expose the patient to any harmful radiation allowing to capture and observe the organ motion over extended time intervals. It also shows good spatial resolution, however, this comes at the cost of increased scan time, leading to motion artifacts in the acquired images. While there exist accelerated 2d and volumetric MRI acquisition schemes (e.g. UNFOLD [42], SENSE [43], as well as combinations thereof, among other approaches), they often provide significantly inferior image resolution and signal-to-noise ratio (SNR). To overcome this trade-off between acquisition speed and image quality, learning-based methods have been proposed, among them 4D-MRI. While MRI does not natively provide a high enough spatial resolution and temporal frame rate required for 4D motion modeling, it is possible to retrospectively sort images of different planes that were acquired over tens of minutes into coherent 3-dimensional stacks of different respiratory phases, yielding a 4-dimensional data set with good spatial as well as temporal resolution. This technique is generally known as 4D-MRI [38]. Other 4D-MRI techniques, as well as hybrid ultrasound-MRI based techniques [44] have been proposed since then, but this chapter describes the original technique proposed in [38].

14.2.1 Acquisition Protocol

In contrast to some of the techniques described in the literature a sagittal slice orientation is chosen. Reason being that sagittal slicing allows tracking the vascular structures during complete breathing cycles with minimal out-of-plane motion. Moreover, the sagittal slice orientation requires the smallest field of view and therefore leads to the fastest temporal frame rate.

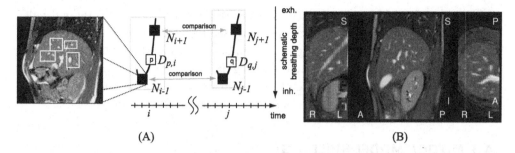

Figure 14.2 (A) Retrospective matching of the acquired MR images and (B) an example reconstruction from orthogonal views.

The volume of interest is covered by sagittal slices (Fig. 14.1A), further called data slices $D_{p,i}$ with p the position of the sagittal slice and i the acquisition time. The key concept of the proposed acquisition scheme is to interleave the acquisition of data slices with a dedicated so-called navigator slice N_j at a fixed spatial position, with j indicating the acquisition time (Fig. 14.1B). These navigator slices will be used retrospectively to derive a gating measure that determines the state of the liver on a certain data frame.

14.2.2 Retrospective Stacking

In order to produce sequences of 3D volumes from the acquired 2D images, one has to find those frames which show the liver in a similar state on different planes, in order to stack them together. The corresponding data frames can be determined by comparing the embracing navigator frames that were acquired immediately before and after the respective data frames in the interleaved sequence, see Fig. 14.1B. The underlying assumption is that two data frames show the same state of the liver if their embracing navigator frames are sufficiently similar. Since only navigator frames are directly compared and not the data frames themselves, data frames can be stacked even when showing the liver at distant positions. Moreover, any difference in amplitude, phase or the amount of deformation between data and navigator frames can be handled as long as the state of the liver can be unambiguously defined by the preceding and the subsequent navigator frames.

To describe the similarity measure for navigator slices in detail, a data frame $D_{p,i}$ that was acquired at location p and time i is considered (Fig. 14.2A). Its embracing navigator frames are N_{i-1} and N_{i+1}. To find a temporally corresponding data frame at a different slice position q, the embracing navigator frames (N_{i-1}, N_{i+1}) are compared to the preceding and the subsequent navigator frames (N_{j-1}, N_{j+1}) of each acquired candidate frame $D_{q,j}$. Among the many different possible similarity measures, best results were achieved by quantifying the positions of prominent vascular structures using template matching, and using these displacements to find similar navigation frames.

The above 4D-MRI technique allows reconstructing 4D data sets that show detailed deformations of an organ during free breathing, see Fig. 14.2B for an example. While variations beyond a regular breathing cycle cannot be observed in 4D CT, they are addressed specifically with the above method. Irregularities such as drifts and deformations are recognized and handled by the retrospective image-based sorting method. Lastly, in contrast to 4D CT, volunteer studies over long acquisition sessions are possible.

14.3 MOTION MODEL BUILDING

In earlier chapters of this book, we have seen various approaches to model the variation of shapes. In this chapter, we move on to modeling the dynamics of shape changes, which goes beyond standard shape modeling. The dynamics of an object is defined as the short term changes in the shape of that object. In the case of face modeling, one can think of modeling expressions of a facial shape [5,3], which means one is interested in the changes of a facial shape with respect to a reference or neutral face shape. In our case, this would correspond to modeling the respiratory motion of abdominal organ shapes with respect to, say, the corresponding organ shape in exhalation state [37,32].

The mindset is somewhat similar to shape modeling, while a shape model is superimposed by a model of shape changes. However, there are some modeling decisions which have to be taken when modeling shape motion. If the respiratory motion over a population of organ shapes shall be modeled, there are usually a different amount of sample shape changes per subject and therefore one has to think about the underlying distribution and its estimation. Furthermore, the case application may differ from shape modeling or fitting. Usually, respiratory motion models find application in estimating and reconstructing the shape changes based on so-called surrogates, which refers to any signal that allows to infer the shape changes. In our case, surrogates might be model points which are tracked in ultrasound images, or a position signal of a magnetic tracker which is mounted on the epigastrium. We will come to such applications in detail later on.

In the following, we formalize the motion model, where we focus on the application of respiratory motion estimation of the liver, and where motion data for learning is captured from 4D-MRI data.

14.3.1 Statistical Shape Model

One distinguishes between the modeling of shape and the modeling of shape motion. In the shape model, the variation among a population of shapes originating from different individuals is considered, whereas in the motion model, the shape changes over time relative to a reference shape is investigated.

For each subject s, there is a finite dimensional reference shape $\mathbf{v}^s \in \mathbb{R}^{3n}$ composed of n model points. In our setting, the shape $\mathbf{v}^s$ represents the right liver lobe in the exhalation state and is derived from a representative exhalation MR image of s. For the moment, we assume that the exhalation shapes are in correspondence. Furthermore, the shapes are assumed to be Gaussian distributed $p(\mathbf{v}^s | \mathbf{v}_\mu, \Sigma_\mathbf{v}) \sim \mathcal{N}(\mathbf{v}_\mu, \Sigma_\mathbf{v})$ where

$$\mathbf{v}_\mu = \frac{1}{S} \sum_s \mathbf{v}^s, \quad \Sigma_\mathbf{v} = \frac{1}{S-1} \sum_s (\mathbf{v}^S - \mathbf{v}_\mu) \otimes (\mathbf{v}^s - \mathbf{v}_\mu) \tag{14.1}$$

are the maximum likelihood parameter estimates of $p(\mathbf{v}^s | \mathbf{v}_\mu, \Sigma_\mathbf{v})$, S is the number of subjects and $\otimes$ is the outer-product. Thus, a shape can be parametrized by $\mathbf{v}^\alpha = \mathbf{v}_\mu + \sum_{i=1}^M \alpha_i \psi_i$, where ψ_i are orthogonal basis vectors of $\Sigma_\mathbf{v}$ weighted by the model parameters α_i and M denotes the number of basis vectors.

14.3.2 Statistical Motion Model

In addition to the shape variation among a population, the relative shape changes over time is modeled. Since each subject is observed for a different amount of time τ, one assumes a mixture of Gaussian distributions over the shape changes $\mathbf{x} \in \mathbb{R}^{3n}$

$$p(\mathbf{x}) = \sum_s^S p(s) p(\mathbf{x}|s) = \sum_s^S \pi^s p(\mathbf{x}|s), \tag{14.2}$$

where $\sum_s^S \pi^s = 1, \pi^s \in (0, 1), \forall s = 1, ..., S$. Each component distribution is assumed to be Gaussian $p(\mathbf{x}^s) \sim \mathcal{N}(\mathbf{x}_\mu^s, \Sigma_{\mathbf{x}^s})$ with

$$\mathbf{x}_\mu^s = \frac{1}{\tau^s} \sum_t^{\tau^s} \mathbf{x}_t^s, \quad \Sigma_{\mathbf{x}^s} = \frac{1}{\tau^s - 1} \sum_t^{\tau^s} (\mathbf{x}_t^s - \mathbf{x}_\mu^s) \otimes (\mathbf{x}_t^s - \mathbf{x}_\mu^s). \tag{14.3}$$

The first two moments of the mixture $p(\mathbf{x})$ are estimated by

$$\mathbf{x}_\mu = \sum_s^S \pi^s \mathbf{x}_\mu^s, \quad \Sigma_\mathbf{x} = \sum_s^S \pi^s \left(\Sigma_{\mathbf{x}^s} + \left(\mathbf{x}_\mu^s - \mathbf{x}_\mu \right) \otimes \left(\mathbf{x}_\mu^s - \mathbf{x}_\mu \right) \right), \tag{14.4}$$

where $\pi^s = \tau^s / \sum_s^S \tau^s$ is the weighting of the component distribution with respect to the number of temporal samples per subject (see more details about finding modes of Gaussian mixtures in [7]). If the number of samples for each subject is equal, Eqs. (14.2)–(14.4) reduce to the case of a single Gaussian.

The variation of the shape changes is finally parametrized by $\mathbf{x} = \mathbf{x}_\mu + \sum_{i=1}^N \beta_i \phi_i$, where ϕ_i are N orthogonal basis vectors of $\Sigma_\mathbf{x}$. With the combination of shape and motion model, a shape to a particular time point can be synthesized by

$$\mathbf{v}_\beta^\alpha = \underbrace{\mathbf{v}_\mu + \sum_{i=1}^{M} \alpha_i \psi_i}_{\text{shape model}} + \underbrace{\mathbf{x}_\mu + \sum_{i=1}^{N} \beta_i \phi_i}_{\text{motion model}}, \tag{14.5}$$

where α_i and β_i are coefficients of the shape and the motion model respectively. Note that for certain applications, the first term (the shape model) can be replaced by a fixed shape $\mathbf{v}^s$ which then merely specifies the topology of the motion model. This simpler case will be the starting point for the applications later in this chapter, before applications with shape and motion models are presented.

14.3.3 Inference on Statistical Models

A major application of motion models is the estimation of the present shape change given some observations. Suppose we are given online ultrasound images depicting *parts* of the changing shape of interest. The goal in this setting would be to estimate the full changing shape given these observed parts based on the already seen shape changes, i.e. the motion model. This can be tailored to observations which are even independent of the topology of the model. Suppose that only the position of a magnetic tracker placed on the epigastrium is given. If the correlation between such a position signal and the motion model is known, the present shape change can be inferred.

14.3.3.1 Sparse Reconstruction

We have seen how to derive linear models of shape changes assuming the samples are Gaussian distributed.[1] In the following, we will use an important property of the Gaussian distribution to infer the shape change of a full shape given sparsely observed changing model points. Before that, we shortly recap Bayes' theorem for applying it to Gaussian random variables.

Bayes' Theorem

The problem of reconstructing shape change from partially observed data can be formulated in a probabilistic way. The goal is to estimate the expected shape change, having given only parts of it. Consider the two fundamental rules of probability,

$$\textbf{sum rule} \quad p(X) = \sum_Y p(X, Y), \tag{14.6}$$

$$\textbf{product rule} \quad p(X, Y) = p(Y|X)p(X), \tag{14.7}$$

---

[1] Since only the first two moments of the Gaussian mixture distribution are estimated the resulting distribution reduces to a Gaussian distribution.

where X and Y are two random variables, $p(X, Y)$ is the joint probability of X and Y and $p(Y|X)$ is the conditional probability of Y given X. Bayes' theorem relates the two conditional probabilities $P(X|Y)$ and $P(Y|X)$ applying the sum and product rule

$$p(X|Y) = \frac{p(Y|X)p(X)}{p(Y)}. \tag{14.8}$$

As we treat each displacement $\mathbf{x}_x \in \mathbf{x}$ at position $x \in \mathbf{v}$ as a Gaussian random variable, the joint distribution $p(\mathbf{x})$ is a multivariate Gaussian distribution. We now partition the variables $\mathbf{x}_x$ into two sets where $\mathbf{x}_a$ are all variables which are observed and $\mathbf{x}_b$ are the remaining variables. Since the marginal distributions $p(\mathbf{x}_a)$ as well as $p(\mathbf{x}_b)$ are Gaussian the conditional distribution

$$p(\mathbf{x}_b|\mathbf{x}_a) = \frac{p(\mathbf{x}_a|\mathbf{x}_b)p(\mathbf{x}_b)}{p(\mathbf{x}_a)} \tag{14.9}$$

is again a Gaussian and given in closed form. In Section 14.6, we will elaborate how Eq. (14.9) can be applied to the problem of reconstruction from partial information.

14.3.3.2 Statistical Model Regression

Let $\mathbf{v}$ be an instance of a model according to Eq. (14.5), where the scripts for model coefficients are dropped for simplicity. The goal in model-based reconstruction is to obtain an estimation of the complete shape vector $\mathbf{v}$ from a vector of partially observed information $\mathbf{r} \in \mathbb{R}^l$, $l < 3n$. Finding the maximum of the posterior distribution $p(\mathbf{v}|\mathbf{r})$, or in other words, the probability of $\mathbf{v}$ given the partial information $\mathbf{r}$, is a regression problem having the statistical model as prior information.

This concept of inferring the shape changes given surrogate data can be drawn even further to surrogates which are not bound to the topology of the motion model. Suppose we are given a position signal of a magnetic tracker which is placed on the epigastrium whose dynamics are naturally correlated to respiration. Let an attribute vector $a \in \mathbb{R}^d$ be such a signal at a particular time point. Consider an observed finite set $A = \{(a_0, \beta_0), ..., (a_n, \beta_n)\} \subset \mathbb{R}^d \times \mathbb{R}^N$ of n pairs of i.i.d. vectors a_i and motion model coefficient vectors β_i. Let further assume that there exists a function $f : \mathbb{R}^d \rightarrow \mathbb{R}^N$ which maps the attribute vectors to the coefficient vectors, while we only observe noisy instances of β such that $\beta \sim \mathcal{N}(f(a), \sigma_\epsilon \mathbf{I})$. The derivation of such a function f is a regression problem as well. However, no statistical model is available as prior and therefore some generic smoothness assumptions are usually taken.

Gaussian Process Regression

Gaussian process regression is a non-linear regression method where such a generic smoothness prior can be applied. Let $f \in \mathcal{GP}(0, k)$ be a Gaussian process with the covariance function $k : \mathbb{R}^d \times \mathbb{R}^d \to \mathbb{R}$. Assuming a Gaussian likelihood, the posterior distribution $p(f|A)$ is given in closed form [33] and is again a Gaussian process $\mathcal{GP}(\mu_A, k_A)$ with

$$\mu_A(a) = \mathbf{K}_{a,A}^T (\mathbf{K}_{A,A} + \sigma_\epsilon \mathbf{I})^{-1} \mathbf{B}, \tag{14.10}$$

$$k_A(a, a') = k(a, a') - \mathbf{K}_{a,A}^T (\mathbf{K}_{A,A} + \sigma_\epsilon \mathbf{I})^{-1} \mathbf{K}_{a',A}, \tag{14.11}$$

where $\mathbf{K}_{a,A} = (k(a_i, a))_{i=1}^n \in \mathbb{R}^n$, $\mathbf{K}_{A,A} = (k(a_i, a_j))_{i,j=1}^n \in \mathbb{R}^{n \times n}$ and $\mathbf{B} = (\beta_0, \ldots, \beta_n)^T \in \mathbb{R}^{n \times N}$. The expectation of an unseen output β^*, given an attribute a^*, yields Eq. (14.10).

14.4 ESTABLISHMENT OF CORRESPONDENCE

Before a population-based statistical model can be built from the 4D-MRI registration results, correspondence must be established between all subjects. In other words, a common topology must be defined for the data to be modeled. Let the vector $\mathbf{v}_s(t)$ describe the liver of subject s at time t, with points located on the surface as well as within the liver, obtained from image registration of the 4D-MRI data. This can be achieved for example using a B-spline based non-rigid registration algorithm [45, 46]. The dependency on s and t hints that a distinction must be made between two types of correspondence: *intra-subject* and *inter-subject* correspondence. Intra–subject correspondence is associated with all time steps t for a single subject, while inter-subject correspondence concerns the establishment of a common data format, or topology, among all subjects. The concept of correspondence is closely related to image and surface registration, and the latter is typically used to establish correspondence between images or shapes. Automatic registration techniques constitute a large area of research in medical image processing. In fact, automatic registration is of highest importance to the clinical success of statistical models, with potential time and cost savings both for building patient-specific models as well as applying population-based models to unseen subjects. However, fully automatic registration is also a very challenging problem, which is the reason why many approaches rely on manual or semi-automatic techniques.

14.4.1 Inter-Subject Correspondence

The goal for inter-subject correspondence in motion modeling is to establish mechanical correspondence [37], which means that the correspondence criterion should be based on how similar points move during respiration. As elaborated in the introduction, this motion is driven by the diaphragm and guided by the abdominal wall as a "tube" defining the direction where the organ can move. The process starts at selecting a reference image at full expiration for each subject, here called exhalation master. Manual establishment of correspondence amounts to selecting a number of landmarks that correspond according to the correspondence criterion, on the exhalation master of each

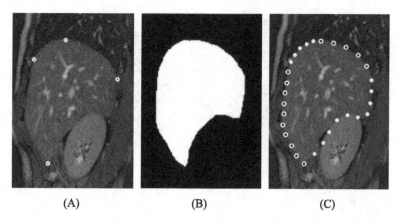

(A)   (B)   (C)

Figure 14.3 (A) Four manually selected landmarks for correspondence on a selected sagittal slice of the exhalation master: anterior and posterior points of the liver where the diaphragm touches the abdominal wall, the most superior point adjacent to the diaphragm and the most inferior point. (B) Manually defined mask. (C) Automatically generated resampling of the liver contour based on mask and landmarks. *(From [29].)*

subject. Through further processing of a relatively small number of manually selected points, a dense set of corresponding surface points can be obtained. A binary mask of the organ of interest can be incorporated into the process, as depicted in Fig. 14.3. For volumetric data, this procedure is performed on each slice. This yields a set of surface points $\check{\mathbf{v}}_s$ for each exhalation master.

In a next step, all surface shapes have to be aligned, for example using generalized procrustes analysis (GPA) [8], and the mean of the aligned points, $\boldsymbol{\mu}^\circ$, can be used to define a regular grid of interior points. In GPA, a rigid transformation T_s, R_s is found for each subject, to move it into a common coordinate system. For each subject s, GPA finds a translation and rotation matrix according to

$$\arg \min_{T_s, R_s} \sum_{s=1}^{n} \left\| T_s R_s \check{\mathbf{v}}_s^\circ - \boldsymbol{\mu}^\circ \right\|_2, \tag{14.12}$$

to move all samples into a common coordinate system. The "goodness-of-fit" criterion is usually the l_2-norm. The procedure is iterative and repeated until convergence. The algorithm outline is the following:

1. Arbitrarily choose a reference shape among all shapes.
2. Superimpose all shapes $\check{\mathbf{v}}_s^\circ$ to current reference shape.
3. Compute the mean shape $\boldsymbol{\mu}^\circ$ of the current set of superimposed shapes.
4. Compute the l_2-norm between the reference and all shapes, set reference to mean shape $\boldsymbol{\mu}^\circ$ and continue to step 2.

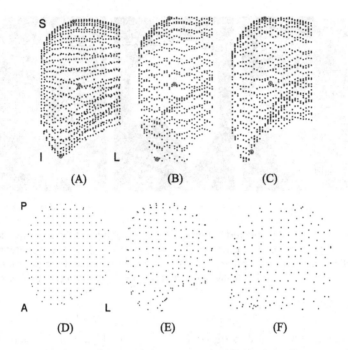

Figure 14.4 (A) Mean exhalation shape and three points highlighted. (B, C) Master exhalation shape of two selected subjects with the same corresponding points highlighted. (D–F) Slice through shape that shows internal grid points for the above examples.

This interior grid can finally be warped to each $\check{\mathbf{v}}_s^\circ$. This way, the mechanical correspondence is transferred from the surface to the inside of the liver, and an exhalation master shape $\check{\mathbf{v}}_s$ with points on the surface as well as in the interior is obtained for each subject. Fig. 14.4 depicts the final result of such an approach, on the mean shape as well as on two subjects.

14.4.2 Inter-Subject Correspondence by Registration

The present approach can be refined in a sense that the correspondence among the individual shapes is established automatically by non-rigid shape registration. As we have given for each volunteer a manually segmented label map of the liver structure (cf. Fig. 14.3B) a dense shape $\check{\mathbf{v}}_s \in \mathbb{R}^{3n}$ can be obtained using standard marching cubes. Correspondence is established by an iterative group-wise registration of the shapes in order to reduce a bias of the mean shape $\bar{\mathbf{v}}$ to a specific exhalation master shape $\check{\mathbf{v}}$:

$$\bar{\mathbf{v}}^1 = \check{\mathbf{v}}_{s=m}, \quad \bar{\mathbf{v}}^{i+1} = \frac{1}{S}\sum_s \bar{\mathbf{v}}^i + \Delta\mathbf{v}_s^i, \tag{14.13}$$

where m is randomly chosen and $\Delta \mathbf{v}_s^i$ is obtained by the Gaussian process registration method of [20] such that $\bar{\mathbf{v}}^i + \Delta \mathbf{v}_s^i \approx \check{\mathbf{v}}_s$. In the following, by $\check{\mathbf{v}}_s$ we refer to the registered shapes $\bar{\mathbf{v}} + \Delta \mathbf{v}_s$, if nothing else is mentioned.

The mean shape $\bar{\mathbf{v}}$ is equidistantly sampled in the inside to add several iterior points to the topology. With thin-plate-spline interpolation these points can be transferred to each shape $\mathbf{v}_s$. To recap, for each subject we have now an exhalation master shape $\check{\mathbf{v}}_s$, which is in correspondence with the population mean shape $\bar{\mathbf{v}}$. Having given the registration results of the 4DMR images, we will see that for each time point a shape $\mathbf{v}_s^t$ can be derived.

14.4.3 Intra-Subject Correspondence

Recall that correspondence is linked to both the subjects s and to time t. The latter means that *temporal* correspondence must be established along the motion sequence for each subject. However, given the inter-subject correspondence from the previous sections, this is now as simple as warping the points of the master shape $\check{\mathbf{v}}_s$ to all individual time steps using the deformation field obtained from the non-rigid registration. Using this approach, the location of every point in the master shape is known over time. In other words, dense motion information $\mathbf{v}_s(t)$ of the liver is obtained for each subject.

The final result of the correspondence step is an exhalation master vector

$$\check{\mathbf{v}}_s = (\check{x}_{s,1}, \check{y}_{s,1}, \check{z}_{s,1}, \ldots \check{x}_{s,m}, \check{y}_{s,m}, \check{z}_{s,m})^T \in \mathbb{R}^{3m} \tag{14.14}$$

for each subject, as well as all n_t respiratory states $\mathbf{v}_s(t) \in \mathbb{R}^{3m}, t \in [1, \ldots, n_t]$ per subject in dense correspondence.

14.5 STATISTICAL MOTION MODELING

With all data samples in dense correspondence, principal component analysis (PCA) is used to transform the 4D-MRI registration results into a representation where each dimension is independent and distributed according to a normal distribution. First, we discuss the most common case, where only motion is modeled [39,28–30,32,31]. This amounts to Eq. (14.5) where the shape model term is replaced with a fixed exhalation master $\check{\mathbf{v}}$,

$$\mathbf{v}_\beta = \check{\mathbf{v}} + \mathbf{x}_\mu + \sum_{i=1}^{N} \beta_i \phi_i. \tag{14.15}$$

Furthermore, often just a single Gaussian component is used for the motion model.

The data used for computing the motion model is the offset between each sample and the exhalation master,

$$\mathbf{x}_s(t) = \mathbf{v}_s(t) - \check{\mathbf{v}}_s. \tag{14.16}$$

The arithmetic mean is estimated using

$$\boldsymbol{\mu}_s = \frac{1}{n} \sum_{i=1}^{n} \mathbf{x}_s(t),$$ (14.17)

and the mean-free data of all subjects is gathered in the data matrix,

$$\mathbf{X} = [\mathbf{x}_1(1) - \boldsymbol{\mu}, \ldots, \mathbf{x}_1(n) - \boldsymbol{\mu}_1, \ldots, \mathbf{x}_{n_s}(1) - \boldsymbol{\mu}_1, \ldots, \mathbf{x}_{n_s}(n) - \boldsymbol{\mu}_{n_s}] \in \mathbb{R}^{3m \times (n_s \cdot n_t)}.$$ (14.18)

Standard Principal Component Analysis is performed on $\mathbf{X}$, as elaborated in much detail in earlier chapters, to obtain the matrix of eigenvectors $\Phi = [\phi_1^T, \ldots, \phi_N^T]$, the coefficient vector $\mathbf{b} = (\beta_1, \ldots, \beta_N)^T$ for each sample as well as the mean displacement $\mathbf{x}_\mu$ for the motion model in Eq. (14.15).

14.6 BAYESIAN RECONSTRUCTION FROM SPARSE DATA

The probabilistic framework of PCA provides an efficient tool to reconstruct missing information based on the statistics of the modeled data. In machine learning language, this is called a *regression* problem. Recall Bayes' rule,

$$p(Y) = \sum_X p(Y|X)p(X).$$ (14.19)

Let X be the observation vector and Y the set of model parameters Θ. The conditional probability

$$p(\mathbf{x}|\Theta)$$ (14.20)

is called the *likelihood* of $\mathbf{x}$ given the set of model parameters Θ. It describes how likely an observed data vector is for different settings of the parameter vector Θ. The value $p(\Theta)$ gives the probability of the model parameters *before* a sample has yet been drawn and is called *prior probability* of Θ. Accordingly, the quantity $p(\Theta|\mathbf{x})$ gives the probability of a set of model parameters *after* a sample $\mathbf{x}$ has been drawn and is therefore called *posterior probability* of Θ.

Here, the goal is to obtain an estimation of the complete data vector for the organ at time t, $\mathbf{x}_s(t)$, from a vector of partially observed information $\mathbf{r}(t) \in \mathbb{R}^l$, $l < 3m$. For the ease of notation, the subscripts are dropped and $\mathbf{x}$ and $\mathbf{r}$ are used instead. In [6], a Bayesian approach for model-based reconstruction from partial information is derived as follows. Let $\mathbf{L} : \mathbb{R}^{3m} \mapsto \mathbb{R}^l$, $l << 3n$ be a linear transformation matrix that governs the mapping of a complete data vector to its partial observation,

$$\mathbf{r} = \mathbf{L}\mathbf{x}.$$ (14.21)

In general, $\mathbf{L}$ is not an injective mapping and thus the solution of Eq. (14.21) with respect to $\mathbf{x}$ is not uniquely defined. Likewise, we can define the reduced version of the model matrix Φ, scaled with the eigenvalues,

$$\mathbf{Q} = (\mathbf{q}_1, \ldots, \mathbf{q}_n) = \mathrm{diag}(\sigma_i) \cdot \mathbf{L}\Phi \in \mathbb{R}^{l \times (n-1-k)}. \tag{14.22}$$

In terms of the unknown model parameter vector $\mathbf{b} = (\beta_1, \ldots, \beta_N)^T$, Bayes' rule becomes

$$p(\mathbf{b} \mid \mathbf{r}) = \frac{p(\mathbf{r} \mid \mathbf{b}) \cdot p(\mathbf{b})}{p(\mathbf{r})}. \tag{14.23}$$

If we assume that the measurement is subject to uncorrelated Gaussian noise of variance σ_N^2, the posterior probability becomes

$$p(\mathbf{b} \mid \mathbf{r}) \propto \exp\left(-\frac{1}{2\sigma_N^2}\|\mathbf{Q}\mathbf{b} - \mathbf{r}\|^2\right). \tag{14.24}$$

We are looking for the model parameters $\mathbf{b}$ that maximize this posterior, which amounts to minimizing the following energy function,

$$E(\mathbf{b}) = \|\mathbf{Q}\mathbf{b} - \mathbf{r}\|^2 + \eta \cdot \|\mathbf{b}\|^2, \tag{14.25}$$

where η is a regularization parameter that allows to trade-off between the data fitting error and the prior probability of $\mathbf{c}$.

After substituting $\mathbf{Q}$ with its singular value decomposition $\mathbf{Q} = \hat{\mathbf{U}}\mathbf{W}\hat{\mathbf{V}}^T$, we obtain

$$\hat{\mathbf{V}}\mathbf{W}^2\hat{\mathbf{V}}^T\mathbf{b} + \eta\mathbf{b} = \hat{\mathbf{V}}\mathbf{W}\hat{\mathbf{U}}^T\mathbf{r}, \tag{14.26}$$

and the following series of transformations leads to the final solution:

$$\mathbf{W}^2\hat{\mathbf{V}}^T\mathbf{b} + \eta\hat{\mathbf{V}}^T\mathbf{b} = \mathbf{W}\hat{\mathbf{U}}^T\mathbf{r}, \tag{14.27}$$

$$\mathrm{diag}(w_i^2 + \eta) \cdot \hat{\mathbf{V}}^T\mathbf{b} = \mathbf{W}\hat{\mathbf{U}}^T\mathbf{r}, \tag{14.28}$$

$$\hat{\mathbf{V}}^T\mathbf{b} = \mathrm{diag}\left(\frac{w_i}{w_i^2 + \eta}\right)\hat{\mathbf{U}}^T\mathbf{r}, \tag{14.29}$$

$$\mathbf{b} = \hat{\mathbf{V}}\mathrm{diag}\left(\frac{w_i}{w_i^2 + \eta}\right)\hat{\mathbf{U}}^T\mathbf{r}. \tag{14.30}$$

The final estimation of the complete data vector is computed according to

$$\mathbf{x} = \mathbf{U}\mathbf{b} + \mu. \tag{14.31}$$

The exhalation master shape is added to obtain the absolute position of the organ, $\mathbf{v} = \mathbf{x} + \check{\mathbf{v}}$.

The matrix $\mathbf{Q}$ is of size $l \times \hat{N}$, with $\hat{N} \ll N$ the number of principal components used for reconstruction. Therefore, Eq. (14.30) can usually be solved in real time.

14.7 APPLICATIONS OF POPULATION-BASED STATISTICAL MOTION MODELS TO MOTION RECONSTRUCTION

In the remainder of this chapter, a number of papers are discussed that leverage the described reconstruction method on 4D-MRI and ultrasound data.

In [28], the Bayesian approach to population-based statistical respiratory motion modeling is described. 4D-MRI liver images were acquired of 12 healthy subjects over roughly one hour on 22 to 30 sagittal slices, at a temporal resolution of 2.6–2.8 Hz. Deformation fields are extracted using non-rigid registration, and a topology for both inter- and intra-subject correspondence are defined using a set of manually selected landmarks on one exhalation master image per subject. The left liver lobe was kept out of the analysis because of the influence of heart motion. From the resulting 12 topologically equivalent 3D liver volumes and their dense spatio-temporal motion data, nine respiratory cycles from the beginning, the middle and the end of the acquisition session are used, each consisting of 8–20 time points. Using this data, 12 leave-one-out population-based models were built. Each of the models was used to evaluate the performance on the left-out subject. Fig. 14.5 shows the first two principal components of the model, and Fig. 14.6A shows a plot of the principal component's cumulative variance. To measure how well the model is able to describe the most extreme respiratory state (full inhalation), the vertices of a typical unseen full inhalation state are projected into each of the leave-one-out models, as a measure of model expressiveness (Fig. 14.6B).

Partially observed data, for example from ultrasound imaging or implanted electromagnetic beacons, is simulated by selecting a point at the inferior tip of Couinaud segment VI, at the diaphragm and near the center of the right liver lobe in each subject. Eq. (14.30) is used to estimate the model parameters of a 20 min sequence for each subject, and the corresponding shape is reconstructed using Eq. (14.31). An average error of 1.2 mm is reported.

Besides quasi-periodic superior/inferior (SI) motion, some data sets further contain secondary modes of motion due to cardiac cycle motion, digestive activity, gravity and muscle relaxation. Some of them cause a drift of the organ. It is shown that including organ drift in the model leads to a significant decrease of the reconstruction error in the presence of such motion, as depicted in Fig. 14.7.

In a clinically realistic scenario, where the goal is to destroy tumor cells under free breathing using motion-compensated radiotherapy, 3D fiducial information is not generally available. More commonly, 2D planar projections of fiducials from a beam-eye view X-ray, portal imager or fluoroscope are acquired in an image-guided therapy setting. Furthermore, there may be a considerable amount of measurement noise which

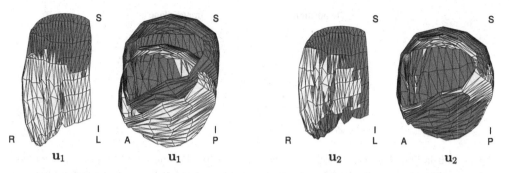

Figure 14.5 Coronal and sagittal views of the mean liver shape deformed in direction of the first principal component ϕ_1 (*left*) and the second principal component ϕ_2 (*right*) of our respiratory model. The white and gray surfaces represent deformations of plus and minus 3 standard deviations, respectively. (*From [28].*)

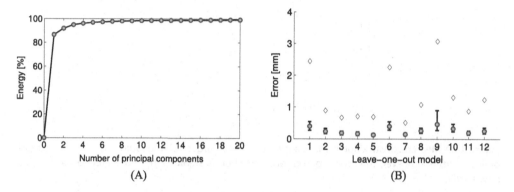

Figure 14.6 (A) Cumulative variance plot of the statistical population model for the liver in [28]. (B) Median, 25th and 75th percentiles, and maximum (diamond) of the projection error at inhalation. (*From [28].*)

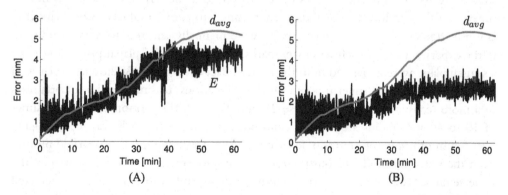

Figure 14.7 Mean reconstruction error for a 60 min sequence subject to organ drift of up to $d_{avg} = 5.6$ mm (averaged over the entire liver). (A) With drift model that only contains states from the beginning of the acquisition session. (B) Without drift information. (*From [28].*)

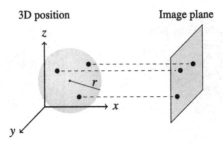

Figure 14.8 In [29], three close model vertices were selected. The motion of all grid points within a sphere of radius $r = 15$ mm around their centroid was reconstructed, based on the planar projection of the fiducials. *(From [29].)*

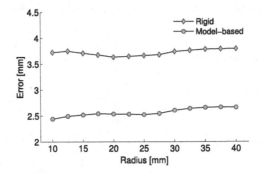

Figure 14.9 Mean reconstruction error as a function of the target volume radius. Both errors increase only very little, however, the model-based reconstruction is significantly more precise for any size. *(From [29].)*

directly impacts the reconstruction accuracy of the statistical model. In [29], these constraints are further studied (Fig. 14.8). Leave-one-out models were generated from a superset of the previously studied data, containing a total of 20 subjects. On each exhalation master shape, three close landmarks were manually selected, to serve as fiducials in the experiments. For each set of landmarks, two-dimensional planar projections onto a sagittal plane were generated from the motion data over 5 min. Additionally, Gaussian noise ($\sigma = 2$ mm) was superimposed on the projections, to simulate the uncertainties in a more realistic setting. From these 2D input points, the motion of target volumes of 10 to 40 mm radius around the centroid of the three fiducials was reconstructed over 5 min of respiratory motion, for each subject. To show that the non-rigid motion from the statistical model reconstruction is an improvement over a simple model, the same sequence was also reconstructed using a rigid body motion model, as depicted in Fig. 14.9. On average, the statistical model outperformed the rigid model by more than 1 mm. The next experiment investigated the influence of the camera projection angle with respect to the patient, as shown in Fig. 14.10A. It was found that projections

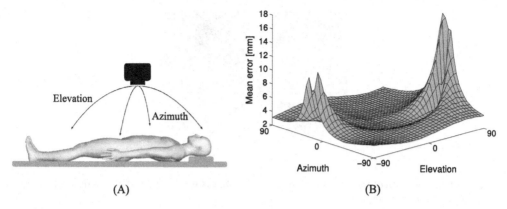

Figure 14.10 (A) In [29], a large number of camera positions was simulated to study its importance for motion reconstruction. (B) This surface shows the mean error over all subjects as a function of the camera position. It can be nicely seen that the reconstruction accuracy drops significantly when the elevation *el* approaches the values $+90°$ and $-90°$. *(From [29].)*

perpendicular to the main direction of respiratory motion yield better reconstructions. Conversely, if the elevation angle gets closer to $+90$ or -90 degrees, the observable motion almost disappears in the projected image, and leads to much higher reconstruction errors. Fig. 14.10B depicts the error as a function of the azimuth and elevation angle.

Another study investigated the difference between 3D, 2D or only 1D surrogate information available to the reconstruction algorithm [30]. While the ability to obtain motion information in 3D would be the preferred case, it is also the most unrealistic of all, since implanted electromagnetic beacons are not a clinically established technique. More commonly, T1-weighted MR images are acquired in an MR-guided high intensity focused ultrasound (MRgHIFU) setting for example, since such images are necessary for temperature monitoring, and are as such readily available for tracking purposes. In addition to 2D images, a 1D pencil beam signal of the diaphragm might additionally be acquired at a fast rate, and small cost in scan time, to increase the temporal resolution of the surrogates. In other scenarios where temperature monitoring might not be needed, such as conventional radiotherapy, 2D ultrasound imaging might be a more cost-efficient method for feature tracking. All these possible settings were simulated in a number of experiments where 3D, 2D or only 1D fiducial information is made available to the reconstruction algorithm, and the location of all model vertices was estimated based on this information [30]. A single fiducial at the diaphragm was manually selected on the 4D-MRI navigator image for each of the 20 subjects. Normalized cross-correlation was used to track this position in 20 min worth of navigator images. Fig. 14.11 depicts the navigator slice of one subject together with the tracked positions of the diaphragm. The results are depicted in Fig. 14.12 and show that even when only 1D surrogate information is available, the statistical model is able to recon-

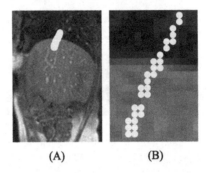

(A)   (B)

Figure 14.11 Selection of tracked organ position using template matching that was used as input for reconstruction (A). Close-up view (B). *(From [30].)*

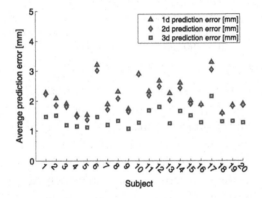

Figure 14.12 Average errors for reconstruction using a single point near the diaphragm in 1D, 2D and 3D. In the 3D case, the actual 3D position of the fiducial was used instead of the tracking results. *(From [30].)*

struct all model vertices with an average error of less than 3 mm for most subjects. The study also considers a gated scenario. In gating, a radiation or imaging device is only triggered when the respiratory state is close to a known reference position, typically full expiration, to minimize the influence of respiratory motion. Fig. 14.13 shows the reconstruction errors as a function of the gating window size. It can be seen that even for a smaller gating window size, the difference between gating only and gating combined with model-based reconstruction is significant.

All previously described approaches rely on synthetic fiducials ultimately generated from the acquired MR images, in one form or another. To address the question how well such a population-based statistical motion model performs on surrogates from 2D ultrasound images, a modified, MR-compatible phased array US transducer [26] is used in [32] to acquire images of the liver under free breathing inside the MR bore, while 4D-MRI is acquired simultaneously (Figs. 14.14, 14.15). The ultrasound images are used

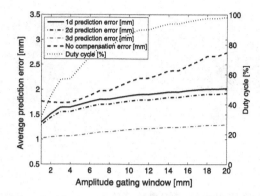

Figure 14.13 Comparison of reconstruction error with amplitude gating. Model-based gating outperforms traditional amplitude gating for any gating window size. *(From [30]).*

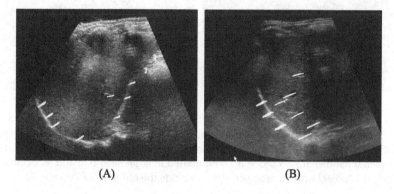

Figure 14.14 Ultrasound tracking of five cycles for two subjects. (A) Tracked points of subject 1 are distributed mainly along the liver surface, as the available acoustic window did not expose many vessel structures inside the organ. (B) Trajectories for subject 4 could be extracted at the diaphragm and also for vessels within the liver. The acoustic shadow occluding about half of the liver is a typical artefact due to absorption by the ribs. *(From [32].)*

for feature tracking, providing the surrogate signal for the reconstruction algorithm, whereas 4D-MRI provides the ground-truth motion for validation.

A mean error of 2.4 mm (standard deviation of 1.7 mm) is reported over all 8 subjects, validated on the 4D-MRI ground-truth data acquired simultaneously. In addition, two observers manually annotated three vessel locations in the MR images, and the prediction error was additionally validated using this manual ground-truth. The reason for this is that the 4D-MRI registration results might itself be biased due to the inaccuracy of image registration. An average error of 1.85 mm (standard deviation 1.2 mm) is reported using the manual ground-truth. In addition, it is shown how a statistical mo-

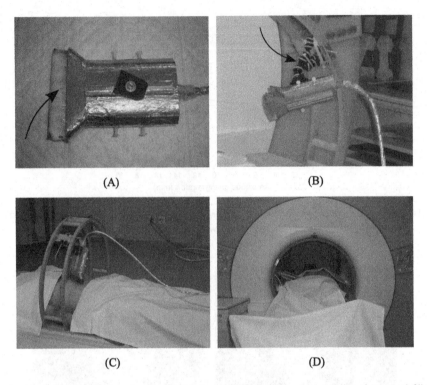

Figure 14.15 Overview of the experimental setup. (A) US probe accommodated in a gel-filled bag and its dedicated holder (EM shielded). (B) US probe attached to an MR-compatible orbital ring using an articulated handler. (C) US probe accommodated in a gel-filled bag and its dedicated holder (EM shielded). (D) Subject with US probe orbital ring inside the MR scanner. *(From [32].)*

tion model can be combined with a temporal predictor, in order to estimate the entire organ from the same small number of surrogates at time $t + t_\Delta$, where t_Δ considers the typical system lag of a radiosurgery device, which is reported to be at least in the order of 100 ms. A neural network is used to learn the mapping of the US surrogates from t to $t + t_\Delta$ from a short sequence of tracking data, and the forward-predicted surrogates are used to reconstruct the entire organ at $t + t_\Delta$. Errors between 2.3 mm ($t_\Delta = 50$ ms) and 2.7 mm ($t_\Delta = 400$ ms) are reported.

All previous approaches model respiratory motion independent of the temporal evolution of the signal. The advantage is that the reconstruction is independent of the specific breathing pattern, such as frequency and amplitude. On the other hand, the information about signal evolution can be incorporated into the reconstruction, as it is done in [31], where the presented Bayesian reconstruction concept is incorporated into a bilinear model that models and reconstructs entire respiratory cycles, rather than individual respiratory states.

14.8 RECONSTRUCTION BY REGRESSION

We now move on to topologically independent surrogates for reconstructing the full shape changes [17]. This has the advantage that one can reduce the involving ultrasound imaging and interest point tracking to a lighter setup. Given an attribute signal which is correlated with the organ motion, the goal is to estimate the motion model parameters β of the current respiratory state with Eq. (14.10). Based on that, a shape change can be synthesized using Eq. (14.5).

Following [17], first a high-temporal resolution respiratory cycle is synthesized in order to give a feeling about the capabilities of motion model regression. This is followed by the evaluation of the estimation performance of this method.

In both experiments, a straightforward Gaussian process model is used, applying a Gaussian kernel as covariance function

$$k_g(x, x') = \theta_0^2 \exp\left(-\frac{\|x - x'\|^2}{2\theta_1^2}\right),\tag{14.32}$$

where θ_0 is a scaling parameter and θ_1 is a length scale or smoothness parameter.

14.8.1 Average Breathing Cycle

We first show a study presented in [17] where respiratory motion of the liver in general was analyzed. A motion model out of the motion samples among $V = 9$ volunteers was built, while 99.9% of the variance was kept. For each temporal sample shape S_t at time point t, an attribute $c \in [0, 2\pi]$ was considered which indicates the cycle state of t within a respiratory cycle.[2] Such a rather abstract attribute can be applied to synthesize an average respiratory cycle of the liver shape. For the regression, 8000 pairs of cycle attributes resp. motion model coefficients were randomly picked among all volunteers.

In Fig. 14.16, an example right liver lobe and its displacements within this average respiratory cycle is visualized. Note that here a semantic and non-linearly captured respiratory cycle of a shape is synthesized, where *not* simply the most dominant principal component of the motion model is varied. Thus, for each patient an average respiratory cycle can be generated e.g. for planning [22]. While the source 4DMRI has a framerate of 2.8 Hz the temporal resolution can now be upsampled to an arbitrary high framerate. In this example, 100 samples were synthesized which corresponds to approximately 25 Hz.

---

[2] This cycle attribute was computed using a greedy cycle detection algorithm which is based on the average vertical coordinates of the shape changes $\mathbf{x}_t$.

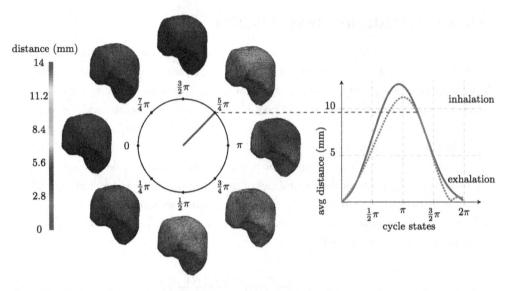

Figure 14.16 Example liver shape S^v deformed by an average respiratory cycle $c = [0, 2\pi]$. The shape is colored with respect to the absolute value of the displacement of a point $|\Delta p| \in [0, 14]$ mm. In the right plot, the average distance to the exhalation master is plotted (*blue* – population mean, *orange dotted* – example of a patient specific cycle). (For interpretation of the references to color in this figure legend, the reader is referred to the web version of this chapter.)

14.8.2 Motion Model Prediction

In the motion estimation experiment of [17] a surrogate signal was simulated which indicates the depth of the diaphragm measured from the abdominal skin for example by a 1D US sensor. A 1D signal was defined which is generated by a ground truth model point in the region of the diaphragm. Let $s : [0, \tau] \to \mathbb{R}^3$ be the 3D signal of absolute coordinates of this point at time point $t \in [0, \tau]$. To get invariant to the absolute positioning of the subject, let us project the signal into its dominant mode of variation

$$\mathcal{F}[s] = (s - \mu_s)\psi_0 + \epsilon, \tag{14.33}$$

where $\mu_s = \frac{1}{\tau} \int_0^\tau s(t) dt$ is the signal mean value, $\epsilon \sim \mathcal{N}(0, \sigma_\epsilon)$ is additive noise, which was set to $\sigma_\epsilon = 2$ mm and ψ_0 is the orthonormal eigenfunction corresponding to the largest eigenvalue λ_0 of the equation $\int_0^\tau \text{cov}(s_i, s_j)\psi_0(s_i)ds_i = \lambda_0\psi_0(s_j)$. Here, cov is the covariance function of the signal s.

In this evaluation, the motion estimation performance of the method can be shown given the simulated signal $\mathcal{F}[s]$. For each volunteer, a leave-one-out (L1O) motion model was generated, where only motion samples from the other volunteers were considered. 99.9% of the variance was kept yielding L1O models of 20 to 22 principal modes. Note that for each temporal sample, additionally $\mathcal{F}[s]$ was computed for the

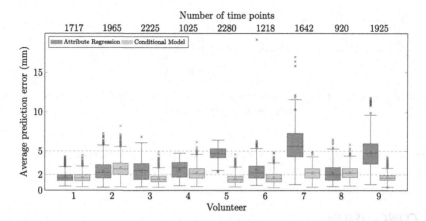

Figure 14.17 For each L1O experiment, the average corresponding point difference between the ground truth and the predicted shape is visualized. We compare our method *Attribute Regression* with the *Conditional Model* [2,28]. The upper x-axis indicates how many time points the motion has been predicted.

later usage as an attribute (Eq. (14.33)). As a shape model, the model with 2571 surface and 368 interior points was used which had been constructed as described in Section 14.4.2. To derive the corresponding model parameters, the shapes of the volunteers are projected into the shape model.

For the Gaussian process regression, 8000 $\mathcal{F}[s]$-attributes resp. ground truth coefficient vector pairs were randomly picked, again only from the other volunteers. The kernel parameters were set to $\theta_1 = 3$, while the exact value of $\theta_0 = 5000$, had only minor influence to the estimation performance. In Fig. 14.17, for each volunteer the average estimation error is plotted. The estimation error is robustly kept below 5 mm, whereas the median stays around 2 to 3 mm. For radiotherapy these are reasonable error bounds.

Let us compare the topology–independent method to the sparse reconstruction method presented above where the simulated 3D point signal s serves as surrogate data. The estimation is performed by estimating the mean of a statistical motion model which is conditioned on s, cf. Eq. (14.9) and [2]. For a fair comparison, we added to s an isotropic Gaussian noise $\mathcal{N}(0, \sigma_\epsilon/\sqrt{3})$. The conditional model performs equally well, while it better generalizes in the experiment with volunteer 7 and 9. Certainly, the topology–independent model with only 9 volunteers is not capable to generalize to the respiratory motion of these two subjects. This can be confirmed when comparing to the results with patient specific models (Fig. 14.18). Here, for each volunteer, a motion model was built where only samples from the volunteer of interest were considered. For the regression, 700 attribute/coefficient pairs were randomly picked and the kernel parameter was adjusted to $\theta_1 = 5$. For all volunteers including for volunteer 7 and 9, the average error was considerably improved to less than 1 mm. In Fig. 14.16 on the right,

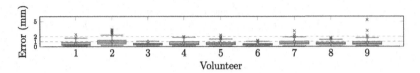

Figure 14.18 Evaluation of the average estimation error for each patient specific experiments, where $\theta_1 = 5,700$ training samples have been picked and 210 tests have been performed.

a patient specific average respiratory cycle is plotted for a comparison to the population mean cycle.

14.9 CONCLUSION

In this chapter, an overview of population-based statistical motion models for respiratory motion was laid out. The 4D-MRI imaging technique was introduced first, along with registration techniques for extracting the motion information from subjects over several minutes. This data is the foundation for the discussed motion models that nicely represent large amounts of respiratory motion data in a compact mathematical framework. The statistical nature of these models allows to reconstruct the entire organ from only a small number of surrogate signals, be it from 1D signals such as a respiratory belt or a pencil beam, from 2D signals such as single slice imaging or projections, or from 3D signals such as implanted electromagnetic beacons. Both manual and automatic techniques for establishing correspondence among a number of subjects were discussed, as well as the Bayesian framework for sparse reconstruction, and a number of applications from the literature were presented. Finally, a generalization based on Gaussian processes was presented that eliminates the need to obtain a direct mapping between surrogates and their corresponding model vertices, allowing for topology-independent reconstruction.

ACKNOWLEDGMENTS

This work was funded by the Swiss National Science Foundation projects P300P2-164647 and CR-SII2-127549.

REFERENCES

[1] AAPM-91, Report of Task Group 76: The Management of Respiratory Motion in Radiation Oncology, Technical Report 91, American Association of Physicists in Medicine, 2006.
[2] Thomas Albrecht, Marcel Lüthi, Thomas Gerig, Thomas Vetter, Posterior shape models, Med. Image Anal. 17 (8) (2013) 959–973.
[3] Brian Amberg, Pascal Paysan, Thomas Vetter, Weight, sex, and facial expressions: on the manipulation of attributes in generative 3D face models, in: Advances in Visual Computing, Springer, 2009, pp. 875–885.

[4] James M. Balter, Kristy K. Brock, W. Dale Litzenberg, Daniel L. McShan, Theodore S. Lawrence, Randall Ten Haken, Cornelius J. McGinn, Kwok L. Lam, Laura A. Dawson, Daily targeting of intrahepatic tumors for radiotherapy, Int. J. Radiat. Oncol. Biol. Phys. 52 (1) (2002) 266–271.

[5] Volker Blanz, Thomas Vetter, A morphable model for the synthesis of 3D faces, in: Proceedings of the 26th Annual Conference on Computer Graphics and Interactive Techniques, ACM Press/Addison-Wesley Publishing Co., 1999, pp. 187–194.

[6] Volker Blanz, Thomas Vetter, Reconstructing the complete 3D shape of faces from partial information, Informationstech. Tech. Inform. 44 (6) (2002) 295–302.

[7] Miguel Á. Carreira-Perpiñán, Mode-finding for mixtures of Gaussian distributions, IEEE Trans. Pattern Anal. Mach. Intell. 22 (11) (2000) 1318–1323.

[8] Ian L. Dryden, Kanti V. Mardia, Statistical Shape Analysis, vol. 4, J. Wiley, Chichester, 1998.

[9] Tami Freeman, Radiotherapy: challenges old and new, http://medicalphysicsweb.org/cws/article/opinion/35930, 2008.

[10] Rohini George, Theodore D. Chung, Sastry S. Vedam, Viswanathan Ramakrishnan, Radhe Mohan, Elisabeth Weiss, Paul J. Keall, Audio-visual biofeedback for respiratory-gated radiotherapy: impact of audio instruction and audio-visual biofeedback on respiratory-gated radiotherapy, Int. J. Radiat. Oncol. Biol. Phys. 65 (3) (2006) 924–933.

[11] Joseph Hanley, Marc M. Debois, Dennis Mah, Gikas S. Mageras, Adam Raben, Kenneth Rosenzweig, Borys Mychalczak, Lawrence H. Schwartz, Paul J. Gloeggler, Wendell Lutz, C. Clifton Ling, Steven A. Leibel, Zvi Fuks, Gerald J. Kutcher, Deep inspiration breath-hold technique for lung tumors: the potential value of target immobilization and reduced lung density in dose escalation, Int. J. Radiat. Oncol. Biol. Phys. 45 (3) (1999) 603–611.

[12] Steve B. Jiang, Technical aspects of image-guided respiration-gated radiation therapy, Med. Dosim. 31 (2) (2006) 141–151.

[13] P.J. Keall, V.R. Kini, S.S. Vedam, R. Mohan, Motion adaptive X-ray therapy: a feasibility study, Phys. Med. Biol. 46 (1) (2001) 1–10.

[14] P.J. Keall, G. Starkschall, H. Shukla, K.M. Forster, V. Ortiz, C.W. Stevens, S.S. Vedam, R. George, T. Guerrero, R. Mohan, Acquiring 4D thoracic CT scans using a multislice helical method, Phys. Med. Biol. 49 (10) (2004) 2053–2067.

[15] Paul J. Keall, Gig S. Mageras, James M. Balter, Richard S. Emery, Kenneth M. Forster, Steve B. Jiang, Jeffrey M. Kapatoes, Daniel A. Low, Martin J. Murphy, Brad Murray, Chester R. Ramsey, Marcel B. van Herk, S. Sastry Vedam, John W. Wong, Ellen Yorke, The management of respiratory motion in radiation oncology report of AAPM Task Group 76, Med. Phys. 33 (10) (2006) 3874–3900.

[16] Rojano Koshani, James M. Balter, James A. Hayman, George T. Henning, Marcel van Herk, Short-term and long-term reproducibility of lung tumor position using active breathing control (abc), Int. J. Radiat. Oncol. Biol. Phys. 65 (5) (2006) 1553–1559.

[17] Christoph Jud, Frank Preiswerk, Philippe C. Cattin, Respiratory Motion Compensation with Topology Independent Surrogates, in: ICART: Imaging and Computer Assistance in Radiation Therapy: a workshop held as part of MICCAI 2015 in Munich, Germany, HAL, 2015, pp. 97–103.

[18] Hideo D. Kubo, Bruce C. Hill, Respiration gated radiotherapy treatment: a technical study, Phys. Med. Biol. 41 (1) (1996) 83–91.

[19] A. Lomax, Intensity modulated methods for proton therapy, Phys. Med. Biol. 44 (1999) 185–205.

[20] Marcel Lüthi, Christoph Jud, Thomas Vetter, A unified approach to shape model fitting and non-rigid registration, in: Machine Learning in Medical Imaging, Springer, 2013, pp. 66–73.

[21] James Mechalakos, Ellen Yorke, Gikas S. Mageras, Agung Hertanto, Andrew Jackson, Ceferino Obcemea, Kenneth Rosenzweig, C. Clifton Ling, Dosimetric effect of respiratory motion in external beam radiotherapy of the lung, Radiother. Oncol. 71 (2) (2004) 191–200.

[22] Nadia Möri, Christoph Jud, Rares Salomir, Philippe Cattin, Leveraging respiratory organ motion for non-invasive tumor treatment devices: a feasibility study, Phys. Med. Biol. 61 (11) (2016) 4247–4267.

[23] Christopher Nelson, George Starkschall, Peter Balter, Mathew J. Fitzpatrick, John A. Antolak, Naresh Tolani, Karl Prado, Respiration-correlated treatment delivery using feedback-guided breath hold: a technical study, Med. Phys. 32 (1) (2005) 175–181.

[24] K. Ohara, T. Okumura, M. Akisada, T. Inada, T. Mori, H. Yokota, M.J.B. Calaguas, Irradiation synchronized with respiration gate, Int. J. Radiat. Oncol. Biol. Phys. 17 (4) (1989) 853–857.

[25] Anders N. Pedersen, Stine Korreman, Hakan Nystrom, Lena Specht, Breathing adapted radiotherapy of breast cancer: reduction of cardiac and pulmonary doses using voluntary inspiration breath-hold, Radiother. Oncol. 72 (1) (2004) 53–60.

[26] Lorena Petrusca, Philippe Cattin, Valeria De Luca, Frank Preiswerk, Zarko Celicanin, Vincent Auboiroux, Magalie Viallon, Patrik Arnold, Francesco Santini, Sylvain Terraz, et al., Hybrid ultra-sound/magnetic resonance simultaneous acquisition and image fusion for motion monitoring in the upper abdomen, Invest. Radiol. 48 (5) (2013) 333–340.

[27] M.H. Phillips, E. Pedroni, H. Blattmann, T. Boehringer, A. Coray, S. Scheib, Effects of respiratory motion on dose uniformity with a charged particle scanning system, Phys. Med. Biol. 37 (1) (1992) 223–233.

[28] Frank Preiswerk, Patrik Arnold, Beat Fasel, Philippe Cattin, A Bayesian framework for estimating respiratory liver motion from sparse measurements, in: Abdominal Imaging. Computational and Clinical Applications, Springer, 2012, pp. 207–214.

[29] Frank Preiswerk, Patrik Arnold, Beat Fasel, Philippe Cattin, Robust tumour tracking from 2D imaging using a population-based statistical motion model, in: 2012 IEEE Workshop on Mathematical Methods in Biomedical Image Analysis (MMBIA), 2012, pp. 209–214.

[30] Frank Preiswerk, Patrik Arnold, Beat Fasel, Philippe Cattin, Towards more precise, minimally-invasive tumour treatment under free breathing, in: Engineering in Medicine and Biology Society (EMBC), 2012 Annual International Conference of the IEEE, IEEE, 2012, pp. 3748–3751.

[31] Frank Preiswerk, Philippe Cattin, A bilinear model for temporally coherent respiratory motion, in: Abdominal Imaging. Computational and Clinical Applications, Springer, 2014, pp. 221–228.

[32] Frank Preiswerk, Valeria De Luca, Patrik Arnold, Zarko Celicanin, Lorena Petrusca, Christine Tanner, Oliver Bieri, Rares Salomir, Philippe Cattin, Model-guided respiratory organ motion prediction of the liver from 2D ultrasound, Med. Image Anal. 18 (5) (2014) 740–751.

[33] Carl Edward Rasmussen, Gaussian Processes for Machine Learning, 2006.

[34] Gregory C. Sharp, Steve B. Jiang, Shinichi Shimizu, Hiroki Shirato, Prediction of respiratory tumour motion for real-time image-guided radiotherapy, Phys. Med. Biol. 49 (3) (2004) 425–440.

[35] Hiroki Shirato, Shinichi Shimizu, Tatsuya Kunieda, Kei Kitamura, Marcel van Herk, Kenji Kagei, Takeshi Nishioka, Seiko Hashimoto, Katsuhisa Fujita, Hidefumi Aoyama, Kazuhiko Tsuchiya, Kohsuke Kudo, Kazuo Miyasaka, Physical aspects of a real-time tumor-tracking system for gated radiotherapy, Int. J. Radiat. Oncol. Biol. Phys. 48 (4) (2000) 1187–1195.

[36] Bernhard W. Stewart, Christopher P. Wild (Eds.), World Cancer Report 2014, International Agency for Research on Cancer, France, 2014.

[37] Martin von Siebenthal, Analysis and Modelling of Respiratory Liver Motion Using 4DMRI, PhD thesis, ETH, Zurich, 2008, Diss. ETH No. 17613.

[38] Martin von Siebenthal, Gabor Szekely, Urs Gamper, Peter Boesiger, Antony Lomax, Philippe Cattin, 4D MR imaging of respiratory organ motion and its variability, Phys. Med. Biol. 52 (2007) 1547–1564.

[39] Martin von Siebenthal, Gabor Szekely, Antony Lomax, Philippe Cattin, Inter-subject modelling of liver deformation during radiation therapy, in: MICCAI, in: Lect. Notes Comput. Sci., vol. 4791, 2007, pp. 659–666.

[40] Martin von Siebenthal, Gabor Szekely, Antony J. Lomax, Philippe C. Cattin, Systematic errors in respiratory gating due to intrafraction deformations of the liver, Med. Phys. 34 (9) (2007) 3620–3629.

[41] C.X. Yu, D.A. Jaffray, J.W. Wong, The effects of intra-fraction organ motion on the delivery of dynamic intensity modulation, Phys. Med. Biol. 43 (1) (1998) 91–104.

[42] B. Madore, G. Glover, N.J. Pelc, and others, Unaliasing by Fourier-encoding the overlaps using the temporal dimension (UNFOLD), applied to cardiac imaging and fMRI, Magnetic Resonance in Medicine 42 (5) (1999) 813–828.

[43] K.P. Pruessmann, M. Weiger, M.B. Scheidegger, P. Boesiger, and others, SENSE: sensitivity encoding for fast MRI, Magnetic Resonance in Medicine 42 (5) (1999) 952–962.

[44] F. Preiswerk, M. Toews, C. Cheng, and others, Hybrid MRI-Ultrasound acquisitions, and scannerless real-time imaging, Magnetic Resonance in Medicine 42 () (2016) 952–962.

[45] D. Rueckert, L.I. Sonoda, C. Hayes, and others, Nonrigid registration using free-form deformations: application to breast MR images, IEEE transactions on medical imaging 18 (8) (1999) 712–721.

[46] M. Modat, G.R. Ridgway, Z.A. Taylor, and others, Fast free-form deformation using graphics processing units, Comput. Methods Programs Biomed. 98 (3) (2010) 278–284.

CHAPTER 15

Statistical Shape and Appearance Models for Bone Quality Assessment

Patrik Raudaschl, Karl Fritscher

Institute for Biomedical Image Analysis (IBIA), University for Health Sciences, Medical Informatics and Technology (UMIT), Hall in Tirol, Austria

Contents

15.1 Introduction		410
15.1.1 Motivation		410
15.1.2 Methods for Bone Quality Assessment		410
15.2 Fundamentals of Statistical Shape and Appearance Models		414
15.2.1 Obtaining Shapes from Medical Images		414
15.2.2 Representing the Shape of Anatomical Structures		415
15.2.2.1 Point Sets and Meshes		415
15.2.2.2 Signed Distance Maps		415
15.2.2.3 Deformation Fields		415
15.2.3 Representing the Appearance of Anatomical Structures		415
15.2.3.1 Active Appearance Model (AAM)		416
15.2.3.2 Direct Appearance Model (DAM)		416
15.2.3.3 InShape Models		416
15.2.4 Establishing the Correspondences Between Shapes		416
15.2.4.1 Pre-Alignment Based on Linear Transforms		416
15.2.4.2 Alignment Using Nonlinear Transformations		417
15.2.5 Statistical Analysis of Shape and Appearance Variability		418
15.2.5.1 Principal Component Analysis		418
15.2.5.2 Kernel Principal Component Analysis (KPCA)		418
15.2.5.3 Independent Component Analysis (ICA)		418
15.2.5.4 Local Nonlinear Dimensionality Reduction		419
15.3 Approaches for Bone Quality Assessment		419
15.3.1 Approaches Based on Statistical Shape and Appearance Models		419
15.3.1.1 Segmentation and 3D Reconstruction		419
15.3.1.2 Fracture Risk Assessment		421
15.3.1.3 Bone Fracture Detection		425
15.3.1.4 Treatment Planning		427
15.3.2 Bone Quality Assessment Based on FE Analysis		428
15.4 Discussion and Conclusion		432
15.4.1 Creation of Statistical Shape and Appearance Models – Potentials and Pitfalls		432
15.4.2 Applications of Bone Quality Assessment		433
15.4.3 Conclusion and Outlook		436
References		436

Statistical Shape and Deformation Analysis
DOI: 10.1016/B978-0-12-810493-4.00018-3

15.1 INTRODUCTION

15.1.1 Motivation

Bone quality is influenced by all geometrical as well as material factors of bone contributing to fracture resistance [1]. Especially material factors are negatively influenced by a decrease of Bone Mineral Density (BMD) leading to osteopenia and finally osteoporosis. Consequently, the risk for fractures of the hip, vertebra and forearm is significantly increased. It has been estimated that approximately one in three women and one in twelve men over the age of 50 years are affected by osteoporosis [2] leading to 9 million fractures annually worldwide. Of these, 1.6 million fractures are located at the hip, 1.7 million at the forearm and 1.4 million are clinical vertebral fractures [3]. In 1990 it was expected that by 2050, the worldwide incidence of hip fractures in men will increase by 310% and in women by 240% [4]. Fortunately, currently it seems that this prediction will not be realized. However, the burden that sustains for the healthcare system due to osteoporotic fractures and its related consequences is severe and increasing rapidly due to rising life expectancy. In the European Osteoporosis Report 2010 the costs of osteoporosis, including pharmacological interventions have been estimated at €37 billion and the total health burden was estimated at 1,180,000 lost QALYs (Quality Adjusted Life Years) [5]. These numbers are motivating researchers and scientists across the globe to develop new methods for bone quality assessment, fracture risk prediction and detection.

15.1.2 Methods for Bone Quality Assessment

Bone strength can be described as the maximum load that a bone can carry before mechanical failure. Basically, structural integrity of bones depend on three different factors: the total bone mass, the bone geometry and the tissue material [1,6]. Bone mass, determined by BMD, is the most frequently used biomechanical parameter for the diagnosis of osteoporosis and the therewith-related fracture risk estimation in clinical routine. The World Health Organization (WHO) defined that a patient with BMD values more than 2.5 standard deviations (SD) below the mean is considered to have osteoporosis and patients with BMD values between 1.0 and 2.5 SD below the mean to have osteopenia correlating with an increased risk for developing osteoporosis.

The most common modalities for measuring BMD are Dual X-ray Absorptiometry (DXA) [7] and Quantitative Computed Tomography (QCT) [8]. The basis of DXA are two X-ray beams with different levels of energy that are aimed at the patient's bone of interest. The principle of QCT is to determine physical density by mass/volume for each voxel. In contrast to DXA which measures an area projected mass (kg/m^2), QCT measures the physical density of each volume element (kg/m^3) [9] in vertebrae or peripheral parts of the body like leg or forearm (= peripheral

QCT). By using QCT approach mechanical bone properties like bone and bending strength can be modeled more accurately [10,11] compared to DXA. However, radiation exposure of QCT is considerably higher compared to DXA [12]. Quantitative ultrasound [13] and Magnetic Resonance Imaging [14] are also approaches that are used for bone quality analysis, although they are not very common in clinical practice. In general, the type of imaging modality used for bone quality assessment preliminary depends on its availability and the patient's anamnesis. However, measuring area or volumetric BMD is still by far the most prevalent method for osteoporosis detection. Unfortunately, using BMD as a single parameter for bone quality assessment and fracture detection/prediction has some major limitations. A similar BMD in young and old people does not mean that they have the same fracture risk and a higher BMD is not always concurring with increased bone strength [15]. Furthermore, BMD of a fractured femur or within a specific region of the cancellous bone cannot be measured. Moreover, in [16] Stone et al. revealed that the definition of osteoporosis based on the T-score ≤ -2 is arbitrary and contentious because less than 10–44% of most type of fractures can be assigned to low bone mass. Siris et al. also found out that the significance of the defined T-score levels of the WHO are not adequate to quantify osteoporosis [17]. Concluding it can be said that bone density is only partly responsible for bone strength due to the poor specificity of BMD in predicting actual fractures [18]. Hence, different approaches, which try to overcome these limitations, have been introduced.

The program Hip Structure Analysis (HSA) tries to predict femoral neck strength from bone mineral data [19]. Femoral neck geometry is derived from raw bone mineral image data in order to estimate hip strength. HSA structural parameters are highly correlated with areal BMD, however, with respect on the prediction of fracture risk, HSA has not shown much improvement compared to areal BMD [20]. The WHO developed the Fracture Risk Assessment Tool (FRAX), which is in clinical use for assessing an individuals' risk of fracture since 2008 [21]. FRAX is based on a mathematical model that includes density measures as well as clinical risk factors (e.g. age, weight, alcohol consummation, etc.). However, also FRAX suffers from different limitations. For example, racial and ethnic differences are not sufficiently considered. Furthermore, hip density is used as modeling input. This, however, may not predict fractures at other locations accurately [15]. Since FRAX is based on performing statistics on clinical data, in populations with insufficient clinical fracture data the confidence for risk estimation is limited. Another alternative approach for fracture risk assessment is QFracture, which is using a similar approach like FRAX [22]. However, instead of using BMD as a parameter, QFracture includes some additional clinical parameters like endocrine disorders or Typ-2 diabetes but suffers similar limitations as FRAX. Hence, there is a further need for methods that improve bone quality analysis and associated questions since the previously mentioned techniques do not deeply consider structural properties

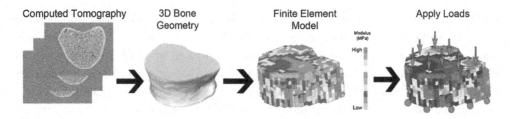

Figure 15.1 Basic steps for creating patient-specific finite element models. *(Figure taken from [23].)*

and anatomical shape, which are also important factors considering structural integrity of bones.

The Finite Element (FE) analysis is an approach for analyzing bone quality and related questions based on using morphological as well as material properties of bones. Basically, FE analysis subdivides a larger problem into smaller and simpler components (= finite elements, FEs). For modeling the whole problem (e.g. bone models), FEs are assembled into a larger system of differential equations. Thus the relationship between stress and strain within each element can be described and a stress distribution within the structure can be calculated when a load is applied. The stress analysis using FE models is the key component for bone quality modeling, fracture risk assessment or fracture fixation modeling. Hence, FE models can be used to analyze how loads are distributed within a structure and how this structure responds to an applied load. In order to create patient-specific FEMs several steps are necessary (Fig. 15.1) [23]. Firstly, the structure of interest is segmented in clinical images (i.e. CT or QCT). The density at each voxel is determined based on the relationship of voxel intensities and mineral density, which is determined using a calibrated phantom. Empirical relationships between bone tissue density and bone tissue Young's modulus are taken from biomechanical studies [24,25]. The density distribution of the resulting 3D model is represented as a distribution of bone stiffness.

Beside FE analysis, the usage of methods coming from the field of Computational Anatomy (CA) for bone quality assessment has become a very active field of research during the last 10 years. The basic concept of CA is to establish correspondences between shapes by using a template that is deformed by using a group of suitable transformations. By statistically analyzing these correspondences and the underlying transformations, Statistical Models (SMs) which assess shape and appearance variations of anatomical structures – also referred to as Statistical Shape and Appearance Models (SSM/SAMs) – can be created. Hence, relevant relationships between shape and appearance variations on the one hand and bone quality on the other hand can be identified. In Fig. 15.2 the different stages for creating a SSM/SAM of a particular example is shown.

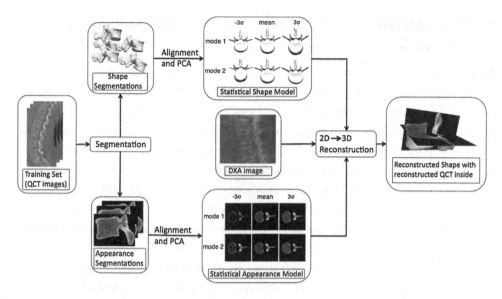

Figure 15.2 A particular pipeline for shape and appearance model creation. Additionally, the typical application of 2D–3D reconstruction is schematically depicted. *(Figure taken from [26].)*

An overview on statistical modeling approaches that are used in the context of osteoporosis detection is given by Castro-Mateos et al. [26].

Generally, SMs are widely used in medical imaging, especially if bones are the object of interest [27]. In case of osteoporosis, the main goals of SMs beside bone quality assessment for treatment planning purposes [28,29] are fracture detection [30–32] and fracture risk prediction [33–38]. Detailed knowledge about local bone quality and the therewith-related fracture risk is important for initiating preventative measures. For most model-based bone quality assessment approaches, accurate image segmentation as well as 3D shape reconstruction based on 2D images using 2D/3D registration methods are of great relevance and SSM/SAMs have also been successfully applied for these purposes (e.g. [30,39–41]).

In the remainder of this chapter, the basic methodology behind the creation of SSM/SAMs will be outlined in Section 15.2. In Section 15.3 different approaches based on SSM/SAMs are described that are used for bone quality assessment, fracture detection and fracture prediction. Furthermore, in this section FEM-based approaches for the analysis of bone quality will be presented. In Section 15.4 different approaches are discussed and the pros and cons will be outlined. Finally, in this section a conclusion and an outlook will be given.

15.2 FUNDAMENTALS OF STATISTICAL SHAPE AND APPEARANCE MODELS

SSM/SAMs are widely used for describing the variability of objects. SSMs were introduced by Cootes et al. in 1995 [42] and SAMs in 2001 also by Cootes et al. [43]. A Point Distribution Model (PDM) has been used to represent an object as a set of labeled points by their mean positions and a small set of modes of variation, which describe the variability of the object's shapes. By learning a linear model, new shapes can be described as a sum of the mean and a weighted combination of the modes of variation. By using this approach, new models can be computed rapidly. For this reason several works accrued based on the work of Cootes et al., including many of bones [32,44–48]. The three major questions that arise when working with SSM/SAMs are the following:

1. How can a specific shape description be obtained from medical images?
2. How can a shape representation be obtained that is suitable for a SSM?
3. How can an appearance representation be obtained that is suitable for a SAM?

In the following, these questions will be used to introduce the basic concept of SSM/SAMs.

15.2.1 Obtaining Shapes from Medical Images

Basically, shapes can be distinguished from its surroundings by its boundary, independent of its location in space. In (medical) imaging the task of determining the boundary of one or more objects is referred to as "segmentation". Automated, accurate and reliable segmentation of an object in an appropriate timeframe is still one of the big unsolved problems in (medical) imaging and a hot topic in the research community. In what follows, the most common segmentation approaches are listed. More details on image segmentation can be found in e.g. [49–51].

1. Segmentation-based on gray level features.
2. Segmentation-based on texture features.
3. Deformable models and level set methods.
4. Atlas-based segmentation methods.
5. Learning-based methods using artificial intelligence and prior knowledge.

Segmentation of medical images is a precondition for the creation of SSMs but it is also one of their major applications (i.e. model-based segmentation) [52,53]. Segmentation typically provides a binarized version of the input image, in which voxels that are part of a particular structure of interest are labeled as "inside" ("1"). All remaining voxels are labeled as "background" ("0").

15.2.2 Representing the Shape of Anatomical Structures

The binary labels resulting from segmentation define a specific shape. However, in most cases these binarized labels are further processed into alternative shape representations, which are more appropriate to establish correspondences between different subjects. In the following, the most common shape descriptors will be presented that are used for creating SSM/SAMs.

15.2.2.1 Point Sets and Meshes

A common way of representing shapes for SSM is to represent the contour of an object by a set of points that are distributed over the object's surface. Cootes et al. [54] have introduced point sets for the creation of SSMs. Single points of a point set often correspond to anatomical landmarks. The shape itself is then represented as a vector that contains the coordinates of all points representing the surface of the shape [52]. If connectivity information of points within the point set is stored additionally, this representation type is called a mesh.

15.2.2.2 Signed Distance Maps

Golland et al. have proposed to use signed distance maps for representing shape [55,56]. When using signed distance maps, any point inside an object is encoded as the (signed) distance from this point to the closest point on the surface of the object to be described.

15.2.2.3 Deformation Fields

Deformation fields are a common way of representing elastic transformations. Frangi et al. have used deformation fields to characterize a shape [57]. Correspondences between two objects are established by Free-Form-Deformation (FFD) using B-Splines [58]. Based on the resulting deformation field, each point of the template object can be mapped to the corresponding point of a new object. Hence, a new shape can be completely described by using the template of a reference shape in combination with the respective deformation field.

15.2.3 Representing the Appearance of Anatomical Structures

Besides shape, the texture and its variation of a structure entails additional information for statistical modeling. Commonly the combination of shape and textural information is referred to as the appearance of an object. With this combined information, especially pathological processes and pharmacological influences can be analyzed more deeply. In the following, common approaches for the representation of the appearance will be described.

15.2.3.1 Active Appearance Model (AAM)

Cootes et al. introduced Active Appearance Models (AAMs) in 1998 [59,60]. Hence, SM can be built which can describe shape variations of an object and the intensities of all pixels/voxels within the object of interest. To build a SM of the appearance a shape-normalized version of the object of interest has to be created (commonly based on using the mean shape $\overline{S}$ of a structure). Each training shape S has to be deformed to $\overline{S}$ and the corresponding textures T for each training subject have to be warped accordingly. The gray values have to be extracted and entered into a vector $T \in \mathbb{R}^L$, where L is the number of pixels/voxels inside $\overline{S}$. Consequently, the appearance can be represented by $A = \{(S, T)\}$.

15.2.3.2 Direct Appearance Model (DAM)

Direct Appearance Models (DAMs) were introduced by Hou et al. in 2001 and are based on the assumption that the appearance of a structure should be uniquely determined by the texture of an object, but not vice versa [61]. Hence, DAMs only use texture information to predict the shape of an object. Consequently, shape and texture are modeled by using two different models. Shape parameters are predicted from texture parameters by using linear regression. More details to DAMs can be found in [61].

15.2.3.3 InShape Models

Fritscher et al. introduced an approach to combine shape information based on deformation fields with gray-level information of an anatomical structure [41]. By using (diffeomorphic) Demons registration [62] in combination with a specific representation of the appearance of the training images, vector fields and shape normalized representations of an object of interest are used to represent the appearance of an anatomical structure. More details on *InShape* models can be found in [41].

15.2.4 Establishing the Correspondences Between Shapes

The statistical analysis of shape and appearance variations commonly requires correspondences between the subjects of a training set. In order to compare anatomical structures coming from different subjects, the shape needs to be superimposed in order to compare them. For this purpose, it is necessary to establish correspondences between corresponding regions on the respective structures. The establishment of the correspondences is commonly accomplished be using image registration methods. In the following, the most common methods for finding correspondences are described briefly.

15.2.4.1 Pre-Alignment Based on Linear Transforms

A registration is based on a similarity transform and is typically performed in the pre-alignment step. If correspondences between different landmarks are already explicitly defined by a human expert who is setting the landmarks, point sets can be pre-aligned in a common coordinate frame using the Procrustes Analysis [63]. For calculating the Procrustes distance, point sets are scaled and centered so that the sum of squared distances of all points in each point set is unity. Based on this, a similarity transform can be computed [64]. If correspondences are not known, approaches like the Iterative Closest Point (ICP) algorithm introduced by Besl et al. [65], or some of its numerous extensions like Softassign [64] or (multi-scale) EM-ICP [66] can be used.

15.2.4.2 Alignment Using Nonlinear Transformations

For anatomical structures with high shape variability and for modeling approaches which represent the shape of an object by using a deformation field in combination with a shape template like statistical deformation models [67] or *InShape* models [38, 40], a nonlinear transformation is essential. Depending on the type of representation used, different approaches exist:

15.2.4.2.1 Non-Rigid Alignment of Point Sets

Chui et al. [68] have presented an extension of the ICP algorithm for non-rigid point set registration by combining the Softassign algorithm [64] with Thin Plate Splines in order to parameterize the spatial mapping. Jian et al. have introduced a non-rigid registration approach which is also using soft margins [69]. More details on non-rigid alignment of point sets can be found in Section 4 in [52].

15.2.4.2.2 Non-Rigid Alignment of Volumetric Representations of Shape and Appearance

If for the creation of SSM/SAMs original image datasets [67], volumetric shape representations like binary labels [70], distance maps [71] or combinations thereof [28] are used, other alignment approaches have to be applied. These representations are used for non-rigid registration and correspondence establishment and not specifically for building statistical models. A popular and efficient approach for volumetric elastic registration that has been used for the construction of SSM/SAMs are Free Form Deformations (FFD) introduced by Rueckert et al. in [58]. Using FFD, registration is achieved by deforming a "virtual" mesh of control points based on using B-Splines. This approach can be combined with arbitrary similarity metrics like Sum of Squared Distances or Mutual Information [72]. Another non-parametric registration approach is the diffeomorphic Demons approach introduced by Vercauteren et al. [62]. This approach is based on the original Demons algorithm introduced by Thirion in [73]. Object boundaries are modeled as membranes on which demons are located. A detailed review on algorithms for deformable registration can be found in [74].

15.2.5 Statistical Analysis of Shape and Appearance Variability

The statistical analysis of a training set of superimposed shapes is a crucial step for the creation of SSM/SAMs. Statistical analysis of the aligned training shapes does not only reduce the intrinsic dimensionality of the training data, but first and foremost allows assessing the variability of an anatomical structure of interest and interpreting the nature of the underlying variations. In the following sections, common methods for dimensionality reduction used for statistical shape and appearance modeling are described.

15.2.5.1 Principal Component Analysis

The Principal Component Analysis (PCA) is one of the most frequently used methods used for the analysis of statistical shape and appearance variations [75]. The basic idea of PCA is to retain the highest possible amount of variation of a (training) set of objects, while reducing the dimensionality of the interrelated variables. This is achieved by transforming the input data to a set of uncorrelated principal components (PCs). PCs are ordered in descending order of component variance, so that the first n PCs usually contain most of the variation present in the input data. Higher PCs cover less relevant variations and/or noise that is/are present in the dataset. In order to create a SM, such input data could typically be m superimposed shapes $s_1, \ldots s_m$, e.g. represented by landmark coordinates. The PCA uses the fact that a $p \times p$ (where $p = 2n$) symmetric, non-singular matrix, like the covariance matrix S can be reduced to a diagonal matrix L by pre- and post-multiplying it by an orthonormal matrix U such that $U'SU = L$ [76]. The diagonal elements $l_1, l_2, \ldots, l_p$ of L are the eigenvalues of S, whereas the columns of U are the eigenvectors of S. As a result, correlated variables of the input data $s_1, s_2, \ldots, s_m$ are transformed into p new uncorrelated variables $z_1, z_2, \ldots, z_p$ [76]. The values for z can be obtained by $z_i = U'[x - \bar{x}]$, where x and $\bar{x}$ are the vectors of observations on the original variables (x) and their respective mean ($\bar{x}$).

15.2.5.2 Kernel Principal Component Analysis (KPCA)

PCA is a linear dimensionality reduction method. More complicated structures, however, cannot be appropriately represented in a linear subspace. For such purposes, a nonlinear extension of the PCA, so called Kernel PCA (KPCA), can be applied. Similar to popular methods like Support Vector Machines [77], KPCA [78] is using the so-called "Kernel Trick" in order to construct nonlinear mappings: Instead of computing the eigenvectors based on the covariance matrix of the input data, KPCA is using a Kernel matrix K. More information on KPCA can be found in [78–81].

15.2.5.3 Independent Component Analysis (ICA)

A different approach to overcome the normality constraint concerning the distribution of the training data is to apply Independent Component Analysis (ICA) [82]. ICA does not assume Gaussian distribution of the input data. In principal, ICA can be seen as an approach that identifies a linear transformation, which maximizes the non–Gaussianity of the training data.

15.2.5.4 Local Nonlinear Dimensionality Reduction

Local dimensionality reduction approaches are aiming at preserving properties of small neighborhoods around input data points. However, the basic principle behind these methods is that – by preserving local properties of the input data – the global layout is retained as well [83]. Members of this group of local dimensionality reduction algorithms are Locally Linear Embedding (LLE) [84], Laplacian Eigenmaps [85], Hessian Locally Linear Embedding [86] and Local Tangent Space Analysis [87].

15.3 APPROACHES FOR BONE QUALITY ASSESSMENT

In this section, different approaches for bone quality assessment will be presented.

15.3.1 Approaches Based on Statistical Shape and Appearance Models

In the following, different applications using SSM/SAMs for bone quality assessment will be presented.

15.3.1.1 Segmentation and 3D Reconstruction

The first step of the SSM/SAM modeling pipeline is typically a 2D/3D segmentation or reconstruction of the bone of interest. The usage of SSMs for these purposes is common. For example, Fritscher et al. introduced combined shape-intensity prior models (i.e. *InShape* models; see also Section 15.2.3.3) for 3D segmentation of the femur [41]. Shape-intensity models have been created by using a hierarchical registration approach based on mutual information and Demons registration [73]. PCA has been used for dimensionality reduction. For segmentation of new, unseen images, *InShape* models have been combined with Geodesic Active Contours (GACs) [88]. For this purpose, GACs have been used in combination with Leveton's MAP approach to combine model-based segmentation and registration introduced in [89]. An exemplary segmentation result of a femur can be seen in Fig. 15.3. The mean distance between automatic segmentation results before and after manual correction averaged at 0.20 mm ($\sigma = 0.063$) and the average similarity index was 0.992 ($\sigma = 0.006$).

Seim et al. have used a SSM for segmentation of human pelvic bones from CT images [90]. Anatomical and geometrical features have been used to subdivide the segmented pelvic bone into patches (i.e. regions of the mesh with the topology of a disc). Point-to-point correspondences have been established for all 50 training shapes. Segmentation of a new, unseen dataset has been consisted of several steps: Firstly, a coarse

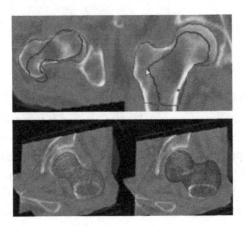

Figure 15.3 CT-scan of the femur with initial surface (red) and segmentation result (green) in 2D (top) and 3D (bottom). (For interpretation of the references to color in this figure legend, the reader is referred to the web version of this chapter.) *(Figure taken from [41].)*

initialization has been performed by placing the average shape of the pelvis model within the CT image using the Generalized Hough Transform (GHT) [91]. Subsequently, the SSM has been adapted by a transformation and variation of its shape modes. Finally, a FFD step based on optimal graph searching [92] is applied in order to consider restrictive parts of the SSM. A leave-one-out evaluation strategy has been used to analyze the developed segmentation approach using 50 datasets. The results have shown that the GHT is able to find an initial placement of the shape within the CT data reliably. The average surface distance after the adaption of the SSM was 1.2 ± 0.3 mm and 0.7 ± 0.3 mm when using the FFD step.

SSM/SAMs can also be used to build hierarchical multi-object models. Hence, the inter-relationship between adjacent structures can be considered within the model. For example, Yokota et al. have used a hierarchical SSM of joint structure for an automated segmentation of the femur and pelvis of diseased hips using 44 datasets [93]. Shape and pose variations of the femur relative to the pelvis have been embedded in a combined SSM of femur and pelvis. Subsequently, the combined SSM has been hierarchically divided into individual SSMs of femur and pelvis, and a partial combined SSM only including proximal femur and acetabulum (i.e. overlapping parts of the shape are identical in individual and combined model). Using individual SSMs of femur and pelvis segmentation resulted in an average contour distance of 2.26 mm in the joint area. The combined SSM reached an average distance of 2.00 mm and the hierarchical SSM 1.78 mm. The results show that using the combined SSM outperformed independent SSMs for segmentation.

For the reconstruction of a 3D shape from 2D radiographs or DXA images, a 3D shape reconstruction is used. By this means, additional morphometric as well as structural information about a bone can be acquired. Commonly, a method based on model-based 2D/3D registration is used for this purpose. Whitmarsh et al. have introduced an approach for reconstructing 3D shape and BMD distribution of the proximal femur from a single 2D DXA image [94]. For this purpose, a SM of the combined 3D shape and BMD distribution has been built using QCT images. First, a SM has been created based on the work of Frangi et al. [57]: By non-rigidly registering the QCT volumes onto a segmented reference subject, the image database has been pre-aligned. In order to model variations of density distribution, individual QCTs have been deformed to the mean shape and a density model has been built from the resulting volumes. The relationship between shape and density model has been established by performing a third PCA using the model parameters of shape and density models as input [59]. The average error of the global shape accuracy was 1.1 mm. The developed approach allows an accurate and detailed 3D analysis of the femur from using 2D DXA images, which reduces costs as well as radiation dosage applied to the patient.

Zheng et al. have used a SSM for 3D reconstruction of the pelvis from a single standard X-ray radiograph [95]. The approach is based on hybrid 2D/3D deformable registration in combination with a landmark-to-ray registration using SSM-based 2D/3D reconstruction. The final reconstruction has been computed by matching projections of contours extracted from a 3D model derived from the SSM to the image contours extracted from the X-ray images. Based on 24 training datasets a SSM has been created using PDMs for shape representation using a PCA for dimensionality reduction. For segmentation of the pelvis in the X-ray images, a semi-automated approach based on the Livewire algorithm introduced by Mortensen and Barrett [96] has been used. Dominant points, which approximate the contours, have been assessed automatically by using an approach of Phillips and Rosenfeld [97]. A surface-based 3D/3D matching between reconstructed models and associated ground truth models (i.e. CT-based reconstructions) have been used for estimating unknown scales. 14 pelvis datasets (X-ray and ground truth CT) have been used for evaluation of the approach. The authors observed an average reconstruction error of 1.6 mm.

Lamecker et al. introduced an atlas-based 3D reconstruction approach of the hip based on using single radiographs [98]. A SSM has been generated from 23 datasets of the pelvic bone, which have been segmented manually. PCA was used for dimensionality reduction. The key part of the approach is an algorithm, which optimizes a similarity measure using the silhouettes of the object in the projections. For minimization, a gradient-descent evolution approach has been used in combination with a multi-resolution registration approach.

15.3.1.2 Fracture Risk Assessment

Prediction of a subject specific fracture risk can act as a marker for starting a specific osteoporosis treatment or even perform preventive surgery (e.g. after contra-lateral hip fracture) in order to decrease the risk for osteoporotic fractures [99]. The simple fact that an existing vertebral fracture increases the risk of another vertebral fracture is an easy but common way for "quantifying" fracture risk. Approaches that are more sophisticated assess shape and/or texture properties of a bone from 2D or 3D images for estimating fracture risk by using regression models on a set of predictive features.

Baker–LePain et al. have used an ASM of the hip to predict incident hip fractures based on radiographs [100]. For this purpose, the first 10 modes of the ASM, which accounted for 95% of the variance of the proximal femur shape have been used. The association between ASM modes and incident hip fractures has been assessed by using logistic regression. The accuracy of fracture discrimination that could be obtained by combining ASM modes with femoral neck BMD resulted in an AUC of 0.835 compared to an AUC of 0.675 when only using femoral neck BMD. The results have also shown that variations of the relative size of the femoral head and neck are vital determinants of incident hip fractures.

Goodyear et al. have introduced a method based on SSM/SAMs for predicting fractures of the proximal femur using DXA scans [36]. The shape of the 2D-model was defined by 72 landmark points, which have been placed semi-automatically. An affine registration approach has been used for alignment of the shapes and a PCA has been applied for dimensionality reduction. Logistic regression has been used to evaluate if variables were suitable fracture predictors. ROC analysis has shown that the use of shape mode 2 and appearance mode 6 in combination with BMD performed 3% better than using only BMD or any other parameter (i.e. modes) as a single predictor.

An approach for assessing fracture risk of the proximal femur in QCT images has been developed by Whitmarsh et al. [33]: A statistical model of shape and bone mineral density distribution of the proximal femur have been used for fracture risk assessment using QCT images. A 3D-SSM in combination with a separate texture model has been used in order to reconstruct patient-specific shape and density of the proximal femur. For the creation of the shape model, a training set of pre-segmented CT images has been used. After creating surface meshes of all femora, a reference image volume has been selected. All other femora have been registered to this reference volume using an intensity–based similarity registration followed by a multi-scale intensity-based Thin Plate Spline (TPS) registration. For TPS registration, the vertices of the surface meshes have been used as the control points on the reference image. In order to remove a potential bias introduced by an arbitrary selection of a reference, a second iteration of TPS registration using the average shape of all training subjects as a reference has been performed. The vertices of the aligned surface meshes have then been used as input for the shape models. The deformed image volumes have acted as an input for the texture model representing the bone mineral density distribution of the proximal femur. PCA has been applied to assess shape and texture variations. Based on the reconstructed model parameters, Fisher Linear Discriminant Analysis (FLDA) has been used to distinguish between a group of subjects suffering from contralateral femoral fractures and a healthy control group. The results have shown that volumetric BMD and cortical thickness is one of the most important discriminators between fracture and non-fractured femora. In addition, the authors have also identified different morphological properties like hip axis length and neck–shaft angle that can potentially act as fracture risk predictors.

In [35,101] Schuler et al. have used a statistical *InShape* model to assess the individual fracture risk by predicting the individual fracture load of a femur. Using 96 femur specimen datasets the fracture load of each femur has been obtained in the course of biomechanical tests and QCT images of all specimen have been acquired. In order to predict the fracture load, *InShape* models (see also Section 15.2.3.3) for different regions of interest have been created. The resulting model parameters have been used as input for Principal Component Regression in order to predict the individual fracture load of the femora. Using model parameters of 30 PCs, a correlation coefficient of $R = 0.91$ between predicted and real fracture loads was achieved, whereas $R = 0.81$ when using only BMD for fracture load prediction.

In [38] Fritscher et al. have applied a SAM in order to predict biomechanical parameters, which are suitable to assess the strength of an individual proximal femur. In these experiments, Fritscher et al. have aimed at predicting femoral failure load using 26 CT datasets. For this purpose, a combined representation of shape and spatial intensity distribution (*InShape* representation, Section 15.2.3.3) of an object of interest has been created by applying an atlas-based non-rigid registration on a training set of CT images. The resulting *InShape* representations have acted as input for PCA. A subset of n relevant PCs has been identified by using a correlation-based feature selection method. Subsequently, PC scores of the selected PCs have been regressed against fracture load that has been obtained in biomechanical tests using Support Vector Regression (SVR). The best subset of PCs and the optimal parameter values for SVR have been identified by using a stratified cross-validation scheme. By using this approach, various geometric and structural properties that can be used to predict biomechanical bone parameters accurately and reliably have been identified. Fig. 15.4 illustrates the major differences between femora with low (top) and high failure load (bottom). Whereas some PCs rather describe variations of the femoral shape (neck thickness, shape of the head and shape of the greater trochanter), others mainly depict variations of cancellous bone in the femoral shaft and head regions (PC 16 and PC 8).

In [37] Bredbenner et al. have developed a method for fracture risk prediction based on statistical shape and density modeling of the proximal femur using QCT images (40 fracture cases and 410 non-fracture cases). After segmentation, surface meshes have been created for all femora. Surface meshes have then been registered to the CT scan axes and intersected at the inferior-most part of the lesser trochanter to create femur surfaces proportionally sized for individual subject hip anatomy. An average femur has been generated by averaging vertex positions for all femora and subsequently warped onto each

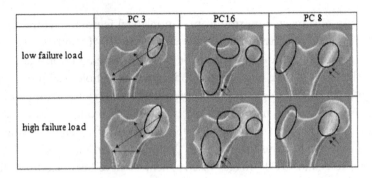

Figure 15.4 Variations of the proximal femur described by the 3 principal components with highest correlation to failure load. *(Figure taken from [38].)*

individual femur surface using vector fields calculated between corresponding surface vertices (→ 450 femur mesh models). Each femur model has been superimposed on CT images of the respective individual and image intensities have been determined at the spatial location of each node. Using PCA, a combined statistical shape and density model (SSDM) that describes shape and density variabilities has been generated. In order to analyze the complex structural variations between fracture and non-fracture cases, separate mean models for fractured and non-fractured datasets have been created. Based on these SMs, differences in shape geometry and mid-plane BMD between fracture and non-fracture cases have been analyzed. For fracture risk assessment, different classification models based on predictor variables like areal BMD, SSDM, age and body mass index have been created. One key finding was that the average fracture model of the fractured group has generally been larger than the average model (especially in the distal neck and greater trochanter region) of the non-fractured group. With respect to fracture prediction, the authors have shown that sensitivity and specificity of fracture prediction increases when using SSDM compared to using only areal BMD: Areal BMD-based fracture prediction model identified 10% of the fracture cases and 91.3% of the non-fracture cases. Age-adjusted SSDM-based prediction approach has performed clearly better – 55% of fracture cases and 94.7% of non-fracture cases were identified correctly.

Another way of analyzing bone quality is the prediction of bone loss. In [102] Varzi et al. have developed a method that predicts rapid bone loss from a single pQCT bone scan within weeks after a spinal cord injury. For building a SSM, a semi-automated landmark placement has been performed (25 femur datasets: 44 points; 23 tibia datasets: 30 points). Baseline scans have been taken within 5 weeks post-surgery, follow up scans 4, 8 and 12 months post-surgery. A Procrustes analysis has been applied for shape alignment. Subsequently, PCA has been performed to identify different characteristics of variability from the average shape. 7 PCs, which describe 74.8% of the total variance, have been selected for further analyses of the femur. The results have shown that PC4 significantly correlated with 12–month percentage changes in total or trabecular BMD. In addition, the authors have observed a significant negative correlation between PC3 and percentage total BMD. Based on these insights, the approach can potentially be used for an early detection and treatment of osteoporotic spinal cord injuries. Furthermore, the method has the potential to reduce the incident of fractures and their implicated consequences.

Li et al. have performed hip fracture risk estimation based on a QCT atlas [103]. A hip QCT atlas has been constructed based on 37 patients with hip fractures and 38 patients without fractures using inter–subject image registration. Again, PCA has been used to identify modes that are related to hip fractures. It could be shown that PCA–based hip fracture risk estimation using the QCT atlas (AUC = 0.880) outperformed traditional BMD–based estimation (AUC = 0.782–0.871).

15.3.1.3 Bone Fracture Detection

SSM/SAMs have also been used for fracture detection and classification. This is specifically important for the detection of vertebral compression fractures, which are typically linked to osteoporosis and often remain undetected since they commonly occur when doing everyday things like coughing, twisting or sneezing. By detecting such fractures at an early stage, treatment can be initiated earlier.

Whitmarsh et al. have estimated hip fracture discrimination by using 3D reconstructions from DXA images [94] in [30]. The BMD distribution and a 3D SSM has been registered onto DXA scans of fracture and non–fracture patients. The first model parameter, the resulting scaling values and the femoral neck areal BMD have been processed by Linear Discriminant Analysis. Discriminating accuracy has been evaluated by generating the ROC. The results indicated that the developed SSM–based approach has an improved power of discrimination between fracture and non–fracture patients compared to only using BMD.

De Bruijne et al. have introduced a conditional SM for vertebral fracture quantification using radiographs [31]. Conditional SMs allow the reconstruction of the shape of an object by including neighboring structures in the analysis. A PDM (see also Section 15.2.2.1) has been used to model shape variations of individual vertebrae and a conditional PDM has been applied to reconstruct the most likely shape of a vertebra based on the shape of another vertebra in the image. After individual SSMs have been constructed, covariance matrices of paired vertebral shape variation for all pairs of vertebra in an image have been built. Subsequently, for each pair of vertebra in a new image the predictor shape is aligned with the model and a shape regression is performed. Since the pairwise shape reconstruction approach results in several shape estimates for each vertebra, the resulting individual estimates have been combined into a single, optimal estimate for each vertebra using a weighted summation of individual estimates (i.e.

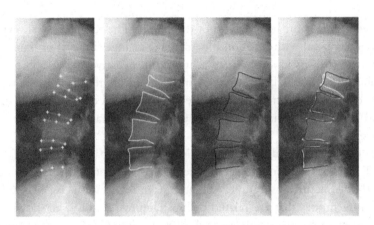

Figure 15.5 Fracture quantification procedure. From left to right: Original lateral X-ray image of a lumbar spine with one severe fracture (pluses indicate the points to measure the anterior, middle and posterior heights annotated by a radiologist). Manual contour annotation. Reconstructed unfractured shapes for each of the vertebra. The difference between the two shapes is a measure of vertebra abnormality. *(Figure taken from [31].)*

weights express the degree of belief in each of the individual estimates). Finally, the difference between the actual segmented shape and the final reconstruction has been used as a measure of abnormality and thus of fracture severity [31]. In Fig. 15.5 the basic approach for fracture quantification is shown.

282 lumbar spine radiographs have been used for evaluation. The patient-specific conditional SM obtained very good fracture detection rates (accuracy = 96.7%) and also outperformed test scenarios in which only a SM of a single vertebra was used (accuracy = 94.8%).

An approach that is not only based on shape analysis, but also includes texture models for vertebral fracture detection has been developed by Roberts et al. [32]. The presented approach has used DXA VFA (vertebral fracture assessment) images of 360 patients as input data. For semi-automatic vertebra segmentation a variant of an AAM-based approach has been applied [59]. The underlying texture models have been extended by adding measures of corner and commensurate edge strength based on the Harris corner detector [104]. Three different vertebral appearance models and classifiers have been constructed by pooling data of nearby vertebrae (L4–L1, T12–T10 and T9–T7). Vertebral fracture detection has consisted of the following steps: Firstly, vertebrae of an unseen image have been segmented using an AAM approach. Subsequently, the corresponding appearance model has been fitted to the previously computed shape and its texture. The parameters of the appearance model have then been used to train a model for automated fracture classification. Manual vertebrae classifications of two radiologists have been used as gold standard. The developed approach has shown that the appearance-based classifier improved specificity and sensitivity compared to the standard height ratio morphome-

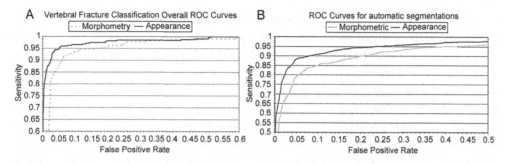

Figure 15.6 (A) ROC curves of fracture detection based on manual segmentations (morphometric method and appearance parameter linear discriminant are compared). (B) ROC curves using semi-automatic segmentation based on AAM. *(Figure taken from [32].)*

try (see Fig. 15.6A). Fig. 15.6B shows a comparison of the classification accuracy that has been obtained by using manual and semi-automated segmentation. A sensitivity loss of 7% was observed when semi-automatic segmentation was used instead of manual segmentations. However, according to the authors, the classification sensitivity is still adequate for computer-assisted vertebral fracture detection. Furthermore, this approach can also be used to assess the severity of a fracture.

15.3.1.4 Treatment Planning

For fracture reduction purposes, choosing implant type and implant anchorage is primarily depending on fracture type and local bone quality. Hence, detailed information, about the (local) bone quality of a patient gives valuable information for surgeons, which can be used for surgical planning (e.g. [105]).

Huber et al. have developed and investigated texture discriminators for the analysis of trabecular bone in proximal femur radiographs [29]. More specifically, the correlation between different texture features and the femurs' anchorage strength has been analyzed. 14 femoral radiographs and BMD measurements of the femoral heads as well as the biomechanical parameter failure load have been used in this study. SAMs ("*In-Shape* models") [38,106] have been applied for the definition of patient-specific regions of interest (ROIs). These ROIs have been used for further texture analysis based on third order gray level co-occurrence matrices (TOGLCM) [107], morphological gradients [108], Minkowski dimensions [109], Minkowski functionals (MF) [110], Gaussian Markov random fields [111] and the Scaling Index Method (SIM) [112,113]. Using a multi-regression analysis, the most predictive texture features have been assessed. The results have shown that the best predictive multi-regression model for anchorage strength described by load to fracture has included the texture methods MF, SIM and TOGLCM but excluded BMD. Highest correlations have been found for regions in the femoral head and lower neck area.

Knowledge about the local bone quality of a patient is also of great value in order to decide about potential preventative surgical interventions at the contralateral femur after a femoral fracture. In [28] Fritscher et al. have analyzed trabecular bone of different regions in the femoral head using 28 CT images and radiographs of femur specimen for local bone quality assessment. In this work, statistical *InShape* models have been used for several tasks: for automated segmentation of individual femurs, for automated adaption/placement of specific ROIs and for automated analysis and classification of the individual intensity distribution values. In the first step, *InShape* representations of the femurs have been created (see also Section 15.2.3.3). Based on these *InShape* representations the size and shape of a ROI used for further analysis have been determined. The local bone quality of a new, unseen image has been assessed by creating an *InShape* representation of the new dataset and projecting this representation into a lower dimensional space. Several approaches for dimensionality reduction have been used and investigated in this work (i.e. PCA, ICA and LLE). Further analysis has then been performed by regressing the *InShape* model parameters against the peak torque to breakaway (BT) of cancellous bone that has been acquired invasively using the so-called "Densiprobe" tool [114] and BMD. The results have shown that dimensionality reduction with PCA and LLE outperformed ICA for the predication of BMD. A correlation of $R > 0.83$ between model parameters and BMD could be acquired using CT images and $R = 0.87$ ($p < 0.01$) could be obtained using radiographs. Better regression results for 2D radiographs are possibly caused by the fact that specimen data with standardized projection geometry has been used. For both image modalities, model-based BT prediction outperformed the prediction based on BMD, demonstrating the potential of this approach for local bone quality assessment.

15.3.2 Bone Quality Assessment Based on FE Analysis

As mentioned in Section 15.1.2, besides SSM/SAM-based approaches especially methods employing FEMs are used for the assessment of bone quality. FEMs are typically generated from 2D or 3D images and consist of FEs for which properties like elastic modulus, material strength or material stiffness must be defined. In the following, some important applications will be presented briefly.

Naylor et al. have used FEMs to determine bone strength and have analyzed the association with hip fracture risk in a longitudinal study [115]. Segmentation of the femur has been performed semi-automatically. For building the FEMs, the patient-specific thickness of the proximal femur has been defined using DXA scans. The thickness has been used to convert areal BMD into volumetric BMD. Hence, it has been assumed that the femur is a plate with constant thickness. Material properties have been derived by using the empirical equations of Morgan et al. [116,117]. Using FEMs, a fall on the greater trochanter has been simulated. The peak impact force (F_{Peak}) is dependent on body weight and height. Femoral strength has been defined based on the von Mises failure criteria and a stress ratio has been calculated at each pixel of the DXA scan. The authors have defined the FE-derived femoral strength as the onset impact force that causes the stress ratio in that area exceeding one. A stress ratio greater than one indicates that the element has failed due to the peak impact force. FE-derived femoral strength F_{FE} has been defined as $F_{FE} = F_{Peak}\beta$, where β is a scaling factor that gradually decreases the stress ratio threshold. Subsequently, the load-to-strength ratio (LSR) has been calculated (= attenuated impact force). For evaluation of the FEM, 56 cadaveric femoral specimens have been used for experiments that simulated sideway falls. The authors came to the result that combining BMD and LSR improves the AUC and discrimination sensitivity of BMD alone by 0.03 and ~4%, respectively. Concluding, the FEM-based approach of Naylor et al. has the potential to discriminate incident hip fracture cases from non-fracture cases without using femoral BMD, FRAX score and prior fractures.

Yang et al. have developed an approach to predict hip fractures by approximating femoral strength based on patient-specific FEMs [118]. For this large cohort study, body weight and height as well as DXA scans of 1941 post-menopausal women (668 of which suffered from incident hip fractures) have been used. The FEM has been created from the DXA scans by using the approach of Naylor et al. [115] (see previous paragraph). By using a Cox regression model the case-cohort design has assessed the association of estimated femoral strength with hip fracture. The performance of FE-estimated femoral strength for fracture risk prediction has been compared to hip BMD and FRAX. An association of the FE-estimated femoral strength with an incident hip fracture was strong (Harrell's C index of 0.770) and outperformed total hip BMD (0.759; $p < 0.05$) and FRAX scores (0.711–0.743; $p < 0.0001$) as a predictor.

Similarly, Qasim et al. have developed a patient-specific FEM that also estimates femur strength as a hip fracture risk predictor [119]. Furthermore, they have also analyzed the effect of methodological determinants like imaging protocols and FE-modeling techniques. The study was based on 100 Caucasian women of whom 50 experienced a hip fracture. All patients have received a DXA scan (femoral neck and total femur BMD was computed) as well as a bilateral proximal femur CT. After femur segmentation, the bone surface has been meshed with tetrahedral elements by using two different methods: a mesh morphing procedure based on anatomical landmarks [120] and a standard automatic meshing algorithm (ICEM CFD 14.0, Ansys Inc., PA, USA). In order to simulate different physiological loading models, a femoral reference system has been generated. This has been created by defining landmarks in the proximal femur scan and creating an approximated FE model of a full femur (based on using a statistical shape model-guided fit). Four different FEMs have been generated to estimate the influence of different meshing approaches and femur orientation methods. Furthermore, 12 different stance loading scenarios as well as 10 different side loading scenarios have been investigated using different boundary conditions. FE-strength has been predicted for a femoral head fracture following a well-established maximum principal strain criterion [121–124]. For statistical analysis, univariate logistic regression models have been used to determine the suitability of areal BMD, T-score, MPhyS (Minimum Physiological Strength: minimal FE-strength of the 12 stance loading conditions) and MPatS (Minimum Pathological Strength: minimal FE-strength of the ten fall loading conditions) to classify fracture and non-fracture cases. The results demonstrate the importance of using the whole femur in order to develop a robust FEM-based tool for hip fracture risk assessment. Apart from this, it could be observed that the mesh morphing algorithm approach has decreased the FE-strength estimation power used to discriminate between fracture and non-fracture cases. This demonstrates the importance of evaluating the effect of different meshing algorithms on the performance of FE-based fracture risk assessment. The results of this work showed that the predictive power of the FE-estimated strength increases when estimating the physiological reference system of the femur using a SSM.

Thevenot et al. have generated 3D FEMs of the hip from single 2D radiographs [125] and predicted failure load. A set of geometrical parameters from hip radiographs have been used in combination with segmented CT training datasets in order to establish relationships between geometrical parameters and the boundary of the femurs. Based on these relationships, the 3D shape of the femur, represented by a mesh, has been generated. Material properties have been evaluated using trabecular structure analysis of the X-ray [126]. For estimation of failure load, a bilinear elastoplastic FE analysis has been performed based on the work of Koivumäki et al. [127]. The authors have generated 3D FEMs of the proximal femur using 2D radiographs, resulting in an average reconstruction error of 1.77 mm ($\pm$1.17 mm). Furthermore, the estimated failure load values correlated well with experimental results ($r^2 = 0.64$).

SSM/SAMs and FEMs have not only been used as alternative approaches for Bone Quality Assessment, but have also been elegantly combined to efficiently create patient-specific FEMs. For example, Sarkalkan et al. have developed a very successful combination of these two techniques [128]. They have used a SSM/SAM for an automated estimation of proximal femur fracture load using 2D FEMs. 70 DXA scans have been used for generating the SM. For all scans, manual expert segmentations have also been available. In this approach, SSM and SAM have been used for building the FEM. Hence, for the creation of the SSM, 70 landmarks have been manually positioned on each proximal femur of the training dataset. Subsequently, a Procrustes analysis has been performed for the alignment on a common coordinate system and a PCA has been used for dimensionality reduction. The SAM has been created based on an approach presented by Stegmann et al. [129]. For segmentation of unseen DXA images, an AAM-based approach has been used based on the work of Cootes et al. [43]. The generation of the patient-specific FEMs for individual fracture load estimation has consisted of several manual and automatic steps: At first, the mean shape of the SSM/SAM has been moved over the unseen DXA image manually. Then the AAM-based search algorithm

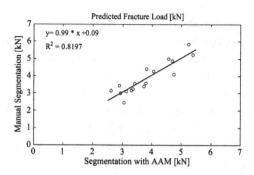

Figure 15.7 Estimated fracture loads using manually and automatically segmented FEMs for 17 evaluation datasets. *(Figure taken from [128].)*

has been used to obtain the best fit of the SSM/SAM to the DXA image and a surface mesh of the SSM/SAM-derived femur model has been created. Material properties of each 2D element have been defined by averaging the gray values of the pixels within the element. These mean gray values have been used to calculate areal BMD, which has then been converted to volumetric BMD [115]. Loads and boundary conditions have been defined in a way that a fall on the grater trochanter could be simulated [115]. FEMs have been solved using a nonlinear implicit solver. To explain 98% of variations of shape 16 PCs were needed and 50 PCs accounted for 98% of the appearance variations. For explaining 98% of shape and appearance in the combined model 44 PCs were used. The mean point-to-curve error was 1.25 ± 0.65 mm of the training dataset based on a leave-one-out test and 1.42 ± 0.75 mm for the 17 evaluation datasets. In Fig. 15.7 the estimated fracture loads using the manually and automatically segmented FE models are compared. The differences primarily originate from segmentation error, which were in the range of the intra-observer variability of manual segmentations (1.03 ± 0.48 mm).

Väänänen et al. have developed an approach combining SAMs and FEMs for reconstructing 3D shape, density and cortical thickness from DXA images. This information acted as a basis for the construction of 3D FEMs of the proximal femur [130]. For this purpose, 3 different sets of images (set 1: 34 cadaver femur DXA and CT images, set 2: 35 clinical hip CT images, set 3: 12 patients with clinical DXA and CT) have been used. CTs have been segmented manually and landmarks have been placed on femur and pelvis in order to create 3D SAMs. By registering SAMs with DXA images, the 3D shape of the femur has been reconstructed and orientation and modes of variations of the SAMs have been adjusted by minimizing the sum of absolute differences between model and image. Subsequently, FEMs have been built from the reconstructed shapes. The mean point-to-surface distance between CT image and reconstructed shape was 1.0 mm for set 1 and 1.4 mm for set 3 (leave-one-out test). Furthermore, the results showed that noise within DXA images had only a small effect on the accuracy of the 3D shape reconstruction. The reconstructed volumetric BMD showed a mean absolute difference of 140 mg/cm^3 for set 1 and 185 mg/cm^3 for set 3. Concluding it can be said that the presented approach provides accurate reconstruction of 3D shapes of the proximal femur, which can act as a basis for FE model creation.

Similarly, Nicolella et al. have developed a parametric FEM of the proximal femur using a Statistical Shape and Density Model (SSDM) [131]. By using a small number of SSDM parameters the shape and density distribution has been described. A specific fall loading has been simulated using the FEM. The 3D FEM was reconstructed from a new, unseen femur with an average spatial error of 0.016 mm and an average bone density error of 0.037 g/cm^3. Furthermore, the effect of each single eigenmode of the SSDM on the predicted force displacement behavior of the femur has been analyzed for the simulated fall condition. For example, the mean SSDM resulted in a predicted maximum force of 1590 N. Shape and density variations of the first eigenmode increased the predicted peak force (i.e. −1 SD: 1996 N and +1 SD: 1727 N).

Bryan et al. have used a SM of the whole femur for fracture risk estimation of the femoral neck [132]. The SM has incorporated geometric and material property variations and has been used for generating FE femur models. The analyses have shown that the percentage of cortical bone in the proximal femur can be used as a discriminator between failed and non-failed groups.

15.4 DISCUSSION AND CONCLUSION

15.4.1 Creation of Statistical Shape and Appearance Models – Potentials and Pitfalls

In general, SSM/SAMs have the potential for being a highly useful tool for different areas in the field of bone quality assessment. On the one hand, they offer the possibility to provide anatomically correct segmentation results. In addition, they can be used to create 3D shape reconstructions, from 2D input images, which can directly be used for further morphological and structural analysis using model-based approaches or FE models. On the other hand, SSM/SAMs can also be directly used for clinical applications like surgical planning and fracture risk assessment in a computationally efficient way. As described in Section 15.2, a large number of different methods have been introduced for creating SSM/SAMs. These approaches do not only provide a lot of potential, but also require a significant amount of expert knowledge to choose an adequate method for a given application. This might be one reason why only a rather small number of clinical studies have yet used the full potential of SSM/SAMs. Additionally, for certain parts of the model building process like the creation of correspondences between datasets, it is still unclear, which approaches are optimal for different applications. Due to the complexity of the correspondence problem, there is no general solution that works best for different structures to be modeled. Hence, it is of utmost importance to put increasing effort in the comparison and evaluation of the different approaches using standard quality measures. Only by this means, it is possible to choose adequate tools for the alignment of shapes in order to establish correct correspondences between datasets. The same can be said about available representations of shape. While landmarks are probably the most widely used approach to represent shapes, they also have some disadvantages. Due to the need for user interaction when placing landmarks, a certain amount of subjectivity is introduced in the model building process [57,133]. When it comes to dimensionality reduction, PCA is by far the most popular approach for SSM/SAM creation. While being easy to use and computationally efficient, PCA also makes certain assumptions concerning the distribution of the input data (e.g. Gaussian distribution of input data), which might not be adequate for certain applications. Consequently, other (nonlinear) approaches like Kernel-PCA, ICA or manifold-learning methods like LLE or Laplacian Eigenmaps might be a better choice. However, compared to PCA they are more complex to handle and again require a significant amount of expert knowledge for a correct and successful application.

At last, the quality and quantity of the input data is a highly relevant factor when creating SSMs. Correctly segmented images on the one hand as well as a sufficient number of training datasets on the other hand are preconditions for the creation of statistically meaningful models. Unfortunately, however, large training sets are rarely publicly available. Consequently, several strategies for artificial enlargement of training datasets have been proposed [134].

In summary, it can be said that – despite their great potential – SSM/SAM-based methods are still not well established in clinical practice. This is most probably due to the complexity of the approach compared to clinically established methods for bone quality assessment like areal BMD or FRAX. However, this complexity is mainly restricted to the creation of the models and less problematic for the application of the models. Consequently, it would be highly beneficial if SSM/SAMs for the certain structures on the one hand and easy-to-use tools for model-based bone quality assessment on the other hand would be publicly available. By means of this, the dissemination of model-based approaches for bone quality assessment in clinical practice could potentially be improved. At the same time, a larger clinical dissemination would be a pre-condition for a comprehensive evaluation and further optimization of model-based methods for a large number of potential approaches in the field of bone quality assessment.

15.4.2 Applications of Bone Quality Assessment

The basis for creating SSM/SAMs are segmented objects that can be used as an input for model creation or as a basis for further analysis. Accurate, automated segmentation, however, is an extremely difficult task due to image noise and artifacts, poor contrast and the lack of inter-institutional, standardized imaging protocols for different applications. By incorporating prior knowledge into the segmentation process, model-based approaches can overcome some of these problems. Using a model-based approach, missing, distorted or degraded structures in images can often be recovered based on using shape and appearance statistics provided by the underlying model. Especially for bony structures, which often show less shape variability than soft-tissue organs, model-based segmentation approaches provide high qualitative results. For example, in [128] Sarkalkan et al. quantified a mean error for proximal femur segmentation between 1.2 and 1.4 mm using a SSM/SSA/AAM-based segmentation approach. Intra-observer variability of manual segmentations was only slightly lower (i.e. 1.0 mm). However, it also has to be said that the accuracy of the segmentation varies for different structures depending on the complexity of the shapes and the amount of shape and appearance variability for the structure throughout a population. For example, when using model-based approaches the segmentation quality for vertebrae, which have a rather complex shape and are located close or even connected to other bony structures (e.g. ribs) is generally a bit lower than the segmentation quality for femur or humerus. This is especially true in the presence of large shape alterations of the vertebrae, e.g. caused by a scoliotic spine or incident vertebral fractures. In addition, the initialization process of SSM/SAMs often requires user interaction in order to place the model in an unseen image. It could be shown that the initial placement of the model can have a significant influence on the segmentation quality [135].

These limitations are also relevant for 3D bone reconstruction. Apart from the complexity of shape, however, also the number and angle of 2D radiographs that are available for the reconstruction have a significant influence on the reconstruction result. Using two orthogonal views leads to significantly better reconstruction results than using a single view configuration [136]. In addition, the underlying registration approach and especially the metric that is used to fit the model to the image are of utmost importance to achieve correct reconstruction results [137]. When using optimal components, however, highly qualitative reconstruction results can be obtained. For example, the approach of Whitmarsh et al. resulted in an average reconstruction error for the proximal femur of 1.1 mm using a single 2D DXA image [94]. Zheng et al. obtained an average mean reconstruction error of 1.2 mm using two views for each femur [138].

Apart from segmentation and shape reconstruction, SSM/SAMs have been successfully used for several clinical applications. Especially the usage of SSM/SAMs for fracture risk prediction [33–35,38], local bone quality assessment [28] or bone loss prediction [102] lead to very promising results, which can be directly used for surgical planning purposes and fracture prevention. The fact that the underlying approaches do not require any user interaction and can be applied in reasonable amount of time (approximately 5–15 min) further increases their practical relevance.

Apart from finding a suitable solution for representing appearance variations and establishing correspondences, the selection of appropriate strategies for attribute selection and the evaluation of suitable approaches for (nonlinear) dimensionality reduction, classification and regression is of utmost importance for a successful application of SAMs for bone quality assessment. PCA for dimensionality reduction as well as Logistic Regression [36,37,100], Support Vector Machines [77] and Principal Component Regression [35,101] for classification and regression have been the most frequently used methods in this field. However, more in–depth analysis will be required to find the optimal combination of machine and manifold-learning tools for specific applications. In addition, it also has to be said that still more training data is required in order to fully account for problems like non–standardized imaging protocols or image noise. Again, the limited availability of such large training sets on the one hand and the high (technical and financial) effort that is required to create a high-quality training dataset on the other hand are two of the main reasons why model-based approaches are not yet used for bone quality assessment in clinical practice.

Since SSM/SAMs are used in order to create patient-specific FEMs in an efficient and practically applicable way, the lack of suitable training datasets probably also explains why FE-based methods are not yet used in clinical practice. For certain applications, however, the combination of SSM/SAMs with FEMs has already proven to be very successful [125,128,130–132]. By using model-based 2D/3D registration, FEM-based fracture prediction could even be used in combination with (multiple) 2D images, which reduces costs for the clinics and the amount of ionizing radiation for the patient. Besides of creating patient-specific FEMs, SSM/SAMs can also be used to create a large number of representative shape and appearance instances which can potentially act as a surrogate for patient specific models [139]. Without using SSM/SAMs, FEMs have been created based on manual or automated segmentation of 2D or 3D images of individual subjects [140]. Such approaches, however, are very time consuming and significantly reduce the clinical relevance of FEM-based approaches. In addition, similar to SSM/SAM approaches, different methodological determinants for estimating femur strength with FEM clearly influence the result. Especially the applied imaging protocol as well as the selected FEM modeling and meshing approach have an important influence on fracture discriminating reliability [119]. Up to now, there is no study which performed a quantitative comparison of SSM/SAM vs. FEM-based approaches for applications in bone quality assessment. However, it can be said that e.g. for fracture load prediction SAM and FEM-based approaches seem to be in the same quantitative range concerning accuracy. For example, the SAM-based approach presented in [101] by Schuler et al. has led to similar results as several FEM-based studies (e.g. FEM-based [141–143]). The SAM-based approach of Schuler et al. showed a correlation between real and predicted fracture load of $R = 0.925$ (96 CT datasets), whereas the FEM-based approach of Keyak et al. resulted in a correlation coefficient of $R = 0.949$ (18 CT datasets). However, it has to be mentioned that FEM-based approaches generally have considerably higher hardware requirements and are computationally more expensive than approaches based on SSM/SAMs.

15.4.3 Conclusion and Outlook

In conclusion it can be said that SSM/SAMs have successfully been applied for various tasks that provide the basis for patient-specific bone quality assessment like image segmentation or 3D shape reconstruction. In addition, however, they can also directly be used for clinical applications like surgical planning or fracture risk prediction. In the future, the usage of higher image resolutions will allow an increasingly detailed analysis of the trabecular bone structure even in clinical settings, which will further increase the potential of statistical models of shape and appearance opening up new fields of applications in bone-related research. Especially for model-based assessment of biomechanical properties and fracture prediction, recent advancements like manifold learning and deep learning could potentially be the basis for a further increase of their predictive quality. Finally, modern additive manufacturing techniques like 3D printing enable the production of patient-specific bone substitutions and also provide the basis for using SSM/SAMs as designing and modeling tool for such approaches. Apart from this, current hardware developments and the usage of GPGPU computing can be used to further increase the applicability of model-based bone quality assessment by allowing to create statistical models more efficiently. This is also true for related approaches like FEM, which offer several possibilities to be elegantly combined with model-based methods. Especially for applications which require the creation of patient-specific models in order to be applicable in an efficient way, hybrid approaches that combine FEMs and SSM/SAMs provide accurate results without requiring any user interaction. With regard to applicability, the creation and usage of larger training datasets for the creation SSM/SAMs will be another important factor. Only by having a solid database, statistical methods can provide results that are statistically significant and practically applicable at the same time.

Summing up, the potential of using SSM/SAMs for different applications related to bone quality assessment is extremely high. However, an increasing amount of effort and the willingness to cooperate with other scientific fields will be essential in order to make SSM/SAMs a clinically applicable tool for diagnosis and treatment of bone-related diseases.

REFERENCES

[1] E. Donnelly, Methods for assessing bone quality, Clin. Orthop. Relat. Res. 469 (2011) 2128–2138.

[2] R. Keen, Burden of osteoporosis and fractures, Curr. Osteoporos. Rep. 1 (2) (2003) 66–70.

[3] O. Johnell, J. Kanis, An estimate of the worldwide prevalence and disability associated with osteoporotic fractures, Osteoporos. Int. 12 (2006) 1726–1733.

[4] B. Gullberg, O. Johnell, J. Kanis, World-wide projections for hip fracture, Osteoporos. Int. 7 (5) (1997) 407–413.

[5] E. Hernlund, A. Svedbom, M. Ivergard, J. Compston, Osteoporosis in the European Union: medical management, epidemiology and economic, Arch. Osteoporos. (2013) 8–136.

[6] M. van der Meulen, K. Jepsen, B. Mikic, Understanding bone strength: size isn't everything, Bone 29 (2001) 101–104.

[7] G. Blake, I. Fogelman, The clinical role of dual energy X-ray absorptiometry, Eur. J. Radiol. 71 (2009) 406–414.

[8] J. Adams, Quantitative computed tomography, Eur. J. Radiol. 71 (2009) 415–424.

[9] M. Leonard, J. Shults, D. Elliot, V. Stallings, B. Zemel, Interpretation of whole body dual X-ray absorption measures in children: comparison with peripheral quantitative computed tomography, Bone 34 (2004) 1044–1052.

[10] T. Jämsä, P. Jalovaara, Z. Peng, H. Väänänen, J. Tuukkanen, Comparison of three-point bending test and peripheral quantitative computed tomography analysis in the evaluation of the strength of mouse femur and tibia, Bone 23 (2) (1998) 155–161.

[11] H. Schiessl, H. Frost, W. Jee, Estrogen and bone-muscle strength and mass relationships, Bone 22 (1998) 1–6.

[12] C. Njeh, T. Fuerst, D. Hans, G. Blake, H. Genant, Radiation exposure in bone mineral density assessment, Appl. Radiat. Isot. 50 (1) (1999) 215–236.

[13] M.-A. Krieg, R. Barkmann, S. Gonnelli, A. Stewart, D. Bauer, L. Del Rio Barquero, J. Kaufman, R. Lorenc, P. Miller, W. Olszynski, C. Poiana, A.-M. Schott, E. Lewiecki, D. Hans, Quantitative ultrasound in the management of osteoporosis: the 2007 ISCD Official Positions, J. Clin. Densitom. 11 (1) (2008) 163–187.

[14] S. Majumdar, H. Genant, A review of the recent advances in magnetic resonance imaging in the assessment of osteoporosis, Osteoporos. Int. 5 (2) (1995) 79–92.

[15] A.A. Licata, Bone density, bone quality, and FRAX: changing concepts in osteoporosis management, Am. J. Obstet. Gynecol. 208 (2) (2013) 92–96.

[16] K.L. Stone, D.G. Seeley, L. Lui, J.A. Cauley, K. Ensrud, W.S. Browner, M.C. Nevitt, S.R. Cummings, BMD at multiple sites and risk of fracture of multiple types: long-term results from the study of osteoporotic fractures, J. Bone Miner. Res. 18 (11) (2003) 1947–1954.

[17] E.S. Siris, P.D. Miller, E. Barrett-Connor, K.G. Faulkner, L.E. Wehren, T.A. Abbott, M.L. Berger, A.C. Santora, L.M. Sherwood, Identification and fracture outcomes of undiagnosed low bone mineral density in postmenopausal women: results from the National Osteoporosis Risk Assessment, JAMA 286 (22) (2001) 2815–2822.

[18] T. Hillier, J. Cauley, J. Rizzo, K. Pedula, K. Ensrud, D. Bauer, L.-Y. Lui, K. Vesco, D. Black, M. Donaldson, E. LeBlanc, S. Cummings, WHO absolute fracture risk models (FRAX): do clinical risk factors improve fracture prediction in older women without osteoporosis?, J. Bone Miner. Res. 26 (8) (2011) 1774–1782.

[19] T. Beck, C. Ruff, K. Warden, W. Scott, G. Rao, Predicting femoral neck strength from bone mineral data. A structural approach, Invest. Radiol. 25 (1) (1990) 6–18.

[20] K. Ohnaru, T. Sone, K. Tanaka, K. Akagi, Y.-I. Ju, H.-J. Choi, T. Tomomitsu, M. Fukunaga, Hip structural analysis: a comparison of DXA with CT in postmenopausal Japanese women, SpringerPlus 2 (2013) 331.

[21] World Health Organization Collaborating Centre for Metabolic Bone Diseases, University of Sheffield, FRAX – Fracture Risk Assessment Tool, available from: http://www.shef.ac.uk/FRAX/, 2016.

[22] J. Hippisley-Cox, C. Coupland, Derivation and validation of updated QFracture algorithm to predict risk of osteoporotic fracture in primary care in the United Kingdom: prospective open cohort study, BMJ, Br. Med. J. 344 (2012) e3427.

[23] C. Hernandez, E. Cresswell, Understanding bone strength from finite element models: concepts for non-engineers, Clin. Rev. Bone Miner. Metabol. (2016) 1–6.

[24] J. Keyak, I. Lee, H. Skinner, Correlations between orthogonal mechanical properties and density of trabecular bone: use of different densitometric measures, J. Biomed. Mater. Res. 28 (11) (1994) 1329–1336.

[25] R. Crawford, C. Cann, T. Keaveny, Finite element models predict in vitro vertebral body compressive strength better than quantitative computed tomography, Bone 33 (4) (2003) 744–750.

[26] I. Castro-Mateos, J. Pozo, T. Cootes, J. Wilkinson, R. Eastell, A. Frangi, Statistical shape and appearance models in osteoporosis, Curr. Osteoporos. Rep. 12 (2014) 163–173.

[27] N. Sarkalkan, H. Weinans, A. Zadpoor, Statistical shape and appearance models of bones, Bone 60 (2014) 129–140.

[28] K. Fritscher, A. Grünerbl, M. Hanni, N. Suhm, C. Hengg, R. Schubert, Trabecular bone analysis in CT and X-ray images of the proximal femur for the assessment of local bone quality, IEEE Trans. Med. Imaging 28 (10) (2009) 1560–1575.

[29] M. Huber, J. Carballido-Gamio, K. Fritscher, R. Schubert, M. Haenni, C. Hengg, S. Majumdar, T. Link, Development and testing of texture discriminators for the analysis of trabecular bone in proximal femur radiographs, Med. Phys. 36 (11) (2009) 5089–5098.

[30] T. Whitmarsh, K. Fritscher, L. Humbert, L. Del-Rio Barquero, R. Schubert, A. Frangi, Hip fracture discrimination using 3D reconstructions from dual-energy X-ray absorptiometry, in: From Nano to Macro, 2011 I.E. International Symposium, Biomed. Imaging (2011) 1189–1192.

[31] M. de Bruijne, M. Lund, L.B. Tankó, P. Pettersen, M. Nielsen, Quantitative vertebral morphometry using neighbor-conditional shape models, Med. Image Anal. 11 (2007) 503–512.

[32] M. Roberts, E. Pacheco, R. Mohankumar, T. Cootes, J. Adams, Detection of vertebral fractures in DXA VFA images using statistical models of appearance and a semi-automatic segmentation, Osteoporos. Int. 21 (12) (2010) 2037–2046.

[33] T. Whitmarsh, K. Fritscher, L. Humbert, A statistical model of shape and bone mineral density distribution of the proximal femur for fracture risk assessment, in: Medical Image Computing and Computer-Assisted Intervention (MICCAI), 2011, pp. 393–400.

[34] T. Whitmarsh, K. Fritscher, L. Humbert, L. Del Rio Barquero, T. Roth, C. Kammerlander, M. Blauth, R. Schubert, A. Frangi, Hip fracture discrimination from dual-energy X-ray absorptiometry by statistical model registration, Bone 51 (2012) 896–901.

[35] B. Schuler, K. Fritscher, V. Kuhn, F. Eckstein, T. Link, R. Schubert, Assessment of the individual fracture risk of the proximal femur by using statistical appearance models, Med. Phys. 37 (6) (2010) 2560–2571.

[36] S. Goodyear, R. Barr, E. McCloskey, S. Alesci, R. Aspden, D. Reid, Can we improve the prediction of hip fracture by assessing bone structure using shape and appearance modelling?, Bone 53 (2013) 188–193.

[37] T.L. Bredbenner, R.L. Mason, L.M. Havill, E.S. Orwoll, D.P. Nicolella, Fracture risk predictions based on statistical shape and density modeling of the proximal femur, J. Bone Miner. Res. 29 (9) (2014) 2090–2100.

[38] K. Fritscher, B. Schuler, T. Link, F. Eckstein, N. Suhm, M. Hänni, C. Hengg, R. Schubert, Prediction of biomechanical parameters of the proximal femur using statistical appearance models and support vector regression, Int. Conf. Med. Image Comput. Comput. Interv. 11 (1) (2008) 568–575.

[39] K. Fritscher, M. Peroni, P. Zaffino, M. Spadea, R. Schubert, G. Sharp, Automatic segmentation of head and neck CT images for radiotherapy treatment planning using multiple atlases, statistical appearance models, and geodesic active contours, Med. Phys. 41 (5) (2014) 51910.

[40] K. Fritscher, A. Grünerbl, R. Schubert, InShape modeling – combined shape-intensity models for level set segmentation, in: CARS – Comput. Aided Radiol. Surgery, 2006.

[41] K. Fritscher, A. Grünerbl, R. Schubert, 3D image segmentation using combined shape-intensity prior models, Int. J. Comput. Assisted Radiol. Surg. 1 (6) (2007) 341–350.

[42] T. Cootes, C. Taylor, J. Graham, Active shape models – their training and applications, Comput. Vis. Image Underst. 61 (1995) 38–59.

[43] T. Cootes, G. Edwards, C. Taylor, Active appearance models, IEEE Trans. Pattern Anal. Mach. Intell. 23 (6) (2001) 681–685.

[44] M. Roberts, T. Cootes, E. Pacheco, J. Adams, Quantitative vertebral fracture detection on DXA images using shape and appearance models, Acad. Radiol. 14 (10) (2007) 1166–1178.

[45] C. Lindner, S. Thiagarajah, J. Wilkinson, G. Wallis, T. Cootes, Accurate bone segmentation in 2D radiographs using fully automatic shape model matching based on regression-voting, in: Medical Image Computing and Computer-Assisted Intervention, Proceedings, Part II, 2013, pp. 181–189.

[46] M. Roberts, T. Cootes, J. Adams, Vertebral morphometry: semiautomatic determination of detailed shape from dual-energy X-ray absorptiometry images using active appearance models, Invest. Radiol. 41 (12) (2006) 849–859.

[47] N. Navab, J. Hornegger, M. Wells, F. Frangi (Eds.), Medical Image Computing and Computer-Assisted Intervention, 18th International Conference, Proceedings, Part II, 2015.

[48] C. Lindner, S. Thiagarajah, J. Wilkinson, K. Panoutsopoulou, A. Day-Williams, T. Cootes, G. Wallis, Investigation of association between hip osteoarthritis susceptibility loci and radiographic proximal femur shape, Arthritis Rheumatol. 67 (8) (2015) 2076–2084.

[49] D. Pham, C. Xu, J. Prince, Current methods in medical image segmentation, Annu. Rev. Biomed. Eng. 2 (2000) 315–337.

[50] N. Sharma, L. Aggarwal, Automated medical image segmentation techniques, Med. Phys. 35 (1) (2010) 3–14.

[51] C. Xu, J. Prince, D. Pham, Image segmentation using deformable models, in: Handbook of Medical Imaging, 2000, pp. 129–174.

[52] T. Heimann, H. Meinzer, Statistical shape models for 3D medical image segmentation: a review, Med. Image Anal. 13 (4) (2009) 543–563.

[53] M. Stegmann, Generative Interpretation of Medical Images, PhD, Technical University of Denmark, 2004.

[54] T. Cootes, C. Taylor, D. Cooper, J. Graham, Training models of shape from sets of examples, in: BMVC92, 1992.

[55] P. Golland, M. Shenton, R. Kikinis, Small sample size learning for shape analysis of anatomical structures, in: Medical Image Computing and Computer-Assisted Intervention (MICCAI), 2000, pp. 72–82.

[56] P. Golland, W. Grimson, M. Shenton, R. Kikinis, Detection and analysis of statistical differences in anatomical shape, Med. Image Anal. 9 (1) (2005) 69–86.

[57] A. Frangi, D. Rueckert, J. Schnabel, W. Niessen, Automatic construction of multiple-object three-dimensional statistical shape models: application to cardiac modeling, IEEE Trans. Med. Imaging 21 (9) (2002) 1151–1166.

[58] D. Rueckert, L. Sonoda, C. Hayes, D. Hill, M. Leach, D. Hawkes, Non-rigid registration using free-form deformations: application to breast MR images, IEEE Trans. Med. Imaging 188 (1999) 721.

[59] T. Cootes, G. Edwards, C. Taylor, Active appearance models, in: Eur. Conf. Comput. Vis., 1998, pp. 484–498.

[60] G. Edwards, T. Cootes, C. Taylor, Face recognition using active appearance models, in: Comput. Vision – ECCV'98, 1998, pp. 581–595.

[61] X. Hou, S. Li, H. Zhang, Q. Cheng, Direct appearance models, in: Proc. 2001 IEEE Comput. Soc. Conf. Comput. Vis. Pattern Recognition, vol. 1, 2001, pp. 828–833.

[62] T. Vercauteren, X. Pennec, A. Perchant, N. Ayache, Non-parametric diffeomorphic image registration with the demons algorithm, in: Med. Image Comput. Comput. Assist. Interv., 2007, pp. 319–326.

[63] J.C. Gower, Generalized procrustes analysis, Psychometrika 40 (1975) 33, http://dx.doi.org/10.1007/BF02291478.

[64] A. Rangarajan, H. Chui, F. Bookstein, The Softassign Procrustes matching algorithm, Inf. Process. Med. Imag. (1997) 29–42.

[65] J. Besl, N. McKay, A method for registration of 3-D shapes, IEEE Trans. Pattern Anal. Mach. Intell. 14 (2) (1992) 239–256.

[66] S. Granger, X. Pennec, Multi-scale EM-ICP: a fast and robust approach for surface registration, in: European Conference on Computer Vision, 2002, pp. 418–432.

[67] D. Rückert, A. Frangi, J. Schnabel, Automatic construction of 3-D statistical deformation models of the brain using nonrigid registration, IEEE Trans. Med. Imaging 22 (8) (2003) 1014–1025.

[68] H. Chui, A. Rangarajan, A new point matching algorithm for non-rigid registration, Comput. Vis. Image Underst. 89 (2003) 114–141.

[69] B. Jian, B. Vemuri, Robust point set registration using Gaussian mixture models, IEEE Trans. Pattern Anal. Mach. Intell. 33 (2011) 1633–1645.

[70] A.F. Frangi, W.J. Niessen, D. Rueckert, J.A. Schnabel, Automatic 3D ASM construction via atlas-based landmarking and volumetric elastic registration, in: M.F. Insana, R.M. Leahy (Eds.), Information Processing in Medical Imaging: 17th International Conference, IPMI 2001, Davis, CA, USA, June 18–22, 2001, Springer, Berlin, Heidelberg, 2001, pp. 78–91.

[71] N. Paragios, M. Rousson, V. Ramesh, Non-rigid registration using distance functions, Comput. Vis. Image Underst. 89 (2) (2003) 142–165.

[72] P. Viola, W. Wells, Alignment of maximization of mutual information, Int. J. Comput. Vis. 22 (1) (1997) 61–97.

[73] P. Thirion, Image matching as a diffusion process: an analogy with Maxwell's demons, Med. Image Anal. 2 (3) (2004) 243–260.

[74] A. Sotiras, C. Davatzikos, N. Paragios, Deformable medical image registration: a survey, IEEE Trans. Med. Imaging 32 (2013) 1153–1190.

[75] I. Joliffe, Principal Component Analysis, 2nd ed., Springer, 2004.

[76] J. Jackson, A User's Guide to Principal Components, Wiley, 2003.

[77] C. Cortes, V. Vapnik, Support-vector networks, Mach. Learn. 20 (1995) 273–297.

[78] S. Mika, A. Smola, K. Müller, M. Scholz, G. Rätsch, B. Schölkopf, Kernel PCA and de-noising in feature spaces, Conf. Neural Inf. Process. Syst. 4 (5) (1998) 1–7.

[79] L. Van Der Maaten, E. Postma, H. Van Den Herik, Dimensionality reduction: a comparative review, J. Mach. Learn. Res. 10 (2009) 1–41.

[80] Y. Rathi, Statistical shape analysis using kernel PCA, Proc. SPIE (2006) 60641B.

[81] C. Twining, C. Taylor, Kernel principal component analysis and the construction of non-linear active shape models, in: BMVC92, 2001, pp. 23–32.

[82] A. Hyvärinen, J. Karhunen, E. Oja, S. Haykin, Independent Component Analysis, John Wiley and Sons Inc., 2001.

[83] L. Van Der Maaten, E. Postma, Dimensionality reduction: a comparative review, J. Mach. Learn. Res. 10 (2009) 1–41.

[84] S. Roweis, L. Saul, Nonlinear dimensionality reduction by locally linear embedding, Science 90 (5500) (2000) 2323–2326.

[85] M. Belkin, P. Niyogi, Laplacian eigenmaps for dimensionality reduction and data representation, Neural Comput. 15 (2003) 1373–1396.

[86] D. Donoho, C. Grimes, Hessian eigenmaps: locally linear embedding techniques for high-dimensional data, Proc. Natl. Acad. Sci. USA 100 (2003) 5591–5596.

[87] Z. Zhang, H. Zha, Principal manifolds and nonlinear dimensionality reduction via tangent space alignment, J. Shanghai Univ. 8 (4) (2004) 406–424.

[88] V. Caselles, R. Kimmel, G. Sapiro, Geodesic active contours, Int. J. Comput. Vis. 22 (1) (1997) 61–79.

[89] M. Leventon, E. Grimson, O. Faugeras, Statistical shape influence in geodesic active contours, Comput. Vis. Pattern Recognit. 1 (2000) 316–323.

[90] H. Seim, D. Kainmueller, M. Heller, H. Lamecker, S. Zachow, H. Hege, Automatic segmentation of the pelvic bones from CT data based on a statistical shape model, in: Eurographics Workshop on Visual Computing for Biomedicine (VCBM), 2008, pp. 93–100.

[91] J. Illingworth, J. Kittler, A survey of the hough transform, Comput. Vis. Graph. Image Process. 44 (1) (1988) 87–116.

[92] K. Li, X. Wu, D.Z. Chen, M. Sonka, Optimal surface segmentation in volumetric images – a graph-theoretic approach, IEEE Trans. Pattern Anal. Mach. Intell. 28 (1) (2006) 119–134.

[93] F. Yokota, T. Okada, M. Takao, N. Sugano, Y. Tada, Y. Sato, Automated segmentation of the femur and pelvis from 3D CT data of diseased hip using hierarchical statistical shape model of joint structure, Med. Image Comput. Comput. Assist. Interv. 12 (Pt 2) (2009) 811–818.

[94] T. Whitmarsh, L. Humbert, M. de Craene, L. Del Rio Barquero, A. Frangi, Reconstructing the 3D shape and bone mineral density distribution of the proximal femur from dual-energy X-ray absorptiometry, IEEE Trans. Med. Imaging 30 (12) (2011) 2101–2114.

[95] G. Zheng, Statistical shape model-based reconstruction of a scaled, patient-specific surface model of the pelvis from a single standard AP x-ray radiograph, Med. Phys. 37 (4) (2010) 1424–1439.

[96] E. Mortensen, W. Barrett, Interactive segmentation with intelligent scissors, Graph. Models Image Process. 60 (5) (1998) 349–384.

[97] T.-Y. Phillips, A. Rosenfeld, A method of curve partitioning using arc-chord distance, Pattern Recognit. Lett. 5 (4) (1987) 285–288.

[98] H. Lamecker, T. Wenckebach, H.-C. Hege, Atlas-based 3D-shape reconstruction from X-ray images, in: 18th International Conference on Pattern Recognition (ICPR'06), vol. 1, 2006, pp. 371–374.

[99] E. Basafa, R. Murphy, Y. Otake, M. Kutzer, Subject-specific planning of femoroplasty: an experimental verification study, J. Biomech. 48 (1) (2015) 59–64.

[100] J. Baker-LePain, K. Luker, J. Lynch, N. Parimi, M. Nevitt, N. Lane, Active shape modeling of the hip in the prediction of incident hip fracture, J. Bone Miner. Res. 26 (3) (2011) 468–474.

[101] B. Schuler, K. Fritscher, V. Kuhn, F. Eckstein, R. Schubert, Using a statistical appearance model to predict the fracture load of the proximal femur, Proc. SPIE 7261 (2009) 72610W.

[102] D. Varzi, S.A.F. Coupaud, M. Purcell, D.B. Allan, J.S. Gregory, R.J. Barr, Bone morphology of the femur and tibia captured by statistical shape modelling predicts rapid bone loss in acute spinal cord injury patients, Bone 81 (2015) 495–501.

[103] W. Li, J. Kornak, T. Harris, Y. Lu, X. Cheng, T. Lang, Hip fracture risk estimation based on principal component analysis of QCT atlas: a preliminary study, Proc. SPIE 7262 (2009).

[104] C. Harris, M. Stephens, A combined corner and edge detector, in: Proc. of Fourth Alvey Vision Conference, 1988, pp. 147–151.

[105] B. Reggiani, L. Cristofolini, E. Varini, M. Viceconti, Predicting the subject-specific primary stability of cementless implants during pre-operative planning: preliminary validation of subject-specific finite-element models, J. Biomech. 40 (2007) 2552–2558.

[106] K. Fritscher, B. Schuler, A. Grünerbl, M. Hänni, K. Schwieger, N. Suhm, R. Schubert, Assessment of femoral bone quality using co-occurrence matrices and adaptive regions of interest, Proc. SPIE 6514 (2007).

[107] H. Anys, D. He, Evaluation of textural and multipolarization radar features for crop classification, IEEE Trans. Geosci. Remote Sens. 33 (1995) 1170–1181.

[108] J. Serra, Image Analysis and Mathematical Morphology, Academic, London, 1982.

[109] P. Maragos, Fractal signal analysis using mathematical morphology, Adv. Electron. Electron Phys. 88 (1994) 199–246.

[110] H.F. Boehm, T. Link, R. Monetti, V. Kuhn, F. Eckstein, C. Raeth, M. Reiser, Analysis of the topological properties of the proximal femur on a regional scale: evaluation of multi-detector CT-scans for the assessment of biomechanical strength using local Minkowski functionals in 3D, Proc. SPIE 6144 (2006).

[111] D. Clausi, B. Yue, Comparing cooccurrence probabilities and Markov random fields for texture analysis of SAR Sea ice imagery, IEEE Trans. Geosci. Remote Sens. 42 (2004) 215–228.

[112] H. Boehm, C. Raeth, R. Monetti, D. Mueller, D. Newitt, S. Majumdar, E. Rummeny, G. Morfill, T. Link, Local 3D scaling properties for the analysis of trabecular bone extracted from high-resolution magnetic resonance imaging of human trabecular bone: comparison with bone mineral density in the prediction of biomechanical strength in vitro, Invest. Radiol. 38 (5) (2003) 269–280.

[113] D. Mueller, T. Link, R. Monetti, J. Bauer, H. Boehm, V. Seifert-Klauss, E. Rummeny, G. Morfill, C. Raeth, The 3D-based scaling index algorithm: a new structure measure to analyze trabecular bone architecture in high-resolution MR images in vivo, Osteoporos. Int. 17 (10) (2006) 1483–1493.

[114] M. Mueller, C. Hengg, M. Hirschmann, D. Schmid, C. Sprecher, L. Audige, N. Suhm, Mechanical torque measurement for in vivo quantification of bone strength in the proximal femur, Injury 43 (10) (2012) 1712–1717.

[115] K.E. Naylor, E.V. McCloskey, R. Eastell, L. Yang, Use of DXA-based finite element analysis of the proximal femur in a longitudinal study of hip fracture, J. Bone Miner. Res. 28 (5) (2013) 1014–1021.

[116] E.F. Morgan, T.M. Keaveny, Dependence of yield strain of human trabecular bone on anatomic site, J. Biomech. 34 (5) (2001) 569–577.

[117] E.F. Morgan, H.H. Bayraktar, T.M. Keaveny, Trabecular bone modulus–density relationships depend on anatomic site, J. Biomech. 36 (7) (2003) 897–904.

[118] L. Yang, L. Palermo, D. Black, R. Eastell, Prediction of incident hip fracture with the estimated femoral strength by finite element analysis of DXA scans in the study of osteoporotic fractures, J. Bone Miner. Res. 29 (12) (2014) 2594–2600.

[119] M. Qasim, G. Farinella, J. Zhang, X. Li, L. Yang, R. Eastell, M. Viceconti, Patient-specific finite element estimated femur strength as a predictor of the risk of hip fracture: the effect of methodological determinants, Osteoporos. Int. 27 (9) (2016) 2815–2822.

[120] L. Grassi, N. Hraiech, E. Schileo, M. Ansaloni, M. Rochette, M. Viceconti, Evaluation of the generality and accuracy of a new mesh morphing procedure for the human femur, Med. Eng. Phys. 33 (1) (2011) 112–120.

[121] E. Schileo, F. Taddei, L. Cristofolini, M. Viceconti, Subject-specific finite element models implementing a maximum principal strain criterion are able to estimate failure risk and fracture location on human femurs tested in vitro, J. Biomech. 41 (2) (2008) 356–367.

[122] C. Falcinelli, E. Schileo, L. Balistreri, F. Baruffaldi, B. Bordini, M. Viceconti, U. Albisinni, F. Ceccarelli, L. Milandri, A. Toni, F. Taddei, Multiple loading conditions analysis can improve the association between finite element bone strength estimates and proximal femur fractures: a preliminary study in elderly women, Bone 67 (2014) 71–80.

[123] E. Schileo, E. Dall'Ara, F. Taddei, A. Malandrino, T. Schotkamp, M. Baleani, M. Viceconti, An accurate estimation of bone density improves the accuracy of subject-specific finite element models, J. Biomech. 41 (11) (2008) 2483–2491.

[124] E. Schileo, L. Balistreri, L. Grassi, L. Cristofolini, F. Taddei, To what extent can linear finite element models of human femora predict failure under stance and fall loading configurations?, J. Biomech. 47 (14) (2014) 3531–3538.

[125] J. Thevenot, J. Koivumaki, V. Kuhn, F. Eckstein, T. Jamsa, A novel methodology for generating 3D finite element models of the hip from 2D radiographs, J. Biomech. 47 (2) (2014) 438–444.

[126] K. Rudman, R. Aspden, J. Meakin, Compression or tension? The stress distribution in the proximal femur, Biomed. Eng. Online 5 (2006) 12.

[127] J. Koivumäki, J. Thevenot, P. Pulkkinen, V. Kuhn, T. Link, F. Eckstein, T. Jämsä, Ct-based finite element models can be used to estimate experimentally measured failure loads in the proximal femur, Bone 50 (4) (2012) 824–829.

[128] N. Sarkalkan, J. Waarsing, P. Bos, H. Weinans, A. Zadpoor, Statistical shape and appearance models for fast and automated estimation of proximal femur fracture load using 2D finite element models, J. Biomech. 47 (12) (2014) 3107–3114.

[129] M. Stegmann, Active appearance models: theory, extensions and cases, Inform. Math. Modell. 262 (2000).

[130] S. Vaananen, L. Grassi, G. Flivik, J. Jurvelin, H. Isaksson, Generation of 3D shape, density, cortical thickness and finite element mesh of proximal femur from a DXA image, Med. Image Anal. 24 (1) (2015) 125–134.

[131] D. Nicolella, T. Bredbenner, Development of a parametric finite element model of the proximal femur using statistical shape and density modelling, Comput. Methods Biomech. Biomed. Eng. 15 (2) (2012) 101–110.

[132] R. Bryan, P.B. Nair, M. Taylor, Use of a statistical model of the whole femur in a large scale, multi-model study of femoral neck fracture risk, J. Biomech. 42 (13) (2009) 2171–2176.

[133] ZheEn Zhao, Eam Khwang Teoh, A new scheme for automated 3D 5PDM6 construction using deformable models, Image Vis. Comput. 26 (2) (2008) 275–288.

[134] J. Koikkalainen, T. Tolli, K. Lauerma, K. Antila, E. Mattila, M. Lilja, J. Lotjonen, Methods of artificial enlargement of the training set for statistical shape models, IEEE Trans. Med. Imaging 27 (11) (2008) 1643–1654.

[135] G. Guglielmi, F. Palmieri, M. Placentino, F. D'Errico, L. Stoppino, Assessment of osteoporotic vertebral fractures using specialized workflow software for 6-point morphometry, Eur. J. Radiol. 70 (1) (2009) 142–148.

[136] J. Yao, R. Taylor, Assessing accuracy factors in deformable 2D/3D medical image registration using a statistical pelvis model, in: Proceedings of the Ninth IEEE International Conference on Computer Vision – vol. 2, IEEE Computer Society, Washington, DC, 2003, p. 1329.

[137] P. Steininger, K. Fritscher, G. Kofler, B. Schuler, M. Hänni, K. Schwieger, R. Schubert, Comparison of different metrics for appearance-model-based 2D/3D-registration with X-ray images, in: Bildverarbeitung für die Medizin 2008: Algorithmen – Systeme – Anwendungen, Springer, Berlin, Heidelberg, 2008, pp. 122–126.

[138] G. Zheng, S. Gollmer, S. Schumann, X. Dong, T. Feilkas, M. Gonzalez Ballester, A 2D/3D correspondence building method for reconstruction of a patient-specific 3D bone surface model using point distribution models and calibrated X-ray images, Med. Image Anal. 13 (6) (2009) 883–899.

[139] M. Taylor, R. Bryan, F. Galloway, Accounting for patient variability in finite element analysis of the intact and implanted hip and knee: a review, Int. J. Numer. Methods Biomed. Eng. 29 (2) (2013) 273–292.

[140] S. Poelert, E. Valstar, H. Weinans, A. Zadpoor, Patient-specific finite element modeling of bones, Proc. Inst. Mech. Eng. H 227 (4) (2013) 464–478.

[141] J. Keyak, Improved prediction of proximal femur fracture load using nonlinear finite element models, Med. Eng. Phys. 23 (2001) 165–173.

[142] M. Bessho, I. Ohnishi, J. Matsuyama, T. Matsumoto, K. Imai, K. Nakamura, Prediction of strength and strain of the proximal femur by a CT-based finite element method, J. Biomech. 40 (8) (2007) 1745–1753.

[143] D. Cody, G. Gross, F. Hou, H. Spencer, S. Goldstein, D. Fyhrie, Femoral strength is better predicted by finite element models than QCT and DXA, J. Biomech. 32 (10) (1999) 1013–1020.

CHAPTER 16

Statistical Shape Models of the Heart: Applications to Cardiac Imaging

Concetta Piazzese*, M. Chiara Carminati†, Mauro Pepi*, Enrico G. Caiani‡
*Centro Cardiologico Monzino IRCCS, Milan, Italy
†Paul Scherrer Institut, Villigen, Switzerland
‡Dipartimento di Elettronica, Informazione e Bioingegneria, Politecnico di Milano, Italy

Contents

16.1	Introduction	445
16.2	The heart	446
	16.2.1 Cardiac Anatomy	446
	16.2.2 Cardiac Cycle	448
16.3	Cardiac Imaging Techniques	449
	16.3.1 Cardiac Magnetic Resonance	449
	16.3.1.1 Cine MR Imaging	*449*
	16.3.2 Real-Time Three-Dimensional Echocardiography	451
	16.3.3 Computed Tomography	451
16.4	Statistical Shape Models	453
	16.4.1 Statistical Shape Model Construction	455
	16.4.2 Statistical Shape Model Deformation	457
	16.4.3 Applications to Cardiac Imaging	458
	16.4.3.1 LV	*458*
	16.4.3.2 RV	*459*
	16.4.3.3 LV-RV	*460*
	16.4.3.4 Atria and Other Structures	*461*
	16.4.3.5 Whole Heart	*462*
16.5	Discussion	462
	References	473

16.1 INTRODUCTION

Cardiovascular diseases are the major cause of death in Europe with the 47% of mortality in women and men every year [81]. In the clinical routine, cardiac magnetic resonance imaging, 3D echocardiography and computed tomography are the most frequently used imaging modalities to assess global volumes and regional function, including stroke volume, ejection fraction and wall thickness. Therefore all these techniques play a crucial role in supporting the correct diagnosis.

Even if manual tracing on medical images remains the gold standard (GS), different automatic or semi-automatic segmentation algorithms have been proposed. Furthermore, recent advances in 3D imaging techniques have increased the implementation of medical image based 3D cardiac models for computational simulations, cardiac mechanics analysis or therapy guidance in procedures such as radiofrequency ablation.

To this respect, statistical shape models (SSMs) have become a powerful tool for cardiac image segmentation and analysis. The adoption of such models that incorporate prior shape knowledge benefits from the fact that the shape of the cardiac chambers with normal or altered function is approximately known. Consequently, the desired structure is detected by matching a predefined geometric shape to the locations of some features automatically or semi-automatically extracted.

In this chapter the application of SSMs in cardiac imaging to quantify the anatomy and the function of the heart will be reviewed. After an introduction section on the anatomy of the heart and imaging modalities, the discussion will be focused on the generation and the use of SSMs to segment the left ventricle (LV), which has been the object of most researches, the right ventricle (RV), both right and left ventricles, the four chambers or the entire heart. Furthermore, different imaging modalities will be analyzed in their ability to infer shape and motion information using SSMs.

16.2 THE HEART

16.2.1 Cardiac Anatomy

The heart is a muscular organ responsible for pumping blood to the lungs and through-out the body at a maintained pressure by repeated, rhythmic contractions [27]. Each day, the heart beats more than 100,000 times pumping about 5 liters of blood per minute [94]. Situated obliquely in the middle mediastinum of the thorax [125], it is cone-shaped with the size of a closed fist (Fig. 16.1 left). The superior end of the heart, where it connects to various veins and arteries (i.e. the aorta or pulmonary arteries and veins, and the vena cavae), is referred to as the base, whereas the blunt inferior tip of the heart is known as the apex.

The heart is enclosed by the pericardium, a double-walled layer that consists of an inner serous membrane called the epicardium and an outer fibrous membrane (Fig. 16.1 right). Underneath the epicardium there is the second layer of the wall of the heart, the myocardium, the layer of cardiac muscle tissue responsible for contraction of the heart. The third and innermost layer of the wall of the heart is the endocardium that acts as a lining for the myocardium and covers the chordae tendineae of the heart valves and the valves themselves [97].

Internally, the heart is divided into four chambers (Fig. 16.1): two upper chambers, i.e., the right atrium (RA) and the left atrium (LA), and two lower chambers, i.e. the RV and the LV. In the normal subject, one-third of the heart lies to the right of the midline and two-thirds to the left [126]. Also, the walls of the LV are three to six times thicker than those of the RV [34,119], because the LV pumps blood to most of the

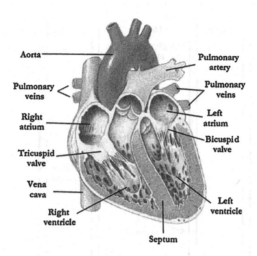

Figure 16.1 Illustration of the heart cut along the frontal plane showing cardiac interior anatomy with the four chambers, valves, septum and part of vessels. Adapted from Rizzo [97].

body (systemic circulation) and it has to sustain and overcome higher haemodynamic pressures than the RV in order to maintain an adequate cardiac output.

The RA has a predominantly smooth interior. The openings of the superior and inferior vena cavae, from which the deoxygenated blood from the body arrives to the heart, are on the roof and floor of the atrium, respectively. The valve connecting the RA to the RV is known as the tricuspid valve.

The LA has smooth walls like the RA. The posterior wall has different openings on either side corresponding to upper and lower pulmonary veins that receive oxygenated blood from the lungs. The mitral opening is located in the inferior part of the LA and, guarded by the mitral valve, allows blood to flow to the LV.

The RV is a lower chamber situated anteriorly immediately behind the sternum [47]. It is separated from the LV by the interventricular septum. The RV has a tripartite structure and can be described in terms of three components: the inlet, which consists of the tricuspid valve, chordae tendineae and papillary muscles; the trabeculated apical myocardium; and the infundibulum, or conus, which corresponds to the smooth myocardial outflow region [38]. The tricuspid valve is one of the upper openings, connecting the RV to the RA.

The LV, considered the primary pump of the heart, is located just below the LA and has a truncated conical ellipsoid shape (bullet shape). The endocardial surface of the LV is less trabeculated than that of the RV [56] and the muscular ridges tend to be relatively thin. The interventricular septum bulges into the cavity of the RV. Consequently, in cross-sectional images, the left ventricular lumen appears circular whereas the lumen of the RV appears crescentic [72].

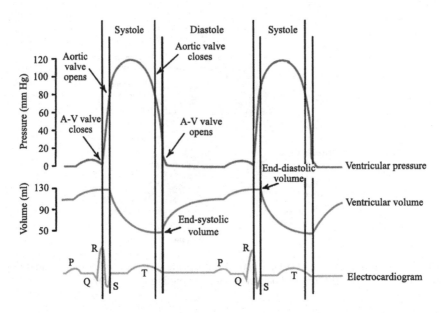

Figure 16.2 Events occurring during two complete cardiac cycles with ventricular pressure, ventricular volume, electrocardiogram and valve opening and closing related to different blood pressure. Adapted from Guyton and Hall [45].

The bicuspid valve, also known as mitral valve, represents the opening of the inflow tract. The LV communicates with the aorta, the largest vessel in the body that aims to transport oxygenated blood from the LV to every organ, through the opening guarded by the aortic semilunar valve. Like the pulmonary valve, the aortic valve is composed by three symmetric, semilunar-shaped cusps [50].

16.2.2 Cardiac Cycle

The cardiac cycle includes the events that occur from the beginning of one heartbeat to the beginning of the next one (Fig. 16.2).

During these events, the heart contracts and relaxes while blood flows through atria and ventricles. Each cycle is initiated by spontaneous generation of an action potential in the sinus node, located in the wall of the RA [45]. The conducting system then propagates and distributes this electrical impulse to stimulate contractile cells to pump blood in the right direction at the proper time. The cardiac cycle can be roughly summarized in two basic phases: diastole and systole. Systole is the term used to refer to a phase of contraction (the LV and the RV contract and eject blood into the aorta and pulmonary vein, respectively) and diastole is the term for a phase of relaxation (ventricles are relaxed and blood passively flows from the atria into ventricles through the atrioventricular valves) [97]. Alteration to one of these mechanisms could generate problems resulting in several cardiac pathologies.

The electrical events in the heart are powerful enough to be detected by electrodes on the surface of the body. A recording of these events is the electrocardiogram (ECG).

16.3 CARDIAC IMAGING TECHNIQUES

Cardiac imaging refers to different technologies or methods that can be used to obtain images and information about the heart. These advanced techniques help the clinicians to assess cardiac structures and function as well as abnormalities due to pathologies [111, 53,59,109,115] and to establish a correct diagnosis.

Cardiac imaging modalities can be divided in invasive and non-invasive [93]. In contrast to invasive techniques, which require to insert catheters inside the heart, the non-invasive imaging technologies are easier to be performed, cost effective and can be used to detect various heart conditions, ranging from myocardial infarction and fibrosis to suspected coronary and valvular diseases, abnormalities that impair the ability of the heart to pump blood [56].

16.3.1 Cardiac Magnetic Resonance

Cardiac magnetic resonance (MR) imaging had a rapidly increasing role in cardiology. It is considered the GS modality for the quantification of cardiac volume and it is routinely used for clinical evaluations [5]. The basic principles of cardiac MR imaging are essentially the same used for other structures of the body when assessed with MR imaging techniques.

When the heart is inside a magnetic field, it is possible to acquire different types of information to analyze and study distinct structures (blood, myocardium, fluids and masses) [14]. Two important sequences used in clinical practice are the spin-echo (SE) and the gradient-echo (GE) sequences [96]. Both sequences are often used to acquire the heart dynamics because their fast signal decay (GE decay time is faster than SE decay time) reduces artefacts (blurring or ghosting) due to the myocardial contraction. The recently introduced steady state free precession (SSFP) sequences, whose commercial names change company by company, are similar to GE ones but provide an excellent contrast between blood and myocardium and a good signal-to-noise ratio. Also, the SSFP sequences are less vulnerable to unwanted flow effects caused by the wash-in and wash-out of the pulsatile blood during the acquisition of a 2D plane [36,90].

16.3.1.1 Cine MR Imaging

Cine MR images are a set of high-quality data reconstructed by information collected over the course of multiple cardiac cycles during patient breath holding. Acquisition are usually based on SSFP sequences and gated with ECG.

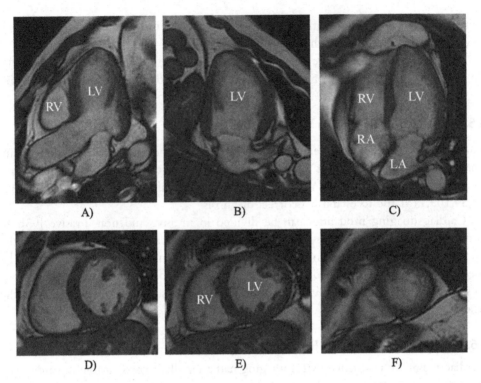

Figure 16.3 Cine images of the LV; A) 3-ch LAX view, B) 2-ch LAX view, C) 4-ch LAX view and D), E) and F) short axes at basal, middle and apical level.

In clinical practice, a cine MR examination contains for each slice between 20 to 30 frames/cardiac cycle [8] and it is performed in a very short time interval [58]. The images, acquired in a rapid succession during different phases of the cardiac cycle, can be displayed as a continuous movie loop.

As recommended by the American Heart Association (AHA), for a correct evaluation of the cardiac function, the oblique planes along the main axes of the heart should also be scanned [19]. For this reason, the standard acquired cardiac protocol includes: a stack of multiple closely spaced short-axis (SAX) slices (6 to 8 mm thick), (Fig. 16.3D, E and F), the horizontal long-axis (LAX) or four-chambers (4-ch) view and vertical LAX or two-chambers (2-ch) view (Fig. 16.3B and C). In addition to standard planes, it is possible to obtain other views, for example the three chamber (3-ch) view in which it is possible to evaluate the section of the ascending aorta (Fig. 16.3A).

The whole image set provides a full coverage of the LV and RV, thus potentially allowing the estimation of fundamental diagnostic indices such as end-diastolic (ED) and end-systolic (ES) volumes, stroke volume, ejection fraction (EF), wall thickness and myocardial mass.

16.3.2 Real-Time Three-Dimensional Echocardiography

Three-dimensional echocardiography (3DE) is a technique for real-time imaging of cardiac structures and function [101]. Its innovative technology allowed to overcome some limitations of the 2D echocardiography approach, such as the mental integration of 2D images to reach a 3D reconstruction of the heart. Furthermore, the transition from slow and labor-intense offline reconstruction to real-time volumetric imaging led to a wide diffusion of 3DE as an imaging tool for everyday clinical practice [61].

The probes or ultrasonic transducers used for echocardiography are able to generate mechanical vibrations and then record the reflected/backscattered echoes generated by the propagation of the pulse. In particular, the transducer is composed of piezoelectric sensors, sensors that can generate an electric signal when deformed. By steering the pulse in different directions (azimuthal and elevation), a volume can be acquired. Nowadays, a typical 3D transducer contains many 2D matrix arrays (Fig. 16.10A) for a total of 2000–3000 electrically individual elements (Fig. 16.10B) arranged in rows and column [10]. Scan lines, generated both azimuthally and elevationally, scan a pyramidal sector from which is possible to obtain 3D volume data (Fig. 16.10C).

Echocardiographic systems allow three acquisition modes: real time (narrow), zoom (magnified) and wide angle. The real-time mode displays a pyramidal data set while the zoom mode displays a smaller, magnified region of the pyramidal scan. The wide-angle mode acquires a large area because it merges 4 narrower pyramidal scans obtained over 4 consecutive heartbeats and for this reason it requires the ECG gating. Even if the larger cardiac structures are scanned with wide-angle mode, acquired images have a low resolution if compared to those obtained with narrow-angle 3D mode [48]. Nowadays, using state-of-the-art equipment, wide-angle mode acquisition is feasible in a single beat with improvements in both spatial and temporal resolution.

Information from volumetric data can be displayed in three ways: slice rendering (Fig. 16.4), in which 2D images are generated by cutting the volume in any desired cross-section (even from physically unavailable cross-sections); volume rendering, generated using the ray casting method; surface rendering, in which a geometrical description (e.g. a triangulated mesh) of cardiac structure is generated by manual or automatic boundary outlining.

3DE is nowadays the recommended imaging modality to assess the LV function and mitral valve anatomy and stenosis [62,2,52,42]. This technology is also used to study the RV, both atria and the tricuspid valve [99,86,79,9].

16.3.3 Computed Tomography

Computed tomography (CT) is a non-invasive imaging modality that allows to obtain different section of the heart by irradiating the patient with a rotating X-ray source [103]. CT imaging of the heart is mainly used for the quantification of coronary artery calcification and less frequently, for minimally invasive coronary angiography.

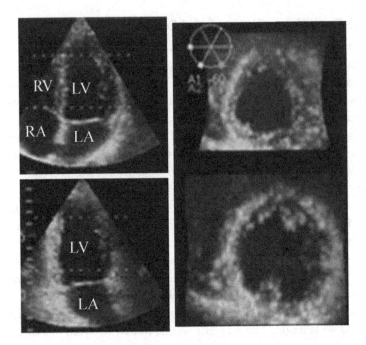

Figure 16.4 Visualization of the 3D data sliced in any desired cross-section. Adapted from Badano et al. [10].

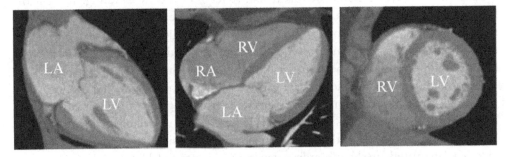

Figure 16.5 Visualization of the standard views of cardiac chambers (64-slice scanner). Adapted from Schoenhagen et al. [102].

The basic principle of CT is to continuously generate different X-rays at different angles. As the patient is automatically moved through the scanner gantry, the X-ray beam passes through the body and its intensity is attenuated according to the density of tissues encountered. The resulting rays are then collected by fixed solid-state detector located in the gantry surrounding the patient and transformed into an image (Fig. 16.5) in which attenuation values are defined with respect to the attenuation of the water and expressed in Hounsfield units.

CT technology has continuously improved since its introduction into clinical practice. With electron beam imaging, characterized by the absence of moving parts, CT images can be obtained in 50–100 ms which is rapid enough to essentially freeze cardiac motion. With helical imaging, characterized by multi-row detectors, from 1 to 16 slices can be obtained at each gantry rotation increasing the acquisition time to 125–500 ms [55].

Beyond a high temporal resolution, CT have high spatial resolution (0.5 mm) so that coronary arteries can be visualizes both with and without contrast enhancement.

Different clinical studies proved the ability of the CT to provide similar information as cardiac MR [63,29]. However, its use in clinical practice is still limited due to the radiation dose that should be the lowest as possible. Also, clinical indications for cardiac CT must always take radiation exposure into account [30,6].

16.4 STATISTICAL SHAPE MODELS

Segmentation includes the processing of an image in order to extract some features or an area of interest in order to separate them from the background. In cardiology, medical data are usually segmented to obtain and quantify clinically important parameters, including ED and ES volume, EF, wall motion, wall thickness, stroke volume, ischemic dilation and myocardial mass [16,80,130,57,17]. Manual delineation is still considered the GS for evaluating cardiac images but it requires a lot of time, expert knowledge and it is affected by inter/intra-observer variability [69]. For these reasons, different computer-aided or fully automated segmentation algorithms have been proposed in the last decade [91] for the segmentation of biomedical images. Even if the processing time has been greatly improved, the automatic or semi-automatic contour extraction is still not considered as accurate as manual contour tracing [73].

According to the classification proposed by Petitjean et al. [91], segmentation techniques can be divided primarily in two groups: based on no or weak priors and based on strong priors.

Prior information can consist in weak or strong assumptions such as simple spatial relationships between objects (for instance, the RV is to the left of the LV) or anatomical assumptions making use of the geometry of the objects, respectively. The amount of a priori knowledge used during the segmentation process has a big impact on the detection accuracy and on the level of user interaction required [54,13,107]. Segmentation methods with weak prior knowledge are usually combined with user interaction at different strengths (mostly moderate ones) and include image-driven techniques, such as thresholding, region-based and edge-based techniques or pixel classification and deformable models. Moreover, due to the none or little prior knowledge incorporated about the organ to assess, their use is limited to simple structures.

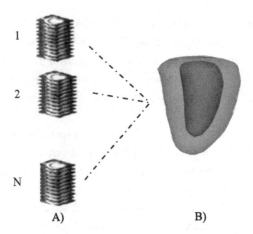

Figure 16.6 Generation of the SSM (B) from a training set of 2D images (A). Adapted from Roohi and Zoroofi [98].

Low-level image processing techniques are usually developed "ad-hoc" for a specific object showing weak robustness to segment target regions that change significantly from one subject to another (for example in pathological patients). In this scenario, it is easy to understand the rapid spread and success of methods relying on strong prior knowledge. Embedding a priori information (i.e. geometric shape, relationships between object) increases stability against local image artefacts and perturbations as well as accuracy and robustness against inter- and intra-subject shape variability [120]. Furthermore, such strong priors can relax the need for user intervention but at the expense of collecting a large training set [21].

Methods using strong prior information, commonly known as SSM techniques, can be categorized [91] as: shape-driven deformable models, atlas-based models and active shape models (ASM).

Shape-driven deformable models evolve following an energy minimization procedure [44] and include a set of a priori learned template shapes [88,70]. Atlas-based models are based on the rigid or non-rigid registration of template images with the data to segment [66,131,87,132,11].

ASMs could be defined as an extension of atlas-based models which encodes shape variability across a training set [24]. The trained model is then used to segment new unseen image with the advantage that it can assume only the plausible shapes included in the dataset.

Segmentation techniques based on ASM attempt to match a predefined geometric shape to the locations of the features extracted from an image. This is accomplished by a two-step procedure: generation of the ASM from a training dataset (Fig. 16.6) and development of an ASM search algorithm to correctly deform the model on the desired structures in the images.

16.4.1 Statistical Shape Model Construction

Building an SSM that contains information about shapes and their variations is not a simple task. A shape (2D or 3D) is usually represented by a set of points, commonly referred to as landmarks, lying on a surface [28]. A set of points containing information about the connectivity between landmarks is called mesh or point distribution model (PDM), as named by Cootes et al. [20]. In order to model the knowledge of the dataset, the training samples are aligned (Fig. 16.7A) in a common coordinate space by applying the generalized Procrustes algorithm that minimizes the mean squared distance between two shapes that have a one-to-one point correspondence [37]. Usually, every set of points in the database is rigidly or affinely registered to a shape arbitrarily chosen as reference.

Samples without a well-defined landmarks correspondence should be aligned (mesh-to-mesh registration) with other matching algorithms. The most popular ones are the iterative closest point (ICP) algorithm [15] and the Softassign Procrustes [95] that tries to find the transformation between surfaces with a potentially different number of vertices. To improve robustness of ICP, Granger and Pennec [41] combined it with the expectation maximization (EM) algorithm, while Toldo et al. [117] proposed the integration of the generalized Procrustes analysis into an ICP framework.

A similarity transformation is not sufficient for a database that includes samples with large shape variations. In these cases, ICP could lead to wrong correspondences. For this reason, other techniques, such as non-rigid registration using B- or Octree splines, minimization of a cost function based on similarity, unsupervised clustering of structure and prior information, should be used [110,33,106,32].

When the training dataset used to build the SSM does not include meshes but rather segmented volumes, other approaches are necessary to compute the alignment step. In particular, a mesh-to-volume registration is usually performed when a deformable surface model is adapted to the segmented binary volumes and the correspondences between the vertex locations of the deformable template are defined. Instead of adapting a template mesh to the training data, it is also possible to register a volumetric atlas (volume-to-volume registration) by propagating the landmarks placed on the atlas to the training data [128,35].

Usually training databases incorporate a lot of knowledge the most of which is superfluous. To discard redundant and noisy parameters, a statistical analysis is performed to limit the information to meaningful features. Principal component analysis (PCA) has been established as the most used multivariate statistical technique since it is able to reduce a complex dataset, described by several dependent variables, to a lower di-

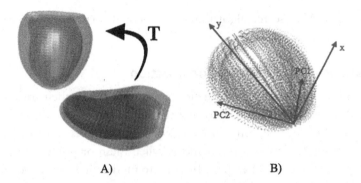

A)   B)

Figure 16.7 Steps to construct SSM; A) alignment of a training sample to a reference shape and B) database dimensionality reduction by using PCA. Adapted from Parajuli et al. [89].

mensional space by keeping only significant information, expressed as a set of new orthogonal variables called principal components (Fig. 16.7B) or modes [1].

Other decomposition methods include independent component analysis (ICA) or kernel PCA. ICA, also known as Blind Source Separation [49], does not require a Gaussian distribution of the data and provides statistically independent projections. Kernel PCA [104] is an extension of the PCA in which the database is non-linearly decomposed avoiding to create invalid shapes which is a common problem of PCA when training samples vary too much.

Reducing the dimensionality of the training set by using PCA is nowadays a must-have step in any model-based segmentation framework [67,85], while application of ICA can be found in a limited number of works [122,112].

An ASMs can be extended to Active Appearance Models (AAM) by including also the texture (intensities or colors across the image patch) variations [22]. In this case, in order to model the texture information, a set of labeled patches are used to train the model. Usually variation of shape and texture are combined together so to be described by common parameters but, since shape and texture have different units, the parameters need to be processed in order to make them commensurable.

The statistical quality of an SSM can be evaluated using the measures introduced by Davies [26]: specificity, generalization ability and compactness. Compactness is a measure of the variance of the model. Specificity is the ability of the model to generate valid shapes that are similar to those in the training set [121]. It is assessed qualitatively by generating a population of N random shapes using the model and comparing them to the closest sample in the training set. Generalization ability is the capability of an SSM to represent unseen instances (new shapes) and it is assessed using leave-one-out test: a test shape excluded from the training set is reconstructed using an SSM built using the remaining shapes [75]. All these measures are strongly related to the number of modes (or parameters) used by the model. The best statistical quality is associated to small values of specificity, generalization ability and compactness.

Recently, Bai et al. [12] used the surface reconstruction error as a measure to assess the representative power of a statistical model. This measure is similar to the generalization ability (with the only difference that the reconstructed shape was used to train the model) and quantify how accurately a training shape is reconstructed using the model.

16.4.2 Statistical Shape Model Deformation

In general, SSMs are deformed to adapt them to the structures to be detected in medical images. As for the low-level segmentation techniques, the first step of an ASM matching algorithm requires an initial estimate of the model pose. Even if the manual initialization of the model pose remains the preferred solution, Lekadir and Yang [65] delivered a fully automatic estimation of the initial pose based on graph search and inter-landmark conditional probabilities achieving reasonably good localization.

After placing the SSM on the image, segmentation is performed by iteratively estimating rotation, translation and scaling. These parameters are usually computed with the Procrustes algorithm [40] that minimizes the mean squared distance between the detected features and the model. The process leads to obtain a vector of residual displacements that are analyzed so to be sure to warping the mesh excluding landmark configurations that are not present in the dataset. This is a double-edged sword because when an SSM is built from small databases with limited number of shapes (a very common situation in medical application due to the small number of available data) and low samples variability, too stringent constraints are imposed to the model deformation process [127]. To overcome this problem, Albà et al. [3] recently proposed to segment pathological patients with a model built from normal population only. By transforming the input image data onto the normal space, the ASM search is performed only in a reference template representing a normal heart, thus without the need to train and use an SSM for each specific pathology.

Another possibility is to create an SSM using artificial image of the ventricles [118, 105]. As reported by Koikkalainen et al. [60], the two best methods to enlarge a training set are the non-rigid movement, in which locally deformations are generated, and the PCA and FEM technique [25], in which new shapes are created by globally deforming the original training set. Tobon-Gomez et al. [116] evaluated the ability of SSMs trained on a database composed by simulated cardiac MR images to segment ventricular cavities from real datasets of RV and LV. The results showed that when proper ground-truth meshes with point correspondence are not available, SSMs trained on simulated data can be deformed on real cardiac MR images with a little loss of accuracy.

Feature points or contours used to deform the mean shape are detected on new images by using different local search algorithms based on image-driven techniques such as thresholding, region-growing, clustering, pixel or voxel classification [84,71,44, 46,78].

In the following, a brief description of some relevant applications for modeling and segmenting cardiac structures using an SSM approach is provided. An overview of all sorted publications can be found in Table 16.1.

16.4.3 Applications to Cardiac Imaging
16.4.3.1 LV

In the work of van Assen et al. [123] a 3D ASM was generated from 2D manually segmented contours. In particular, the correspondence between landmarks was achieved by sampling the 2D curves at equidistant angles with respect to the LV long axis. PCA was used to compute modes of variation and a fuzzy c-means clustering was applied to segment SAX MR and CT images. Detected endocardium and epicardium borders were then processed to extract scaling, rotation and translation parameters required to update the model.

In the work of Santiago et al. [100], the basic approach to construct an ASM was improved in order to be able to construct a model from 3D MRI volumes with a different number of slices. O'Brien et al. [82] built a 4D ASM by processing shape, spatial and temporal variation with a kernel PCA. ASMs have also been used to detect and localize abnormally contracting regions of the myocardium [113].

High-resolution multi-slice CT data were used by Swoboda et al. [114] to create an ASM of the LV and use it to segment simulated and in-vivo angiograms. In particular, during the training phase the common one-to-one points correspondence problem was solved with a back-propagation of the landmarks on a mean shape.

An inter-modality SSM approach was proposed by Caiani et al. [18] in which an intrinsically 3D ASM was trained on a large database of surfaces extracted from 3D echocardiographic images of the LV. The model was then used to segment cardiac MR images of pathological and healthy subjects (Fig. 16.8). In a more recent work, Piazzese et al. [92] used the same inter-modality SSM approach to evaluate the segmentation accuracy of different 3D ASMs built by varying the temporal information (number of cardiac frames) included in the database and the type of registration used during the training samples alignment.

In cardiac field, ASMs have been extended to active appearance models (AAM) to ensure to have a realistic solution and a precise match to the image texture [23] by including also a modeling of the gray video-intensity levels [7,77,51]. Recently, Leiner et al. [64] created a 4D AAM of the LV from a set of CT images and computed mesh updating parameters (angle of rotation, translation vector and a scaling factor) with a weighted sum of Mahalanobis distances function.

Fetal echocardiographic images were used in the work of Vargas-Quintero et al. [124] to build an AAM based on the Hermite transform (HT), a common mathematical

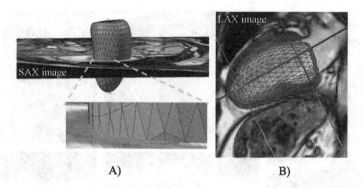

Figure 16.8 The ASM, trained on intrinsically 3D surfaces extracted from 3DE images of the LV, is deformed on the base of features extracted from SAX cardiac MR images until a stable condition is reached (A). The final result superimposed on a LAX cardiac MR image (B). Adapted from Caiani et al. [18].

tool for image analysis applications to locally decompose an image using a Gaussian function and orthogonal polynomials.

A hybrid LV model combining ASM and AAM technique was proposed by Zhang et al. [127] by training a 3D and a 3D + time model on cine cardiac MR images. In particular, SAX and LAX views, corrected for breathing motion and fused in a single 4D dataset, were manually traced and used as database to generate two different models: a 4D model that contains temporal information and a 3D model that includes shape/texture data of all cardiac phases. For each ventricle, the 4D model was first used to achieve robust preliminary segmentation on all cardiac phases simultaneously with a full hybrid ASM/AAM process and a 3D model was then applied to each phase to improve local accuracy while maintaining the overall robustness of the 4D segmentation.

Temporal information was also used by Gopal et al. [39] who exploited a trained AAM to segment the LV in MR images at the ED phase and then mapped the resulting ED model to the ES phase with a deformable superquadratic approach.

16.4.3.2 RV

For the RV, Mansi et al. [74] created an ASM using cine MR images from which to quantify the regional impact of valve regurgitation and heart growth in patients with tetralogy of Fallot. PCA was computed to estimate variables, predictors and local effects of RV regurgitation. A two-dimensional extension of the traditional PCA was used by ElBaz and Fahmy [31] to capture the inter-profile relations among shape's landmarks and the variations between the different training shapes.

In the works of Grosgeorge et al. [44], Sedai et al. [105] and Moolan-Feroze et al. [78] the RV was segmented on cardiac MR images using an ASM approach in combination with a pixel classification based method or shape regressions. Grosgeorge

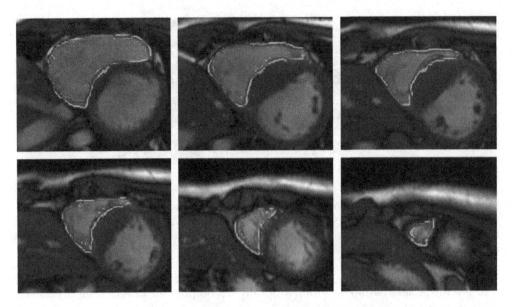

Figure 16.9 Manual tracings (red) and segmented contours (green) obtained by incorporating the SSM into the graph-cut technique. Adapted from Grosgeorge et al. [44]. (For interpretation of the references to color in this figure, the reader is referred to the web version of this chapter.)

et al. [44] incorporated the ASM, trained on manual delineation of 2D SAX images, into the graph-cut technique so to guide the RV segmentation (Fig. 16.9).

The learned RV shape information was included in the Markov Random Field formulation in the work of Moolan-Feroze et al. [78] to iteratively define the pose of the model on the image and extract the segmented RV contours.

The training set used to create an ASM of the RV was artificially enlarged by generating synthetic data in the approach proposed by Sedai et al. [105]. A cascade of shape regressors were then trained on the augmented database and applied to segment the RV cavity in new SAX MR images by determining the probability of each pixel composing the edge detected on a Gaussian distance map.

Oghli et al. [83] created two ASM of the RV (one for the apical and basal slices and one for the mid-ventricular slices) and integrated them as a force term into a deformable model formulation for curve evolution to segment the RV cavity in patients with arrhythmogenic right ventricular dysplasia.

16.4.3.3 LV-RV

Examples of biventricular segmentation by using an SSM can be found in the model-based approaches proposed by Mitchell et al. [77], Ordas et al. [84], Lötjönen et al. [68], Albà et al. [4], Mitchell et al. [76].

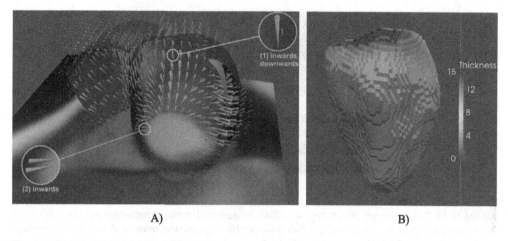

Figure 16.10 Statistical motion model superimposed on the 2D image (A) and LV myocardial wall thickness map (B) used to apply the statistical parametric mapping. Adapted from Bai et al. [12].

Recently, Bai et al. [12] constructed a model of the LV and RV by using a large database (1000+) of 3D cardiac MR images. To overcome the time consuming process of manual segmenting the training data, a semi-automatic atlas-based approach was used to segment the images. The 3D shapes obtained with the marching cubes algorithm were used to train two SSMs: an SSM to study the variation of cardiac shapes and a statistical motion model to assess the distribution of cardiac motion across the spatio-temporal domain (Fig. 16.10A). Furthermore, a statistical parametrical mapping, an approach widely used in neuroimaging to analyze each voxel (or each vertex) using statistical test and present the resulting statistical parameters on an image or on a mesh, was performed to evaluate the regional myocardial wall thickness (Fig. 16.10B).

16.4.3.4 Atria and Other Structures

To the best of our knowledge, few works investigated a strategy to create an SSM of the LA. Stender et al. [108] developed an ASM strategy to segment the LA in both CT and MR images. In particular, each modality-specific ASM was trained on manual contours and deformed on the new images on the base of different gradient features extracted.

The complex geometry of the LA was analyzed and segmented separately by Zheng et al. [129] that proposed a part-based LA model including the chamber, the appendage, four major PVs, and right middle PVs (Fig. 16.11).

A complete and modular model of the cardiac valves including the anatomy of the aortic, mitral, tricuspid and pulmonary valves was proposed by Grbić et al. [43].

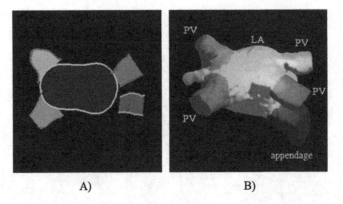

Figure 16.11 Part based left atrium segmentation: A) detected contours superimposed on a 2D mask; B) resulting mesh of the LA (cyan), the PVs (orange, blue, green and magenta) and the appendage (dark red). Adapted from Zheng et al. [129]. (For interpretation of the references to color in this figure, the reader is referred to the web version of this chapter.)

16.4.3.5 Whole Heart

A complete heart model comprising atria and ventricles can be found in the work of Lötjönen et al. [67]. The 3D SSM was created from manually segmented MR images. In particular, a triangulated surface, modeled with PCA and ICA, was iteratively deformed on both SAX and LAX views using landmark probability distribution, probability atlas and non-rigid registration.

CT angiographic images were used by Unberath et al. [121] to build the first open-source dynamical model (3D + t) of the heart (descending aorta, four chambers, and left ventricular myocardium). In this approach, in order to preserve the point-to-point correspondence among samples, the training images are registered using a similarity transform and mutual information and then segmented using an atlas-based segmentation.

A similar approach was exploited by Haak et al. [46] to derive an ASM containing the LV, RV, LA, RA and aorta from CT angiographic images (Fig. 16.12A). The model was then used to segment the heart in TEE images (Fig. 16.12B) with three-stage segmentation scheme: initial pose estimation, global pose and shape updating, individual cavities pose and shape updating. At each iteration, the model was consecutively fitted to tissue probability maps estimated by using tissue/blood classification based on a gamma mixture model.

16.5 DISCUSSION

This chapter presented the most relevant applications based on SSM techniques applied to cardiac images to model and segment the heart.

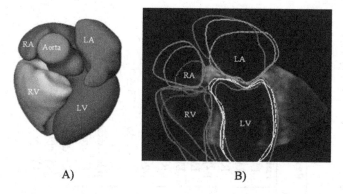

A)   B)

Figure 16.12 Whole heart mean shape of the ASM trained on CT angiographic images (A) and final segmented contours obtained with the ASM approach (dotted lines) and two manual observers (solid and dashed lines) superimposed on 3D TEE volume (B). Adapted from Haak et al. [46].

Since their introduction, SSMs gained more and more attention, becoming nowadays one of the most powerful technique used in the cardiac field. By using a collection of data, it is possible to train an SSM and use it to segment complex cardiac structures. Realistic shape solutions are guaranteed by the shape variability of the training set that constrain the model's deformation during the matching process.

Different variations have been applied to the first ASM algorithm presented by Cootes et al. [21] so to adapt and apply it to different cardiac structures. Most of the SSM approaches present in literature are applied to the LV for its crucial role in developing the correct diagnosis and also for its simple shape. Only a small amount of SSM-based works are focused on other cardiac chambers or on the whole heart segmentation. For such structures, a multi-part SSM, in which the complex shape is decomposed into sub-structures (i.e. the atrium is divided in chamber, vessels and appendage), revealed a more accurate detection.

By including the temporal information, a generic SSM is transformed into a dynamic SSM to detect cardiac structures in all the phases of the cardiac cycle and to derive regional parameters (i.e. local displacement and thickness, regional curves) useful to assess regional wall motion abnormalities.

Grey level variations, combination with image-driven techniques are other possible elements that can be included into an SSM to improve the segmentation accuracy. In general, the choice of the SSM technique to implement depends on the structure to be segmented and on the available training dataset.

The typical problems of the SSM technique such as a small training set, that impose too stringent constraints to the model deformation process, or the point-to-point correspondence problem can be overcome by adopting one of the solution proposed in literature.

Table 16.1 Overview of SSM applications in cardiac field. SSM, statistical shape model; AAM, active appearance model; PCA, principal component analysis; ICA, independent component analysis; LV, left ventricle; RV, right ventricle; LV-RV, bi-ventricle; LA, left atrium; SAX, short-axis; LAX, long-axis; MR, magnetic resonance; CT, computed tomography; TTE, transthoracic echocardiographic; TEE, transesophageal echocardiographic; pat, pathological subjects; he, healthy subjects; pat-he, pathological and healthy subjects; p2p/ p2c/p2s/s2s, point-to-point/point-to-contour/point-to-surface/surface-to-surface distance; rms, root-mean-square; DM, Dice similarity coefficient; HD, Hausdorff distance; CC, correlation coefficient; endo, endocardium; epi, epicardium

SSM applications

References	Techniques	Structure	Training set	User interaction	Test cases	Results
Mitchell et al. [76]	ASM + AAM	LV-RV	102 SAX MR images (pat-he)	manual tracing of training set	60 SAX MRI (pat-he)	**rms error:** LV endo = 1.71 ± 0.82; LV epi = 1.75 ± 0.83; RV endo = 2.46 ± 1.39; **p2p distance [mm]:** LV endo = 0.22 ± 1.90; LV epi = −0.01 ± 1.92; RV endo = −0.32 ± 2.80;
Mitchell et al. [77]	AAM	LV	MR: leave-one-out; TEE: 64 TTE images	manual tracing of training set	56 SAX MR images (pat-he); 65 TTE images	**p2s distance (MR) [mm]:** LV endo = 2.75 ± 0.86; LV epi = 2.63 ± 0.76; **p2s distance (TEE) [mm]:** LV endo = 3.90 ± 1.38;
Ordas et al. [84]	IOF-ASM	LV-RV	21 SAX MR images (pat-he)	manual tracing of training set	74 SAX MR images (pat-he)	**p2c distance [mm]:** LV endo = 1.80 ± 1.74; LV epi = 1.52 ± 2.01; RV endo = 1.20 ± 1.47;
Üzümcü et al. [122]	ICA-based SSM	LV/LV-RV	51 LAX MR images; 89 X-ray angiography images; 150 SAX MR images	manual tracing of training set	–	–

continued on next page

Table 16.1 (continued)

SSM applications

References	Techniques	Structure	Training set	User interaction	Test cases	Results
Lötjönen et al. [67]	SSM	whole heart	leave-one-out cross-validation	manual tracing of training set	25 SAX + LAX MR images (he)	**s2s distance (PCA) [mm]:** whole model = 3.31 ± 1.17; **s2s distance (ICA) [mm]:** whole model = 3.32 ± 0.89; **s2s distance [mm]** reported for LV, RV, LA, RA and epicardium separately
Tölli et al. [118]	ASM	whole heart	leave-one-out cross-validation	manual tracing of training set	25 SAX + LAX MR images (he); artificial enlarged training set	**segmentation error [mm]** reported for the different structures using the standard PCA method and the artificial training set enlargement
Ordas et al. [85]	ASM	whole heart + aorta	100 multi-slice CT images (pat)	manual tracing of training set and points selection	–	–
Andreopoulos and Tsotsos [7]	3D AAM + 2D ASM	LV	three-fold cross-validation	manual tracing of training set	7980 SAX MR images (pat)	**volume error:** endo = 6.10 ± 6.44; epi = 9.82 ± 8.97;
Koikkalainen et al. [60]	ASM	whole heart	leave-one-out cross-validation	manual tracing of training set	25 SAX + LAX MR images (he); different artificial enlarged training set	**s2s distance [mm]** and **DM** reported for the different structures using the standard PCA method and the artificial training set enlargement

continued on next page

Table 16.1 (*continued*)
SSM applications

References	Techniques	Structure	Training set	User interaction	Test cases	Results
Lekadir and Yang [65]	ASM + graph search + outlier correction	LV	leave-one-out	manual tracing of training set	20 SAX + LAX MR images (pat-he)	**segmentation error [mm]:** 2D = 1.13 ± 0.32; 3D = 1.11 ± 0.25;
van Assen et al. [123]	ASM + fuzzy inference	LV	53 SAX MR images (pat-he)	manual tracing of training set	25 reconstructed SAX CT images (pat); 15 SAX MR images (he)	**p2p distance [mm]:** CT endo $= 1.31 - 2.60(1.85)$; CT epi $= 1.02 - 2.97(1.60)$; MR endo $= 1.34 - 2.05(1.72)$; MR epi $= 1.27 - 1.85(1.55)$;
Suinesiaputra et al. [113]	ICA-based statistical model detection	LV	44 SAX MR images (he)	manual tracing of training set	45 SAX MR images (pat)	percentage values of **accuracy, sensitivity and specificity** reported for the different slice levels (basal, middle and apical)
Grbić et al. [43]	constrained multi-linear PCA-ICA shape model	valves	three-fold cross-validation	points selection	640 4D CT images (pat)	**p2s distance [mm]:** aortic valve $= 1.22 \pm 0.38$; mitral valve $= 1.32 \pm 0.57$; pulmonary valve $= 1.35 \pm 0.9$; tricuspic valve $= 1.40 \pm 1.41$; **Euclidean distance [mm]:** aortic valve $= 2.65 \pm 1.50$; mitral valve $= 2.75 \pm 1.19$; pulmonary valve $= 3.50 \pm 2.53$; tricuspic valve $= 3.59 \pm 2.55$;

continued on next page

Table 16.1 (*continued*)

SSM applications

References	Techniques	Structure	Training set	User interaction	Test cases	Results
Zhang et al. [127]	ASM + AAM	LV-RV	five-fold cross-validation	manual tracing of training set	50 SAX MR images (pat-he)	**p2s distance (healthy) [mm]:** LV endo = 1.67 ± 0.30; LV epi = 1.81 ± 0.40; RV endo = 2.13 ± 0.39; **DM (healthy):** LV endo = 89.9 ± 2.7; LV epi = 91.8 ± 1.9; RV endo = 85.2 ± 2.9; **p2s distance (TOF subjects) [mm]:** LV endo = 1.71 ± 0.45; LV epi = 1.97 ± 0.58; RV endo = 2.92 ± 0.73; **DM (TOF subjects):** LV endo = 89.6 ± 3.2; LV epi = 90.6 ± 3.1; RV endo = 84.2 ± 4.2;
Mansi et al. [74]	non-parametric SSM	RV	ten-fold cross-validation	manual refinement of the fitting (if necessary)	49 SAX MR images (pat)	**p2p distance [mm]:** 1.2 ± 0.4;

continued on next page

Table 16.1 (*continued*)
SSM applications

References	Techniques	Structure	Training set	User interaction	Test cases	Results
O'Brien et al. [82]	ASM + global contour optimization	LV	x SAX MR images (pat), with $x = 1, \ldots, 33$	manual tracing of training set	$(33 - x)$ SAX MR images (pat)	average **p2c distance [mm]** reported for the different levels of training and testing data
ElBaz and Fahmy [31]	AAM + 2D PCA + multi-stage searching	RV	five-fold cross-validation	–	90 SAX MR images (pat)	**p2c distance [mm]:** 0.73 ± 0.37; **DM:** 0.78 ± 0.06;
Gopal et al. [39]	AAM + deformable super-quadric model + correspondence model	LV	leave-one-out	manual tracing of training set	35 SAX MR images (pat)	**Mean perpendicular distance [mm]:** endo ED = 2.50 ± 1.06; endo ES = 4.20 ± 1.93; epi ED = 2.73 ± 1.61; epi ES = 3.27 ± 1.62; **DM:** endo ED = 0.88 ± 0.04; endo ES = 0.73 ± 0.07; epi ED = 0.90 ± 0.05; epi ES = 0.87 ± 0.05;
Leiner et al. [64]	AAM (3D + t)	LV	10 CT images	model initialization	several time-series (10 volumes per series)	**Mean Euclidean distance** reported separately for each time-series

continued on next page

Table 16.1 (*continued*)
SSM applications

References	Techniques	Structure	Training set	User interaction	Test cases	Results
Tobon-Gomez et al. [116]	AAM	LV-RV	400 simulated images (pat-he); 40 SAX MR images (pat-he)	real data: manual tracing of training set and points selection	simulated AAM: 400 simulated images (pat-he), 95 SAX MR images (pat-he); real AAM: 95 SAX MR images (pat-he)	**p2s distance [mm]** presented as box plots for LV endo, LV epi and RV endo.
Grosgeorge et al. [44]	ASM + graph-cut	RV	16 SAX MR images (pat)	manual tracing of training set and points selection	16 SAX MR images (pat)	**p2c distance [mm]:** 2.32 ± 1.57; **DM:** 0.83 ± 0.15;
Inamdar and Ramdasi [51]	AAM	LV	10 SAX MR images (he)	–	–	–
Stender et al. [108]	ASM	LA	10 CT/10 MR images	manual tracing of training set	20 CT/20 MR images	**DM:** $CT = 0.83 \pm 0.05$; $MR = 0.83 \pm 0.06$;
Caiani et al. [18]	ASM	LV	205 3D TTE images (pat-he)	selection of points	24 SAX MR images (pat-he)	**CC:** volume $= 0.97$ $EF = 0.91$
Oghli et al. [83]	Shape based deformable model	RV	30 SAX MR images (pat)	manual tracing of training set	30 ptz SAX MR images (pat)	**Sensitivity (%):** 93.9 ± 5.82; **Specificity (%):** 89.5 ± 3.52;

continued on next page

Table 16.1 (continued)

SSM applications

References	Techniques	Structure	Training set	User interaction	Test cases	Results
Zheng et al. [129]	ASM + Marginal Space Learning	LA	four-fold cross-validation	–	434 small C-ARM CT volumes; 253 small C-ARM CT volumes;	**s2s distance [mm]:** small C-ARM CT volumes = 1.57 ± 0.70; large C-ARM CT volumes = 1.65 ± 0.75;
Bai et al. [12]	SSM + statistical motion model + statistical parametric mapping	LV-RV	1093 3D MR images (he)	manual tracing of training set (20 subjects) and selection of points	–	–
Haak et al. [46]	ASM	whole heart + aorta	151 CT angiography (pat)	points selection	20 TEE (pat)	**p2s distance [mm]:** LV = 2.82; RV = 5.17; LA = 3.10; RA = 5.70; Ao = 3.34; **DM:** LV = 90.0; RV = 79.5; LA = 87.4; RA = 64.8; Ao = 70.1;
Santiago et al. [100]	Robust ASM	LV	leave-one-out	manual tracing of training set	20 SAX MR images (pat)	**p2c distance [mm]:** 1.3 ± 0.7; **DM:** 0.88 ± 0.07;

continued on next page

Table 16.1 (*continued*)
SSM applications

References	Techniques	Structure	Training set	User interaction	Test cases	Results
Sedai et al. [105]	ASM with shape Regression	RV	15 SAX MR images (pat)	points selection	15 SAX MR images (pat)	**DM:** 0.87 ± 0.06; **HD [mm]:** 6.20 ± 2.50;
Swoboda et al. [114]	ASM	LV	20 3D CT angiographic images (he)	manual tracing of training set	simulated: leave-one-out; in-vivo: 3 in-vivo angiograms	**p2s distance [mm]:** simulated = 2.61 ± 0.65; in-vivo = 2.57 ± 0.34.
Unberath et al. [121]	ASM (3D + t)	heart	20 CT angiographic images (he)	manual tracing of training set	–	–
Albà et al. [4]	AAM	LV–RV	20 SAX MR images (he)	points selection	40 SAX MR images (pat)	**p2s distance [mm]:** pulmonary hypertension = 2.60 ± 0.34; hypertrophic cardiomyopathy = 2.57 ± 0.46;
Ma et al. [71]	ASM + GMM	LV	SAX MR images (pat): 13 basal slices, 45 mid slices, 18 apical slices	manual tracing of training set	10 SAX MR images (pat)	**average perpendicular distance [mm]:** endo = 2.26; epi = 2.13; **DM:** endo = 0.88; epi = 0.93; **percentage of good contours (%):** endo = 91.54; epi = 96.29;

continued on next page

Table 16.1 (*continued*)
SSM applications

References	Techniques	Structure	Training set	User interaction	Test cases	Results
Moolan-Feroze et al. [78]	ASM	RV	16 SAX MR images (pat)	manual tracing of training set	16 SAX MR images (pat)	**DM:** ED phase = 0.87 ± 0.09; ES phase = 0.78 ± 0.17; **HD [mm]:** ED phase:9.72 ± 4.97; ES phase:9.71 ± 4.68;
Piazzese et al. [92]	ASM	LV	435 3D TTE images (pat-he)	selection of points	90 SAX MR images (pat-he)	CC reported separately for 15 SSMs built using different strategies
Vargas-Quintero et al. [124]	AAM con Hermite transform	LV	98 fetal echocardiographic images	manual rotation and scaling of the training images	45 fetal echocardiographic images	**p2c distance** and **DM** reported separately for 5 patients

In conclusion, SSMs are a powerful tool to segment different cardiac structures in medical images. The results of the model-based approaches proposed in these years proved the robustness of the technique to the recent advances in imaging techniques and therefore reinforced the idea that in the future SSM will continue to be extensively used to guide cardiac image segmentation.

REFERENCES

[1] H. Abdi, L.J. Williams, Principal component analysis, Wiley Interdiscip. Rev.: Comput. Stat. 2 (4) (2010) 433–459.

[2] K. Addetia, V. Mor-Avi, R.M. Lang, Impact of three-dimensional echocardiography, Argent. J. Cardiol. 82 (2) (2014) 145–156.

[3] X. Albà, K. Lekadir, C. Hoogendoorn, M. Pereanez, A.J. Swift, J.M. Wild, A.F. Frangi, Reusability of statistical shape models for the segmentation of severely abnormal hearts, in: Statistical Atlases and Computational Models of the Heart-Imaging and Modelling Challenges, Springer International Publishing, 2014, pp. 257–264.

[4] X. Albà, M. Pereañez, C. Hoogendoorn, A.J. Swift, J.M. Wild, A.F. Frangi, K. Lekadir, An algorithm for the segmentation of highly abnormal hearts using a generic statistical shape model, in: IEEE Trans. Med. Imaging, 2016, pp. 845–859.

[5] K. Alfakih, T. Bloomer, S. Bainbridge, G. Bainbridge, J. Ridgway, G. Williams, M. Sivananthan, A comparison of left ventricular mass between two-dimensional echocardiography, using fundamental and tissue harmonic imaging, and cardiac MRI in patients with hypertension, Eur. J. Radiol. 52 (2) (2004) 103–109.

[6] E.S. Amis Jr., P.F. Butler, K.E. Applegate, S.B. Birnbaum, L.F. Brateman, J.M. Hevezi, F.A. Mettler, R.L. Morin, M.J. Pentecost, G.G. Smith, K.J. Strauss, R.K. Zeman, American College of radiology white paper on radiation dose in medicine, J. Am. Coll. Radiol. 4 (2007) 272–284.

[7] A. Andreopoulos, J.K. Tsotsos, Efficient and generalizable statistical models of shape and appearance for analysis of cardiac MRI, Med. Image Anal. 12 (3) (2008) 335–357.

[8] M. Anuloi, J. Brorzino, D.R. Peterson, Medical Imaging: Principles and Practices, CRC Press, 2012.

[9] A.M. Anwar, M.L. Geleijnse, O.I. Soliman, J.S. McGhie, R. Williams, A. Nemes, A.E. van den Bosch, T.W. Galema, F.J. Ten Cate, Assessment of normal tricuspid valve anatomy in adults by real-time three-dimensional echocardiography, Int. J. Cardiovasc. Imaging 23 (6) (2007) 717–724.

[10] L.P. Badano, R.M. Lang, J.L. Zamorano, Textbook of Real-Time Three Dimensional Echocardiography, Springer, 2011.

[11] W. Bai, W. Shi, H. Wang, N.S. Peters, D. Rueckert, Multi-atlas based segmentation with local label fusion for right ventricle MR images, in: Workshop in Medical Image Computing and Computer Assisted Intervention, 2012.

[12] W. Bai, W. Shi, A. de Marvao, T.J. Dawes, D.P. O'Regan, S.A. Cook, D. Rueckert, A bi-ventricular cardiac atlas built from 1000+ high resolution MR images of healthy subjects and an analysis of shape and motion, Med. Image Anal. 26 (1) (2015) 133–145.

[13] I. Ben Ayed, K. Punithakumar, S. Li, A. Islam, J. Chong, Left ventricle segmentation via graph cut distribution matching, Med. Image Comput. Comput. Assist. Interv. 12 (Pt 2) (2009) 901–909.

[14] M.A. Bernstein, K.F. King, X.J. Zhou, Handbook of MRI Pulse Sequences, illustrated ed., Elsevier, 2004.

[15] P.J. Besl, N.D. McKay, A method for registration of 3-D shapes, IEEE Trans. Pattern Anal. Mach. Intell. 14 (2) (1992) 239–256.

[16] F. Beygui, A. Furber, S. Delépine, G. Helft, J.P. Metzger, P. Geslin, J.J. Le Jeune, Routine breath-hold gradient echo MRI–derived right ventricular mass, volumes and function: accuracy, reproducibility and coherence study, Int. J. Card. Imaging 20 (6) (2004) 509–516.

[17] E.V. Buechel, T. Kaiser, C. Jackson, A. Schmitz, C.J. Kellenberger, Normal right- and left ventricular volumes and myocardial mass in children measured by steady state free precession cardiovascular magnetic resonance, J. Cardiovasc. Magn. Reson. 11 (2009) 19.

[18] E.G. Caiani, A. Colombo, M. Pepi, C. Piazzese, F. Maffessanti, R.M. Lang, M.C. Carminati, Three-dimensional left ventricular segmentation from magnetic resonance imaging for patient-specific modelling purposes, Europace 16 (Suppl. 4) (2014), iv96–iv101.

[19] M.D. Cerqueira, N.J. Weissman, V. Dilsizian, A.K. Jacobs, S. Kaul, W.K. Laskey, D.J. Pennell, J.A. Rumberger, T. Ryan, M.S. Verani, American Heart Association Writing Group on Myocardial Segmentation and Registration for Cardiac Imaging, Standardized myocardial segmentation and nomenclature for tomographic imaging of the heart. A statement for healthcare professionals from the Cardiac Imaging Committee of the Council on Clinical Cardiology of the American Heart Association, Circulation 105 (4) (2002) 539–542.

[20] T.F. Cootes, C.J. Taylor, D. Cooper, J. Graham, Training models of shape from sets of examples, in: Proc. British Machine Vision Conference, Springer, 1992.

[21] T.F. Cootes, C.J. Taylor, D.H. Cooper, J. Graham, Active shape models – their training and application, Comput. Vis. Image Underst. 61 (1) (1995) 38–59.

[22] T.F. Cootes, G.J. Edwards, C.J. Taylor, Active appearance models, in: H. Burkhardt, B. Neumann (Eds.), Proc. ECCV, vol. 2, 1998.

[23] T.F. Cootes, G.J. Edwards, C.J. Taylor, Comparing active shape models with active appearance models, in: T. Pridmore, D. Elliman (Eds.), Proc. British Machine Vision Conference, vol. 1, 1999, pp. 173–182.

[24] T.F. Cootes, E.R. Baldock, J. Graham, An introduction to active shape models, in: Image Processing and Analysis, 2000, pp. 223–248.

[25] T.F. Cootes, C.J. Taylor, Statistical Models of Appearance for Computer Vision, Technical report, University of Manchester, Wolfson Image Analysis Unit, Imaging Science and Biomedical Engineering, Manchester M13 9PT, United Kingdom, 2004.

[26] R.H. Davies, Learning shape: optimal models for analysing shape variability, Ph.D. thesis, University of Manchester, 2002.

[27] M. Deo, Modelling the Role of the Purkinje System in Cardiac Arrhythmias, University of Calgary, 2008.

[28] I.L. Dryden, K.V. Mardia, Statistical Shape Analysis, Wiley & Sons, 1998.

[29] G. Dwivedi, H. Al-Shehri, R.A. deKemp, et al., Scar imaging using multislice computed tomography versus metabolic imaging by F-18 FDG positron emission tomography: a pilot study, Int. J. Cardiol. 168 (2013) 739–745.

[30] A.J. Einstein, K.W. Moser, R.C. Thompson, M.D. Cerqueira, M.J. Henzlova, Radiation dose to patients from cardiac diagnostic imaging, Circulation 116 (2007) 1290–1305.

[31] M.S. ElBaz, A.S. Fahmy, Active shape model with inter-profile modeling paradigm for cardiac right ventricle segmentation, Med. Image Comput. Comput. Assist. Interv. 15 (Pt 1) (2012) 691–698.

[32] L. Ferrarini, H. Olofsen, W.M. Palm, M.A. van Buchem, J.H.C. Reiber, F. Admiraal-Behloul, GAMEs: growing and adaptive meshes for fully automatic shape modeling and analysis, Med. Image Anal. 11 (3) (2007) 302–314.

[33] M. Fleute, S. Lavallée, R. Julliard, Incorporating a statistically based shape model into a system for computer-assisted anterior cruciate ligament surgery, Med. Image Anal. 3 (3) (1999) 209–222.

[34] R. Foale, P. Nihoyannopoulos, W. McKenna, A. Kleinebenne, A. Nadazdin, E. Rowland, G. Smith, A. Klienebenne, Echocardiographic measurement of the normal adult right ventricle, Br. Heart. J. 56 (1) (1986) 33–44.

[35] A.F. Frangi, D. Rueckert, J.A. Schnabel, W.J. Niessen, Automatic construction of multiple-object three-dimensional statistical shape models: application to cardiac modeling, IEEE Trans. Med. Imaging 21 (9) (2002) 1151–1166.

[36] D.T. Ginat, M.W. Fong, D.J. Tuttle, S.K. Hobbs, R.C. Vyas, Cardiac imaging: Part 1, MR pulse sequences, imaging planes, and basic anatomy, Am. J. Roentgenol. 197 (4) (2011) 808–815.

[37] C. Goodall, Procrustes methods in the statistical analysis of shape, J. R. Stat. Soc. B 53 (2) (1991) 285–339.

[38] D.A. Goor, C.W. Lillehei, Congenital malformations of the heart, in: D.A. Goor, C.W. Lillehei (Eds.), Congenital Malformations of the Heart: Embryology, Anatomy, and Operative Considerations, 1st ed., Grune & Stratton, New York, NY, 1975, pp. 1–37.

[39] S. Gopal, Y. Otaki, R. Arsanjani, D. Berman, D. Terzopoulos, P. Slomka, Combining active appearance and deformable superquadric models for LV segmentation in cardiac MRI, in: SPIE Medical Imaging, International Society for Optics and Photonics, 2013.

[40] J. Gower, Generalized Procrustes analysis, Psychometrika 40 (1975) 33–51.

[41] S. Granger, X. Pennec, Multi-scale EM-ICP: a fast and robust approach for surface registration, in: European Conference on Computer Vision, vol. 2353, 2002, pp. 418–432.

[42] P.A. Grayburn, N.J. Weissman, J.L. Zamorano, Quantitation of mitral regurgitation, Circulation 126 (2012) 2005–2017.

[43] S. Grbić, R. Ionasec, D. Vitanovski, I. Voigt, Y. Wang, B. Georgescu, N. Navab, D. Comaniciu, Complete valvular heart apparatus model from 4D cardiac CT, Med. Image Comput. Comput. Assist. Interv. 13 (Pt 1) (2010) 218–226.

[44] D. Grosgeorge, C. Petitjean, J.N. Dacher, S. Ruan, Graph cut segmentation with a statistical shape model in cardiac MRI, Comput. Vis. Image Underst. 117 (9) (2013) 1027–1035.

[45] A.C. Guyton, J.E. Hall, Guyton Hall and Textbook of Medical Physiology, 11th ed., Elsevier Saunders, Philadelphia, 2006.

[46] A. Haak, G. Vegas-Sánchez-Ferrero, H.W. Mulder, B. Ren, H.A. Kirişli, C. Metz, G. van Burken, M. van Stralen, J.P. Pluim, A.F. van der Steen, T. van Walsum, J.G. Bosch, Segmentation of multiple heart cavities in 3-D transesophageal ultrasound images, IEEE Trans. Ultrason. Ferroelectr. Freq. Control 62 (6) (2015) 1179–1189.

[47] S.Y. Ho, P. Nihoyannopoulos, Anatomy, echocardiography, and normal right ventricular dimensions, Heart 92 (Suppl. 1) (2006) i2–i13.

[48] J. Hung, R. Lang, F. Flachskampf, S. Shernan, M.L. McCulloch, D.B. Adams, J. Thomas, M. Vannan, T. Ryan, 3D echocardiography: a review of the current status and future directions, J. Am. Soc. Echocard. 20 (2007) 213–233.

[49] A. Hyvarinen, J. Karhunen, E. Oja, Independent Component Analysis, John Wiley & Sons, 2001.

[50] P. Iazzo, Handbook of Cardiac Anatomy, Physiology, and Devices, 2009.

[51] R.S. Inamdar, D.S. Ramdasi, Active appearance models for segmentation of cardiac MRI data, in: 2013 International Conference on Communications and Signal Processing (ICCSP), 2013, pp. 96–100.

[52] L.D. Jacobs, I.S. Salgo, S. Goonewardena, L. Weinert, P. Coon, D. Bardo, O. Gerard, P. Allain, J.L. Zamorano, L.P. de Isla, V. Mor-Avi, R.M. Lang, Rapid online quantification of left ventricular volume from real-time three-dimensional echocardiographic data, Eur. Heart J. 27 (4) (2006) 460–468.

[53] A. Jain, H. Tandri, H. Calkins, D.A. Bluemke, Role of cardiovascular magnetic resonance imaging in arrhythmogenic right ventricular dysplasia, J. Cardiovasc. Magn. Reson. 10 (1) (2008) 32, http://dx.doi.org/10.1186/1532-429X-10-32.

[54] K. Jaspers, H.G. Freling, K. van Wijk, E.I. Romijn, M.J. Greuter, T.P. Willems, Improving the reproducibility of MR-derived left ventricular volume and function measurements with a semi-automatic threshold-based segmentation algorithm, Int. J. Card. Imaging 29 (3) (2013) 617–623.

[55] G.C. Kagadis, in: U. Joseph Schoepf (Ed.), CT of the Heart. Principles and Applications, Med. Phys. 32 (5) (2005) 1452.

[56] B. Kastler, MRI of Cardiovascular Malformations, 1st ed., Springer, Berlin/Heidelberg, 2011.

[57] F. Khalifa, G. Beache, M. Nitzken, G. Gimel'farb, A. El-Baz, Automatic analysis of left ventricle wall thickness using short-axis cine CMR images, in: IEEE International Symposium on Biomedical Imaging: From Nano to Macro, 2011, pp. 1306–1309.

[58] H.W. Kim, A. Faraneh-Far, R.J. Kim, Cardiovascular magnetic resonance in patients with myocardial infarction, J. Am. Coll. Cardiol. 55 (2010) 1–16.

[59] A.L. Knauth, K. Gauvreau, A.J. Powell, M.J. Landzberg, E.P. Walsh, J.E. Lock, P.J. Lock, T. Geva, Ventricular size and function assessed by cardiac MRI predict major adverse clinical outcomes late after tetralogy of Fallot repair, Heart 94 (2) (2008) 211–216.

[60] J. Koikkalainen, T. Tölli, K. Lauerma, K. Antila, E. Mattila, M. Lilja, J. Lötjönen, Methods of artificial enlargement of the training set for statistical shape models, IEEE Trans. Med. Imaging 27 (11) (2008) 1643–1654.

[61] R.M. Lang, V. Mor-Avi, J.M. Dent, C.M. Kramer, Three-dimensional echocardiography: is it ready for everyday clinical use? JACC Cardiovasc. Imaging 2 (2009) 114–117.

[62] R.M. Lang, V. Mor-Avi, L. Sugeng, P.S. Nieman, D.J. Sahn, Three-dimensional echocardiography: the benefits of the additional dimension, J. Am. Coll. Cardiol. 48 (2006) 2053–2069.

[63] J.B. le Polain de Waroux, A.C. Pouleur, C. Goffinet, et al., Combined coronary and late-enhanced multidetector-computed tomography for delineation of the etiology of left ventricular dysfunction: comparison with coronary angiography and contrast-enhanced cardiac magnetic resonance imaging, Eur. Heart J. 29 (2008) 2544–2551.

[64] B.J. Leiner, J. Olveres, B. Escalante-Ramrez, F. Armbula, E. Vallejo, Segmentation of 4D cardiac computed tomography images using active shape models, in: Proceedings of SPIE Medical Imaging, 2012.

[65] K. Lekadir, G.Z. Yang, Optimal feature point selection and automatic initialization in active shape model search, Med. Image Comput. Comput. Assist. Interv. 29 (Pt 1) (2008) 434–441.

[66] M. Lorenzo-Valdés, G.I. Sanchez-Ortiz, A.G. Elkington, R.H. Mohiaddin, D. Rueckert, Segmentation of 4D cardiac MR images using a probabilistic atlas and the EM algorithm, Med. Image Anal. 8 (3) (2004) 255–265.

[67] J. Lötjönen, S. Kivisto, J. Koikkalainen, D. Smutek, K. Lauerma, Statistical shape model of atria, ventricles and epicardium from short- and long-axis MR images, Med. Image Anal. 8 (3) (2004) 371–386.

[68] J. Lötjönen, V.M. Järvinen, B. Cheong, E. Wu, S. Kivistö, J.R. Koikkalainen, J.J. Mattila, H.M. Kervinen, R. Muthupillai, F.H. Sheehan, K. Lauerma, Evaluation of cardiac biventricular segmentation from multiaxis MRI data: a multicenter study, J. Magn. Reson. Imaging 28 (3) (2008) 626–636.

[69] S.E. Luijnenburg, D. Robbers-Visser, A. Moelker, H.W. Vliegen, B.J. Mulder, W.A. Helbing, Intraobserver and interobserver variability of biventricular function, volumes and mass in patients with congenital heart disease measured by CMR imaging, Int. J. Card. Imaging 26 (1) (2010) 57–64.

[70] M. Lynch, O. Ghita, P.F. Whelan, Left-ventricle myocardium segmentation using a coupled level-set with a priori knowledge, Comput. Med. Imaging Graph. 30 (4) (2006) 255–262.

[71] Y. Ma, D. Wang, Y. Ma, R. Lei, M. Dong, K. Wang, L. Wang, An effective approach for automatic LV segmentation based on GMM and ASM, in: International Conference on Neural Information Processing, Springer International Publishing, 2016, pp. 663–672.

[72] V. Mahadevan, Anatomy of the heart, Surgery (Oxford) 22 (6) (2004) 121–123.

[73] A. Mahnken, G. Mühlenbruch, R. Koos, S. Stanzel, P. Busch, M. Niethammer, R. Günther, J. Wildberger, Automated vs. manual assessment of left ventricular function in cardiac multidetector row computed tomography: comparison with magnetic resonance imaging, J. Eur. Radiol. 16 (7) (2006) 1416–1423.

[74] T. Mansi, I. Voigt, B. Leonardi, X. Pennec, S. Durrleman, M. Sermesant, H. Delingette, A.M. Taylor, Y. Boudjemline, G. Pongiglione, N. Ayache, A statistical model for quantification and prediction of cardiac remodelling: application to tetralogy of Fallot, IEEE Trans. Med. Imaging 30 (9) (2011) 1605–1616.

[75] M. Mayya, S. Poltaretskyi, C. Hamitouche, J. Chaoui, Mesh correspondence improvement using regional affine registration: application to statistical shape model of the scapula, IRBM 36 (4) (2015) 220–232.

[76] S.C. Mitchell, B.P. Lelieveldt, R.J. van der Geest, H.G. Bosch, J.H. Reiber, M. Sonka, Multistage hybrid active appearance model matching: segmentation of left and right ventricles in cardiac MR images, IEEE Trans. Med. Imaging 20 (5) (2001) 415–423.

[77] S. Mitchell, J. Bosch, B. Lelieveldt, R. van der Geest, J. Reiber, M. Sonka, 3-D active appearance models: segmentation of cardiac MR and ultrasound images, IEEE Trans. Med. Imaging 21 (9) (2002) 1167–1178.

[78] O. Moolan-Feroze, M. Mirmehdi, M. Hamilton, Right ventricle segmentation using a 3D cylindrical shape model, in: 2016 IEEE 13th International Symposium on Biomedical Imaging (ISBI), IEEE, 2016, pp. 44–48.

[79] V. Mor-Avi, C. Yodwut, C. Jenkins, H. Kühl, H.J. Nesser, T.H. Marwick, A. Franke, L. Weinert, J. Niel, R. Steringer-Mascherbauer, B.H. Freed, L. Sugeng, R.M. Lang, Real-time 3D echocardiographic quantification of left atrial volume: multicenter study for validation with CMR, JACC Cardiovasc. Imaging 5 (8) (2012) 769–777.

[80] C. Nguyen, E. Kuoy, S. Ruehm, M. Krishnam, Reliability and reproducibility of quantitative assessment of left ventricular function and volumes with 3-slice segmentation of cine steady-state free precession short axis images, Eur. J. Radiol. 84 (7) (2015) 1249–1258, http://dx.doi.org/10.1016/j.ejrad.2015.03.019.

[81] M. Nichols, N. Townsend, P. Scarborough, R. Luengo-Fernandez, J. Leal, A. Gray, M. Rayner, European Cardiovascular Disease Statistics 2012, European Heart Network/European Society of Cardiology, Brussels/Sophia Antipolis, 2012.

[82] S.P. O'Brien, O. Ghita, P.F. Whelan, A novel model-based 3D+time left ventricular segmentation technique, IEEE Trans. Med. Imaging 30 (2) (2011) 461–474.

[83] M.G. Oghli, V. Dehlaghi, A.M. Zadeh, A. Fallahi, M. Pooyan, Right ventricle functional parameters estimation in arrhythmogenic right ventricular dysplasia using a robust shape based deformable model, J. Med. Signals Sens. 4 (3) (2014) 211–222.

[84] S. Ordas, L. Boisrobert, M. Huguet, A. Frangi, Active shape models with invariant optimal features (IOF-ASM) application to cardiac MRI segmentation, in: Computers in Cardiology, 2003, pp. 633–636.

[85] S. Ordas, E. Oubel, R. Leta, F. Carreras, A.F. Frangi, A statistical shape model of the heart and its application to model-based segmentation, in: Proc. SPIE, Med. Imaging, vol. 6511, 2007.

[86] E. Ostenfeld, F.A. Flachskampf, Assessment of right ventricular volumes and ejection fraction by echocardiography: from geometric approximations to realistic shapes, Echo Res. Practice 2 (1) (2015) R1–R11.

[87] Y. Ou, J. Doshi, G. Erus, C. Davatzikos, Multi-atlas segmentation of the right ventricle in cardiac MRI, in: Proceedings of MICCAI RV Segmentation Challenge, 2012.

[88] N. Paragios, A level set approach for shape-driven segmentation and tracking of the left ventricle, IEEE Trans. Med. Imaging 22 (6) (2003) 773–776.

[89] N. Parajuli, A. Lu, J. Duncan, Left ventricle classification using active shape model and support vector machine, in: Statistical Atlases and Computational Models of the Heart, Imaging and Modelling Challenges, Springer International Publishing, 2015, pp. 154–161.

[90] F.S. Pereles, V. Kapoor, J.C. Carr, et al., Usefulness of segmented true FISP cardiac pulse sequence in evaluation of congenital and acquired adult cardiac abnormalities, Am. J. Roentgenol. 177 (5) (2001) 1155–1160.

[91] C. Petitjean, J.N. Dacher, A review of segmentation methods in short axis cardiac MR images, Med. Image Anal. 15 (2) (2011) 169–184.

[92] C. Piazzese, M.C. Carminati, A. Colombo, R. Krause, M. Potse, A. Auricchio, L. Weinert, G. Tamborini, M. Pepi, R.M. Lang, E.G. Caiani, Segmentation of the left ventricular endocardium from magnetic resonance images by using different statistical shape models, J. Electrocardiol. 49 (3) (2016) 383–391.

[93] S.K. Prasad, R.G. Assomull, D.J. Pennell, Recent developments in non-invasive cardiology, BMJ, Br. Med. J. 329 (7479) (2004) 1386–1389.

[94] C. Ramesh, G. Priya, K. Jyothi, E. Victoria, Effectiveness of twin therapeutic approaches on pain and anxiety among patients following cardiac surgery, Nitte Univ. J. Health Sci. 3 (4) (2013) 34–39.

[95] A. Rangarajan, H. Chui, F.L. Bookstein, The softassign Procrustes matching algorithm, in: Proc. IPMI, in: LNCS, vol. 1230, Springer, 1997.

[96] P. Reimer, P.M. Parizel, J.F.M. Meaney, F.A. Stichnoth, Clinical MR Imaging, Springer Verlag, Berlin, 2010.

[97] D. Rizzo, Fundamentals of Anatomy and Physiology, 2009.

[98] S.F. Roohi, R.A. Zoroofi, 4D statistical shape modeling of the left ventricle in cardiac MR images, Int. J. Comput. Assisted Radiol. Surg. 8 (3) (2013) 335–351.

[99] L.G. Rudski, W.W. Lai, J. Afilalo, L. Hua, M.D. Handschumacher, K. Chandrasekaran, S.D. Solomon, E.K. Louie, N.B. Schiller, Guidelines for the echocardiographic assessment of the right heart in adults: a report from the American Society of Echocardiography endorsed by the European Association of Echocardiography, a registered branch of the European Society of Cardiology, and the Canadian Society of Echocardiography, J. Am. Soc. Echocard. 23 (7) (2010) 685–713.

[100] C. Santiago, J.C. Nascimento, J.S. Marques, Robust 3D active shape model for the segmentation of the left ventricle in MRI, in: Pattern Recognition and Image Analysis IbPRIA'15, Springer, 2015, pp. 283–290.

[101] P.M. Sapin, K.M. Schroder, A.S. Gopal, M.D. Smith, A.N. DeMaria, D.L. King, Comparison of two- and three-dimensional echocardiography with cineventriculography for measurement of left ventricular volume in patients, J. Am. Coll. Cardiol. 24 (1994) 1054–1063.

[102] P. Schoenhagen, C.J. Schultz, S.S. Halliburton, Cardiac CT Made Easy: An Introduction to Cardiovascular Multidetector Computed Tomography, CRC Press, 2014.

[103] U.J. Schoepf, CT of the Heart: Principles and Applications, Humana Press, Totowa, NJ, 2005.

[104] B. Schölkopf, A. Smola, K.R. Müller, Nonlinear component analysis as a kernel eigenvalue problem, Neural Comput. 10 (5) (1998) 1299–1319.

[105] S. Sedai, P. Roy, R. Garnavi, Segmentation of right ventricle in cardiac MR images using shape regression, in: International Workshop on Machine Learning in Medical Imaging, 2015, pp. 1–8.

[106] C.R. Shelton, Morphable surface models, Int. J. Comput. Vis. 38 (1) (2000) 75–91.

[107] M. Souto, L.R. Masip, M. Couto, J.J. Suárez-Cuenca, A. Martínez, P.G. Tahoces, J.M. Carreira, P. Croisille, Quantification of right and left ventricular function in cardiac MR imaging: comparison of semiautomatic and manual segmentation algorithms, Diagnostics 3 (2013) 271–282.

[108] B. Stender, O. Blanck, B. Wang, A. Schlaefer, Model-based segmentation of the left atrium in CT and MRI scans, in: International Workshop on Statistical Atlases and Computational Models of the Heart, Springer, Berlin/Heidelberg, 2013, pp. 31–41.

[109] P. Stolzmann, H. Scheffel, P.T. Trindade, A.R. Plass, L. Husmann, S. Leschka, M. Genoni, B. Marincek, P.A. Kaufmann, H. Alkadhi, Left ventricular and left atrial dimensions and volumes: comparison between dual-source CT and echocardiography, Invest. Radiol. 43 (5) (2008) 284–289.

[110] G. Subsol, J.P. Thirion, N. Ayache, A scheme for automatically building three-dimensional morphometric anatomical atlases: application to a skull atlas, Med. Image Anal. 2 (1) (1998) 37–60.

[111] L. Sugeng, V. Mor-Avi, L. Weinert, J. Niel, C. Ebner, R. Steringer-Mascherbauer, F. Schmidt, C. Galuschky, G. Schummers, R.M. Lang, H.J. Nesser, Quantitative assessment of left ventricular

size and function: side-by-side comparison of real-time three-dimensional echocardiography and computed tomography with magnetic resonance reference, Circulation 114 (2006) 654–661.

[112] A. Suinesiaputra, A.F. Frangi, M. Üzümcü, J.H.C. Reiber, B.P.F. Lelieveldt, Extraction of myocardial contractility patterns from short-axes MR images using independent component analysis, in: Computer Vision and Mathematical Methods in Medical and Biomedical Image Analysis, vol. 3117, Springer, 2004, pp. 75–86.

[113] A. Suinesiaputra, A.F. Frangi, T.A. Kaandorp, H.J. Lamb, J.J. Bax, J.H. Reiber, B.P. Lelieveldt, Automated detection of regional wall motion abnormalities based on a statistical model applied to multislice short-axis cardiac MR images, IEEE Trans. Med. Imaging 28 (4) (2009) 595–607.

[114] R. Swoboda, J. Scharinger, C. Steinwender, Model-based 3-D LV shape recovery in biplane X-ray angiography: a-priori information learned from CT, in: 2015 Computing in Cardiology Conference (CinC), IEEE, 2015, pp. 101–104.

[115] J.D. Thomas, Z.B. Popović, Assessment of left ventricular function by cardiac ultrasound, J. Am. Coll. Cardiol. 48 (10) (2006) 2012–2025.

[116] C. Tobon-Gomez, F.M. Sukno, C. Butakoff, M. Huguet, A.F. Frangi, Automatic training and reliability estimation for 3D ASM applied to cardiac MRI segmentation, Phys. Med. Biol. 57 (13) (2012) 4155–4174.

[117] R. Toldo, A. Beinat, F. Crosilla, Global registration of multiple point clouds embedding the generalized Procrustes analysis into an ICP framework, in: 3DPVT 2010 Conference, 2010.

[118] T. Tölli, J. Koikkalainen, K. Lauerma, J. Lötjönen, Artificially enlarged training set in image segmentation, Med. Image Comput. Comput. Assist. Interv. 9 (Pt 1) (2006) 75–82.

[119] B.L. Troy, J. Pombo, C.E. Rackley, Measurement of left ventricular wall thickness and mass by echocardiography, Circulation 45 (1972) 602–611.

[120] A. Tsai, A.J. Yezzi, A.S. Willsky, A shape-based approach to the segmentation of medical imagery using level sets, IEEE Trans. Med. Imaging 22 (2) (2003) 137–154.

[121] M. Unberath, A. Maier, D. Fleischmann, J. Hornegger, R. Fahrig, Open-source 4D statistical shape model of the heart for X-ray projection imaging, in: IEEE 12th International Symposium on Biomedical Imaging, ISBI, 2015, pp. 739–742.

[122] M. Üzümcü, A.F. Frangi, J.H. Reiber, B.P. Lelieveldt, Independent component analysis in statistical shape models, in: Proceedings of SPIE Medical Imaging, 2003.

[123] H.C. van Assen, M.G. Danilouchkine, M.S. Dirksen, J.H. Reiber, B.P. Lelieveldt, A 3-D active shape model driven by fuzzy inference: application to cardiac CT and MR, IEEE Trans. Inf. Technol. Biomed. 12 (5) (2008) 595–605.

[124] L. Vargas-Quintero, B. Escalante-Ramírez, L.C. Marín, M.G. Huerta, F.A. Cosio, H.B. Olivares, Left ventricle segmentation in fetal echocardiography using a multi-texture active appearance model based on the steered Hermite transform, Comput. Methods Programs Biomed. 137 (2016) 231–245.

[125] R.H. Whitaker, Anatomy of the heart, Medicine 42 (8) (2014) 406–408.

[126] B.R. Wilcox, A.C. Cook, R.H. Anderson, Surgical Anatomy of the Heart, Cambridge University Press, Cambridge, 2005.

[127] H. Zhang, A. Wahle, R.K. Johnson, T.D. Scholz, M. Sonka, 4-D cardiac MR image analysis: left and right ventricular morphology and function, IEEE Trans. Med. Imaging 29 (2) (2010) 350–364.

[128] Z. Zhao, E.K. Teoh, A novel framework for automated 3D PDM construction using deformable models, in: Proc. SPIE, vol. 5747, 2005, pp. 303–314.

[129] Y. Zheng, D. Yang, M. John, D. Comaniciu, Multi-part modeling and segmentation of left atrium in C-arm CT for image-guided ablation of atrial fibrillation, IEEE Trans. Med. Imaging 33 (2) (2014) 318–331.

[130] L. Zhong, Y. Su, S.Y. Yeo, R.S. Tan, D.N. Ghista, G. Kassab, Left ventricular regional wall curvedness and wall stress in patients with ischemic dilated cardiomyopathy, Am. J. Physiol., Heart Circ. Physiol. 296 (3) (2009) H573–H584.

[131] X. Zhuang, K. Rhode, S. Arridge, R. Razavi, D. Hill, D. Hawkes, S. Ourselin, An atlas-based segmentation propagation framework locally affine registration–application to automatic whole heart segmentation, Med. Image Comput. Comput. Assist. Interv. 11 (Pt 2) (2008) 425–433.

[132] M.A. Zuluaga, M.J. Cardoso, S. Ourselin, Automatic right ventricle segmentation using multi-label fusion in cardiac MRI, in: Workshop in Medical Image Computing and Computer Assisted Intervention, 2012.

INDEX

Symbols

2D–3D reconstruction algorithm, 332
3D Morphable model (3DMM), 116, 121

A

Active appearance models (AAMs), 14, 119, 416, 456
Active shape models (ASMs), 103, 454
Affine transformations, 38, 52, 224
Alzheimer's disease, 352, 369
Anatomical landmarks, 13, 37, 47, 275, 343, 415, 429
AP image, 334
Appearance, 120, 415, 431
Appearance models, 19, 101, 119, 426
Appearance of individual landmarks, 92
Applications to cardiac imaging, 458
Areal BMD, 411, 424, 428
Assignment problem, 100
Atlas, 141, 189, 272, 319, 322, 340, 363, 455
Atlas-based segmentation, 319, 462
Atrial fibrillation (AF), 276, 291, 325
Atrium
    left (AL), 275, 291, 446
    right (RL), 321, 446
Automatic segmentation of hippocampal subfields (ASHS), 361

B

Background, 118, 125, 285, 414
Background model, 125
Basel face model (BFM), 120
Basis, 7, 116, 156, 174, 231, 381, 410, 431, 433
Basis functions, 49, 168, 180, 197, 353
    Fourier (sinusoidal), 140
Bayesian statistics, 195
Bending energy, 54, 237
    TPS, 71
Best position (of landmark point), 11
Binary volumes, 60, 285, 331
Biomechanical modeling, 58
Bladder, 302
Bone mineral density (BMD), 410, 422
Bone quality, 410, 424

Bone quality assessment, 411
    applications of, 433
    approaches for, 419
    methods for, 410
    patient-specific, 436
Bones, 126, 259, 274, 291, 310, 338, 410, 419
    fracture risk of, 195
    nasal, 232
    strength of, 410
Boundary, 40, 54, 91, 103, 138, 153, 259, 274, 322, 333, 362, 366, 414, 430
Boundary conditions (BCs), 59, 62, 304, 309, 429
    pelvic, 310
Brain imaging, 352
Brain structures, *see* hippocampus

C

Cadaveric femur, 338
Candidate points, 103
    optimal, 104
Canonical variate analysis (CVA), 231
Cardiac imaging, 449
    applications, 458
C-arm images, 344
Chambers, 56, 320, 446
Clinical applications, 432
Clinical practice, 411, 449
Closest points, 248
    iterative (ICP), 250, 417, 455
Clusters, 99, 207
Coefficients, 357
Color model, 116
Combined model, 166, 186, 420, 431
    deformation of the, 185
Combined shape-texture model, 18
Common atlas, 324
Common coordinate system, 330, 341, 389, 430
Common space and model generalization, 47
Compactness, 5, 79, 156, 179, 456
Complete graph (CG), 93
Composite principal nested spheres (CPNS), 149
Computation, 44
Computational anatomy (CA), 193, 352, 412
    Bayesian treatment, 196

Configuration space, 234, 262
Connections, 91, 105, 142
 total number of, 93
Constrained local models (CLMs), 18
Contours, 4, 47, 69, 306, 332, 415, 421, 457
Control points, 49, 151, 197, 282, 340, 417, 422
Coordinates, 44, 68, 220, 353, 415
 2D or 3D, 218
Correspondence, 4, 50, 76, 153, 260, 271, 330, 432, 455
 2D, 119
 3D shape, 85
 dense, 119, 391
 establishment of, 388, 412
 inter-subject, 388
 intra-subject, 391
 open-shape, 70
 pairwise, 75
 shape, 68
 with regression, 279
Correspondence criterion, 388
Correspondence information, 131
Correspondence model, 262, 289
 optimized, 261
Correspondence optimization, 153, 279
Correspondence point positions, 287
Correspondence points, 259
Correspondence positions, 259, 262, 271, 276, 289
Cortical surfaces, 273
Cost function, 71, 263, 278
 minimization of, 455
 shape, 268
Covariance function, 121, 169, 388, 401
 and mean, 183
 sum of two, 177
Covariation, 232
CPU time, 82
Creation of the heat map, 236
CT angiographic images, 462
CT images, 25, 419, 453, 458
Curved manifolds, 139, 151, 158
Curves, 5, 75, 152, 237
 2D, 237, 458
 3D, 237

D

Data driven Markov chain Monte Carlo (DDMCMC), 116
Deformable models, 414, 453

Deformation component, 176, 363
Deformation fields, 53, 172, 208, 272, 363, 394, 415
Deformation manifold, 44
Deformation models, 34, 141, 185, 330, 346
 biophysical, 40
 general-purpose, 48
 global, 39
 heuristic, 40
 local, 53
 statistical, 40
Deformation population, 48
Deformations, 34, 120, 141, 160, 168, 176, 319, 352, 370, 380, 395, 457
 large, 54, 203, 308
Dense displacement fields (DDFs), 35
Destinations, 97
Diaphragm, 388, 397
Digitally reconstructed radiographs (DRRs), 339
Dimensionality reduction, 7, 15, 418, 428, 433
Dimensions, 139, 148
Direct appearance model (DAM), 416
Dirichlet process, 206
Dirichlet process mixture model, 208
Displacement field, 37, 58, 251, 360
Displacements, 36, 142, 222, 308, 383
Distance metric, 44
Distance-weighted discrimination (DWD), 145, 152
Distribution
 final proposal, 129
 probability, 151, 167
 proposal, 117, 122
Dual X-ray absorptiometry (DXA), 410
DXA images, 420, 431
DXA scans, 428

E

Early mild cognitive impairment (EMCI), 365
Echocardiography, 451
 3D, 445
Efficient approaches, 148, 417
Eigenfaces approach, 119
Eigenvalues, 7, 41, 102, 157, 180, 227, 269, 393
 largest, 7, 102, 157, 403
Eigenvectors, 7, 41, 102, 227, 269, 418
Ellipsoid, 150, 159, 229, 272, 357
 first order (FOE), 358
Entropy, 57, 154, 259, 277

Estimated positions, 11
Euclidean spaces, 138
Euclidean vector space, 44
Euclideanization, 142, 147, 158
Euclideanized, 139, 144, 149
Expectation-maximization (EM), 118, 203, 313,
    455
Experiments, 82, 186, 343, 396
Explanatory variables, 279

F

Faces, 116, 242
    3D, 115
FE models, 308, 412, 432
Feature point detections, 9, 23, 123, 130
FEM-based approaches, 413, 435
Femoral head, 336, 422, 427
Femur, 83, 331, 338, 419
FFD registration, 54, 322
Fiducials, 37, 396
    2D planar projections of, 394
    3D, 330
Finite element analysis (FEA), 36, 58, 306
First order ellipsoid (FOE), 358
Fitting process, 125, 253, 322
Flexibility, 53, 125, 187
Fracture risk, 411, 421
Fracture risk assessment tool (FRAX), 411, 429
Fractures, 208, 410, 423
Fréchet mean, 45, 146
Free form deformation (FFD), 53, 417
    backward, 342
    forward, 341
Functional data, 277
Functionality, 218, 239, 250
Functions
    objective, 19, 143, 196, 355
    spherical, 355
    vector-valued, 277

G

Gaussian kernel, 43, 51, 170, 176, 251, 267, 401
Gaussian process, 116, 166, 204, 388
Gaussian process model (GPM), 121, 170, 252, 401
Gaussian process morphable models (GPMMs),
    166
Gender, 364
General linear models (GLMs), 352, 362
Generalization, 151, 161, 166, 274

Generalization ability, 47, 187, 456
Generalized Hough transform (GHT), 25, 420
Generalized Procrustes analysis (GPA), 6, 225, 286,
    389, 455
Generalized rotation, 158
Geodesic active contours (GACs), 419
Geodesic distances, 44, 147, 268, 287, 360
Geometric information, 4, 97, 274, 287
Geometric morphometrics, 218
Geometric object properties (GOPs), 139
Gibbs samplers, 210
Graph coloring, 107
Graphical lasso (GL), 96
Ground-truth PDMs, 79
Groups, 231

H

Hamiltonian Monte Carlo (HMC), 196
Healthy control (HC), 361
Heart, 320, 446
    atria, 292
Hip fractures, 410, 425
Hip structure analysis (HSA), 411
Hippocampal surface, 357, 364, 370
    3D, 352
Hippocampus, 138, 149, 155, 158, 351, 352, 357,
    359, 369
Hybrid landmark and surface registration, 173

I

Illumination, 116
Illumination models, 119
Image analysis, 67, 91, 133
    3D, 93, 104
    face, 116, 126
    landmark-based, 92, 103
    medical, 34, 67, 89, 116, 126, 157, 194
Image calibration, 329
Image coordinate frame, 6
Image-guided interventions, 62
Image interpretation, 27
Image registration, 10, 35, 57, 193, 388
    groupwise, 9
    medical, 35, 62
Image segmentation, 94, 262, 284, 414
Image voxels, 36, 47
Independent component analysis (ICA), 418, 456
Individual landmarks, 16, 90, 101, 107

Information
    temporal, 458
    underlying surface, 239
Initialization, 13, 18, 26, 103, 115, 123, 260, 286,
        322, 355, 420
Input images, 115, 414
    2D, 432
InShape models, 416
    statistical, 423, 428
Interpolation, 35, 55
Intersection, 61, 274

K
KD-Trees, 248
Kernel-based approaches, 48, 264
Kernel functions, 46
Kernels, 51, 121, 170, 178, 364

L
Landmark-based shape representation, 92
    error, 72
Landmark candidate points, 103, 109
Landmark clustering, 98
Landmark configurations, 91, 220, 226
    mirrored, 222
Landmark connections, 91
Landmark coordinates, 234, 418
Landmark data, 219, 352
Landmark detection, 90
    shape-based, 91, 101
Landmark information, 174, 225
Landmark points, 6, 11, 18, 68, 124, 167, 181
Landmark samples, 92
    aligned, 93
Landmark sliding, 73, 82
Landmark-sliding algorithm, 73
Landmarks, 4, 54, 68, 89, 173, 182, 219, 321, 359,
        417, 455
    adjacent, 94, 105
    bilateral, 223
    dummy, 100
    identified, 79
    large number of, 94
    manually placed, 219, 242
    manually selected, 394
    pairs of, 92
Late mild cognitive impairment (LMCI), 364
Likelihood functions, 115, 172, 195, 313
Liver, 323, 381

Loadings, 58
Local gray-value models, 12
Local regions, 54
Locally affine registration method (LARM), 56,
        322
Locally affine transformation (LAT), 54
Locally linear embedding (LLE), 419
Lumber back pain (LBP), 206

M
Machine learning approaches, 11, 110
Mahalanobis distances, 8, 22, 105, 223
Manifold data, 148
Manifolds, 44, 139, 146, 246, 364
Markov chain, 118, 200
Markov chain Monte Carlo (MCMC), 196
Material properties, 61, 245, 308, 428
Medial models, 143
Medical image computing, 54, 58
Medical image segmentation, 414
Medical imaging, 6, 13, 25, 67, 69, 413
    3D, 85
    technologies, 292
Meshes, 53, 94, 220, 236, 308, 415, 455
    face, 121
Metropolis–Hastings algorithm, 122, 204
Minimal cost path, 104
    2D, 106
Minimum description length (MDL), 5, 76, 154,
        261
Minimum spanning tree (MST), 76
    rooted, 79
Missing landmarks, 221, 240
Model adaptation, 115
Model-based approaches, 331, 432, 460
Model-based bone quality assessment, 433
Model-based segmentation, 419
    deformable, 322
Model building, 12, 18, 21
Model matching performance, 17
Model quality, 166
Model selection problem, 261
Model space, 131, 253
Model surface, 315
Modeling deformations, 40, 168
    on multiple scale levels, 177
    statistical, 44
Modeling of shape motion, 384
Modeling smooth deformations, 170

Modeling symmetric variations, 171
Models
    atlas-based, 454
    bias, 185
    conditional, 403
    leave-one-out, 394
    mesh/shape, 251
    occlusion-aware, 126
    population-based, 317, 394
    symmetric, 180
Morphable models, 119, 130, 166
    Gaussian process, 174
Motion, 34, 265
    3D, 311
    along a curve, 152
    variance of the, 308
Motion model, 384
MR images, 305
    cardiac, 459
    T1-weighted, 397
    T2-weighted, 302
MRF problem, 107
Multi-atlas segmentation (MAS), 324
Multiple path propagation and segmentation
        (MUPPS), 324
Multiscale kernels, 178

N

Non-rigid registration, 172, 343, 391, 417, 454
Normal controls, 156, 290
Normalized mutual information (NMI), 46, 54
Normals distribution models, 140

O

Object boundary, 90, 94, 106, 143, 154, 417
Object class, 3
Object detection, 24
Object of interest, 3, 16, 24, 91, 416
    bones, 413
    boundary of the, 91, 94, 105
    contour of the, 12, 20, 24
    global shape of the, 9
    patch of the, 16
    position, orientation and/or scale of the, 10, 109
    properties of the, 106
    search for the, 19
    shape of the, 5, 95, 101
    translation, scaling and rotation of the, 92
Object spaces, 358

Object surface, 140, 353
Observed object, 90
Occlusion-aware morphable model, 126
Occlusions, 24, 115
Optimal solution, 91, 106, 261
Optimal transportation-based graph, 98
Optimization, 115, 259, 262, 355
Ordinary differential equations (ODEs), 45, 208
Organ motion, 317, 380, 400
Organs, 34, 67, 126, 259, 308, 380, 448
    deformation of, 34, 58, 384
    motion of, 379
Orientation, 10, 22, 115, 337, 431
Orthopedics, 259
Osteoporosis, 27, 410
Outliers, 11, 115, 223

P

Parameter space, 353
Parametric appearance model (PAM), 116
Parametric models, 166, 209, 259, 272
Particle-based modeling (PBM), 262
Particle configuration, 264
Particle positions, 263, 281
Particle systems, 259, 282
    class, 288
Particles, 260
Patches, 333
Pathological condition, 44, 325
Pathological deformations, 166, 184
Pathologies, 156, 184, 321, 449
PBM algorithm, 261, 274
    extension of the, 276
    generalized, 277
PBM approach, 261
PBM code library, 288
PBM model, 282
PBM optimization, 271
Pelvic bone, 302, 421
Pelvis, 304, 331, 420
Performance evaluation, 68
Point clouds, 36, 248
Point distribution models (PDMs), 68, 139, 331,
        414, 455
Point sets, 259, 315, 415
Point-to-point correspondences, 419, 462
Poisson disk sampling, 249
Polynomial time, 104
Population, 48, 139, 282, 384, 456

Population-based model generation method, 317
Population-based statistical motion models, 394
Positional information, 262, 276
Predictor, 42, 232, 366, 429
Predictor of interest, 364
Principal component analysis (PCA), 5, 40, 103,
    119, 138, 167, 223, 271, 331, 391, 418,
    455
Principal nested spheres (PNS), 148
Prior models, 166, 182, 419
Probabilistic framework, 392
    fully, 115
Probabilistic PCA (PPCA), 120
Probabilistic setting, 115
Procrustes distance, 228, 237, 417
Prostate, 160, 302
    gland, 302
Proximal femur, 9, 330, 421, 428
p-Value, 366

Q

Quantitative computed tomography (QCT), 410
    images, 421

R

Radial basis functions (RBFs), 283
Random field theory (RFT), 364
    peak, 368, 369
Random samples, 156, 176, 195
Receiver operating characteristic (ROC), 158, 425
Reconstructed femoral model, 336
Reconstructed pelvic model, 335
Reconstruction, 258, 289, 400
    2D–3D, 329
    3D, 419, 425, 451
    3D shape, 413, 432
    accurate, 332, 357, 432
    by regression, 400
    degree one, 357
    hierarchical 2D–3D, 332
    model-based, 387
    pairwise shape, 425
    shape, 121, 131, 434
Reconstruction accuracies, 338, 396
Reconstruction error, 394
    average mean, 434
Reconstruction process, 283, 341
    2D–3D, 333, 335
Rectum, 302

Reference coordinate frame, 9
Reference frame, 17
Reference image, 44, 54, 272, 388, 422
Reference objects, 141
Regions of interest (ROI), 36, 259, 302, 423
Registered surface models, 352
Registration, 10, 36, 181, 242, 253, 313, 358, 390,
    416, 458
Registration problem, 47, 172
Regression model, 46, 280, 366
Regression modeling, 44
    shape, 274
Relative warp analysis (RWA), 227
Rendering, 122
    3D surface, 321
    inverse, 119
    slice, 451
    surface, 451
    volume, 451
Reparameterizations, 141, 153
Resolution, 21, 246
Respiratory motion, 379
Response images, 19
Rigid registration, 338, 454
Rigid transformation, 38, 282, 316
Root mean square distance (RMSD), 93
Rotation, 6, 37, 69, 92, 158, 286, 317, 335, 357,
    458
Rotation angle, 38, 158, 458

S

Sample set, 261, 268
SAX MR images, 458
Scale, 6, 53, 92, 124, 139, 176, 224, 276, 357
Scale factors, 150
Schizophrenia, 158
Scientific and clinical applications, 262
SD objective function, 360
Search image, 11
Segmentation, 36, 115, 127, 157, 274, 303, 414,
    453
    automated, 285, 453
    manual, 285, 427, 435
Segmentation quality, 434
Segmentation results, 320, 361
Semilandmarks, 237
    transfer, 242
Semilandmarks on symmetric structures, 241
Shape analysis of registered surface models, 352

Shape and appearance variations, 412
Shape correspondence algorithm, 68, 79
Shape descriptors, 352, 415
Shape model matching, 10, 23
Shape representations, 96, 149, 414
    3D object, 94
    alternative, 415
    multivariate, 258
    parametric, 259
    specific, 259
    volumetric, 417
Shape spaces, 81
Shape variables, 227
ShapeWorks, 259
    shape modeling workflow, 284
    Studio, 289
Signal, 363, 384, 400
Signed distance maps, 415
Simplified finite element models, 303
Skeletal models, 142
Skeletal points, 143
Skeletal surface, 143
Skull-registration, 181
Slices, 314, 389, 450, 458
    2D, 186, 314
Sliding, 237
Smoothness parameter, 177, 401
Sources, 97
Space
    curved, 139
    latent, 121, 231
Sparse linear combination for shape modeling, 103
Spatial relationships, 90, 110
Specificity, 156, 180, 456
SPHARM expansion, 355
SPHARM models, 358
SPHARM surface modeling, 353
Spheres, 139, 355
    2D, 147
    products of, 149
Spherical coordinates, 353
Spherical harmonic models, 140
Spherical harmonics (SPHARM), 60, 121, 129, 140, 353
    classic, 352
    expansion, 122, 356
Spherical image, 360
Spherical parameterization, 141, 259, 353

Standard deviation (SD), 8, 105, 130, 176, 210, 272, 317, 410
Standard PCA, 227
Statistical analysis, 47, 143, 218, 363, 416, 455
Statistical appearance models (SAMs), 14
Statistical approaches, 40, 151
Statistical deformation model, 40, 41, 44, 63, 417
    building, 47, 53
    concept of, 48
Statistical deformation model (SDM), 35, 42
    2D–3D reconstruction, 338
    b-spline-based, 339
Statistical deformation models, nested, 40
Statistical model, 5, 40, 115, 313, 331, 387, 412, 422, 457
    construction of the, 44, 417
    deformation, 34
    inference on, 386
    of computational anatomy, 196
    of deformation, 34
    of the skull, 174
Statistical model regression, 387
Statistical modeling, 69, 319, 415
Statistical modeling of finite element analysis, 63
Statistical motion modeling, 391
Statistical motion models (SMMs), 35, 50, 302, 311, 312, 381, 385, 461
Statistical parametric map (SPM), 364
    distribution analysis (SPM-DA), 352
    distribution analysis (SPM-DA) method, 366
Statistical shape and appearance models (SSM/SAMs), 412
Statistical shape and density model (SSDM), 424, 432
Statistical shape model construction, 331, 455
Statistical shape model deformation, 457
Statistical shape modeling, 258
Statistical shape models (SSMs), 5, 165, 258, 384, 415, 446
    3D reconstruction, 421
    computing, 182
    construction of, 331
    mixed-subject, 318, 319
    revisited classical, 167
    subject-specific, 317
Statistical skull model, 184
Subfields, 359
Subject-specific probability density function (SSPDF), 318

Subjects, 13, 34, 55, 62, 196, 273, 317, 384, 415, 454
  specific, 44
Subspaces, 63, 151, 228
Substructures, 54
Surface-based morphometry (SBM), 352
Surface boundary, 274
  open, 275
Surface constraint, 266, 285
Surface correspondence, 260, 358
Surface meshes, 61, 106, 244, 260, 307, 423
  sample-specific, 282
Surface meshing, 60
Surface models, 329
  3D, 321
Surface registration, 61, 173, 359, 388
Surface registration and shape models, 250
Surface representations, 263, 356
Surface signals, 363, 370
Surfaces, 36, 59, 91, 143, 169, 218, 237, 260, 307, 353, 388, 415, 449
  3D closed, 356
  closed, 68, 274
  implicit, 259, 285
  open, 68, 274

T

Target image, 47, 57, 106, 115, 322
Target surface, 172, 236
Template, 24, 70, 154, 196, 219, 241, 282, 352, 360, 412
  landmarks, 71
  new, 75
  parameterization, 363
Test shape-correspondence algorithm, 79
Texture, 11, 245, 415, 456
Texture models, 14, 422
Texture vector, 14
Thin-plate spline (TPS), 51, 282
Threshold, 22, 57, 155, 230
Topology, 60, 68, 143, 260, 282, 355, 386, 419
Topology preservation and landmark-sliding algorithm, 73
Total hip arthroplasty (THA), 330
TPS deformation, 222
Traditional landmarks, 237
Training data, 7, 47, 145, 166, 303, 418, 455
Training datasets, 430

Training images, 6, 18, 46, 339, 416, 462
Training process, 339
Training samples, 21, 46, 103, 138, 455
Training set, 6, 16, 92, 153, 321, 416, 423, 454
Transformation models, 37, 56, 316
  nonrigid, 37, 316
Transformations, 36, 69, 81, 105, 224, 322, 360, 412, 455
Translation, 37, 69, 92, 286, 357, 389, 457
Transportation problem, 98
Triangular meshes, 245
TRUS images, 307
TRUS probe, 302

U

Ubiquitous displacement models, 36
Ultrasound images, 161, 384
  2D, 398
Uniform sampling, 265
Unit sphere, 139, 353

V

Variance, 7, 102, 121, 158, 168, 203, 237, 303, 456
  principal, 155
  total, 150, 155, 180
Variations
  modeling, 184, 421
  texture, 11, 422
Vector space, 228
Velocity vectors, 44, 141
Ventricles, 321, 448, 457
  left, 323, 446
  right, 45, 323, 446
Verification steps, 117
Vertebra, 425
Voxel-based morphometry (VBM), 352
Voxel surface, 353
Voxels, 35, 141, 272, 314, 412

W

Whole heart segmentation (WHS), 319, 463

X

X-ray images, 332, 421
  2D, 339
  2D calibrated, 330
  calibrated, 338

Printed in the United States
By Bookmasters